Lecture Notes in Computer Science 1738

Edited by G. Goos, J. Hartmanis and J. van Leeuwen

Springer
Berlin
Heidelberg
New York
Barcelona
Hong Kong
London
Milan
Paris
Singapore
Tokyo

C. Pandu Rangan V. Raman
R. Ramanujam (Eds.)

Foundations of Software Technology and Theoretical Computer Science

19th Conference
Chennai, India, December 13-15, 1999
Proceedings

Springer

Series Editors

Gerhard Goos, Karlsruhe University, Germany
Juris Hartmanis, Cornell University, NY, USA
Jan van Leeuwen, Utrecht University, The Netherlands

Volume Editors

C. Pandu Rangan
Indian Institute of Technology
Department of Computer Science and Engineering
Chennai 600036, India
E-mail: rangan@iitm.ernet.in

V. Raman
R. Ramanujam
Institute of Mathematical Sciences
C.I.T. Campus, Chennai 600113, India
E-mail: {vraman,jam}@imsc.ernet.in

Cataloging-in-Publication data applied for

Die Deutsche Bibliothek - CIP-Einheitsaufnahme

**Foundations of software technology and theoretical computer
science** : 19th conference, Chennai, India, December 13 - 15, 1999 ;
proceedings / [FST and TCS 19]. C. Pandu Rangan ... (ed.). - Berlin ;
Heidelberg ; New York ; Barcelona ; Hong Kong ; London ; Milan ;
Paris ; Singapore ; Tokyo : Springer, 1999
   (Lecture notes in computer science ; Vol. 1738)
   ISBN 3-540-66836-5

CR Subject Classification (1998): F.1, F.3-4, G.2, F.2, D.2, D.1, D.3

ISSN 0302-9743
ISBN 3-540-66836-5 Springer-Verlag Berlin Heidelberg New York

Typesetting: Camera-ready by author
SPIN: 10749931   06/3142 – 5 4 3 2 1 0   Printed on acid-free paper

Preface

This volume contains the proceedings of the 19th FST&TCS conference (Foundations of Software Technology and Theoretical Computer Science), organized under the auspices of the Indian Association for Research in Computing Science (http://www.imsc.ernet.in/~iarcs).

This year's conference attracted 84 submissions from as many as 25 different countries. Each submission was reviewed by at least three independent referees. After a week-long e-mail discussion, the program committee met on the 7th and 8th of August, 1999, in Chennai and selected 30 papers for inclusion in the conference program. We thank the program committee members and the reviewers for their sincere efforts.

We are fortunate to have five invited speakers this year, providing for a very attractive program: *Martín Abadi* (Bell Labs - Lucent Technologies, Palo Alto, USA), *Lila Kari* (Univ. Western Ontario, Canada), *Jean-Jacques Lévy* (INRIA, Paris, France), *Micha Sharir*, (Univ. Tel Aviv, Israel), and *Seinosuke Toda* (IEC, Tokyo, Japan). Moreover, the conference is preceded by a two-day workshop (December 11–12, 1999) on Data Structures and succeeded by a two-day workshop (December 16-17, 1999) on Foundations of Mobile Computation. The conference also features two joint sessions with the **International Symposium on Automata, Algorithms and Computation** (December 16–18, 1999, Chennai): *Monika Henzinger* (Compaq Systems Research, Palo Alto, USA) is presenting a tutorial on Web algorithmics as the last event of FST&TCS and *Kurt Mehlhorn* (Max-Planck-Institut, Saarbrücken, Germany) is giving a talk on algorithm engineering as the first event of ISAAC. We thank all the invited speakers for agreeing to talk at the conference, as well as for providing abstracts and articles for the proceedings.

The Institute of Mathematical Sciences and the Indian Institute of Technology, both at Chennai, are co-hosting the conference. We thank these Institutes, as well as the others who extended financial support, agencies of the Government of India and private software companies in India.

We thank the members of the organizing committee for making it happen. Special thanks go to the staff of our institutes and to Alfred Hofmann of Springer-Verlag for their help with the proceedings.

September 1999

C. Pandu Rangan (IIT, Madras)
Venkatesh Raman (IMSc, Chennai)
R. Ramanujam

Conference General Chair

C. Pandu Rangan *(IIT, Madras)*

Program Committee

S. Arun-Kumar *(IIT, Delhi)*
V. Arvind *(IMSc, Chennai)*
Zhou Chaochen *(UNU/IIST, Macau)*
Tamal Dey *(IIT, Kharagpur)*
Devdatt Dubhashi *(IIT, Delhi)*
Melvin Fitting *(CUNY, New York)*
Lance Fortnow *(U. Chicago)*
Ramesh Hariharan *(IISc, Bangalore)*
Kohei Honda *(U. Edinburgh)*
Samir Khuller *(U. Maryland, College Park)*
Kamala Krithivasan *(IIT, Madras)*
K. Narayan Kumar *(SMI, Chennai)*
Ming Li *(U. Waterloo)*
Kamal Lodaya *(IMSc, Chennai)*
Paliath Narendran *(SUNY, Albany)*
Rajeev Raman *(King's College, London)*
Venkatesh Raman *(IMSc, Chennai)* **(Co-chair)**
R. Ramanujam *(IMSc, Chennai)* **(Co-chair)**
S. Ramesh *(IIT, Bombay)*
Uday S. Reddy *(U. Illinois at Urbana-Champaign)*
Natarajan Shankar *(SRI, Menlo Park)*
Scott Smolka *(SUNY, Stony Brook)*
Milind Sohoni *(IIT, Bombay)*
Wolfgang Thomas *(RWTH, Aachen)*

Organizing Committee

E. Boopal *(IIT, Madras)*
V. Kamakoti *(ATI Research, Chennai)*
R. Rama *(IIT, Madras)*
K. Rangarajan *(MCC, Chennai)*

List of Reviewers

Luca Aceto
Manindra Agrawal
Stefano Aguzzoli
Rajeev Alur
Andris Ambainis
G.V. Anand
Siva Anantharaman
Lars Arge
S. Arun-Kumar
V. Arvind
Sunil Arya
Arnon Avron
R. Balasubramanian
Paolo Baldan
Mohua Banerjee
Marie-Pierre Beal
Bernhard Beckert
Martin Berger
Luca Bernardinello
Karen Bernstein
Binay Bhattacharya
Dragan Bosnacki
Gerhard Brewka
Gerth Brodal
Alberto Caprara
Walter Carnielli
Mihir Chakraborty
Zhou Chaochen
Siu-Wing Cheng
S.A. Choudum
Judy Crow
Erzsebet Csuhaj-Varju
Deepak D'Souza
Ivan Damgard
Philippe Darondeau
Alexandre David
Rocco De Nicola
S. Debray
Tamal Dey
Volker Diekert
Shlomi Dolev
Manfred Droste
Xiaoqun Du
Devdatt Dubhashi

Didier Dubois
Paul Dunne
Bruno Dutertre
Christian Fermueller
Melvin Fitting
Lance Fortnow
Yuxi Fu
Naveen Garg
Leslie Goldberg
Mordecai Golin
Ganesh Gopalakrishnan
Anupam Gupta
Prosenjit Gupta
Andrew Haas
Reiner Haehnle
Petr Hajek
Ramesh Hariharan
Jifeng He
Matthew Hennessy
Kohei Honda
Jozef Hooman
Peter Hoyer
Hans Huttel
Pino Italiano
R. Jagadeesan
Ravi Janardan
Tomasz Janowski
Lalita Jategaonkar
R.K. Joshi
Yuh-Jzer Joung
Marcin Jurdzinski
Helmut Jurgensen
Deepak Kapur
S. Kar
Rohit Khandekar
Sanjeev Khanna
Samir Khuller
Astrid Kiehn
Josva Kleist
J. Koebler
Jochen Konemann
Roman Kossak
V. Krishnamoorthy
Kamala Krithivasan

Manfred Kudlek
Anil Kumar
K. Narayan Kumar
Orna Kupferman
Dietrich Kuske
Tony Lai
Yassine Lakhnech
Kim Larsen
Daniel Lehmann
Francesca Levi
Francois Levy
Ming Li
Huimin Lin
Kamal Lodaya
Christof Löding
Satyanarayana Lokam
Anna Lubiw
Christopher Lynch
Zahir Maazouzi
Meena Mahajan
B. Mans
Felip Manyà
V. Marek
O. Matz
David McAllester
Massimo Merro
Jochen Messner
Ron van der Meyden
Daniele Micciancio
Joe Mitchel
Tulika Mitra
Swarup Mohalik
Larry Moss
David Mount
Madhavan Mukund
Daniele Mundici
Hema Murthy
David Musser
S. Muthukrishnan
Rajagopal Nagarajan
Seffi Naor
Giri Narasimhan
Paliath Narendran
V. Natarajan

Ashwin Nayak

Mogens Nielsen

Catuscia Palamidessi

Alessandro Panconesi

Rohit Parikh

Giridhar Pemmasani

W. Penczek

Antoine Petit

Anna Philippou

Michele Pinna

Sanjiva Prasad

Volker Priebe

Shaz Qadeer

Xu Qiwen

J. Radhakrishnan

Balaji Raghavachari

S. Rajasekaran

R. Rama

C.R. Ramakrishnan

G. Ramalingam

Rajeev Raman

Venkatesh Raman

R. Ramanujam

S. Ramesh

Edgar Ramos

Abhiram Ranade

J-F Raskin

R. Ravi

Michel Raynal

Uday Reddy

James Riely

C. Roeckl

Denis Roegel

Erik Rosenthal

Abhik Roychoudhury

Harald Ruess

John Rushby

Hassen Saidi

Louis Salvail

Peter Sanders

Davide Sangiorgi

V. Sassone

Rainer Schuler

Sandeep Sen

Anil Seth

Pradyut Shah

Jeff Shallit

N. Shankar

Priti Shankar

M.S. Shanmugam

Sandeep Shukla

R.K. Shyamasundar

Riccardo Silvestri

Steve Sims

Ashish Singhai

Rakesh Sinha

Rani Siromoney

G. Sivakumar

Michiel Smid

Scott Smolka

Milind Sohoni

Oleg Sokolsky

Aravind Srinivasan

Z. Stachniak

V.S. Subrahmanian

K.V. Subrahmanyam

K.G. Subramanian

P.S. Subramanian

Terrance Swift

Lars Thalmann

P.S. Thiagarajan

Thomas Thierauf

Wolfgang Thomas

David Toman

Luca Trevisan

Stavros Tripakis

Dang Van Hung

Kasturi Varadarajan

Moshe Vardi

Helmut Veith

V.N. Venkatakrishnan

G. Venkatesh

Victor Vianu

K. Vidyasankar

Sundar Vishwanathan

Walter Vogler

Igor Walukiewicz

Ming-wei Wang

Pascal Weil

Carsten Weise

Angela Weiss

David Williamson

David Wolfram

Tomoyuki Yamakami

Hongseok Yang

Reza Ziaei

Table of Contents

Invited Talk 1

Recent Developments in the Theory of Arrangements of Surfaces 1
Micha Sharir

Session 1(a)

Dynamic Compressed Hyperoctrees with Application to
the N-body Problem 21
Srinivas Aluru and Fatih E. Sevilgen

Largest Empty Rectangle among a Point Set 34
Jeet Chaudhuri and Subhas C. Nandy

Session 1(b)

Renaming Is Necessary in Timed Regular Expressions 47
Philippe Herrmann

Product Interval Automata: A Subclass of Timed Automata 60
Deepak D'Souza and P. S. Thiagarajan

Session 2(a)

The Complexity of Rebalancing a Binary Search Tree 72
Rolf Fagerberg

Fast Allocation and Deallocation with an Improved Buddy System 84
Erik D. Demaine and Ian J. Munro

Session 2(b)

Optimal Bounds for Transformations of ω-Automata 97
Christof Löding

CTL^+ Is Exponentially More Succinct than CTL 110
Thomas Wilke

Invited Talk 2

A Top-Down Look at a Secure Message 122
Martín Abadi, Cédric Fournet and Georges Gonthier

Session 3

Explaining Updates by Minimal Sums 142
Jürgen Dix and Karl Schlechta

A Foundation for Hybrid Knowledge Bases 155
James J. Lu, Neil V. Murray and Erik Rosenthal

Session 4

Hoare Logic for Mutual Recursion and Local Variables 168
David von Oheimb

Invited Talk 3

Explicit Substitutions and Programming Languages 181
Jean-Jacques Lévy and Luc Maranget

Session 5(a)

Approximation Algorithms for Routing and Call Scheduling in
All-Optical Chains and Rings 201
Luca Becchetti, Miriam Di Ianni and Alberto Marchetti-Spaccamela

A Randomized Algorithm for Flow Shop Scheduling 213
Naveen Garg, Sachin Jain and Chaitanya Swamy

Session 5(b)

Synthesizing Distributed Transition Systems from Global Specifications ... 219
Ilaria Castellani, Madhavan Mukund and P. S. Thiagarajan

Beyond Region Graphs: Symbolic Forward Analysis of Timed Automata .. 232
Supratik Mukhopadhyay and Andreas Podelski

Session 6

Implicit Temporal Query Languages: Towards Completeness 245
Nicole Bidoit and Sandra de Amo

On the Undecidability of Some Sub-classical First-Order Logics 258
Matthias Baaz, Agata Ciabattoni, Christian Fermüller and Helmut Veith

Invited Talk 4

How to Compute with DNA* 269
*Lila Kari, Mark Daley, Greg Gloor, Rani Siromoney and
Laura F. Landweber*

Session 7(a)

A High Girth Graph Construction and a Lower Bound for
Hitting Set Size for Combinatorial Rectangles 283
L. Sunil Chandran

Protecting Facets in Layered Manufacturing 291
*Jörg Schwerdt, Michiel Smid, Ravi Janardan, Eric Johnson and
Jayanth Majhi*

Session 7(b)

The Receptive Distributed π-Calculus (Extended Abstract) 304
Roberto M. Amadio, Gérard Boudol and Cédric Lhoussaine

Series and Parallel Operations on Pomsets 316
Zoltán Ésik and Satoshi Okawa

Session 8

Unreliable Failure Detectors with Limited Scope Accuracy and
an Application to Consensus 329
Achour Mostéfaoui and Michel Raynal

Invited Talk 5

Graph Isomorphism: Its Complexity and Algorithms 341
Seinosuke Toda

Session 9(a)

Computing with Restricted Nondeterminism: The Dependence of
the OBDD Size on the Number of Nondeterministic Variables 342
Martin Sauerhoff

Lower Bounds for Linear Transformed OBDDs and FBDDs
(Extended Abstract) 356
Detlef Sieling

Session 9(b)

A Unifying Framework for Model Checking Labeled Kripke Structures,
Modal Transition Systems, and Interval Transition Systems 369
Michael Huth

Graded Modalities and Resource Bisimulation 381
Flavio Corradini, Rocco De Nicola and Anna Labella

Session 10(a)

The Non-recursive Power of Erroneous Computation 394
Christian Schindelhauer and Andreas Jakoby

Analysis of Quantum Functions (Preliminary Version) 407
Tomoyuki Yamakami

Session 10(b)

On Sets Growing Continuously 420
Bernhard Heinemann

Model Checking Knowledge and Time in Systems with Perfect Recall
(Extended Abstract) 432
Ron van der Meyden and Nikolay V. Shilov

FST&TCS — ISAAC Joint Session Talks

The Engineering of Some Bipartite Matching Programs 446
Kurt Mehlhorn

Author Index 451

Recent Developments in the Theory of Arrangements of Surfaces*

Micha Sharir

[1] School of Mathematical Sciences, Tel Aviv University
Tel Aviv 69978, Israel
[2] Courant Institute of Mathematical Sciences, New York University
New York, NY 10012, USA

Abstract. We review recent progress in the study of arrangements of surfaces in higher dimensions. This progress involves new and nearly tight bounds on the complexity of lower envelopes, single cells, zones, and other substructures in such arrangements, and the design of efficient algorithms (near optimal in the worst case) for constructing and manipulating these structures. We then present applications of the new results to a variety of problems in computational geometry and its applications, including motion planning, Voronoi diagrams, union of geometric objects, visibility, and geometric optimization.

1 Introduction

The combinatorial, algebraic, and topological analysis of arrangements of surfaces in higher dimensions has become one of the most active areas of research in computational geometry during the past decade. In this paper we will review the recent progress in the study of combinatorial and algorithmic problems related to such arrangements.

Given a set Γ of n surfaces in $\mathbb{R}^d$, the arrangement $\mathcal{A}(\Gamma)$ that they induce is the decomposition of $\mathbb{R}^d$ into maximal connected regions (cells) of dimensions $0, 1, \ldots, d$, such that each region is contained in the intersection of a fixed subset of Γ and is disjoint from all the other surfaces. For example, the arrangement of a set L of lines in the plane consists of *vertices* (0-dimensional cells) which are the intersection points of the lines, *edges* (1-dimensional cells) which are relatively open intervals along the lines delimited by pairs of consecutive intersection points, and *faces* (2-dimensional cells) which are the connected components of $\mathbb{R}^2 \setminus \bigcup L$.

Arrangements of lines and of hyperplanes have been studied earlier quite extensively. In fact, Edelsbrunner's book from 1987 [36], one of the earliest textbooks on computational geometry, is essentially devoted to the study of such

* Work on this paper has been supported by NSF Grant CCR-97-32101, by a grant from the U.S.-Israeli Binational Science Foundation, by the ESPRIT IV LTR project No. 21957 (CGAL), and by the Hermann Minkowski–MINERVA Center for Geometry at Tel Aviv University.

arrangements. However, it has recently been realized that the more general arrangements of (curved) surfaces are more significant and have a wider range of applications, because many geometric problems in diverse areas can be reduced to problems involving such arrangements (but not necessarily to arrangements of hyperplanes). We will give here four such examples:

Motion planning. Assume that we have a robot system B with d degrees of freedom, i.e., we can represent each placement of B as a point in d-space. Suppose that the workspace of B is cluttered with obstacles, whose shapes and locations are known. For each combination of a geometric feature (vertex, edge, face) of an obstacle and a similar feature of B, define their *contact surface* as the set of all points in d-space that represent a placement of B in which contact is made between these specific features. Let Z be a point corresponding to a given initial *free* placement of B, in which it does not intersect any obstacle. Then the set of all free placements of B that can be reached from Z via a collision-free continuous motion will obviously correspond to the cell containing Z in the arrangement of the contact surfaces. Thus, the robot motion planning problem leads to the problem of computing a single cell in an arrangement of surfaces in higher dimensions. The *combinatorial complexity* of this cell, i.e., the total number of lower-dimensional faces appearing on its boundary, serves as a trivial lower bound for the running time of the motion planning problem (assuming the entire cell has to be output). It turns out that in most instances this bound can be almost matched by suitable algorithms.

Generalized Voronoi diagrams. Let S be a set of n 'simply-shaped'[1] pairwise-disjoint compact convex objects in d-space, and let ρ be some metric. The *Voronoi diagram* $\mathrm{Vor}_\rho(S)$ of S under the metric ρ is defined, as usual, as the decomposition of d-space into *Voronoi cells* $V(s)$, for $s \in S$, where

$$V(s) = \{x \in \mathbb{R}^d \mid \rho(x,s) \le \rho(x,s') \text{ for all } s' \in S\}.$$

For each $s \in S$ define a function $f_s(x) = \rho(x,s)$, for $x \in \mathbb{R}^d$ (where $\rho(x,s) = \min_{y \in s} \rho(x,y)$). Define $F(x) = \min_{s \in S} f_s(x)$, for $x \in \mathbb{R}^d$; we refer to F as the *lower envelope* of the f_s's. If we project the graph of F onto $\mathbb{R}^d$, we obtain a decomposition of $\mathbb{R}^d$ into connected cells of dimensions $0, \dots, d$, so that over each cell F is attained by a fixed set of functions. This decomposition is nothing but $\mathrm{Vor}_\rho(S)$, as follows directly by definition, and as has already been noted in [39]. In other words, the study of Voronoi diagrams in d dimensions is equivalent to the study of the lower envelope of a collection of surfaces in $\mathbb{R}^{d+1}$. Planar Voronoi diagrams have been studied intensively (see [20,36,60,68]), but very little is known about generalized Voronoi diagrams in higher dimensions.

---

[1] Formally, this means that each object in S is defined as a Boolean combination of a constant number of polynomial equalities and inequalities of constant maximum degree in a constant number of variables; we also refer to such objects as semialgebraic sets of *constant description complexity*.

Transversals. Let S be a set of 'simply-shaped' compact convex objects in $\mathbb{R}^d$. A hyperplane h is called a *transversal* of S if h intersects every member of S. Let $\mathcal{T}(S)$ denote the space of all hyperplane transversals of S. We wish to study the structure of $\mathcal{T}(S)$. To facilitate this study, we apply the geometric duality described, e.g., in [36], which maps hyperplanes to points and vice versa, and which preserves incidences and the above/below relationship between points and hyperplanes. Specifically, this duality maps a point $(a_1, \ldots, a_d)$ to the hyperplane $x_d = a_1 x_1 + \cdots + a_{d-1} x_{d-1} + a_d$, and a hyperplane $x_d = b_1 x_1 + \cdots + b_{d-1} x_{d-1} + b_d$ to the point $(-b_1, \ldots, -b_{d-1}, b_d)$. Let h be the hyperplane $x_d = a_1 x_1 + \cdots + a_{d-1} x_{d-1} + a_d$, and suppose that h intersects an object $s \in S$. Translate h up and down until it becomes tangent to s. Denote the resulting upper and lower tangent hyperplanes by

$$x_d = a_1 x_1 + \cdots + a_{d-1} x_{d-1} + U_s(a_1, \ldots, a_{d-1})$$

and

$$x_d = a_1 x_1 + \cdots + a_{d-1} x_{d-1} + L_s(a_1, \ldots, a_{d-1}),$$

respectively. Then we have

$$L_s(a_1, \ldots, a_{d-1}) \le a_d \le U_s(a_1, \ldots, a_{d-1}).$$

Now if h is a tranversal of S, we must have

$$\max_{s \in S} L_s(a_1, \ldots, a_{d-1}) \le a_d \le \min_{s \in S} U_s(a_1, \ldots, a_{d-1}).$$

In other words, in dual space, $\mathcal{T}(S)$ is the region 'sandwiched' between a lower envelope of a collection of n surfaces and an upper envelope of another collection of n surfaces.

Geometric optimization. Many problems in geometric optimization can be solved by reducing them to problems involving the analysis of substructures in arrangements. One such example is the problem of computing the *width* of a set S of n points in $\mathbb{R}^3$; this is the smallest distance between two parallel planes enclosing S between them. This problem has been studied in a series of recent papers [2,10,27], where more details can be found. Following standard techniques, it suffices to solve the decision problem where we are given a parameter $w > 0$ and wish to determine whether the width of S is $\le w$. For this, we need to test all 'antipodal' pairs of edges of the convex hull C of S (these are pairs of edges that admit parallel supporting planes), and determine whether there exists such a pair whose supporting planes lie at distance $\le w$. In the worst case the number of such pairs may be $\Theta(n^2)$, and the goal is to avoid having to test explicitly all such pairs. This is achieved by a step that partitions the set of all antipodal pairs of edges into a collection of complete bipartite graphs, so that, for each such graph $E \times E'$, it suffices to determine whether there exist $e \in E$, $e' \in E'$, such that the distance between the lines containing e and e', respectively, is $\le w$. This problem can be solved by mapping the lines containing the edges in E' into points in 4-space (lines in three dimensions have four degrees of freedom), and by mapping the lines containing the edges of E into trivariate functions, so that the distance between a line ℓ' in the first set and a line ℓ in the second set is $\leq w$ if and only if the point that represents ℓ' lies below the graph of the function that represents ℓ. We have thus landed in a problem where we are given a collection of points and a collection of surfaces in 4-space, and we need to determine whether there exists a point that lies below the upper envelope of the surfaces. Full details can be found in the papers cited above.

As these examples show, many problems in combinatorial and computational geometry can be rephrased in terms of certain substructures (single cells, lower envelopes, regions enclosed between an upper envelope and a lower envelope, etc.) in arrangements of surfaces in higher dimensions. The past decade has seen a significant expansion in the study of arrangements. This work has been described in the recent book [75] and in several other surveys [45,51,52,73,74].

The main theme in the study of arrangements is to analyze the combinatorial complexity of such substructures, with the goal of obtaining bounds on this complexity that are significantly smaller than the complexity of the full arrangement. A second theme is the design of efficient algorithms for constructing such substructures.

In this study there are three main relevant parameters: the number n of surfaces, their maximum algebraic degree b, and the dimension d. The approach taken here is 'combinatorial', in which we want to calibrate the dependence of the complexity of the various structures and algorithms on the number n of surfaces, assuming that the dimension and maximum degree, as well as any other factor that does not depend on n, is constant. In this way, all issues related to the algebraic complexity of the problem are 'swept under the rug'. These issues should be (and indeed have been) picked up in the complementary 'algebraic' mode of research, where the dependence on the maximum degree b is more relevant; this area is known as computational real algebraic geometry; see [56,69,64] for studies of this kind.

2   Complexity of Lower Envelopes

During the past decade, significant progress has been made on the problem of bounding the complexity of the lower envelope of a collection of multivariate functions. This problem has been open since 1986, when it was shown in [54] that the combinatorial complexity of the lower envelope of n *univariate* continuous functions, each pair of which intersect in at most s points, is at most $\lambda_s(n)$, the maximum length of an (n, s)-*Davenport–Schinzel* sequence. This bound is slightly super-linear in n, for any fixed s (for example, it is $\Theta(n\alpha(n))$ for $s = 3$, where $\alpha(n)$ is the extremely slowly growing inverse of Ackermann's function [54]; see also [12,75]). Since the complexity of the full arrangement of such a collection of functions can be $\Theta(n^2)$ in the worst case, this result shows that the worst-case complexity of the lower envelope is smaller than the overall complexity of the arrangement by nearly a factor of n.

It was then conjectured that a similar phenomenon occurs in higher dimensions. That is, the combinatorial complexity of the lower envelope of a collection $\mathcal{F}$ of n 'well-behaved' d-variate functions should be close to $O(n^d)$ (as opposed to $\Theta(n^{d+1})$, which can be the complexity of the entire arrangement of the function graphs). More precisely, according to a stronger version of this conjecture, this quantity should be at most $O(n^{d-1}\lambda_s(n))$, for some constant s depending on the shape of the functions in $\mathcal{F}$.

This problem was almost completely settled in 1994 [49,72]: Let $\mathcal{F}$ be a collection of (possibly partially-defined) d-variate functions, such that all functions in $\mathcal{F}$ are algebraic of constant maximum degree and, in case of partial functions, the domain of definition of each function is a semi-algebraic set defined by a constant number of polynomial equalities and inequalities of constant maximum degree. (As already mentioned, we refer to such a region as having *constant description complexity*.) Then, for any $\varepsilon > 0$, the combinatorial complexity of the lower envelope of $\mathcal{F}$ is $O(n^{d+\varepsilon})$, where the constant of proportionality depends on ε, d, and on the maximum degree of the functions and of the polynomials defining their domains.[2] Thus, the weak form of the above conjecture has been settled in the affirmative.

Prior to the derivation of this bound, the above conjectures have been confirmed only in some special cases, including the case in which the graphs of the functions are d-simplices in $\mathbb{R}^{d+1}$, where a tight worst-case bound, $\Theta(n^d\alpha(n))$, was established in [37,65]. (The case $d = 1$, involving n segments in the plane, where the bound is $\Theta(n\alpha(n))$, had been analyzed earlier, in [54,80].) There are also some even more special cases, like the case of hyperplanes, where the maximum complexity of their lower envelope is known to be $\Theta(n^{\lfloor(d+1)/2\rfloor})$, by the so-called *Upper Bound Theorem* for convex polytopes [62]. The case of balls also admits a much better bound, using a standard lifting transformation (see [36]).

The new analysis technique (whose details can be found in [75]) has later been extended in various ways, to be described below. These additional results include near-optimal bounds on the complexity of a single cell in an arrangement of surfaces in $\mathbb{R}^d$ and on the complexity of the region enclosed between two envelopes of bivariate functions, as well as to many other combinatorial results involving arrangements, as noted below, and efficient algorithms for constructing lower envelopes and single cells. The new results have been applied to obtain improved algorithmic and combinatorial bounds for a variety of problems; we will mention some of these applications later on.

---

[2] In this paper we will state similar complexity bounds that depend on $\varepsilon > 0$, without saying repeatedly 'for any $\varepsilon > 0$'; the meaning of such a bound is that it holds for any $\varepsilon > 0$, where the constant of proportionality depends on ε (and perhaps also on other problem constants), and usually tends to infinity when ε tends to zero. For algorithmic complexity, the meaning of such a bound is that the algorithm can be fine-tuned, as a function of ε, so that its complexity obeys the stated bound.

3 Algorithms for Lower Envelopes

Once the combinatorial complexity of lower envelopes of multivariate functions has been (more or less) resolved, the next task is to derive efficient algorithms for computing such lower envelopes. One of the strongest specifications of such a computation is as follows. We are given a collection $\mathcal{F}$ of d-variate algebraic functions satisfying the above conditions. We want to compute the lower envelope $E_{\mathcal{F}}$ and store it in some data structure, so that, given a query point $p \in \mathbb{R}^d$, we can efficiently compute the value $E_{\mathcal{F}}(p)$, and the function(s) attaining $E_{\mathcal{F}}$ at p. (This is, for example, the version that is needed for solving the three-dimensional width problem described above.) Of course, we need to assume here an appropriate model of computation, where various primitive operations on a constant number of functions can be each performed in constant time. As mentioned above, the exact arithmetic model in computational real algebraic geometry is an appropriate choice.

This task has recently been accomplished for the case of bivariate functions in several papers [7,22,23,33,72]. Some of these techniques use randomized algorithms, and their expected running time is $O(n^{2+\varepsilon})$, which is comparable with the maximum complexity of such an envelope. The simplest algorithm is probably the one given in [7]. It is deterministic and uses divide-and-conquer. That is, it partitions the set of functions into two subsets of roughly equal size, and computes recursively the lower envelopes E_1, E_2 of these two subsets. It then projects these envelopes onto the xy-plane, to obtain two respective planar maps M_1, M_2, and forms the overlay M of M_1 and M_2. Over each face f of M there are only two functions that can attain the final envelope (the function attaining E_1 over f and the function attaining E_2, so we compute the lower envelope of these two functions over f, and repeat this step for all faces of M. It is easy to see that the cost of this step is proportional to the number of faces of M. In general, overlaying two planar maps of complexity N each, can result in a map whose complexity is $\Theta(N^2)$, which may be $\Omega(n^4)$ in our case. Fortunately, it was shown in [7] that for lower envelopes, the overlay of M_1 and M_2 has complexity $O(n^{2+\varepsilon})$. In other words, the worst-case bound on the complexity of the overlay of (the xy-projections of) two lower envelopes of bivariate functions is asymptotically the same as the bound for the complexity of a single envelope. This implies that the complexity of the above divide-and-conquer algorithm is $O(n^{2+\varepsilon})$.

In higher dimensions, the only result known so far is that lower envelopes of trivariate functions satisfying the above properties can be computed, in the above strong sense, in randomized expected time $O(n^{3+\varepsilon})$ [2]. For $d > 3$, it is also shown in [2] that all vertices, edges and 2-faces of the lower envelope of n d-variate functions, as above, can be computed in randomized expected time $O(n^{d+\varepsilon})$. It is still an open problem whether such a lower envelope can be computed within similar time bounds in the above stronger sense. Another, more difficult problem is to devise *output-sensitive* algorithms, whose complexity depends on the actual combinatorial complexity of the envelope. This problem is really hard even for collections of nonintersecting triangles in 3-space. It would also be interesting to develop algorithms for certain special classes of functions, where better bounds are known for the complexity of the envelope, e.g., for envelopes of piecewise-linear functions (see below for more details).

4   Single Cells

Lower envelopes are closely related to other substructures in arrangements, notably *single cells* and *zones*. (The lower envelope is a portion of the boundary of the bottommost cell of the arrangement.) In two dimensions, it was shown in [46] that the complexity of a single face in an arrangement of n arcs, each pair of which intersect in at most s points, is $O(\lambda_{s+2}(n))$, and so is of the same asymptotic order of magnitude as the complexity of the lower envelope of such a collection of arcs. Again, the prevailing conjecture is that the same holds in higher dimensions. That is, the complexity of a single cell in an arrangement of n algebraic surfaces in d-space satisfying the above assumptions is close to $O(n^{d-1})$, or, in a stronger form, this complexity should be $O(n^{d-2}\lambda_s(n))$, for some appropriate constant s. The weaker version of this conjecture has recently been confirmed in [21]; see also [50] for an earlier result for the three-dimensional case: Let $\mathcal{A}$ be an arrangement of n surface patches in $\mathbb{R}^d$, all of them algebraic of constant description complexity. It was proved in [21,50] that, for any $\varepsilon > 0$, the complexity of a single cell in $\mathcal{A}$ is $O(n^{d-1+\varepsilon})$, where the constant of proportionality depends on ε, d, and on the maximum degree of the surfaces and of their boundaries.

The results of [50] mentioned above easily imply that, for fairly general robot systems with d degrees of freedom, the complexity of the space of all free placements of the system, reachable from a given initial placement, is $O(n^{d-1+\varepsilon})$, a significant improvement over the previous, naive bound $O(n^d)$. In three dimensions, the corresponding algorithmic problem, of devising an efficient (near-quadratic) algorithm for computing such a cell, has recently been solved in [70]. We will say more about this result when we discuss vertical decompositions below. Prior to this result, several other near-quadratic algorithms were proposed for some special classes of surfaces [7,17,47,48]. For example, the paper [48] gives a near-quadratic algorithm for the single cell problem in the special case of arrangements that arise in the motion planning problem for a (nonconvex) polygonal robot moving (translating and rotating) in a planar polygonal region. However, this algorithm exploits the special structure of the surfaces that arise in this case (namely, that any cross-section of such an arrangement at a fixed orientation of the polygon is polygonal), and does not extend to the general case. The algorithm given in [7] also provides a near-quadratic solution for the case where all the surfaces are graphs of totally-defined continuous algebraic bivariate functions (so that the cell in question is xy-monotone).

In higher dimensions, we also mention the special case of a single cell in an arrangement of n $(d-1)$-simplices in $\mathbb{R}^d$. It was shown in [17] that the complexity of such a cell is $O(n^{d-1}\log n)$; a simplified proof was recently given in [77]. This bound is sharper than the general bound stated above; the best lower bound known for this complexity is $\Omega(n^{d-1}\alpha(n))$, so a small gap between the upper and lower bounds still remains.

5   Zones

Given an arrangement $\mathcal{A}$ of surfaces in $\mathbb{R}^d$, and another surface σ_0, the *zone* of σ_0 is the collection of all cells of the arrangement $\mathcal{A}$ that σ_0 crosses, and the complexity of the zone is the sum of complexities of all these cells. The 'classical' Zone Theorem [36,40] asserts that the maximum complexity of the zone of a hyperplane in an arrangement of n hyperplanes in $\mathbb{R}^d$ is $\Theta(n^{d-1})$, where the constant of proportionality depends on d. This has been extended in [14] to the zone of an algebraic or convex surface (of any dimension $p < d$) in an arrangement of hyperplanes. The bound on the complexity of such a zone is $O(n^{\lfloor(d+p)/2\rfloor}\log^c n)$, and $\Omega(n^{\lfloor(d+p)/2\rfloor})$ in the worst case, where $c = 1$ when $d - p$ is odd and $c = 0$ when $d - p$ is even. It is not clear whether the logarithmic factor is really needed, or that it is just an artifact of the proof technique.

The result of [21,50] can easily be extended to obtain a bound of $O(n^{d-1+\varepsilon})$, on the complexity of the zone of an algebraic surface σ_0 (of constant description complexity) in an arrangement of n algebraic surfaces in $\mathbb{R}^d$, as above. Intuitively, the proof proceeds by cutting each of the given surfaces along its intersection with σ_0, and by shrinking the surface away from that intersection, thus leaving a 'tiny' gap there. These modifications transform the zone of σ into a single cell in the arrangement of the new surfaces, and the result of [21] can then be applied. (The same technique has been used earlier in [38], to obtain a near-linear bound on the complexity of the zone of an arc in a 2-dimensional arrangement of arcs.) A similar technique implies that the complexity of the zone of an algebraic or convex surface in an arrangement of n $(d-1)$-simplices in $\mathbb{R}^d$ is $O(n^{d-1}\log n)$ [17,77].

6   Generalized Voronoi Diagrams

One of the interesting applications of the new bounds on the complexity of lower envelopes is to generalized Voronoi diagrams in higher dimensions. Let S be a set of n pairwise-disjoint convex objects in d-space, each of constant description complexity, and let ρ be some metric. The *Voronoi diagram* $\text{Vor}_\rho(S)$ of S under the metric ρ has been defined in the introduction. As shown there, the diagram is simply the minimization diagram of the lower envelope of n d-variate 'distance functions' induced by the objects of S.

In the classical case, in which ρ is the Euclidean metric and the objects in S are singletons (points), one can replace these distance functions by a collection of n hyperplanes in $\mathbb{R}^{d+1}$, so the maximum possible complexity of their lower envelope (and thus of $\text{Vor}_\rho(S)$) is $\Theta(n^{\lceil d/2\rceil})$ (see, e.g., [36]). In more general settings, though, this reduction is not possible. Nevertheless, the new bounds on the complexity of lower envelopes imply that the complexity of the diagram is $O(n^{d+\varepsilon})$. While this bound is nontrivial, it is conjectured to be too weak. For example, this bound is near-quadratic for planar Voronoi diagrams, but the complexity of practically any kind of planar Voronoi diagram is known to be only $O(n)$.

In three dimensions, the above-mentioned bound for point sites and Euclidean metric is $\Theta(n^2)$. It has been a long-standing open problem whether a similar quadratic or near-quadratic bound holds in 3-space for more general objects and metrics (here the new results on lower envelopes give only an upper bound of $O(n^{3+\varepsilon})$). Thus the problem stated above calls for improving this bound by roughly another factor of n. It thus appears to be a considerably more difficult problem than that of lower envelopes, and the only hope of making progress here is to exploit the special structure of the distance functions $\rho(x, s)$.

Fortunately, some progress on this problem was made recently. It was shown in [29] that the complexity of the Voronoi diagram is $O(n^2\alpha(n) \log n)$, for the case where the objects of S are lines, and the metric ρ is a convex distance function induced by a convex *polytope* with a constant number of facets. (Note that the L_1 and L_∞ metrics are special cases of such distance functions. Note also that such a distance function is not necessarily a metric, because it will fail to be symmetric if the defining polytope is not centrally symmetric.) The best known lower bound for the complexity of the diagram in this special case is $\Omega(n^2\alpha(n))$. In another recent paper [24], it is shown that the maximum complexity of the L_1-Voronoi diagram of a set of n points in $\mathbb{R}^3$ is $\Theta(n^2)$. Finally, it is shown in [78] that the complexity of the three-dimensional Voronoi diagram of point sites under general polyhedral convex distance functions is $O(n^2 \log n)$. The most intriguing unsolved problem is to obtain a similar bound for a set S of n lines in space but under the Euclidean metric.

An interesting special case of these problems involves *dynamic Voronoi diagrams* for moving points in the plane. Let S be a set of n points in the plane, each moving along some line at some fixed velocity. The goal is to bound the number of combinatorial changes of $\text{Vor}_\rho(S)$ over time. This dynamic Voronoi diagram can easily be transformed into a 3-dimensional Voronoi diagram, by adding the time t as a third coordinate. The points become lines in 3-space, and the metric is a distance function induced by a horizontal disc (that is, the distance from a point $p(x_0, y_0, t_0)$ to a line ℓ is the Euclidean distance from p to the point of intersection of ℓ with the horizontal plane $t = t_0$). Here too the open problem is to derive a near-quadratic bound on the complexity of the diagram. Cubic or near-cubic bounds are known for this problem, even under more general settings [42,44,72], but subcubic bounds are known only in some very special cases [28].

Next, consider the problem of bounding the complexity of generalized Voronoi diagrams in higher dimensions. As mentioned above, when the objects in S are n points in $\mathbb{R}^d$ and the metric ρ is Euclidean, the complexity of $\text{Vor}_\rho(S)$ is $O(n^{\lceil d/2 \rceil})$. As d increases, this becomes drastically smaller than the naive $O(n^{d+1})$ bound or the improved bound, $O(n^{d+\varepsilon})$, obtained by viewing the Voronoi diagram as a lower envelope in $\mathbb{R}^{d+1}$. The same bound of $O(n^{\lceil d/2 \rceil})$ has recently been obtained in [24] for the complexity of the L_∞-diagram of n points in d-space (it was also shown that this bound is tight in the worst case).

It is thus tempting to conjecture that the maximum complexity of generalized Voronoi diagrams in higher dimensions is close to this bound. Unfortunately, this was recently shown to be false in [15], where a lower bound of $\Omega(n^{d-1})$ is given. The sites used in this construction are convex polytopes, and the distance is either Euclidean or a polyhedral convex distance function. For $d = 3$, this lower bound does not contradict the conjecture made above, that the complexity of generalized Voronoi diagrams should be at most near-quadratic in this case. Also, in higher dimensions, the conjecture mentioned above is still not refuted when the sites are singleton points. Finally, for the general case, the construction of [15] still leaves a gap of roughly a factor of n between the known upper and lower bounds.

7   Union of Geometric Objects

A subproblem related to generalized Voronoi diagrams is as follows. Let S and ρ be as above (say, for the 3-dimensional case). Let K denote the region consisting of all points $x \in \mathbb{R}^3$ whose smallest distance from a site in S is at most r, for some fixed parameter $r > 0$. Then $K = \bigcup_{s \in S} B(s,r)$, where $B(s,r) = \{x \in \mathbb{R}^3 \mid \rho(x,s) \leq r\}$. We thus face the problem of bounding the combinatorial complexity of the union of n objects in 3-space (of some special type). For example, if S is a set of lines and ρ is the Euclidean distance, the objects are n congruent infinite cylinders in 3-space. In general, if the metric ρ is a distance function induced by some convex body P, the resulting objects are the *Minkowski sums* $s \oplus (-rP)$, for $s \in S$, where $A \oplus B = \{x + y \mid x \in A, \ y \in B\}$. Of course, this problem can also be stated in any higher dimension.

Since it has been conjectured that the complexity of the whole Voronoi diagram should be near-quadratic (in 3-space), the same conjecture should apply to the (simpler) structure K (whose boundary can be regarded as a 'cross-section' of the diagram at 'height' r; it does indeed correspond to the cross-section at height r of the lower envelope that represents the diagram). Recently, this conjecture has been confirmed in [18], in the special case where both P and the objects of S are convex polyhedra (see also [53] for an earlier study of a special case, with a slightly better bound). Let us discuss this result in more detail. An earlier paper [16] has studied the case involving the union of k arbitrary convex polyhedra in 3-space, with a total of n faces. It was shown there that the complexity of the union is $O(k^3 + nk \log^2 k)$, and can be $\Omega(k^3 + nk\alpha(k))$ in the worst case. The upper bound was subsequently improved to $O(k^3 + nk \log k)$ [19], which still leaves a small gap between the upper and lower bounds. In the subsequent paper [18], these bounds were improved in the special case where the polyhedra in question are Minkowski sums of the form $s_i \oplus P$, where the s_i's are k pairwise-disjoint convex polyhedra, P is a convex polyhedron, and the total number of faces of these Minkowski sums is n. The improved bounds are $O(nk \log k)$ and $\Omega(nk\alpha(k))$. They are indeed near-quadratic, as conjectured.

Recently, the case where P is a ball (namely, the case of the Euclidean distance) has been solved in [11]. It is shown there that the complexity of the union of the Minkowski sums of n pairwise-disjoint triangles in $\mathbb{R}^3$ with a ball is $O(n^{2+\varepsilon})$, for any $\varepsilon > 0$. In the special case where instead of triangles we have n lines, we obtain that the complexity of the union of n infinite congruent cylinders in 3-space is $O(n^{2+\varepsilon})$.

In higher dimensions, it was recently shown in [24] that the maximum complexity of the union of n axis-parallel hypercubes in d-space is $\Theta(n^{\lceil d/2 \rceil})$, and this improves to $\Theta(n^{\lfloor d/2 \rfloor})$ when all the hypercubes have the same size.

The above instances involve Minkowski sums of a collection of pairwise disjoint convex objects with a fixed convex object. Of course, one may consider the union of arbitrary objects and look for special cases where improved combinatorial bounds can be established. For example, it is conjectured that the union of n arbitrarily-aligned cubes in 3-space has near-quadratic complexity.

8   Vertical Decomposition

In many algorithmic applications, one needs to decompose a d-dimensional arrangement, or certain portions thereof, into a small number of subcells, each having constant description complexity. In a typical setup where this problem arises, we need to process in a certain manner an arrangement of n surfaces in d-space. We choose a random sample of r of the surfaces, for some sufficiently large constant r, construct the arrangement of these r surfaces, and decompose it into subcells as above. Since no such subcell is crossed by any surface in the random sample, it follows by standard ε-net theory [30,55,59] that with high probability, none of these subcells is crossed by more than $O(\frac{n}{r} \log r)$ of the n given surfaces. (For this result to hold, it is essential that each of these subcells have constant description complexity.) This allows us to break the problem into recursive subproblems, one for each of these subcells, solve each subproblem separately, and then combine their outputs to obtain a solution for the original problem. The efficiency of this method crucially depends on the number of subcells. The smaller this number is, the faster is the resulting algorithm. (We note that the construction of a 'good' sample of r surfaces can also be performed deterministically, e.g., using the techniques of Matoušek [61].)

The only general-purpose known technique for decomposing an arrangement of surfaces into subcells of constant description complexity is the *vertical decomposition* technique. In this method, we erect a vertical 'wall' up and down (in the x_d-direction) from each $(d-2)$-dimensional face of the arrangement, and extend these walls until they hit another surface. This results in a decomposition of the arrangement into subcells so that each subcell has a unique top facet and a unique bottom facet, and each vertical line cuts it in a connected (possibly empty) interval. We next project each resulting subcell on the hyperplane $x_d = 0$, and apply recursively the same technique within each resulting $(d-1)$-dimensional projected cell, and then 'lift' this decomposition back into d-space, by extending each subcell c in the projection into the vertical cylinder $c \times \mathbb{R}$, and by cutting the original cell by these cylinders. We continue the recursion in this manner until we reach $d = 1$, and thereby obtain the vertical decomposition of the given arrangement. The resulting subcells have the desired properties. Furthermore, if we assume that the originally given surfaces are algebraic of constant maximum degree, then the resulting subcells are semi-algebraic and are defined by a constant number of polynomials of constant maximum degree (although the latter degree can grow quite fast with d). In what follows, we ignore the algebraic complexity of the subcells of the vertical decomposition, and will be mainly interested in bounding their number as a function of n, the number of given surfaces.

It was shown in [26] that the number of cells in such a vertical decomposition of the entire arrangement is $O(n^{2d-3}\beta(n))$, where $\beta(n)$ is a slowly growing function of n (related to the inverse Ackermann's function). However, the only known lower bound is the trivial $\Omega(n^d)$, so there is a considerable gap here, for $d > 3$; for $d = 3$ the two bounds nearly coincide. Improving the upper bound appears to be a very hard task. This problem has been open since 1989; it seems difficult enough to preempt, at the present state of knowledge, any specific conjecture on the true maximum complexity of the vertical decomposition of arrangements in $d > 3$ dimensions.

The bound stated above applies to the vertical decomposition of an entire arrangement of surfaces. In many applications, however, one is interested in the vertical decomposition of only a portion of the arrangement, e.g., a single cell, the region lying below the lower envelope of the given surfaces, the zone of some surface, a specific collection of cells of the arrangement, etc. Since, in general, the complexity of such a portion is known (or conjectured) to be smaller than the complexity of the entire arrangement, one would like to conjecture that a similar phenomenon applies to vertical decompositions. Recently, it was shown in [70] that the complexity of the vertical decomposition of a single cell in an arrangement of n surface patches in 3-space, as above, is $O(n^{2+\varepsilon})$. As mentioned above, this leads to a near-quadratic algorithm for computing such a single cell, which implies that motion planning for fairly general systems with three degrees of freedom can be performed in near-quadratic time, thus settling a major open problem in the area. A similar near-quadratic bound has been obtained in [7] for the vertical decomposition of the region enclosed between a lower envelope and an upper envelope of bivariate functions. Another recent result [4] gives a bound on the complexity of the vertical decomposition of the first k levels in an arrangement of surfaces in 3-space, which is only slightly worse than the worst-case complexity of this undecomposed portion of the arrangement. A challenging open problem is to obtain improved bounds for the complexity of the vertical decomposition of the region lying below the lower envelope of n d-variate functions, for $d \geq 3$.

Finally, an interesting special case is that of hyperplanes. For such arrangements, the vertical decomposition is a too cumbersome construct, because there are other easy methods for decomposing each cell into simplices, whose total number is $O(n^d)$. Still, it is probably a useful exercise to understand the complexity of the vertical decomposition of an arrangement of n hyperplanes in d-space. A recent result of [43] gives an almost tight bound of $O(n^4 \log n)$ for this problem in 4-space, but nothing significantly better than the general bound is known for $d > 4$. Another interesting special case is that of triangles in 3-space. This has been studied by [34,77], where almost tight bounds were obtained for the case of a single cell ($O(n^2 \log^2 n)$), and for the entire arrangement ($O(n^2 \alpha(n) \log n + K)$, where K is the complexity of the undecomposed arrangement). The first bound is slightly better than the general bound of [70] mentioned above. The paper [77] also derives sharp complexity bounds for the vertical decomposition of many cells in such an arrangement, including the case of all nonconvex cells.

9 Other Applications

We conclude this survey by mentioning some additional applications of the new advances in the study of arrangements. We have already discussed in some detail the motion planning application, and have seen how the new results lead to a near-optimal algorithm for the general motion planning problem with three degrees of freedom. Here we discuss three other kinds of applications: to visibility problems in three dimensions, to geometric optimization, and to transversals.

9.1 Visibility in Three Dimensions

Let us consider a special case of the so-called *aspect graph* problem, which has recently attracted much attention, especially in the context of three-dimensional scene analysis and object recognition in computer vision. The aspect graph of a scene represents all topologically-different views of the scene. For background and a survey of recent research on aspect graphs, see [25]. Here we will show how the new complexity bounds for lower envelopes, with some additional machinery, can be used to derive near-tight bounds on the number of views of polyhedral terrains.

Let K be a *polyhedral terrain* in 3-space; that is, K is the graph of a continuous piecewise-linear bivariate function, so it intersects each vertical line in exactly one point. Let n denote the number of edges of K. A line ℓ is said to *lie over* K if every point on ℓ lies on or above K. Let $\mathcal{L}_K$ denote the space of all lines that lie over K. (Since lines in 3-space can be parametrized by four real parameters, we can regard $\mathcal{L}_K$ as a subset of 4-space.) Using an appropriate parametrization, the *lower envelope* of $\mathcal{L}_K$ consists of those lines in $\mathcal{L}_K$ that touch at least one edge of K. Assuming general position of the edges of K, a line in $\mathcal{L}_K$ (or any line, for that matter) can touch at most four edges of K. We estimate the combinatorial complexity of this lower envelope, in terms of the number of its *vertices*, namely those lines in $\mathcal{L}_K$ that touch four distinct edges of K. It was shown in [49] that the number of vertices of $\mathcal{L}_K$, as defined above, is $O(n^3 \cdot 2^{c\sqrt{\log n}})$, for some absolute positive constant c.

We give here a sketch of the proof. We fix an edge e_0 of K, and bound the number of lines of $\mathcal{L}_K$ that touch e_0 and three other edges of K, with the additional proviso that the three other contact points all lie on one fixed side of the vertical plane passing through e_0. We then multiply this bound by the number n of edges, to obtain a bound on the overall number of vertices of $\mathcal{L}_K$. We first rephrase this problem in terms of the lower envelope of a certain collection of surface patches in 3-space, one patch for each edge of K (other than e_0), and then exploit the results on lower envelopes reviewed above.

The space $\mathcal{L}_{e_0}$ of oriented lines that touch e_0 is 3-dimensional: each such line ℓ can be specified by a triple (t, k, ζ), where t is the point of contact with e_0, and $k = \tan\theta$, $\zeta = -\cot\phi$, where (θ, ϕ) are the spherical coordinates of the direction of ℓ, that is, θ is the orientation of the xy-projection of ℓ, and ϕ is the angle between ℓ and the positive z-axis.

For each edge $e \neq e_0$ of K, let σ_e be the surface patch in $\mathcal{L}_{e_0}$ consisting of all points (t, k, ζ) representing lines that touch e and are oriented from e_0 to e. Note that if $(t, k, \zeta) \in \sigma_e$ then $\zeta' > \zeta$ iff the line (t, k, ζ') passes below e. It thus follows that a line ℓ in $\mathcal{L}_{e_0}$ is a vertex of the lower envelope of $\mathcal{L}_K$ if and only if ℓ is a vertex of the lower envelope of the surfaces σ_e in the $tk\zeta$-space, where the height of a point is its ζ-coordinate. It is easy to show that these surfaces are algebraic of constant description complexity. Actually, it is easily seen that the number s of intersections of any triple of these surfaces is at most 2. The paper [49] studies the special case of lower envelopes of collections of such algebraic surface patches in 3-space, with the extra assumption that $s = 2$. It is shown there that the complexity of the lower envelope of such a collection is $O(n^2 \cdot 2^{c\sqrt{\log n}})$, for some absolute positive constant c, a bound that is slightly better than the general bound stated above. These arguments immediately complete the proof. (This bound has been independently obtained by Pellegrini [66], using a different proof technique.) Recently, de Berg [32] has given a lower bound construction, in which the lower envelope of $\mathcal{L}_K$ has complexity $\Omega(n^3)$, implying that the upper bound stated above is almost tight in the worst case.

We can extend the above result as follows. Let K be a polyhedral terrain, as above. Let $\mathcal{R}_K$ denote the space of all rays in 3-space with the property that each point on such a ray lies on or above K. We define the lower envelope of $\mathcal{R}_K$ and its vertices in complete analogy to the case of $\mathcal{L}_K$. By inspecting the proof sketched above, one easily verifies that it applies equally well to rays instead of lines. Hence we obtain that the number of vertices of $\mathcal{R}_K$, as defined above, is also $O(n^3 \cdot 2^{c\sqrt{\log n}})$.

We can apply this bound to obtain a bound of $O(n^5 \cdot 2^{c'\sqrt{\log n}})$, for some constant $c' > c$, on the number of topologically-different orthographic views (i.e., views from infinity) of a polyhedral terrain K with n edges. We omit here details of this analysis, which can be found in [49]. The paper [35] gives a lower bound construction that produces $\Omega(n^5 \alpha(n))$ topologically-different orthographic views of a polyhedral terrain, so the above bound is almost tight in the worst case. It is also instructive to note that, if K is an arbitrary polyhedral set in 3-space with n edges, then the maximum possible number of topologically-different orthographic views of K is $\Theta(n^6)$ [67].

Consider next the extension of the above analysis to bound the number of perspective views of a terrain. As shown recently in [8], the problem can be reduced to the analysis of $O(n^3)$ lower envelopes of appropriate collections of

5-variate functions. This leads to an overall bound of $O(n^{8+\varepsilon})$, for the number of topologically-different perspective views of a polyhedral terrain with n edges. This bound is also known to be almost tight in the worst case, as follows from another lower bound construction given in [35]. Again, in contrast, If K is an arbitrary polyhedral set with n edges, the maximum possible number of topologically-different perspective views of K is $\Theta(n^9)$ [67].

9.2 Geometric Optimization

In the past few years, many problems in geometric optimization have been attacked by techniques that reduce the problem to a problem involving arrangements of surfaces in higher dimensions. These reduced problems sometimes call for the construction of, and searching in lower envelopes or other substructures in such arrangements; one such example, that of computing the width in three dimensions, has been sketched in the introduction. Hence the area of geometric optimization is a natural extension, and a good application area, of the study of arrangements, as described above. See [9] for a recent survey on geometric optimization.

One of the basic techniques for geometric optimization is the *parametric searching* technique, originally proposed by Megiddo [63]. Roughly speaking, the technique derives an efficient algorithm for the optimization problem from an effieicnt algorithm for the *decision problem*: determining whether the optimal value is at most (or at least) some specified value W. This technique has been used to solve a wide variety of geometric optimization problems, including many of those that involve arrangements. Some specific results of this kind include:

- **Selecting distances in the plane:** Given a set S of n points in $\mathbb{R}^2$ and a parameter $k \le \binom{n}{2}$, find the k-th largest distance among the points of S [3]. Here the problem reduces to the construction and searching in 2-dimensional arrangements of congruent disks.
- **The segment center problem:** Given a set S of n points in $\mathbb{R}^2$, and a line segment e, find a placement of e that minimizes the largest distance from the points of S to e [41]. Using lower envelopes of certain special kinds of bivariate functions, and applying a more careful analysis, the problem can be solved in $O(n^{1+\varepsilon})$ time, improving substantially a previous near-quadratic solution given in [5].
- **Extremal polygon placement:** Given a convex polygon P and a closed polygonal environment Q, find the largest similar copy of P that is fully contained in Q [76]. This is just an extension of the corresponding motion planning problem, where the size of P is fixed. The running time of the algorithm is almost the same as that of the motion planning algorithm given in [57,58].
- **Width in three dimensions** (see also the introduction): Compute the width of a set S of n points in $\mathbb{R}^3$; this is the smallest distance between two parallel planes enclosing S between them. This problem has been studied in a series of papers [2,10,27], and the current best bound is $O(n^{3/2+\varepsilon})$ [10].

The technique used in attacking this and the three following problems reduce them to problems involving lower envelopes in 4 dimensions, where we need to construct and to search in such an envelope.

- **Translational separation of two intersecting convex polytopes:** Given two intersecting convex polytopes A and B, find the shortest translation of A that will make its interior disjoint from B. In case $A = B$, the solution is the width of A. Using a similar technique, this problem can be solved in time $O(m^{3/4+\varepsilon}n^{3/4+\varepsilon} + m^{1+\varepsilon} + n^{1+\varepsilon})$, where m and n are the numbers of facets of A and B, respectively [6].

- **Biggest stick in a simple polygon:** Compute the longest line segment that can fit inside a given simple polygon with n edges. The current best solution is $O(n^{3/2+\varepsilon})$ [10] (see also [2,13]).

- **Smallest-width annulus:** Compute the annulus of smallest width that encloses a given set of n points in the plane. Again, the current best solution is $O(n^{3/2+\varepsilon})$ [10] (see also [2,13], and a recent approximation algorithm in [1]).

- **Geometric matching:** Consider the problem where we are given two sets S_1, S_2 of n points in the plane, and we wish to compute a minimum-weight matching in the complete bipartite graph $S_1 \times S_2$, where the weight of an edge (p, q) is the Euclidean distance between p and q. One can also consider the analogous nonbipartite version of the problem, which involves just one set S of $2n$ points, and the complete graph on S. The goal is to explore the underlying geometric structure of these graphs, to obtain faster algorithms than those available for general abstract graphs.

  It was shown in [79] that both the bipartite and the nonbipartite versions of the problem can be solved in time close to $O(n^{2.5})$. Recently, a fairly sophisticated application of vertical decomposition in 3-dimensional arrangements, given in [4], has improved the running time for the bipartite case to $O(n^{2+\varepsilon})$.

This list is by no means exhaustive.

9.3   Transversals

Let S be a set of n compact convex sets in $\mathbb{R}^d$, each of constant description complexity. The space $\mathcal{T}(S)$ of all hyperplane transversals of S has been defined in the introduction. It was shown there that this space, in a dual setting, is the region in $\mathbb{R}^d$ enclosed between the upper envelope of n surfaces and the lower envelope of n other surfaces, where each surface in the first (resp. second) collection represents the locus of the lower (resp. upper) tangent hyperplanes to an object of S. The results of [7] imply that the complexity of $\mathcal{T}(S)$ is $O(n^{2+\varepsilon})$ in three dimensions. No similarly sharp bounds are known in higher dimensions. The results of [7] concerning the complexity of the vertical decomposition of such a 'sandwiched' region imply that $\mathcal{T}(S)$ can be constructed in $O(n^{2+\varepsilon})$ time.

The problem can be generalized by considering lower-dimensional transversals. For example, in three dimensions we can also consider the space of all line transversals of S. By mapping lines in $\mathbb{R}^3$ into points in 4-space, and by using an appropriate parametrization of the lines, the space of all line transversals of S can also be represented as the region in $\mathbb{R}^4$ enclosed between an upper envelope and a lower envelope of two respective collections of surfaces. Since no sharp bounds are known for the complexity of such a region in 4-space, the exact calibration of the complexity of the space of line transversals in 3-space is still an open problem.

References

1. P. Agarwal, B. Aronov, S. Har-Peled and M. Sharir, Approximation and exact algorithms for minimum-width annuli and shells, *Proc. 15th ACM Symp. on Computational Geometry* (1999), 380–389.
2. P.K. Agarwal, B. Aronov and M. Sharir, Computing envelopes in four dimensions with applications, *SIAM J. Comput.* 26 (1997), 1714–1732.
3. P.K. Agarwal, B. Aronov, M. Sharir and S. Suri, Selecting distances in the plane, *Algorithmica* 9 (1993), 495–514.
4. P.K. Agarwal, A. Efrat and M. Sharir, Vertical decompositions of shallow levels in arrangements and their applications, *Proc. 11th ACM Symp. on Computational Geometry* (1995), 39–50. Also to appear in *SIAM J. Comput.*
5. P.K. Agarwal, A. Efrat, M. Sharir and S. Toledo, Computing a segment-center for a planar point set, *J. Algorithms* 15 (1993), 314–323.
6. P.K. Agarwal, L. Guibas, S. Har-Peled, A. Rabinovitch and M. Sharir, Computing exact and approximate shortest separating translations of convex polytopes in three dimensions, in preparation.
7. P.K. Agarwal, O. Schwarzkopf and M. Sharir, The overlay of lower envelopes in 3-space and its applications, *Discrete Comput. Geom.* 15 (1996), 1–13.
8. P.K. Agarwal and M. Sharir, On the number of views of polyhedral terrains, *Discrete Comput. Geom.* 12 (1994), 177–182.
9. P.K. Agarwal and M. Sharir, Efficient algorithms for geometric optimization, *ACM Computing Surveys* 30 (1998), 412–458.
10. P.K. Agarwal and M. Sharir, Efficient randomized algorithms for some geometric optimization problems, *Discrete Comput. Geom.* 16 (1996), 317–337.
11. P. Agarwal and M. Sharir, Pipes, cigars and kreplach: The union of Minkowski sums in three dimensions, *Proc. 15th ACM Symp. on Computational Geometry* (1999), 143–153.
12. P. Agarwal, M. Sharir and P. Shor, Sharp upper and lower bounds for the length of general Davenport Schinzel sequences, *J. Combin. Theory, Ser. A* 52 (1989), 228–274.
13. P.K. Agarwal, M. Sharir and S. Toledo, New applications of parametric searching in computational geometry. *J. Algorithms* 17 (1994), 292–318.
14. B. Aronov, M. Pellegrini and M. Sharir, On the zone of a surface in a hyperplane arrangement, *Discrete Comput. Geom.* 9 (1993), 177–186.
15. B. Aronov, personal communication, 1995.
16. B. Aronov and M. Sharir, The union of convex polyhedra in three dimensions, *Proc. 34th IEEE Symp. on Foundations of Computer Science* (1993), 518–527.
17. B. Aronov and M. Sharir, Castles in the air revisited, *Discrete Comput. Geom.* 12 (1994), 119–150.
18. B. Aronov and M. Sharir, On translational motion planning of a convex polyhedron in 3-space, *SIAM J. Comput.* 26 (1997), 1785–1803.

19. B. Aronov, M. Sharir and B. Tagansky, The union of convex polyhedra in three dimensions, *SIAM J. Comput.* 26 (1997), 1670–1688 (a revised version of [16]).

20. F. Aurenhammer, Voronoi diagrams—A survey of a fundamental geometric data structure, *ACM Computing Surveys* 23 (1991), 346–405.

21. S. Basu, On the combinatorial and topological complexity of a single cell, *Proc. 39th Annu. IEEE Sympos. Found. Comput. Sci.*, 1998, 606–616.

22. J.D. Boissonnat and K. Dobrindt, Randomized construction of the upper envelope of triangles in $\mathbb{R}^3$, *Proc. 4th Canadian Conf. on Computational Geometry* (1992), 311–315.

23. J.D. Boissonnat and K. Dobrindt, On-line randomized construction of the upper envelope of triangles and surface patches in $\mathbb{R}^3$, *Comp. Geom. Theory Appls.* 5 (1996), 303–320.

24. J.D. Boissonnat, M. Sharir, B. Tagansky and M. Yvinec, Voronoi diagrams in higher dimensions under certain polyhedral distance functions, *Discrete Comput. Geom.* 19 (1998), 485–519.

25. K.W. Bowyer and C.R. Dyer, Aspect graphs: An introduction and survey of recent results, *Int. J. of Imaging Systems and Technology* 2 (1990), 315–328.

26. B. Chazelle, H. Edelsbrunner, L. Guibas and M. Sharir, A singly exponential stratification scheme for real semi-algebraic varieties and its applications, *Proc. 16th Int. Colloq. on Automata, Languages and Programming* (1989), 179–193.

27. B. Chazelle, H. Edelsbrunner, L. Guibas and M. Sharir, Diameter, width, closest line pair, and parametric searching, *Discrete Comput. Geom.* 10 (1993), 183–196.

28. L.P. Chew, Near-quadratic bounds for the L_1 Voronoi diagram of moving points, *Proc. 5th Canadian Conf. on Computational Geometry* (1993), 364–369.

29. L.P. Chew, K. Kedem, M. Sharir, B. Tagansky and E. Welzl, Voronoi diagrams of lines in three dimensions under polyhedral convex distance functions, *J. Algorithms* 29 (1998), 238–255.

30. K.L. Clarkson, New applications of random sampling in computational geometry, *Discrete Comput. Geom.* 2 (1987), 195–222.

31. K.L. Clarkson and P.W. Shor, Applications of random sampling in computational geometry, II, *Discrete Comput. Geom.* 4 (1989), 387–421.

32. M. de Berg, personal communication, 1993.

33. M. de Berg, K. Dobrindt and O. Schwarzkopf, On lazy randomized incremental construction, *Discrete Comput. Geom.* 14 (1995), 261–286.

34. M. de Berg, L. Guibas and D. Halperin, Vertical decomposition for triangles in 3-space, *Discrete Comput. Geom.* 15 (1996), 35–61.

35. M. de Berg, D. Halperin, M. Overmars and M. van Kreveld, Sparse arrangements and the number of views of polyhedral scenes, *Internat. J. Comput. Geom. Appl.* 7 (1997), 175–195.

36. H. Edelsbrunner, *Algorithms in Combinatorial Geometry*, Springer-Verlag, Heidelberg 1987.

37. H. Edelsbrunner, The upper envelope of piecewise linear functions: Tight complexity bounds in higher dimensions, *Discrete Comput. Geom.* 4 (1989), 337–343.

38. H. Edelsbrunner, L. Guibas, J. Pach, R. Pollack, R. Seidel and M. Sharir, Arrangements of curves in the plane: topology, combinatorics, and algorithms, *Theoret. Comput. Sci.* 92 (1992), 319–336.

39. H. Edelsbrunner and R. Seidel, Voronoi diagrams and arrangements, *Discrete Comput. Geom.* 1 (1986), 25–44.

40. H. Edelsbrunner, R. Seidel and M. Sharir, On the zone theorem for hyperplane arrangements, *SIAM J. Comput.* 22 (1993), 418–429.

41. A. Efrat and M. Sharir, A near-linear algorithm for the planar segment center problem, *Discrete Comput. Geom.* 16 (1996), 239–257.
42. J.-J. Fu and R.C.T. Lee, Voronoi diagrams of moving points in the plane, *Internat. J. Comput. Geom. Appl.* 1 (1994), 23–32.
43. L. Guibas, D. Halperin, J. Matoušek and M. Sharir, On vertical decomposition of arrangements of hyperplanes in four dimensions, *Discrete Comput. Geom.* 14 (1995), 113–122.
44. L. Guibas, J. Mitchell and T. Roos, Voronoi diagrams of moving points in the plane, *Proc. 17th Internat. Workshop Graph-Theoret. Concepts Computer Science*, Lecture Notes in Comp. Sci., vol. 570, Springer-Verlag, pp. 113–125.
45. L. Guibas and M. Sharir, Combinatorics and algorithms of arrangements, in *New Trends in Discrete and Computational Geometry*, (J. Pach, Ed.), Springer-Verlag, 1993, 9–36.
46. L. Guibas, M. Sharir and S. Sifrony, On the general motion planning problem with two degrees of freedom, *Discrete Comput. Geom.* 4 (1989), 491–521.
47. D. Halperin, On the complexity of a single cell in certain arrangements of surfaces in 3-space, *Discrete Comput. Geom.* 11 (1994), 1–33.
48. D. Halperin and M. Sharir, Near-quadratic bounds for the motion planning problem for a polygon in a polygonal environment, *Discrete Comput. Geom.* 16 (1996), 121–134.
49. D. Halperin and M. Sharir, New bounds for lower envelopes in three dimensions with applications to visibility of terrains, *Discrete Comput. Geom.* 12 (1994), 313–326.
50. D. Halperin and M. Sharir, Almost tight upper bounds for the single cell and zone problems in three dimensions, *Discrete Comput. Geom.* 14 (1995), 285–410.
51. D. Halperin, Arrangements, in *Handbook of Discrete and Computational Geometry* (J.E. Goodman and J. O'Rourke, Editors), CRC Press, Boca Raton, FL, 1997, 389–412.
52. D. Halperin and M. Sharir, Arrangements and their applications in robotics: Recent developments, *Proc. Workshop on Algorithmic Foundations of Robotics* (K. Goldberg et al., Editors), A. K. Peters, Boston, MA, 1995, 495–511.
53. D. Halperin and C.-K. Yap, Complexity of translating a box in polyhedral 3-space, *Proc. 9th Annu. ACM Sympos. Comput. Geom.* 1993, 29–37.
54. S. Hart and M. Sharir, Nonlinearity of Davenport–Schinzel sequences and of generalized path compression schemes, *Combinatorica* 6 (1986), 151–177.
55. D. Haussler and E. Welzl, ε-nets and simplex range queries, *Discrete Comput. Geom.* 2 (1987), 127–151.
56. J. Heintz, T. Recio and M.F. Roy, Algorithms in real algebraic geometry and applications to computational geometry, in *Discrete and Computational Geometry: Papers from DIMACS Special Year*, (J. Goodman, R. Pollack, and W. Steiger, Eds.), American Mathematical Society, Providence, RI, 137–163.
57. K. Kedem and M. Sharir, An efficient motion planning algorithm for a convex rigid polygonal object in 2-dimensional polygonal space, *Discrete Comput. Geom.* 5 (1990), 43–75.
58. K. Kedem, M. Sharir and S. Toledo, On critical orientations in the Kedem-Sharir motion planning algorithm, *Discrete Comput. Geom.* 17 (1997), 227–239.
59. J. Komlós, J. Pach and G. Woeginger, Almost tight bound on epsilon–nets, *Discrete Comput. Geom.* 7 (1992), 163–173.
60. D. Leven and M. Sharir, Intersection and proximity problems and Voronoi diagrams, in *Advances in Robotics*, Vol. I, (J. Schwartz and C. Yap, Eds.), 1987, 187–228.

61. J. Matoušek, Approximations and optimal geometric divide-and-conquer, *J. Comput. Syst. Sci.* 50 (1995), 203–208.

62. P. McMullen and G. C. Shephard, *Convex Polytopes and the Upper Bound Conjecture*, Lecture Notes Ser. 3, Cambridge University Press, Cambridge, England, 1971.

63. N. Megiddo, Applying parallel computation algorithms in the design of serial algorithms, *J. ACM* 30, 852–865.

64. B. Mishra, Computational real algebraic geometry, in *Handbook of Discrete and Computational Geometry* (J.E. Goodman and J. O'Rourke, Eds.), CRC Press LLC, Boca Raton, FL, 1997, 537–556.

65. J. Pach and M. Sharir, The upper envelope of piecewise linear functions and the boundary of a region enclosed by convex plates: Combinatorial analysis, *Discrete Comput. Geom.* 4 (1989), 291–309.

66. M. Pellegrini, On lines missing polyhedral sets in 3-space, *Proc. 9th ACM Symp. on Computational Geometry* (1993), 19–28.

67. H. Plantinga and C. Dyer, Visibility, occlusion, and the aspect graph, *Internat. J. Computer Vision*, 5 (1990), 137–160.

68. F. Preparata and M. Shamos, *Computational Gemetry: An Introduction*, Springer-Verlag, Heidelberg, 1985.

69. J.T. Schwartz and M. Sharir, On the Piano Movers' problem: II. General techniques for computing topological properties of real algebraic manifolds, *Advances in Appl. Math.* 4 (1983), 298–351.

70. O. Schwarzkopf and M. Sharir, Vertical decomposition of a single cell in a 3-dimensional arrangement of surfaces, *Discrete Comput. Geom.* 18 (1997), 269–288.

71. M. Sharir, On k–sets in arrangements of curves and surfaces, *Discrete Comput. Geom.* 6 (1991), 593–613.

72. M. Sharir, Almost tight upper bounds for lower envelopes in higher dimensions, *Discrete Comput. Geom.* 12 (1994), 327–345.

73. M. Sharir, Arrangements in higher dimensions: Voronoi diagrams, motion planning, and other applications, *Proc. Workshop on Algorithms and Data Structures*, Ottawa, Canada, August, 1995, Lecture Notes in Computer Science, Vol. 955, Springer-Verlag, 109–121.

74. M. Sharir, Arrangements of surfaces in higher dimensions, in *Advances in Discrete and Computational Geometry* (Proc. 1996 AMS Mt. Holyoke Summer Research Conference, B. Chazelle, J.E. Goodman and R. Pollack, Eds.) Contemporary Mathematics No. 223, American Mathematical Society, 1999, 335–353.

75. M. Sharir and P.K. Agarwal, *Davenport-Schinzel Sequences and Their Geometric Applications*, Cambridge University Press, New York, 1995.

76. M. Sharir and S. Toledo, Extremal polygon containment problems, *Comput. Geom. Theory Appls.* 4 (1994), 99–118.

77. B. Tagansky, A new technique for analyzing substructures in arrangements, *Discrete Comput. Geom.* 16 (1996), 455–479.

78. B. Tagansky, *The Complexity of Substructures in Arrangements of Surfaces*, Ph.D. Dissertation, Tel Aviv University, July 1996.

79. P.M. Vaidya, Geometry helps in matching, *SIAM J. Comput.* 18 (1989), 1201–1225.

80. A. Wiernik and M. Sharir, Planar realization of nonlinear Davenport–Schinzel sequences by segments, *Discrete Comput. Geom.* 3 (1988), 15–47.

Dynamic Compressed Hyperoctrees with Application to the N-body Problem*

Srinivas Aluru[1] and Fatih E. Sevilgen[2]

[1] Iowa State University, Ames, IA 50011, USA
aluru@iastate.edu
[2] Syracuse University, Syracuse, NY 13244, USA
sevilgen@ecs.syr.edu

Abstract. Hyperoctree is a popular data structure for organizing multidimensional point data. The main drawback of this data structure is that its size and the run-time of operations supported by it are dependent upon the distribution of the points. Clarkson rectified the distribution-dependency in the size of hyperoctrees by introducing compressed hyperoctrees. He presents an $O(n \log n)$ expected time randomized algorithm to construct a compressed hyperoctree. In this paper, we give three deterministic algorithms to construct a compressed hyperoctree in $O(n \log n)$ time, for any fixed dimension d. We present $O(\log n)$ algorithms for point and cubic region searches, point insertions and deletions. We propose a solution to the N-body problem in $O(n)$ time, given the tree. Our algorithms also reduce the run-time dependency on the number of dimensions.

1 Introduction

Hyperoctrees are used in a number of application areas such as computational geometry, computer graphics, databases and scientific computing. In this paper, we focus our attention on hyperoctrees for any fixed dimension d. These are popularly known as quadtrees in two dimensions and octrees in three dimensions. Hyperoctrees can be constructed by starting with a cubic region containing all the points and subdividing it into 2^d equal sized cubic regions recursively until all the resulting regions have exactly one point.

We term a data structure *distribution-independent* if its size is a function of the number of points only and is not dependent upon the distribution of the input points. We term an algorithm *distribution-independent* if its run-time is independent of the distribution of the points. Hyperoctrees are distribution-dependent because recursively subdividing a cubic region may result in just one occupied region for an arbitrarily large number of steps. In terms of the tree,

---

* This research is supported in part by ARO under DAAG55-97-1-0368, NSF CAREER under CCR-9702991 and Sandia National Laboratories. The content of the information does not necessarily reflect the position or the policy of the U.S. federal government, and no official endorsement should be inferred.

C. Pandu Rangan, V. Raman, R. Ramanujam (Eds.): FSTTCS'99, LNCS 1738, pp. 21–33, 1999.

there can be an arbitrarily long path without any branching. If the properties of interest to the problem at hand depend only on the points, all the nodes along such a path essentially contain the same information. Thus, one can compress such a path into a single node without losing any information.

Such *compressed hyperoctrees* have been introduced by Clarkson [10] in the context of the all nearest neighbors problem. Clarkson presents a randomized algorithm to construct such a tree for n d-dimensional points in $O(c^d n \log n)$ expected time (where c is a constant). Later, Bern [5] proposed an $O((cd)^d n \log n)$ time deterministic algorithm for construction and used centroid decomposition [9] for $O(d \log n)$ time point searches. Callahan and Kosaraju's *fair split tree* achieves similar run-time complexities as the algorithms presented in this paper by allowing non-cubical regions [8]. Dynamizing these and similar data structures has been discussed in [4], [6], [8] and [15] by using auxiliary data structures, such as topology trees [13] or dynamic trees [16]. Further research on related techniques can be found in [2], [3], [11], [14] and [18].

In this paper, we present a dynamic variant of compressed hyperoctrees without using any of the aforementioned complicated auxiliary data structures. We call this new data structure *Distribution-Independent Adaptive Tree (DIAT)*. We present three deterministic algorithms that each construct the DIAT tree in $O(dn \log n)$ time. We present algorithms for point and cubic region searches, point insertion and point deletion that run in $O(d \log n)$ time. We assume an extended model of computation in which the floor, bitwise exclusive-or and logarithm operations can be done in constant time.

We expect that any algorithm that makes use of hyperoctrees can use DIAT trees to eliminate the dependence on distribution and result in faster running times. As evidence, we provide an algorithm to solve the N-body problem in $O(n)$ time (using the standard algebraic model of computation) on a given DIAT tree.

2   The DIAT Tree

We make use of the following terminology to describe DIAT trees: Call a cubic region containing all the points the *root cell*. Define a hierarchy of cells by the following: The root cell is in the hierarchy. If a cell is in the hierarchy, then the 2^d equal-sized cubic subregions obtained by bisecting along each coordinate of the cell are also called cells and belong to the hierarchy. Two cells are *disjoint* if they do not intersect at all or if they are merely adjacent, i.e., they intersect only at the boundaries. Given two cells in the hierarchy, possibly of different sizes, either one is completely contained in the other or they are disjoint. We use the term *subcell* to describe a cell that is completely contained in another. The 2^d subcells obtained by bisecting a cell along each dimension are also called the *immediate subcells* with respect to the bisected cell. Also, a cell is a *supercell* (*immediate supercell*) of any of its subcells (immediate subcells). We use the notation $C \subseteq D$ to indicate that C is a subcell of D and $C \subset D$ to indicate that C is a subcell of D and C is strictly smaller than D. Define the *length of a cell* C, denoted $length(C)$, to be the span of C along any dimension.

The DIAT Tree is essentially a compressed hyperoctree with some pointers attached to it. Each node v in the DIAT tree contains two cells, *large cell of v* and *small cell of v*, denoted by $L(v)$ and $S(v)$ respectively. The large cell is the largest cell that the node is responsible for and the small cell is the smallest subcell of the large cell that contains all points within the large cell. Consider any internal node v in the DIAT tree. To obtain its children, we subdivide $S(v)$ into 2^d equal-sized subcells. Note that subdivision of $S(v)$ results in at least two non-empty subcells. For each non-empty subcell C resulting from the subdivision, there is a child u such that $L(u) = C$. Since each internal node has at least two children, the size of a DIAT tree for n points is $O(n)$.

DIAT trees can be searched in a manner similar to binary search trees. Such algorithms are not efficient on DIAT trees because the height of a DIAT tree is $O(n)$ in the worst-case. For speeding accesses and for efficient maintenance in the dynamic case, we equip the DIAT tree with pointers.

Each internal node in the DIAT tree is pointed by a leaf. An internal node is pointed by the leftmost leaf in its subtree, unless its leftmost leaf is also the leftmost leaf in its parent's subtree. In such a case, the node is pointed by the rightmost leaf in its subtree. The following lemma asserts that each leaf needs to store only one pointer.

Lemma 1. *Each leaf in a DIAT tree points to at most one internal node.*

Proof. A leaf node that is neither the leftmost nor the rightmost child of its parent does not point to any internal node. A leaf that is the leftmost child of its parent points only to its ancestor closest to the root for which this leaf is the leftmost leaf in the ancestor's subtree. Now consider a leaf l that is the rightmost child of its parent. Suppose this leaf has two or more pointers. Consider any two of its pointers, and the nodes pointed by them, say u and v. Since l is the rightmost leaf in the subtrees of u and v, one must be an ancestor of the other. Without loss of generality, let u be the ancestor of v. Since v must be the rightmost child of its parent, it should be pointed by the leftmost leaf in its subtree and not the rightmost leaf l. This contradicts our earlier assumption that l points to v. It follows that each leaf points to at most one internal node.

2.1   Building the DIAT Tree

In the subsequent algorithms to be presented for the construction and manipulation of DIAT trees, we make use of the following result by Clarkson [10]:

Lemma 2. *Let R be the product of d intervals $I_1 \times I_2 \times \ldots \times I_d$, i.e., R is a hyperrectangular region in d dimensional space. The smallest cell containing R can be found in constant time.*

Proof. See [10]. The required smallest cell can be determined in $O(d)$ time, which is constant for any fixed dimension d. The procedure uses floor, logarithm and bitwise exclusive-or operations.

In what follows, we present three algorithms that construct the DIAT tree in $O(n \log n)$ time. While all three algorithms achieve the same time-bound, understanding the first two is necessary as the ideas developed in them are needed to make the tree dynamic.

A Divide-and-Conquer Algorithm Let T_1 and T_2 be two DIAT trees representing two distinct sets S_1 and S_2 of points. Let r_1 (resp., r_2) be the root node of T_1 (resp., T_2). Suppose that $L(r_1) = L(r_2)$, i.e., both T_1 and T_2 are constructed starting from a cell large enough to contain $S_1 \cup S_2$. A DIAT tree T for $S_1 \cup S_2$ can be constructed in $O(|S_1| + |S_2|)$ time by merging T_1 and T_2.

To merge T_1 and T_2, we start at their roots r_1 and r_2. Suppose that at some stage during the execution of the algorithm, we are at node v_1 in T_1 and at node v_2 in T_2 with the task of merging the subtrees rooted at v_1 and v_2. An invariant of the merging algorithm is that $L(v_1)$ and $L(v_2)$ cannot be disjoint. Furthermore, it can be asserted that $L(v_1) \cap L(v_2) \supseteq S(v_1) \cup S(v_2)$. For convenience, assume that a node may be empty. If a node has less than 2^d children, we may assume empty nodes in the place of absent children. The following possibilities arise:

- Case I: v_1 is an empty node. Return the subtree rooted at v_2.
- Case II: $S(v_1) = S(v_2)$. In this case, merge each child of v_1 with the corresponding child of v_2 and return the subtree rooted at v_2.
- Case III: $S(v_1) \subset S(v_2)$. There exists a child u of v_2 such that $S(v_1) \subset L(u)$. Merge v_1 with u and return the subtree rooted at v_2.
- Case IV: $S(v_1)$ and $S(v_2)$ are disjoint. Create a new node v with $L(v) = L(v_1) \cap L(v_2)$. Set $S(v) =$ smallest cell containing $S(v_1) \cup S(v_2)$. Subdivide $S(v)$ to separate $S(v_1)$ and $S(v_2)$ and create two corresponding children of v. Return the subtree rooted at v.

Cases I and III admit symmetric cases with the roles of v_1 and v_2 interchanged and can be handled similarly. The merging algorithm performs a preorder traversal of each tree. In every step of the merging algorithm, we advance on one of the trees after performing at most a constant amount of work. Thus, two DIAT trees with a common root cell can be merged in time proportional to the sum of their sizes. Using the merging algorithm and a recursive divide-and-conquer, a DIAT tree can be built in $O(n \log n)$ time.

Once the tree is built, it remains to assign the pointers from leaves to internal nodes. A recursive algorithm to assign all required pointers is given in Figure 1. The procedure **Assign-Pointers**, when called with a node in the DIAT tree, assigns pointers to all internal nodes in the node's subtree. If the node is the leftmost (resp., rightmost) child of its parent, it returns the leftmost (resp., rightmost) leaf in its subtree. The run-time of the algorithm is proportional to the size of the tree, which is $O(n)$.

Bottom-up Construction of the DIAT Tree Let f be any bijective function that maps the 2^d immediate subcells of a cell to the set $\{1, 2, \ldots, 2^d\}$. Consider

---

Algorithm 1 Assign-Pointers (v)

p = Assign-Pointers(leftmost child of v)
q = Assign-Pointers(rightmost child of v)
for every other child u of v
    Assign-Pointers(u)
If v is the leftmost child of its parent
    $q.pointer = v$, return(p)
If v is the rightmost child of its parent
    $p.pointer = v$, return(q)
Else $p.pointer = v$, return(nil)

Fig. 1. Algorithm for assigning pointers from leaves to internal nodes.

an internal node v and let u_1 and u_2 be two children of v. Define an ordering as follows: u_1 appears earlier in the ordering than u_2 if and only if $f(L(u_1)) < f(L(u_2))$. If u_1 appears earlier in the ordering than u_2, then every node in the subtree of u_1 appears earlier in the ordering than any node in the subtree of u_2. The ordering can be extended to sets of nodes and is a complete order for any set satisfying the property that no node in the set is an ancestor of another. In particular, the leaves of a DIAT tree (or the points) can be ordered. The left-to-right order of the corresponding leaves in the DIAT tree is the same as this ordering, if the children of each node are ordered according to the function f.

To perform a bottom-up construction, first compute the ordering of the points as they should appear in the DIAT tree and compute the tree bottom-up using this order. Similar ordering has been used by Bern et. al. [4] as a way of constructing hyperoctrees for points with integer coordinates. To order the points according to the DIAT tree, we establish a procedure that orders any two points: Given two points p_1 and p_2, compute the smallest subcell containing them. Subdividing this smallest subcell into its immediate subcells separates p_1 and p_2. The ordering of the two points is the same as the ordering of the immediate subcells they belong to and can be determined in $O(1)$ time by Lemma 2.

Given n points and a root cell containing them, the points can be sorted in $O(n \log n)$ time according to the total order just defined by using any optimal sorting algorithm. The DIAT tree is then incrementally constructed using this sorted list of points starting from the single node tree for the first point and the root cell. During the insertion process, keep track of the most recently inserted leaf. Let p be the next point to be inserted. Starting from the most recently inserted leaf, traverse the path from the leaf to the root until we find the first node v such that $p \in L(v)$. Two possibilities arise:

- <u>Case I:</u> $p \notin S(v)$. Create a new node u in place of v where $S(u)$ is the smallest subcell containing p and $S(v)$. Make v (along with its subtree) and the node containing p children of u.

– <u>Case II:</u> $p \in S(v)$. Consider the immediate subcell of $S(v)$ that contains p. The DIAT tree presently does not contain a child for v that corresponds to this subcell. Therefore, the node containing p can be inserted as a child of v.

Once the points are sorted, the rest of the algorithm is identical to a post-order walk on the final DIAT tree with $O(1)$ work per node. The time for a single insertion is not bounded by a constant but all n insertions together require only $O(n)$ time. Combined with the initial sorting of the points, the tree can be constructed in $O(n \log n)$ time. The pointer assignment procedure described Algorithm 1 can be used to assign required pointers.

Construction by Repeated Insertions An algorithm to insert a point in an n-point DIAT tree in $O(\log n)$ time is presented later. Using this algorithm, a DIAT tree can be constructed by repeated insertions in $O(n \log n)$ time.

Theorem 1. *A DIAT tree for n points can be constructed in $O(n \log n)$ time.*

2.2 Querying DIAT Trees

We consider two types of searches: point searches and cell searches. To facilitate fast algorithms, an auxiliary data structure is used in conjunction with the DIAT tree. The auxiliary data structure is a balanced binary search tree (abbreviated BBST) built on the input points using their order of appearance as leaves in the DIAT tree. Given the DIAT tree, the sorted order of points according to the DIAT tree can be read in $O(n)$ time followed by a binary search tree construction on this sorted data taking additional $O(n)$ time. Each node in the BBST represents a point and contains a pointer to the leaf representing the same point in the DIAT tree.

The general idea behind searches is as follows: The searches are always first conducted in the BBST which helps in locating the relevant leaves in the DIAT tree. The DIAT tree itself is then accessed from the leaves.

Point Search To locate a point in the DIAT tree, first locate it in the BBST. If the point does not exist in the BBST, it does not exist in the DIAT tree. Otherwise, the node in the BBST has a pointer to the corresponding leaf in the DIAT tree. The search in the BBST is performed using the aforementioned ordering procedure. The overall search time is $O(\log n)$.

Cell Search Given a cell C, the cell search problem is to locate the node in the DIAT tree representing C. A node v in the tree is said to represent C if $S(v) \subseteq C \subseteq L(v)$, i.e., the points in cell C are exactly the points in the subtree under v. If C does not contain any points, it is returned that the cell does not exist in the DIAT tree.

Consider a given cell C and the bijective function f used in defining the ordering of immediate subcells of a cell. Identify the two immediate subcells C_1

and C_2 of C such that $f(C_1) \leq f(C') \leq f(C_2)$ for any immediate subcell C' of C. Find the corner l of C_1 (and h of C_2) that is not adjacent to any other immediate subcell of C. Locate the first point p_1 that should appear after l and the last point p_2 that should appear before h in the BBST. It is clear that p_1 is the leftmost leaf in the subtree at the node representing C in the DIAT tree and p_2 is the rightmost leaf. Since each node is pointed by either the leftmost or the rightmost leaf in its subtree, one of these leaves leads to the required node. Therefore, a cell search is equivalent to two point searches followed by $O(1)$ work, for a total of $O(\log n)$ run-time.

2.3 Dynamic Operations on DIAT Trees

Point Insertion To insert a point q in the DIAT tree, we first insert it in the BBST and find its predecessor point p and successor point s. The smallest cell that contains p and q and the smallest cell that contains s and q are either the same or one is contained in the other. Let w be the node in the DIAT tree representing the smaller of these two cells. This can be located in $O(\log n)$ time using a cell search. If $q \in S(w)$, the immediate subcell of $S(w)$ containing q does not contain any other point and q is inserted as a child of w. Otherwise, create and insert a new node x as the parent of w where $S(x)$ is the smallest cell containing both q and $S(w)$. The node representing q is then inserted as a child of x.

During insertion, pointers need to be updated to be consistent with the pointer assignment mechanism in DIAT trees. For example, if the newly inserted leaf is a leftmost child of its parent, it has to take over the pointer from the previous leftmost child in the parent's subtree. For all possible cases that may arise during insertion, it can be shown that only a constant number of pointer updates are needed. The total run-time is bounded by $O(\log n)$.

Point Deletion To delete a point p, first search for p in the BBST and identify the corresponding leaf in the DIAT tree. Delete p from the BBST and delete the leaf containing p from the DIAT tree. If the removal of the leaf leaves its parent with only one child, simply delete the parent and assign to the other child the largest cell it is responsible for. Since each internal node has at least two children, the delete operation can not propagate to higher levels in the DIAT tree. Like insertions, deletions also require a constant number of pointer adjustments. Deletions can be performed in $O(\log n)$ time.

Theorem 2. *Point search, cell search, point insertion and point deletion in a DIAT tree of n points can be performed in $O(\log n)$ time.*

3 The N-body Problem

The N-body problem is defined as follows: Given n bodies and their positions, where each pair of bodies interact with a force inversely proportional to the square of the distance between them, compute the force on each body due to all other bodies. A direct algorithm for computing all pairwise interactions requires $O(n^2)$ time. Greengard's fast multipole method [12], which uses a hyperoctree data structure, reduces this complexity by approximating the interaction between clusters of particles instead of computing individual interactions. For each cell in the hyperoctree, the algorithm computes a multipole expansion and a local expansion. The multipole expansion at a cell C, denoted $\phi(C)$, is the effect of the particles within the cell C on distant points. The ϕ's are computed by a bottom-up traversal of the hyperoctree. The local expansion at a cell C, denoted $\psi(C)$, is the effect of all distant particles on the points within the cell C. The ψ's are computed by a top-down traversal (by combining the multipole expansions of *well-separated* cells). Though widely accepted to be $O(n)$, Greengard's algorithm is distribution-dependent and the number of cell-cell interactions proportional to the size of the hyperoctree.

Recently, Callahan et. al. presented a distribution-independent algorithm for solving the N-body problem [7]. The algorithm computes a well-separated decomposition of the particles in $O(n \log n)$ time followed by computing the interactions in $O(n)$ time. In what follows, we present an $O(n)$ algorithm for the N-body problem given the DIAT tree. Though our run-time is the same, there are some important differences between the two algorithms: Almost all the N-body algorithms used by practitioners involve hyperoctrees. DIAT trees contain same type of cells. The regularity in the shapes and locations of the cells makes it easy to perform error calculations.

The starting point of our algorithm is Greengard's fast multipole method [12]. Greengard considers cells of the same length. Two such cells are called *well-separated* if the multipole expansion at one cell converges at any point in the other i.e., they are not adjacent. For the DIAT tree, a capability to deal with different cell lengths is needed. We generalize well-separatedness criteria for any two cells C and D which are not necessarily of the same length. Define a predicate *well-separated*(C, D) to be true if D's multipole expansion converges at any point in C, and false otherwise. If two cells are not well-separated, they are *proximate*. Similarly, two nodes v_1 and v_2 in the DIAT tree are said to be *well-separated* if and only if $S(v_1)$ and $S(v_2)$ are well-separated. Otherwise, we say that v_1 and v_2 are *proximate*.

Our DIAT tree based N-body algorithm is as follows: For each node v in the DIAT tree, we wish to compute the multipole expansion $\phi(v)$ and the local expansion $\psi(v)$. Both $\phi(v)$ and $\psi(v)$ are with respect to the cell $S(v)$. The multipole expansions can be computed by a simple bottom-up traversal in $O(n)$ time. At a node v, $\phi(v)$ is computed by aggregating the multipole expansions of the children of v.

The algorithm to compute the local expansions is given in Figure 2. The computations are done using a top-down traversal of the tree. To compute local expansion at node v, we have to consider the set of nodes that are proximate to its parent, which is the *proximity set*, $P(parent(v))$. The proximity set of the root node contains only itself. We recursively decompose these nodes until each

---

Algorithm 2 Compute-Local-Exp (v)

I. Find the proximity set $P(v)$ and interaction set $I(v)$ for v
    $E(v) = P(parent(v))$
    $I(v) = \emptyset$; $P(v) = \emptyset$
    While $E(v) \neq \emptyset$ do
        Pick some $u \in E(v)$
        $E(v) = E(v) - \{u\}$
    If $well\text{-}separated(S(v), S(u))$
        $I(v) = I(v) \cup \{u\}$
    Else if $S(u)$ is smaller than $S(v)$
        $P(v) = P(v) \cup \{u\}$
    Else $E(v) = E(v) \cup children(u)$

II. Calculate the local expansion at v
    Assign shifted $\psi(parent(v))$ to $\psi(v)$
    For each node $u \in I(v)$
        Add shifted $\phi(u)$ to $\psi(v)$

III. Calculate the local expansions at the children of v with recursive calls
    For each child u of v
        Compute-Local-Exp (u)

Fig. 2. Algorithm for calculating local expansions of all nodes in the tree rooted at v.

node is either 1) well-separated from v or 2) proximate to v and the length of the small cell of the node is smaller than the small cell of v. The nodes satisfying the first condition form the *interaction set* of v, $I(v)$ and the nodes satisfying the second condition are in the proximity set of v, $P(v)$. In the algorithm, the set $E(v)$ contains the nodes that are yet to be processed. As in Greengard's algorithm, local expansions are computed by combining parent's local expansion and the multipole expansions of the nodes in $I(v)$. For the leaf nodes, potential calculation is completed by using the direct method.

Before analyzing the run-time of this algorithm, we need to precisely define the sets used in the algorithm. The set of cells that are proximate to the cell C and having same length as C is called *the proximity set of C* and is defined by $P^=(C) = \{D \mid length(C) = length(D), \neg well\text{-}separated(C, D)\}$. The superscript "=" is used to indicate that cells of the same length are being considered. For a node v in the DIAT tree, define the *proximity set $P(v)$* as the set of all nodes proximate to v and having the small cell smaller and large cell larger than $S(v)$. More precisely, $P(v) = \{w \mid \neg well\text{-}separated(S(v), S(w)),$ $length(S(w)) \leq length(S(v)) \leq length(L(w))\}$. The *interaction set $I(v)$* of v is defined as $I(v) = \{w \mid well\text{-}separated(S(v), S(w)), [\, w \in P(parent(v)) \vee \{\exists u \in P(parent(v)),$ w is a descendant of u, $\neg well\text{-}sep(v, parent(w)),$ $length(S(v)) < length(S(parent(w)))\}]\}$. We use $parent(w)$ to denote the parent of the node w.

In the rest of the section, we prove that our algorithm achieves the running-time of $O(n)$ for any predicate *well-separated* that satisfies the following there conditions:

C1. The relation *well-separated* is symmetric for equal length cells, that is, $length(C) = length(D) \Rightarrow well\text{-}separated(C, D) = well\text{-}separated(D, C)$.

C2. For any cell C, $|P^=(C)|$ is bounded by a constant.

C3. If two cells C and D are not well-separated, any two cells C' and D' such that $C \subseteq C'$ and $D \subseteq D'$ are not well-separated as well.

These three conditions are respected by the various well-separatedness criteria used in N-body algorithms and in particular, Greengard's algorithm. In N-body methods, the well-separatedness decision is solely based on the geometry of the cells and their relative distance and is oblivious to the number of particles or their distribution within the cells. Given two cells C and D of the same length, if D can be approximated with respect to C, then C can be approximated with respect to D as well, as stipulated by Condition C1. The size of the proximity sets of cells of the same length should be $O(1)$ as prescribed by Condition C2 in order that an $O(n)$ algorithm is possible. Otherwise, we can construct an input that requires processing the proximity sets of $O(n)$ such cells, making an $O(n)$ algorithm impossible. Condition C3 merely states that two cells C' and D' are not well-separated unless every subcell of C' is well-separated from every subcell of D'.

Lemma 3. *For any node v in the DIAT tree, $|P(v)| = O(1)$.*

Proof. Consider any node v. Each $u \in P(v)$ can be associated with a unique cell $C \in P^=(S(v))$ such that $S(u) \subseteq C$. This is because any subcell of C which is not a subcell of $S(u)$ is not represented in the DIAT tree. It follows that $|P(v)| \leq |P^=(S(v))| = O(1)$ (by Condition C2).

Lemma 4. *The sum of interaction set sizes over all nodes in the DIAT tree is linear in the number of nodes in the DIAT tree i.e., $\sum_v |I(v)| = O(n)$.*

Proof. Let v be a node in the DIAT tree. Consider any $w \in I(v)$, either $w \in P(parent(v))$ or w is in the subtree rooted by a node $u \in P(parent(v))$. Thus,

$$\sum_v |I(v)| = \sum_v |\{w \mid w \in I(v), w \in P(parent(v))\}|$$
$$+ \sum_v |\{w \mid w \in I(v), w \notin P(parent(v))\}|.$$

Consider these two summations separately. The bound for the first summation is easy; From Lemma 3, $|P(parent(v))| = O(1)$. So,

$$\sum_v |\{w \mid w \in I(v), w \in P(parent(v))\}| = \sum_v O(1) = O(n).$$

The second summation should be explored more carefully.

$$\sum_v |\{w \mid w \in I(v), w \notin P(parent(v))\}| = \sum_w |\{v \mid w \in I(v), w \notin P(parent(v))\}|$$

In what follows, we bound the size of the set $M(w) = \{v \mid w \in I(v), w \notin P(parent(v))\}$ for any node.

Since $w \notin P(parent(v))$, there exists a node $u \in P(parent(v))$ such that w is in the subtree rooted by u. Consider $parent(w)$: The node $parent(w)$ is either u or a node in the subtree rooted at u. In either case, $length(S(parent(w)))$ $\leq length(S(parent(v)))$. Thus, for each $v \in M(w)$, there exists a cell C such that $S(v) \subseteq C \subseteq S(parent(v))$ and $length(S(parent(w))) = length(C)$. Further, since v and w are not well-separated, C and $S(parent(w))$ are not well-separated as well by Condition C3. That is to say $S(parent(w)) \in P^=(C)$ and $C \in P^=(S(parent(w)))$ by Condition C1. By Condition C2, we know that $|P^=(S(parent(w)))| = O(1)$. Moreover, for each cell $C \in P^=(S(parent(w)))$, there are at most 2^d choices of v because $length(C) \leq length(S(parent(v)))$. As a result, $|M(w)| \leq 2^d \times O(1) = O(1)$ for any node w. Thus, $\sum_v |I(v)| = \sum_w |\{v \mid w \in I(v), w \notin P(parent(v))\}| = \sum_w O(1) = O(n)$.

Theorem 3. *Given a DIAT tree for n particles, the N-body problem can be solved in $O(n)$ time.*

Proof. Computing the multipole expansion at a node takes constant time and the number of nodes in the DIAT tree is $O(n)$. Thus, total time required for the multipole expansion calculation is $O(n)$. The nodes explored during the local expansion calculation at a node v are either in $P(v)$ or $I(v)$. In both cases, it takes constant time to process a node. By Lemma 3 and 4, the total size of both sets for all nodes in the DIAT tree is bounded by $O(n)$. Thus, local expansion calculation takes $O(n)$ time. As a conclusion, the running time of the fast multipole algorithm on the DIAT tree takes $O(n)$ time irrespective of the distribution of the particles.

4   Conclusions

In this paper we presented the DIAT tree, a new dynamic data structure for multidimensional point data. We presented construction algorithms for DIAT trees in $O(n \log n)$ time and search and insertion/deletion algorithms in $O(\log n)$ time.

DIAT trees can potentially be used to solve any application that currently uses hyperoctrees. We have presented an optimal algorithm for the N-body problem using DIAT trees. The DIAT trees and algorithms presented improves or matches the results presented in [4], [10] and [15] and provide dynamic handling of the data points. For example, the randomized algorithm for compressed hyperoctree construction in [10] can be replaced by a DIAT tree construction algorithm presented in this paper to achieve $O(n \log n)$ deterministic algorithm for all nearest neighbors problem. Many other applications use hyperoctrees and we believe that faster sequential and parallel algorithms can be designed for them using DIAT trees. We recently developed parallel algorithms for constructing DIAT trees and for solving the N-body problem using them. Both the algorithms are independent of the distribution of points and have rigorously analyzable worst case running times.

Acknowledgments

The authors gratefully acknowledge valuable suggestions made by the reviewers which led to improvements in the presentation of the paper.

References

1. Aluru, S.: Greengard's N-body algorithm is not order N. SIAM Journal on Scientific Computing **17** (1996) 773–776.
2. Arya, S., Mount, D., Netanyahu, N., Silverman, R., Wu, A.Y.: An optimal algorithm for approximate nearest neighbor searching. Proc. ACM-SIAM Symposium on Discrete Algorithms (1994) 573–582.
3. Arya, S., Mount, D., Netanyahu, N., Silverman, R., Wu, A.Y.: An optimal algorithm for approximate nearest neighbor searching in fixed dimensions. Journal of the ACM **45** (1998) 891–923.
4. Bern, M., Eppstein, D., Teng, S.H.: Parallel construction of quadtrees and quality triangulations. Proc. Workshop on Algorithms and Data Structures (1993) 188–199.
5. Bern, M.: Approximate closest-point queries in high dimensions. Information Processing Letters **45** (1993) 95–99.
6. Bespamyatnikh, S.N.: An optimal algorithm for closest-pair maintenance. Discrete Comput. Geom. **19** (1998) 175–195.
7. Callahan, P.B., Kosaraju, S.R.: A decomposition of multidimensional point sets with applications to k-nearest neighbors and N-body potential fields. Journal of the ACM **42** (1995) 67–90.
8. Callahan, P.B., Kosaraju, S.R.: Algorithms for dynamic closest pair and n-body potential fields. Proc. ACM-SIAM Symposium on Discrete Algorithms (1995) 263–272.
9. Chazelle, B.: A theorem on polygon cutting with applications. Proc. Foundations of Computer Science (1982) 339–349.
10. Clarkson, K.L.: Fast algorithms for the All-Nearest-Neighbors problem. Proc. Foundations of Computer Science (1983) 226–232.
11. Cohen, R.F., Tamassia, R.: Combine and conquer. Algorithmica **18** (1997) 51–73.
12. Greengard, L., Rokhlin, V.: A fast algorithm for particle simulations. Journal of Computational Physics **73** (1987) 325–348.
13. Frederickson, G.N.: A data structure for dynamically maintaining rooted trees. Proc. ACM-SIAM Symposium on Discrete Algorithms (1993) 175–194.
14. Mitchell, J.S.B., Mount, D.M., Suri, S.: Query-Sensitive ray shooting. International Journal of Computational Geometry and Applications **7** (1997) 317–347.

15. Schwarz, C., Smid, M., Snoeyink, J.: An optimal algorithm for the on-line closest-pair problem. Algorithmica **12** (1994) 18–29.
16. Sleator, D.D., Tarjan, R.E.: A data structure for dynamic trees. Journal of Computer and System Sciences **26** (1983) 362–391.
17. Teng, S.H.: Provably good partitioning and load balancing algorithms for parallel adaptive N-body simulations. SIAM Journal on Scientific Computing **19** (1998) 635–656.
18. Vaidya, P.M.: An $O(n \log n)$ algorithm for the All-Nearest-Neighbors problem. Discrete Computational Geometry **4** (1989) 101–115.

Largest Empty Rectangle among a Point Set

Jeet Chaudhuri[1] and Subhas C. Nandy[2*]

[1] Wipro Limited, Technology Solns, Bangalore 560034, India
[2] Indian Statistical Institute, Calcutta - 700 035, India

Abstract. This paper generalizes the classical MER problem in 2D. Given a set P of n points, here a maximal empty rectangle (MER) is defined as a rectangle of arbitrary orientation such that each of its four sides coincides with at least one member of P and the interior of the rectangle is empty. We propose a simple algorithm based on standard data structure to locate largest area MER on the floor. The time and space complexities of our algorithm are $O(n^3)$ and $O(n^2)$ respectively.

1 Introduction

Recognition of all maximal empty axes-parallel (isothetic) rectangles, commonly known as MER problem was first introduced in [3]. Here a set of points, say $P = \{p_1, p_2, \ldots, p_n\}$ is distributed on a rectangular floor. The objective is to locate all possible isothetic *maximal empty rectangles* (MER). They proposed an algorithm for this problem with time complexity $O(min(n^2, Rlogn))$, where R denoting the number of reported MERs', may be $O(n^2)$ in the worst case. The time complexity was later improved to $O(R + nlogn)$ [6]. The best result for locating the largest empty isothetic rectangle among a point set without inspecting all MER's appeared in [1]; it uses divide and conquer and matrix searching techniques and its time complexity is $O(nlog^2n)$. The MER problem is later generalized among a set of isothetic obstacles [4], and among a set of non-isothetic obstacles [5].

In this context, it needs to mention that for a given set of points P, the location of all/largest empty r-gon whose vertices coincide with the members in P, can be reported in $O(\gamma_3(P) + r\gamma_r(P))$ time if $r \geq 5$; for r = 3 and 4, it requires $O(\gamma_r(P))$ time [2], where $\gamma_r(P)$ is the number of such empty r-gons. It is also shown that $\gamma_4(P) \geq \gamma_3(P) - \binom{n-1}{2}$, and the expected value of $\gamma_3(P)$ is $O(n^2)$. This provides a lower bound on the number of empty convex quadrilateral in terms of number of empty triangles.

This paper outlines a natural generalization of the classical MER problem. Given n points on a 2D plane, a long standing open problem is to locate an empty rectangle of maximum area. Thus the earlier restriction of the isotheticity of the MER is relaxed. This type of problem often arises in different industrial applications where one needs to cut a largest defect-free rectangular piece from a

---

* This work was done when the author was visiting School of Information Science, Japan Advanced Institute of Science and Technology, Ishikawa 923-1292, Japan.

C. Pandu Rangan, V. Raman, R. Ramanujam (Eds.): FSTTCS'99, LNCS 1738, pp. 34–46, 1999.
© Springer-Verlag Berlin Heidelberg 1999

given metal sheet. We adopt a new algorithmic paradigm, called *grid rotation*, to solve this problem. The time and space complexities of our algorithm are $O(n^3)$ and $O(n^2)$ respectively.

2 Basic Concepts

Let $P = \{p_1, p_2, \ldots, p_n\}$ be a set of n arbitrarily distributed points on a 2D region. Without loss of generality, we may choose the origin of the coordinate system such that all the points lie in the first quadrant. The coordinate of point p_i will be denoted by (x_i, y_i). We shall define a *maximal empty rectangle* (MER) by an ordered tuple as follows :

Definition : A rectangle (of arbitrary orientation) on the floor is called an MER if it is empty, i.e., not containing any member of P, and no other empty rectangle can properly enclose it. Thus, each of the four boundaries of an MER must contain at least one point of P.

The notion of a rectangle assumes that this geometric structure is bounded on all sides. But we take the liberty to loosen this notion to include also those rectangles as MERs that is unbounded on some side, provided, of course, they are empty and cannot be properly contained in some other rectangle. The rectangles and MERs in the conventional sense shall then be referred to specifically as "bounded".

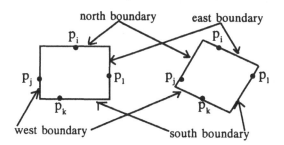

Fig. 1. Definition of an MER

It needs to mention here that, by slightly rotating an empty rectangle with four specified points $\{p_i, p_j, p_k, p_\ell\}$ on its four boundaries, one may get multiple MERs' with the same set of bounding points. (The pictorial representation of the above situations, is dropped due to the space limitation.) This motivates us to define a PMER as follows :

Definition : Consider a set of MERs' with $\{p_i, p_j, p_k, p_\ell\}$ on its four boundaries and their south boundary makes an angle between ϕ and ψ with the positive direction of the x-axis. An MER in this set is said to be *prime MER* (PMER) if there exists no MER in this set which is of larger size than it. A set of MERs, corresponding to a PMER, is represented by a six tuple $\{p_i, p_j, p_k, p_\ell, \phi, \psi\}$.

In order to assign some order among the points in four sides of an MER, consider a corner of the MER having maximum y-coordinate. The side (boundary) of the MER incident to that corner and having non-negative slope will be referred to as its *north boundary*. The other boundaries viz. the *east*, *south* and *west* boundaries are defined based on the *north boundary* (see Figure 1). Actually, this type of nomenclature is misnomer in the context of non-axis parallel rectangles, but it will help to explain our method. Our objective is to locate the largest MER whose each of the four sides is bounded by some point(s) of P.

The algorithmic technique, discussed in this paper, shows the number of PMERs may be $O(n^3)$ in the worst case. Let us mention, once and for all, that henceforth, we shall refer to the terms 'MER' and 'rectangle' interchangeably. Further we may refer north, west, south and east as top, left, bottom and right respectively.

3   Identification of PMERs

In this section, we explain the recognition of all PMERs using a very simple algorithmic technique, called *grid rotation*. Initially, we draw n horizontal lines and n vertical lines through all the members in P. The resulting diagram is a *grid*, but the separation among each pair of vertical (horizontal) lines are not same. For a given P, the initial grid diagram is shown in Figure 2a. During execution of the algorithm these lines will be rotated and will no longer remain horizontal/vertical. So, we shall refer to the lines which are initially horizontal (vertical), as *red lines* (*blue lines*). At any instant of time during the execution of algorithm, the angle θ made by each of the red lines with the x-axis, will be referred to as the *grid angle*.

Consider the set of MERs which are *embedded* in the grid, i.e., the set of MERs whose sides are incident to the grid lines. We maintain these rectangles in a data structure, called *grid diagram*, as follows.

3.1   Grid Diagram

First we aim to represent the grid in Figure 2a using an $n+1 \times n+1$ matrix $\mathcal{M}$, where $n = |P|$. The row numbers 0 to n of this matrix increase from bottom to top while the column numbers 0 to n from left to right. Let $P_y = \{p'_1, p'_2, \ldots p'_n\}$ be the ordered set of points in P in increasing values of their y-coordinates, and $P_x = \{p''_1, p''_2, \ldots p''_n\}$ be the ordered set of points in P in increasing values of their x-coordinates. Let $p \in P$ such that $p = p'_k$ and $p = p''_j$. Then we put $\mathcal{M}(k, j) = 1$. The other entries in $\mathcal{M}$ are initialized to 0. Thus each red (blue) line of the grid is mapped to each row (column) of the matrix $\mathcal{M}$, each point in P is mapped to an 1 entry in $\mathcal{M}$.

Our next step will be to represent the *embedded* rectangles. First we consider only those that are bounded on all sides by some point in P. Let us consider a

rectangle in the grid as in Figure 2a bounded by $\{p_n, p_w, p_s, p_e\}(= \{b, a, g, d\})$
at its north, west, south and east sides respectively. Let $p_n = p'_{\alpha_n}$ and $p_n = p''_{\beta_n}$,
or in other words, p_n corresponds to the α_n-th row and the β_n-th column of the
matrix $\mathcal{M}$. Similarly, the row (column) indices corresponding to p_w, p_s and p_e
are α_w, α_s and α_e (β_w, β_s and β_e) respectively. We represent this MER by (i)
storing the point pair (p_w, p_e) in the (α_s, β_n)-th entry of the matrix $\mathcal{M}$, which
we shall henceforth term as *storing at the south boundary*, and also (ii) storing
the point pair (p_n, p_s) in the (α_e, β_w)-th entry of the same matrix, which we
shall describe as *storing at the west boundary*.

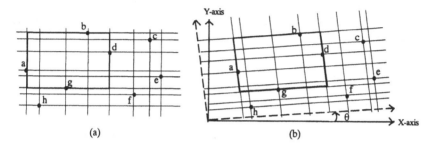

Fig. 2. Demonstration of grid rotation technique using grid diagram

Next we aim to store rectangles that are unbounded in one side. An MER that
is unbounded in the north, will have p_n being $-\infty$ while storing it at the west,
and is stored at the south in $\mathcal{M}(\alpha_s, \beta_s)$, i.e., the entry of $\mathcal{M}$ representing the
point which bounds it at the south. An MER unbounded in the south, likewise,
will have p_s being ∞ while storing it at the west, and is assumed to be bounded
by the 0-th row in the matrix in the south. Thus, we store it at the south at the
entry in the 0-th row and column corresponding to the north bounding point.
The case is absolutely symmetrical for MERs unbounded at either west or east
but not both.

Next we tackle the case where MERs that are unbounded in exactly two
sides. If a rectangle is unbounded in the north and south, we would store the
rectangle at the west boundary only, with p_n and p_s being $-\infty$ and ∞ respectively. Similarly, a rectangle unbounded in the east and west is stored at the
south boundary only, with p_w and p_e being $-\infty$ and ∞ respectively. Also, an
MER that is unbounded in two adjacent sides, for example in north and west, are
stored both in the west and south, in the manner outlined in the last paragraph.

A rectangle that is bounded on only one side can be and is also represented.
An MER that is bounded only in the south, is represented at the south at the
point which bounds it there. The west and east bounding vertices are represented
by $-\infty$ and ∞ respectively. Again, an MER that is bounded only in the north,
is represented at the south at the corresponding entry in the 0-th row of the
matrix. The west and east boundaries are represented as in the last case. We,
however, do not store any of these MERs at the west. In an exact symmetric

manner, we can take care of rectangles that are bounded only in the west or in the east.

Observation 1 *i) Given a fixed grid angle, and a pair of points p_n and p_s, if there exists an MER whose opposite sides contain p_n and p_s respectively, then its bounding blue lines are unique.*

ii) Given a fixed grid angle, and a pair of points p_w and p_e, if there exists an MER whose opposite sides contain p_w and p_e respectively, then its bounding red lines are unique.   □

The matrix $\mathcal{M}$ is initialized by the set of all axis-parallel MERs' which are obtained by invoking the algorithm presented in [6], and it requires $O(R+n\log n)$ time, where R is the total number of MERs whose sides are parallel to the coordinate axes.

3.2   Data Structure

We consider the set of points P as an array whose each element contains its corresponding row and column numbers in the matrix $\mathcal{M}$. Also we maintain an array P' whose elements correspond to the rows in $\mathcal{M}$ in a bottom to top order. An entry of P' is the index of the point in P corresponding to that row. In an exact similar fashion, we have another array P'', whose elements correspond to the columns ordered from left to right, and its each entry stores the index of the point in P, corresponding to that column. Each element of $\mathcal{M}$ now consists of the following fields :

(i) a *value* field containing 1, 2 or 4 depending on whether the corresponding entry represents a point in P, or stores an MER *at its south boundary* or stores an MER *at its west boundary*. The *value* field may also contain a sum of any of these three primary values, to denote that the corresponding cases occur simultaneously. If none of these cases occur, the *value* field contains 0. Now, it may be observed that any point in the set P shall always bound an MER in the south that is unbounded in the north, and bound an MER in the west that is unbounded in the east. In view of this, it is evident that once this field contains 1, it should also contain 2 and 4 respectively. Thus, a matrix entry representing a point in P should contain 7 (1+2+4). A matrix entry representing only an MER stored at its south (west) boundary, will contain 2 (4). A matrix entry representing two MERs, one of them is stored at its south boundary and the other one is stored at its west boundary, will contain a value 6. By Observation 1, there exists no matrix entry representing more than one MERs all stored at their south boundary, or all stored at their west boundary. Thus the possible values of the matrix entries are 2, 4, 6 or 7.

(ii) two pointers $P1$ and $P2$ storing the (indices in P of) two different points which appear on the west and east boundaries of the MER represented by that element *at the south*. Thus, if the *value* field is 4 or 0, $P1$ and $P2$

contain NULL. Again, if the MER represented by an element is unbounded in a particular side, then also the corresponding pointer is set to NULL.

(iii) two more pointers $P3$ and $P4$ storing the (indices in P of) two different points which appear on the north and south boundaries of the MER represented by that element *at the west*. Thus, if the *value* field is 2 or 0, $P3$ and $P4$ contain NULL. Again, if the MER represented by an element is unbounded in a particular side, then also the corresponding pointer is set to NULL.

(iv) A pair of initial grid angles, one for the MER represented *at the south* by this entry, one for the MER represented *at the west*, if at all any/both is/are represented.

3.3   Grid Rotation

Now we describe how the grid diagram changes due to the rotation of the grid. The rotation of the grid implies that for each point, the pair of mutually perpendicular lines, passing through that point, are rotated gradually in an anticlockwise direction, and all at the same rate. Let us imagine the rectangles *embedded* in the grid to be rotating also with the rotation of the grid as shown in Figure 2b. Now, for a very small rotation, although the rectangles change in size, their bounding points nevertheless remain same. However, when a pair of grid lines of the grid actually coincide, some rectangles might degenerate, some rectangles might be formed anew, while some may have its bounding vertices changed. At this stage we need to do the following :

Update the data structure to account for the new set of MERs. For each such MER, we need to store the current grid angle at the appropriate place.

The MERs that remain with the current set of bounding vertices after the rotation, do not need any computation.

For the MERs (defined by a specified set of tuples) which were present in the data structure, but will not remain alive from the current instant of time onwards, we may need to compute the PMER, and update the data structures.

In other words, in a particular orientation of the grid angle, say at $\theta = \phi$, if a set of four points $\{p_n, p_w, p_s, p_e\}$ defines an MER for the first time, and if it remains valid for an interval $\phi \leq \theta \leq \psi$ of the grid angle, then the entry in $\mathcal{M}$ representing that MER is created when the $\theta = \phi$, and the PMER corresponding to the six-tuple $\{p_n, p_w, p_s, p_e, \phi, \psi\}$ is computed when θ becomes equal to ψ during the grid rotation. Recall that we store the initial grid angle ϕ for these set of MERs; it is actually done for this purpose.

Note that, if we gradually rotate the grid by an angle $\frac{\pi}{2}$, we can enumerate all the PMERs that exists on the floor. Our aim is to find the one having maximum area.

Selection of event points

In order to perform the rotational sweep of the grid, we need to know the order in which a pair of grid lines of same color swaps. This requires a sorting of the absolute gradients of the lines obtained by joining each pair of points. During the rotation of the grid, we need to stop $O(n^2)$ times when either the *red lines* or the *blue lines* become parallel to any of those lines. We consider two different sets containing all the lines having positive and negative slopes respectively. The lines in the first (second) set are sorted in increasing order of the angle θ with the x-axis (y-axis) in counter-clockwise direction. Easy to understand, in each set the lines are stored in increased order of their gradients. Finally, these two sets are merged with respect to the angle θ considered for sorting. Needless to say, this requires $O(n^2)$ space for storing the angles of all the lines, and the sorting requires $O(n^2 \log n)$ time in the worst case. But note that, we don't need to store the gradient of all the lines permanently; rather we are satisfied if we get the event points (the angles) in proper order during grid rotation. In the full paper, we will show, using geometric duality, that the event points can be generated using $O(n)$ space.

Some important properties of grid rotation

Next, we come to the most crucial part of generating the new set of MERs and consequently updating $\mathcal{M}$ when a pair of red (blue) lines in the grid swap. This actually causes the swap of a pair of rows (columns) of the matrix $\mathcal{M}$.

Here we need to mention that while processing an event point, if the angle of the joining line of the corresponding pair of points with the x-axis is $< \frac{\pi}{2}$ ($> \frac{\pi}{2}$), it results in a swap of rows (columns) corresponding to those points.

Lemma 1. *While processing an event point corresponding to the line joining a pair of points (p_α, p_β),*

(a) *if the angle of the line joining p_α and p_β with the x-axis is less than $\frac{\pi}{2}$, then the MERs whose neither of the north or south boundaries contain p_α nor p_β, will not be changed with respect to their definition.*

(b) *if the angle of the line joining p_α and p_β with the x-axis is greater than $\frac{\pi}{2}$, then the MERs whose neither of the east or west boundaries contain p_α nor p_β, will not be changed with respect to their definition.*

(c) *if the angle of the line joining p_α and p_β with the x-axis is less than $\frac{\pi}{2}$, and p_α is to the left of p_β, then MERs whose south bounding point is p_α, but p_β does not appear on any of its sides, and MERs whose north bounding point is p_β, but p_α does not appear on any of its sides, will not be changed with respect to their definition.*

(d) *if the angle of the line joining p_α and p_β with the x-axis is greater than $\frac{\pi}{2}$, and p_α is below p_β, then MERs whose east bounding point is p_α, but p_β does not appear on any of its sides, and MERs whose west bounding point is p_β, but p_α does not appear on any of its sides, will not be changed with respect to their definition.*

In view of this lemma, we state the following exhaustive set of MERs which may take birth or die out due to the swap of a pair of rows corresponding to a pair of points, say p_α and p_β (p_α is assumed to be to the left of p_β). A similar set of situations may also hold when a pair of columns swap; we will not mention them explicitly.

Following is the exhaustive set of MERs that may die out due to the swap of two red lines corresponding to p_α and p_β.

A : An MER with p_α and p_β at its the south and north boundaries respectively,
B : A set of MERs with p_α and p_β on south and east boundaries respectively,
C : A set of MERs with p_β on their south boundaries,
D : A set of MERs with p_α and p_β on west and north boundaries respectively,
E : A set of MERs with p_α on their north boundaries.

And following is the exhaustive set of MERs that may possibly result due to the swap of two red lines corresponding to p_α and p_β.

A' : An MER with p_α and p_β at its north and south boundaries respectively,
B' : A set of MERs with p_β and p_α on south and west boundaries respectively,
C' : A set of MERs with p_α on their south boundaries,
D' : A set of MERs with p_α on their north sides and p_β appears on the east,
E' : A set of MERs with p_β on their north boundaries.

In the following section, we shall highlight the necessary actions when a row swap takes place and also indicate how all the cases above are taken care of.

Now, note that the MER in A modifies into the MER in A'.

Further, the MERs in B all collapse to form members of C', and moreover, all the new members in C' are derived from B. In the latter case, to be a bit more explicit, there are actually two sets of MERs. First, the ones that have their north bounding points to the left of p_β, and secondly, the ones that have their north bounding points to the right of p_β. What is to be noted is that these north bounding points of the second class of MERs bound an MER of the first class at the east.

Similarly, the members in C that collapse, result in members of B' if at all they remain, and conversely all the members of B' result from members in C. To be a bit more explicit about the former case: among the MERs that collapse in this case, ones having their north bounding points to the right of p_α only would still exist and degenerate into members of B'. Rest are all destroyed.

Again, the MERs in D degenerate into MERs of E', and all the newly introduced MERs of E' are derived from D. Similarly, all the members of E that collapse, result in members of D' if at all they remain, and every member in D' is derived from a member of E. These observations will guide our actions due to a row swap.

3.4   Updating Grid Diagram

We need to consider the two cases - (A) swap of two adjacent rows, and (B) swap of two adjacent columns, separately. When a new MER takes birth with a specified set of points in its four sides, it is entered in $\mathcal{M}$ along with the initial grid angle. When it disappears, we update the corresponding entries of $\mathcal{M}$, and evaluate the PMER. Also, all the appropriate data structures are updated at each stage. In this subsection, all references to storage of an MER at a matrix entry will imply *at the south*, unless otherwise specified.

Swap of two adjacent rows

Let the line joining (p_α, p_β) is under process which is having the smallest absolute gradient among the set of unprocessed lines, and the gradient of that line is positive. We now study the effect of rotating the grid so that all red lines become parallel to the line joining (p_α, p_β). Let i and $i+1$ be the rows in $\mathcal{M}$ corresponding to the points p_α and p_β, before the rotation. Thus after the swap, the rows corresponding to the points p_α and p_β are $i+1$ and i respectively. The columns corresponding to p_α and p_β be k and ℓ ($k < \ell$) respectively.

Step A : The MER with p_α and p_β at its the south and north boundaries respectively before the swap, is unbounded at its east and west sides. This MER will not exist further. But this gives birth to another unbounded empty rectangle with p_β and p_α at its south and north boundaries respectively. Thus the necessary changes in the *value* fields in $\mathcal{M}(i, \ell)$ and $\mathcal{M}(i+1, k)$ need to be made.

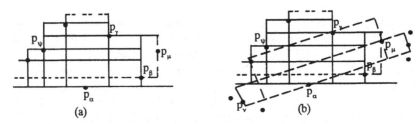

Fig. 3. Illustration of (a) Step B.1, (b) Step B.2

Step B : The MERs with p_α and p_β on their south and east boundaries respectively before the swap (see Figure 3a), will eventually collapse; so, for each of them the corresponding PMER needs to be computed. First proceed along the i-th row, from extreme left and include each encountered MER in a set R till we reach i) an MER whose east vertex is to the left of p_β, or ii) the entry $\mathcal{M}(i, k)$. In the latter case, the corresponding MER is included in R if it is bounded by p_β at the east. Next proceed from the right along the i-th row to include in R the first MER that is bounded by p_β at the east, unless of course, this is the one represented at $\mathcal{M}(i, k)$. The last action is guided by the observation that there can be only one MER that is bounded by p_β to the east, p_α to the south, and by a point to the right of p_α at the north. Now for all the members in R, p_β will no longer remain in their east boundary, and we need to update their east boundaries as follows (see Figure 3).

B.1 Consider a rectangle in R whose north and west bounding vertices are p_γ and p_ψ respectively. Note that before the rotation, there must have been a rectangle with p_γ and p_β at its north and south respectively, and it is also bounded by p_ψ at its west. Further, if the east boundary of this second rectangle contains p_μ, then the rectangle in R we started with, shall degenerate to be bounded by p_μ at the east. Thus, for each member in R, replace the $P1$, $P2$ pointers of their representative entries in the i-th row by the corresponding entries at the same column of $i + 1$-th row. If the member of R is however, the one stored at $\mathcal{M}(i, k)$ and thus unbounded at the north, the $P1$, $P2$ pointers are obtained from those at $\mathcal{M}(i + 1, \ell)$.

B.2 Note that, here a new MER appears with p_μ and p_α on its north and south boundaries respectively. This would, in fact, be the case for each rectangle in R. In other words, the newly obtained east boundary vertices for each of the rectangles in R, would now bound a rectangle in the north that are bounded by p_α in the south. This gives rise to a new set of rectangles R'. But this case can also be tackled much like in B.1 (as shown in Figure 3b). The details are omited in this version.

Step C : Next we consider the set of MERs, each having south boundary containing p_β. Due to the rotation, some of them will be truncated by p_α at their west side. The corresponding PMERs need to be reported and the matrix entries need to be updated to give birth to a new set of MERs.

C.1 The entries in row $i+1$ (corresponding to p_β) that store MERs are observed from extreme right one by one. If the west boundary of such an MER is observed to be to the left of p_α, we replace its west boundary by p_α. The scan terminates as soon as an entry is encountered which does not satisfy the above criterion, or the cell $\mathcal{M}(i + 1, \ell)$ is reached. If the MER stored at $\mathcal{M}(i + 1, \ell)$ has its west boundary to the left of p_α, then it is replaced by p_α.

C.2 Next, we check the entries of the $i + 1$-th row that represent an MER at the south from the extreme left of that row. All the MERs that appear to the left of k-th column, i.e., whose north boundary is defined by a point to the left of p_α, will not remain valid after the current rotation (See Figure 4a). The search continues along that row past the k-th column, to detect the MERs having their west bounding vertices to the left of p_α. These set of MERs will be truncated by p_α to the west. We stop when the first MER with its west boundary to the right of p_α is encountered or the cell $\mathcal{M}(i + 1, \ell)$ is reached.

Step D : We now consider the set of MERs having p_α and p_β at their west and north boundaries respectively. These MERs will now collapse as a result of this swap. This case is exactly similar to that in Step B, and the new set of MERs shall be determined in the same way by traversing along the column ℓ in a downward direction from $\mathcal{M}(i+1, \ell)$, and collecting all the rectangles in a set R, until i) an MER is obtained which is not bounded by p_α to its west side, or ii) the bottom of the column is reached. In the latter case, the MER represented at the 0-th row(and consequently unbounded at the south) is included in R, if it has p_α as its west vertex. Note that, after the current grid rotation, this set of rectangles will no longer be bounded by p_α towards west.

D.1 Next, consider an MER in the set R. It is bounded in the north and west by p_β and p_α respectively. Suppose its south and east sides are bounded by p_θ and p_η respectively. After the rotation, surely this MER is not going to remain bounded at the west by p_α. But observe that before the rotation, in this case, there would be an MER bounded by p_α on the north and p_θ on the south, and it would also have p_η on the east. If this rectangle is bounded by p_δ in the west, then surely the rectangle in R we started with, will have p_δ at its west after the rotation (as shown in Figure 4b). Thus the $P1$, $P2$ pointers in the entries of the ℓ-th column corresponding to the members in R are obtained from those in the same row and k-th column.

D.2 It is to be noted, exactly as in Case B, that as a result of rotation, the point p_δ will also bound a MER at the south that is bounded by p_β at the north. This is however, true for all the MERs in R. So, a new set of MERs R' thus arises, which can be identified much like in Step D.1.

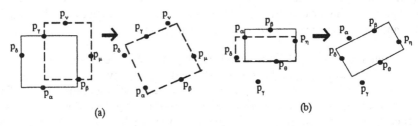

Fig. 4. Illustration of (a) Step C, (b) Step D.4

Step E : Some of the MERs with p_α on their north boundaries, might be affected with p_β entering them due to rotation, and will be either truncated on the east by p_β, or simply destroyed. The situation is similar to that in step C.

Processing for this step involves a traversal along the k-th column in a downward direction. The MERs encountered are eliminated if their south boundary vertices are to the right of p_β, and otherwise truncated by p_β at the east, if the latter is to the left of the east vertex of this MER.

Here, the traversal terminates once (i) an MER is encountered which is bounded by a point in the east that is to the left of p_β or (ii) bottom of the column is reached. In the second case, however, we perform the above processing if the east boundary of the MER represented there is to the right of p_β.

Finally, after the computation of the PMERs, and the necessary updates in $\mathcal{M}$, we swap the two rows i and $i+1$ of the matrix $\mathcal{M}$. This swap requires $O(n)$ unit of time. The row-id of the points p_α and p_β in P will be changed, and p_α and p_β will be swapped in P'.

An important activity to be taken care of is the updating of the representation of an MER at its west, once it is introduced or deleted, provided it is bounded either in east or west. But this is easy upon noting that, when an MER is deleted or introduced, we know the exact set of bounding points.

One crucial point is to be kept in mind while making this update. The steps have to be executed exactly in this order. This is because one step ahead of another may corrupt the values being used by the other. As an example, the updates in row i in Step B depend on the existing entries in row $i + 1$. If we execute Step C ahead of B, it is evident that the entries in row $i+1$ get corrupted.

Swap of two adjacent columns

The data structures, as also our approach is evidently exactly symmetric with respect to the rows and columns. Hence an exactly same process is followed in case of a column swap-over.

Note : By our design, an MER that is unbounded in both north and south is never represented at the south. This could lead to a potential problem because, when two rows swap, this MER will not get updated in our data structures. But note that such an MER can never be changed in terms of its determining vertices if two rows are interchanged. Indeed they can be modified only if two columns swap, and we at once store them at the south once they become bounded either at the south or north.

Symmetrical is the case for MERs that are unbounded at both west and east.

4   Complexity Analysis

As discussed in the preceding sections, our algorithm consists of two phases, (i) deciding the event points, i.e., the grid angles, and (ii) the management of grid diagram during each step of rotation. The first phase requires $O(n^2 \log n)$ time. Now it remains to analyze the time complexity of the second phase.

The construction of initial grid matrix $\mathcal{M}$ requires $O(n^2)$ time in the worst case. Next, we process $O(n^2)$ event points. For each such event point involving a pair of points p_α and p_β, the procedure outlined in the last section traverses atmost a pair of rows and a pair of columns, and does not more than constant time processing at each entry, thereby involving a total of $O(n)$ time. Moreover, all the PMERs are evaluated in our algorithm. Thus, we have the final theorem stating the time complexity of our algorithm.

Theorem 1. *(a) The number of PMERs generated during the entire processing, may be $O(n^3)$ in the worst case, and (b) the time complexity of our algorithm is also $O(n^3)$ in the worst case.*   □

As an $n + 1 \times n + 1$ matrix is maintained throughout the processing, the space complexity is $O(n^2)$.

References

1. A. Aggarwal and S. Suri, *Fast algorithm for computing the largest empty rectangle*, Proc. 3rd Annual ACM Symp. on Computational Geometry, pp. 278-290, 1987.
2. D. P. Dobkin, H. Edelsbrunner and M. H. Overmars, *Searching for empty convex polygons*, Proc. 4th ACM Symp. on Computational Geometry, pp. 224-228, 1988.

3. A. Naamad, D. T. Lee and W. L. Hsu, *On the maximum empty rectangle problem*, Discrete Applied Mathematics, vol. 8, pp. 267-277, 1984.
4. S. C. Nandy, B. B. Bhattacharya and S. Ray, *Efficient algorithms for identifying all maximal empty rectangles in VLSI layout design*, Proc. FSTTCS - 10, Lecture Notes in Computer Science, vol. 437, Springer, pp. 255-269, 1990.
5. S. C. Nandy, A. Sinha and B. B. Bhattacharya, *Location of largest empty rectangle among arbitrary obstacles*, Proc. FSTTCS - 14, Lecture Notes in Computer Science, vol. 880, Springer, pp. 159-170, 1994.
6. M. Orlowski, *A new algorithm for largest empty rectangle problem*, Algorithmica, vol. 5, pp. 65-73, 1990.

Renaming Is Necessary in Timed Regular Expressions

Philippe Herrmann

LIAFA, Université Paris 7
2, place Jussieu F-75251 Paris Cedex 05 France
herrmann@liafa.jussieu.fr

Abstract. We prove that timed regular expressions without renaming are strictly less expressive than timed automata, as conjectured by Asarin, Caspi and Maler in [3], where this extension of regular expressions was introduced. We also show how this result allows us to exhibit an infinite hierarchy of timed regular expressions.

1 Introduction

Among the different models that have been developed in order to describe real-time systems, the timed automata of Alur and Dill [2] are particularly interesting since they provide a timed counterpart of finite state automata, which have been studied intensively by the formal languages community. Thus it is only natural to try to adapt well-known results about finite state automata, of which there are plenty of, to the more general picture of timed automata. A basic result of automata theory being Kleene's theorem [7], stating the equivalence between finite automata and regular expressions, Asarin, Caspi and Maler designed *timed* regular expressions in [3]. They proved that, at the price of augmenting regular expressions with a very natural timed restriction operator, and a somewhat less natural but indispensable conjunction operator, one gets timed regular expressions which are exactly as expressive as timed automata *up to renaming*. That is, every timed regular expression is equivalent to a timed automaton, while every timed automaton can be translated into an equivalent timed regular expression provided a subset of the actions are renamed. Asarin, Caspi and Maler conjectured that renaming is indeed necessary to get the full expressive power of timed automata. In this paper we prove that their conjecture was indeed correct, namely that *one cannot get rid of the renaming*. To achieve this goal, we first give in Sect. 6 an easy proof of the necessity of introducing the conjunction operator, a result that already appeared in the original paper [3], but with a different construction. Then we proceed in Sect. 7 with the much more involved proof of the necessity of renaming. Finally, in Sect. 8, we introduce an infinite hierarchy of timed regular expressions, based on the number of conjunction operators and the use of renaming.

C. Pandu Rangan, V. Raman, R. Ramanujam (Eds.): FSTTCS'99, LNCS 1738, pp. 47–59, 1999.
© Springer-Verlag Berlin Heidelberg 1999

2   Timed Regular Expressions, Timed Automata

2.1   Notations

Let Σ be a finite alphabet and let $\mathbb{R}_{\geq 0}$ denote the set of nonnegative reals. A *timed language* is a subset of $(\Sigma \times \mathbb{R}_{\geq 0})^+$, i.e. a set of *timed words* (note that the notion of empty word does not appear in our definition).

Intuitively, a timed word $w = (a_1, \delta_1)(a_2, \delta_2) \cdots (a_n, \delta_n)$ corresponds to an action a_1 of duration δ_1, followed by an action a_2 of duration δ_2 ... and ending by an action a_n of duration δ_n. The duration of w is denoted by $\delta(w) = \sum_{k=1}^{n} \delta_k$.

To such a timed word we associate a non-decreasing sequence $(t_0, t_1, \cdots, t_n)$ with $t_0 = 0$ and $t_{k+1} = t_k + \delta_{k+1}$ (the intended meaning being that action a_k starts at time t_{k-1} and ends at time t_k). So δ_i will be called the duration of event i, while t_i will be its time stamp. Sequences of durations and sequences of time stamps are in bijection.

A *timed interval* is an interval of $\mathbb{R}_{\geq 0}$ of the form $[a, b]$, $[a, b)$, $(a, b]$, (a, b), $[a, +\infty)$ or $(a, +\infty)$ with a and b in $\mathbb{N}$. We denote $[a, a]$ by a.

Let σ be a mapping from Σ to Σ. The domain and range of σ can be extended to $\Sigma \times \mathbb{R}_{\geq 0}$ by letting $\sigma((a, \delta)) = (\sigma(a), \delta)$ for $a \in \Sigma$. We also denote by σ the generated morphism from $(\Sigma \times \mathbb{R}_{\geq 0})^+$ to $(\Sigma \times \mathbb{R}_{\geq 0})^+$ and call it a *renaming*.

3   Timed Regular Expressions

Timed Regular Expressions or *TREs* were introduced in [3] and are defined inductively as follows (we let ϕ, ϕ_1 and ϕ_2 be *TREs*):

- a is a *TRE*, with $a \in \Sigma$ (atom);
- $\phi_1 \wedge \phi_2$ is a *TRE* (conjunction);
- $\phi_1 \vee \phi_2$ is a *TRE* (disjunction);
- $\phi_1 \cdot \phi_2$ is a *TRE* (concatenation);
- ϕ^+ is a *TRE* (iteration);
- $\langle \phi \rangle_I$ is a *TRE* (restriction), where I is a timed interval.

The interpretation of a *TRE* is given by $[\![\cdot]\!]$ which is a function from the set of *TREs* to the set of timed languages defined inductively by:

- $[\![a]\!] = \{(a, \delta) \mid \delta \in \mathbb{R}_{\geq 0}\}$ for $a \in \Sigma$;
- $[\![\phi_1 \wedge \phi_2]\!] = [\![\phi_1]\!] \cap [\![\phi_2]\!]$;
- $[\![\phi_1 \vee \phi_2]\!] = [\![\phi_1]\!] \cup [\![\phi_2]\!]$;
- $[\![\phi_1 \cdot \phi_2]\!] = \{w_1 w_2 \mid w_1 \in [\![\phi_1]\!], w_2 \in [\![\phi_2]\!]\} = [\![\phi_1]\!] \cdot [\![\phi_2]\!]$;
- $[\![\phi^+]\!] = \bigcup_{k > 0} \{w_1 \ldots w_k \mid w_1, \ldots, w_k \in [\![\phi]\!]\} = \bigcup_{k > 0} [\![\phi]\!]^k$;
- $[\![\langle \phi \rangle_I]\!] = \{w \mid w \in [\![\phi]\!], \delta(w) \in I\}$.

Note that the iteration operator is a Kleene plus, not a Kleene star: for any *TRE* ϕ, $\varepsilon \notin [\![\phi]\!]$. Also we do not use signals, but timed words. We do so in order to simplify subsequent proofs, as pointed out in Sect. 9.

The function *sons* is defined on *TREs* and returns a set of *TREs*: for an atom it returns the emptyset, for the conjunction, disjunction and concatenation it returns the two sub-*TREs*, and for the iteration and restriction it returns the single sub-*TRE*. The *syntax tree* of a *TRE* ϕ is a tree with root ϕ, and each node ψ has $sons(\psi)$ as sons. A *cut* is a set of nodes (i.e. *TREs*) that contains exactly one node from every branch from the root to a leaf.

A $\wedge$-free *TRE* (resp. $\vee$-free) is a *TRE* containing no $\wedge$ (resp. $\vee$) operator.

A *TRE* can be put in disjunctive normal form (disjunction of a finite number of $\vee$-free *TREs*) using the fact that conjunction, concatenation and restriction distribute over disjunction. For the iteration, we check that the following identity on regular expressions extends to timed ones:

$$[\![(\phi_1 \vee \phi_2)^+]\!] = [\![\phi_1^+ \vee \phi_2^+ \vee (\phi_1^+ \cdot \phi_2^+)^+ \vee (\phi_2^+ \cdot \phi_1^+)^+ \vee (\phi_1^+ \cdot \phi_2^+)^+ \cdot \phi_1^+ \vee (\phi_2^+ \cdot \phi_1^+)^+ \cdot \phi_2^+]\!]$$

3.1   Timed Automata

A *timed automaton* or *TA* (see [1]) is a tuple $\mathcal{A} = (\Sigma, Q, S, F, X, E)$ where Σ is a finite set of actions, Q a finite set of states, $S \subseteq Q$ is the set of initial states, $F \subseteq Q$ the set of final states, X a finite set of clocks, E a finite set of transitions of the form (q, q', a, g, ρ) where q and q' are states of Q, a an action, g is a condition on clocks (see below) and ρ a subset of X called the reset set of the transition.

A *condition on clocks* is a possibly empty conjunction of terms of the form $x \in I$ where x is a clock and I a timed interval (we say that $x \in I$ is part of the condition). Recall that the ending points of a timed interval are integers or $+\infty$. A *clock valuation* is a vector $\nu \in (\mathbb{R}_{\geq 0})^X$: hence the valuation of clock x in ν is denoted by $\nu(x)$. For a reset set $\rho \subseteq X$ and a clock valuation ν, we define $Reset_\rho(\nu) \in (\mathbb{R}_{\geq 0})^X$ by $Reset_\rho(\nu)(x) = 0$ if $x \in \rho$ and $\nu(x)$ otherwise. A condition on clocks holds or not for a given clock valuation, and the evaluation of a condition g on a valuation ν is denoted by $g(\nu)$. We define the valuation $\nu + \delta$ with $\delta \in \mathbb{R}_{\geq 0}$ by $(\nu + \delta)(x) = \nu(x) + \delta$.

Let $\pi = q_0 \xrightarrow{e_1} q_1 \xrightarrow{e_2} \cdots \xrightarrow{e_n} q_n$ with $n > 0$ be a path of a *TA* $\mathcal{A}$ (i.e. $q_k \in Q$ for $0 \leq k \leq n$ and the source state of $e_k \in E$ is q_{k-1} while its goal state is q_k for $1 \leq k \leq n$). For such a path, the reset set of e_k will be denoted by ρ_k (we let $\rho_0 = X$ by convention), its condition on clocks by g_k and its action by a_k. The path is *accepting* if $q_0 \in S$ and $q_n \in F$ (i.e. it starts in an initial state and ends in a final state). The *trace* of π is the untimed word $a_1 \cdots a_n$. A *cycle* is a path which starts and ends in the same state. Now a *run* of $\mathcal{A}$ associated with π is a sequence

$$(q_0, \nu_0) \xrightarrow{e_1, \delta_1} (q_1, \nu_1) \xrightarrow{e_2, \delta_2} \cdots \xrightarrow{e_n, \delta_n} (q_n, \nu_n)$$

(where ν_k is a clock valuation and $\delta_k \in \mathbb{R}_{\geq 0}$) which satisfies the additional conditions: $g_k(\nu_{k-1} + \delta_k)$ holds and $\nu_k = Reset_{\rho_k}(\nu_{k-1} + \delta_k)$. The trace of this run is the timed word $(a_1, \delta_1) \cdots (a_n, \delta_n)$. A run is *accepting* if the associated path is accepting and $\nu_0 = \{0\}^X$ (i.e. the clocks are set to 0 initially).

The semantics of a *TA* $\mathcal{A}$ is a timed language $\mathcal{L}(\mathcal{A})$ defined as follows: $w \in \mathcal{L}(\mathcal{A})$ iff w is the trace of an accepting run of $\mathcal{A}$.

4   From *TREs* to *TAs*

This section is an adaptation of the translation from expressions to automata of [3]: recall that we do not allow the empty word as part of our languages.

A *TRE* ϕ can be translated into a *TA* $\mathcal{A}$ such that $[\![\phi]\!] = \mathcal{L}(\mathcal{A})$. To see this, we proceed by structural induction. For the base case, note that an atom $a \in \Sigma$ corresponds to the *TA* $(\{a\}, \{q_0, q_1\}, \{q_0\}, \{q_1\}, \emptyset, \{(q_0, q_1, a, true, \emptyset)\})$. For the induction hypothesis, let ϕ_1 correspond to $(\Sigma_1, Q_1, S_1, F_1, X_1, E_1)$, while ϕ_2 corresponds to $(\Sigma_2, Q_2, S_2, F_2, X_2, E_2)$, where the *TAs* are such that Q_1 and Q_2 as well as X_1 and X_2 are disjoint, and such that no final state has an outgoing transition (this is verified in the base case). Then we have:

- $\phi = \phi_1 \wedge \phi_2$ becomes $(\Sigma_1 \cup \Sigma_2, Q_1 \times Q_2, S_1 \times S_2, F_1 \times F_2, X_1 \uplus X_2, E)$ where E is defined as the smallest set such that $(q_1, q_1', a, g_1, \rho_1) \in E_1$ and that $(q_2, q_2', a, g_2, \rho_2) \in E_2$ implies $((q_1, q_2), (q_1', q_2'), a, g_1 \wedge g_2, \rho_1 \cup \rho_2) \in E$;
- $\phi = \phi_1 \vee \phi_2$ becomes $(\Sigma_1 \cup \Sigma_2, Q_1 \uplus Q_2, S_1 \uplus S_2, F_1 \uplus F_2, X_1 \uplus X_2, E_1 \uplus E_2)$;
- $\phi = \phi_1 \cdot \phi_2$ is translated into $(\Sigma_1 \cup \Sigma_2, Q_1 \backslash F_1 \uplus Q_2, S_1, F_2, X_1 \uplus X_2, E)$ where E is obtained by replacing every transition (q, q', a, g, ρ) of E_1 with $q' \in F_1$ by the set of transitions $\{(q, q'', a, g, X_2) \mid q'' \in S_2\}$ and leaving the other transitions just as they are (the fact that we do not recognize the empty word and that F_1 has no outgoing transition plays a crucial role here). Note that all clocks of X_2 are reset when moving from the automaton of ϕ_1 to the automaton of ϕ_2 and that the clocks in X_1 will never be used again;
- $\phi = \phi_1^+$ is translated into $(\Sigma_1, Q_1, S_1, F_1, X_1, E)$ where E is obtained by adding to E_1 the set of transitions $\{(q, q'', a, g, X_1) \mid q'' \in S_1\}$ for each transition (q, q', a, g, ρ) of E_1 with $q' \in F_1$. Note that all clocks are reset after each iteration;
- $\phi = \langle \phi_1 \rangle_I$ is translated into $(\Sigma_1, Q_1, S_1, F_1, X_1 \uplus \{x\}, E)$ with $x \notin X_1$, where E is obtained by changing every transition (q, q', a, g, ρ) of E_1 with $q' \in F_1$ into a transition $(q, q', a, g \wedge (x \in I), \rho)$ (the fact that no state of F_1 has an outgoing transition is important here);

At each step, we check that no final state has an outgoing transition, and that the constructed *TA* and ϕ have the same semantics (the proofs are left to the reader). The *TA* obtained from ϕ in the way we just described will be denoted by $\mathcal{A}_\phi$.

5   Technical Results

This section presents a few technical lemmas based on ideas from [4] and [6].

Let $\mathcal{A}$ be a *TA*, $\pi = q_0 \xrightarrow{e_1} q_1 \xrightarrow{e_2} \cdots \xrightarrow{e_n} q_n$ be a path of $\mathcal{A}$. Let $\rho_0 = X$ and

$$C_{i,j} = \{I \mid \exists x \text{ such that } x \in \rho_i \backslash (\rho_{i+1} \cup \ldots \cup \rho_{j-1}) \text{ and } (x \in I) \text{ is part of } g_j\}$$

It is the set of timed intervals being part of the condition on clocks g_j involving clocks that are reset in ρ_i and not reset again until at least ρ_j. With π we associate a graph denoted by $G(\pi)$ with $n + 1$ nodes labeled by $0 \ldots n$, and containing for all nodes $i < j$ s.t. $C_{i,j} \neq \emptyset$ the edge $i \xrightarrow{I_{i,j}} j$ with $I_{i,j} = \bigcap_{c \in C_{i,j}} c$.

We will also say that $G(\pi)$ is a graph of $\mathcal{A}$ as an abbreviation.

Almost by definition (recall that $t_k = \sum_{i=1}^{k} \delta_k$):

Lemma 1. *Let π be a path of a TA $\mathcal{A}$ with trace$(\pi) = a_1 \cdots a_n$. Then we have that the timed word $(a_1, \delta_1) \cdots (a_n, \delta_n)$ is the trace of a run associated with π iff for all edges $i \xrightarrow{I_{i,j}} j$ of $G(\pi)$ we have $t_j - t_i \in I_{i,j}$.*

Proof. see [6].

A graph $G(\pi)$ contains a *crossing* denoted by $i \overline{ j k } l$ (where $i < j < k < l$ are nodes of $G(\pi)$) if there is an edge $i \xrightarrow{I_{i,k}} k$ and an edge $j \xrightarrow{I_{j,l}} l$.

The following result gives a characterization of $\wedge$-free *TREs*:

Lemma 2. *Let ϕ be a $\wedge$-free TRE, and let $\mathcal{A}_\phi$ be its associated TA. Then no graph of $\mathcal{A}_\phi$ contains a crossing.*

Proof. Proceed by structural induction on $\wedge$-free *TREs*. The result is trivial for atoms. For disjunction, note that the paths and hence graphs of the corresponding *TA* are obtained by taking the union of the paths of the sub-*TAs*. For concatenation and iteration, the fact that the clocks are reset respectively when moving from the first to the second *TA* or when starting a new iteration shows that the graphs of the resulting automaton are obtained by concatenating certain graphs of the sub-*TAs*. For restriction, edges are only added between the first and the last node of certain graphs (those associated with accepting paths). All these operations cannot create a crossing. The $\wedge$-freeness is important since conjunction corresponds roughly to the 'superposition' of graphs which may very well create a crossing.

Let $\pi = q_0 \cdots q_n$ (resp. $\pi' = q'_0 \cdots q'_n$) be an *accepting* path in a *TA* $\mathcal{A}$ (resp. $\mathcal{A}'$). We write that $\pi' \preceq \pi$ (π' is less constraining than π) iff

- trace$(\pi) = $ trace(π');
- $i \xrightarrow{I'_{i,j}} j$ in $G(\pi')$ implies $i \xrightarrow{I_{i,j}} j$ in $G(\pi)$ with $I_{i,j} \subseteq I'_{i,j}$.

Note that $\preceq$ is a partial order. This definition is motivated by:

Lemma 3. *Let ϕ be a $\vee$-free TRE such that $\mathcal{A}_\phi$ has an accepting path π of length n for which there is a crossing $\overline{0 i j n}$ in $G(\pi)$. Then for all $\phi' \in$ sons(ϕ), $\mathcal{A}_{\phi'}$ has a path π' satisfying $\pi' \preceq \pi$.*

Proof. The different possible cases are:

- $\phi = \phi_1 \wedge \phi_2$: the accepting path π can only be obtained as the synchronous product of an accepting path π_1 in $\mathcal{A}_{\phi_1}$ and an accepting path π_2 in $\mathcal{A}_{\phi_2}$ with the same traces as π. The condition on edges is ensured by the fact that the constraints in $\mathcal{A}_\phi$ are obtained by conjuncting constraints of $\mathcal{A}_{\phi_1}$ and $\mathcal{A}_{\phi_2}$;
- $\phi = \phi_1 \cdot \phi_2$: π is obtained roughly by concatenating a path π_1 of $\mathcal{A}_{\phi_1}$ and a path π_2 of $\mathcal{A}_{\phi_2}$. All the clocks are reset when we 'move' from π_1 to π_2: this contradicts the fact that there is a crossing $\overline{0 \quad i \quad j \quad} n$ in π (recall that the crossing starts in the first node and ends in the last). Hence this case is impossible and we have nothing to prove;
- $\phi = (\phi_1)^+$: $\mathcal{A}_\phi$ is obtained from $\mathcal{A}_{\phi_1}$ by adding edges resetting all clocks. Since there is a crossing $\overline{0 \quad i \quad j \quad} n$ in $G(\pi)$ (starting in the first node and ending in the last), these edges cannot be part of π. Therefore π also appears as an accepting path in $\mathcal{A}_{\phi_1}$;
- $\phi = \langle \phi_1 \rangle_I$: since $\mathcal{A}_\phi$ is obtained from $\mathcal{A}_{\phi_1}$ by adding a condition on a new clock on edges ending in a final state, the path π in $\mathcal{A}_\phi$ already appears in $\mathcal{A}_{\phi_1}$ and the only difference in the corresponding graphs is for the edge from node 0 to node n for which the second condition holds.

The lemma is true for *TREs* which are not $\vee$-free as long as they are not of the form $\phi_1 \vee \phi_2$. Also note that $\phi = \phi_1 \cdot \phi_2$ is impossible.

Corollary 1. *Let ϕ be a $\vee$-free TRE such that $\mathcal{A}_\phi$ accepts a timed word w through a path π of length n for which there is a crossing $\overline{0 \quad i \quad j \quad} n$. Then for all $\phi' \in sons(\phi)$, $w \in [\![\phi']\!]$.*

Proof. We use lemma 3 and lemma 1.

6 Necessity of the Conjunction Operator

In the usual definition of (untimed) regular expressions, there is no conjunction operator such as $\wedge$: regular expressions and finite automata having the same expressive power, it is clear that such an operator is unnecessary. In the timed case, we wish to show that we cannot get rid of the intersection operator $\wedge$ if we want *TREs* to be as expressive as *TAs*. A more involved proof of this result appears in [3] in the case of signals.

Proposition 1. *Let $\phi = (\langle a \cdot b \rangle_1 \cdot c) \wedge (a \cdot \langle b \cdot c \rangle_1)$. There is no $\wedge$-free TRE ϕ' such that $[\![\phi']\!] = [\![\phi]\!]$.*

Proof. Let $\phi = (\langle a \cdot b \rangle_1 \cdot c) \wedge (a \cdot \langle b \cdot c \rangle_1)$. We proceed by contradiction: suppose there exists a $\wedge$-free *TRE* ϕ' such that $[\![\phi]\!] = [\![\phi']\!]$. It is clear that the timed word $w = (a, \frac{1}{2})(b, \frac{1}{2})(c, \frac{1}{2})$ is recognized by $\mathcal{A}_{\phi'}$ through an accepting path π. By lemma 2, we obtain that the graph $G(\pi)$ doesn't contain any crossing $\overline{0 \quad 1 \quad 2 \quad} 3$. Hence we cannot at the same time have an edge $0 \xrightarrow{I_{0,2}} 2$ and an edge $1 \xrightarrow{I_{1,3}} 3$. Suppose there is no edge $0 \xrightarrow{I_{0,2}} 2$. Let us consider node 2 (corresponding to the end of the b event): it may share edges only with nodes 1 and 3, and by lemma 1, the timed intervals $I_{1,2}$ and $I_{2,3}$ (if they exist) must contain $(0, 1)$, since $t_2 - t_1 = t_3 - t_2 = \frac{1}{2}$. Using lemma 1, we see that the timed word $w' = (a, \frac{1}{2})(b, \frac{3}{4})(c, \frac{1}{4})$ is recognized by $\mathcal{A}_{\phi'}$ through π. Hence $w' \in [\![\phi']\!]$, which is a contradiction since $w' \notin [\![\phi]\!]$. The case where there is no edge $1 \xrightarrow{I_{1,3}} 3$ is similar.

7   Necessity of Renaming

In this section we prove the fact that *TREs* alone cannot express the whole power of *TAs*. In fact we need to add renaming, defined earlier in Sect. 2.1. In order to prove that renaming is indeed necessary, which is a problem that was left open in [3], consider the *TA* $\mathcal{A}$ of Fig. 1. It is easy to see that $\mathcal{L}(\mathcal{A}) = \sigma([\![\phi]\!])$ where $\phi = \langle a^+ \cdot b \rangle_1 \cdot a^+ \wedge a^+ \cdot \langle b \cdot a^+ \rangle_1$ and σ maps a and b to a. Note that $\sigma([\![\phi]\!])$ and $[\![\langle a^+ \cdot a \rangle_1 \cdot a^+ \wedge a^+ \cdot \langle a \cdot a^+ \rangle_1]\!]$ are not equal for reasons of synchronization on the b event in ϕ: this is the idea we exploit to prove the necessity of renaming.

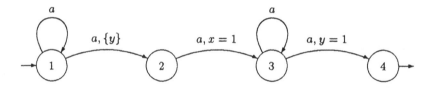

Fig. 1. the *TA* $\mathcal{A}$

Proposition 2. *There is no* TRE ψ *such that* $[\![\psi]\!] = \sigma([\![\phi]\!])$.

Proof. We proceed by contradiction. So let us suppose that there exists a *TRE* ψ such that $[\![\psi]\!] = \sigma([\![\phi]\!])$. We may assume that ψ is in disjunctive normal form, i.e. $\psi = \bigvee_{i=1}^{n} \psi_i$ where the ψ_i's are $\vee$-free. Hence $\mathcal{A}_\psi$ is the juxtaposition of n *TAs* $\mathcal{A}_{\psi_1}, \ldots, \mathcal{A}_{\psi_n}$. Let K be an integer greater than the number of states of $\mathcal{A}_\psi$ (and big enough for all subsequent constructions). At least one of the *TAs*, say $\mathcal{A}_{\psi_1}$ w.l.o.g., accepts the timed word $w = (a^{4K}, (\delta_i)_{1 \leq i \leq 4K})$, where the sequence of durations/time stamps is any sequence satisfying the following:

- $t_0 = 0 < t_1 < t_2 < \cdots < t_{4K}$;
- $t_{2K} = 1$;
- $t_{2K-1} = t_{4K} - 1$;
- $t_{4K-1} < t_1 + 1$.

The constraints on this timed word may be visualized by:

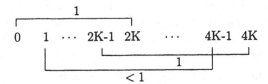

We define the predicate $crossing(\chi)$ where χ is a *TRE*. It is true iff $\mathcal{A}_\chi$ has at least one path π through which w is recognized and such that $G(\pi)$ has a crossing $0 \underline{\ \ i\ \ } \underline{\ \ j\ \ } 4K$. Also let *keep* be the set of *TRE*s satisfying $\neg crossing$ (in particular it contains χ whenever $w \notin [\![\chi]\!]$).

We consider the way ψ_1 is built, that is the syntax tree of this *TRE*: note that since ψ_1 is $\vee$-free, it is also the case for all nodes of the syntax tree. Moreover the number of states of any *TA* associated with a node of the syntax tree is less than K.

We construct a sequence of cuts $(C_i)_{i \geq 0}$ in the following way:

- C_1 is the root of the tree (corresponding to ψ_1);
- $C_{i+1} = (C_i \cap keep) \cup sons(C_i \setminus keep)$ (it is clearly a cut if C_i is a cut because *keep* includes all leaves of any syntax tree).

Lemma 4. *For all $i \geq 1$, for all $\chi \in C_i$, $w \in [\![\chi]\!]$.*

Proof. We proceed by induction.

- base case: $C_1 = \{\psi_1\}$, and by hypothesis $w \in [\![\psi_1]\!]$;
- • induction hypothesis (IH): all members χ of C_i are such that $w \in [\![\chi]\!]$;
  • from C_i to C_{i+1}: let $\chi' \in C_{i+1}$. Either $\chi' \in C_i$ and we use the IH directly, or $\chi' \in sons(\chi)$ with $\chi \in C_i$. But then $crossing(\chi)$ is true, and since χ if $\vee$-free (just as any node of the syntax tree) we apply corollary 1 to prove that $w \in [\![\chi']\!]$. This ends the proof.

Since the syntax tree is finite, there is a rank l for which $C_{l+1} = C_l \cap keep = C_l$. Thus for all $\chi \in C_l$, $\mathcal{A}_\chi$ recognizes w through a path $\pi = q_0 \cdots q_{4K}$ with no crossing $0 \underline{\ \ i\ \ } \underline{\ \ j\ \ } 4K$. And since K is greater than the number of states of any node of the syntax tree, there exists $1 < j_1 < j_2 < 2K - 1$ and $2K < j_3 < j_4 < 4K - 1$ such that $q_{j_1} = q_{j_2}$ and $q_{j_3} = q_{j_4}$ ('left and right cycles').

For $\chi \in C_l$, let $ln(\chi)$ denote the product $(j_2 - j_1)(j_4 - j_3)$ of the lengths of the aforementioned cycles. Let

$$L = \prod_{\chi \in C_l} ln(\chi)$$

Now let $w' = (a^{4K+L}, (\delta'_i)_{1 \leq i \leq 4K+L})$ be any timed word satisfying:

- $t'_0 = 0 < t'_1 < \cdots < t'_{4K+L}$;

- $t'_{2K+L} = 1;$
- $t'_{2K-1} = t'_{4K+L} - 1;$
- $t'_{4K-1+L} < t'_1 + 1.$

The constraints on w' can be obtained from those on w by shifting positions $2K \cdots 4K$ from L places to the right as can be visualized by:

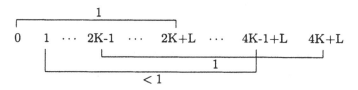

Lemma 5. For all $\chi \in C_l$, $w' \in [\![\chi]\!]$.

Proof. We know that $\mathcal{A}_\chi$ recognizes w through a path π such that $G(\pi)$ has no crossing $\overline{0 \; i \quad j \; 4K}$; also, π is such that there exists $1 < j_1 < j_2 < 2K - 1$ and $2K < j_3 < j_4 < 4K - 1$ with $q_{j_1} = q_{j_2}$, $q_{j_3} = q_{j_4}$, and satisfying the fact that $(j_2 - j_1)(j_4 - j_3)$ divides L. Hence $\pi = \beta_1 \alpha_1 \beta_2 \alpha_2 \beta_3$, with $|\beta_1| = j_1$, $|\alpha_1| = j_2 - j_1$, $|\beta_2| = j_3 - j_2$, $|\alpha_2| = j_4 - j_3$ and $|\beta_3| = 4K - j_4$. Let $\pi_l = \beta_1(\alpha_1)^{1+L/(j_2-j_1)}\beta_2\alpha_2\beta_3$, and let $\pi_r = \beta_1\alpha_1\beta_2(\alpha_2)^{1+L/(j_4-j_3)}\beta_3$. Therefore π_r and π_l are accepting paths of $\mathcal{A}_\chi$ and their traces are the same and equal to a^{4K+L}. We now show that w' is recognized by $\mathcal{A}_\chi$ through π_l or π_r by examining the edges of $G(\pi)$. Since w is recognized through π, we get by using lemma 1:

- for all $0 \xrightarrow{I} i$ in $G(\pi)$,
  - if $i < 2K$ then $(0, 1) \subseteq I$;
  - if $i = 2K$ then $1 \in I$;
  - if $i > 2K$ then $(1, 2) \subseteq I$;
- for all $i \xrightarrow{I} j$ in $G(\pi)$, if $1 \leq i < j \leq 4K - 1$ then $(0, 1) \subseteq I$;
- for all $i \xrightarrow{I} 4K$ in $G(\pi)$,
  - if $i < 2K - 1$ then $(1, 2) \subseteq I$;
  - if $i = 2K - 1$ then $1 \in I$;
  - if $i > 2K - 1$ then $(0, 1) \subseteq I$.

All possible edges are considered. We distinguish two cases:

Case 1. Suppose there is no edge $i \xrightarrow{I} 4K$ with $i \leq 2K - 1$ in $G(\pi)$. In this case, we consider the edges of $G(\pi_l)$. By the construction of π_l from π we used, particularly the fact that the transitions appearing in π_l and π are the same, we get that (only the first case is detailed, the others are similar):

- for all $0 \xrightarrow{I} i$ in $G(\pi_l)$,

- if $i < 2K + L$ then $(0,1) \subseteq I$: simply note that the transition corresponding to node i in $G(\pi_l)$ already appeared in π, and then it was attached to a node $< 2K$ (otherwise it would be shifted farther to the right), hence the constraint either disappears because of a clock reset in the cycle or stays the same as before the iteration of the cycle (that is the same than for an edge $0 \xrightarrow{I} i$ with $i < 2K$ in $G(\pi)$);
  - if $i = 2K + L$ then $1 \in I$ (same constraint);
  - if $i > 2K + L$ then $(1,2) \subseteq I$ (same constraint);
- for all $i \xrightarrow{I} j$ in $G(\pi_l)$, if $1 \leq i < j \leq 4K - 1 + L$ then $(0,1) \in I$ (the constraint either disappears because of a clock reset in the cycle or stays the same);
- for all $i \xrightarrow{I} 4K + L$ in $G(\pi_l)$,
  - $i \leq 2K - 1 + L$ cannot happen (by hypothesis);
  - if $i > 2K - 1 + L$ then $(0,1) \subseteq I$ (same constraint).

By lemma 1, it is easy to see that $\mathcal{A}_\chi$ recognizes w' through π_l.

Case 2. Now suppose there is an edge $i \xrightarrow{I} 4K$ with $i \leq 2K - 1$ in $G(\pi)$: since there is no crossing $\overline{0 \ \underline{i \quad j} \ 4K}$, there is no edge $0 \xrightarrow{I'} j$ with $j > 2K - 1$. In this second and last case, we consider the edges of $G(\pi_r)$ and we get that $\mathcal{A}_\chi$ recognizes w' through π_r. This ends the proof of Lemma 5.

Lemma 6. $w' \in [\![\psi_1]\!]$

Proof. Again we proceed by induction.

- base case: all members χ of C_l are such that $w' \in [\![\chi]\!]$ since we proved Lemma 5;
- induction hypothesis (IH): all members χ' of C_{i+1} satisfy $w' \in [\![\chi']\!]$;
- from C_{i+1} to C_i: let $\chi \in C_i$. If $\chi \in C_{i+1}$, then we use the IH directly. Otherwise, we distinguish four cases:
  - $\chi = \chi' \wedge \chi''$ with $\chi', \chi'' \in C_{i+1}$: here $w' \in [\![\chi']\!]$ and $w' \in [\![\chi'']\!]$ immediately implies $w' \in [\![\chi]\!]$;
  - $\chi = \chi' \cdot \chi''$ with $\chi', \chi'' \in C_{i+1}$: this case is impossible, as we noticed in the demonstration of lemma 3 and by construction of the sequence of cuts;
  - $\chi = (\chi')^+$ with $\chi' \in C_{i+1}$: here $w' \in [\![\chi']\!]$ immediately implies $w' \in [\![\chi]\!]$;
  - $\chi = \langle \chi' \rangle_I$ with $\chi' \in C_{i+1}$: we know $w \in [\![\chi]\!]$ and $\delta(w) \in (1,2)$, and since I is a timed interval this implies $(1,2) \subseteq I$. Thus $\delta(w') \in I$ and $w' \in [\![\chi]\!]$.

This ends the proof of Lemma 6, and leads us to a contradiction since we have $w' \notin \sigma([\![\phi]\!])$ and therefore $[\![\psi]\!] \neq \sigma([\![\phi]\!])$. Thus Proposition 2 is proven.

8   An Infinite Hierarchy of *TREs*

In this section we introduce an infinite hierarchy of *TREs* based on the number of $\wedge$-operators and the use or not of renaming. A *TRE* is n-$\wedge$ if it is constructed with at most n operators $\wedge$. Hence 0-$\wedge$ corresponds to $\wedge$-free. In the syntax tree of a *TRE*, a $\wedge$-node is any node of the form $\phi_1 \wedge \phi_2$. A n-*crossing* for $n > 0$ is a set of pairs of nodes $\{(i_k, j_k) \mid k = 1 \ldots n)\}$ satisfying the following requirements:

- $\displaystyle\bigcap_{k=1}^{n+1} [i_k, j_k] \neq \emptyset$;
- for all $k, l \in [1, n+1]$, $k \neq l$, $[i_k, j_k] \not\subseteq [i_l, j_l]$.

One checks that a 2-crossing is simply a crossing ('2 edges overlapping'). Also if a graph contains no n-crossing it won't contain any $(n+1)$-crossing either. We can now state a generalization of lemma 2:

Lemma 7. *Let ϕ be a n-$\wedge$ TRE, and let $\mathcal{A}_\phi$ be its associated TA. Then no graph of $\mathcal{A}_\phi$ contains any $(n+2)$-crossing.*

Proof. We proceed by induction over n.

- base case (n=0): this is exactly lemma 2;
- induction hypothesis (IH): we suppose that no graph of $\mathcal{A}_\phi$, where ϕ is any n-$\wedge$ *TRE*, contains any $(n+2)$-crossing;
- from n to $n+1$: let ϕ be a $(n+1)$-$\wedge$ *TRE* and consider its syntax tree. Let C be the unique cut consisting of nodes that are either leaves or $\wedge$-nodes and such that none of their ancestors is a $\wedge$-node. Each of the $\wedge$-nodes of C has two sub-trees that have respectively c_1 and c_2 $\wedge$-nodes. We know that $c_1 + c_2 \leq n$, thus we can apply the IH to the *TREs* corresponding to the two sub-trees of each $\wedge$-nodes of C. Now we can check that if ϕ_1 (resp. ϕ_2) is such that no graph of $\mathcal{A}_{\phi_1}$ (resp. $\mathcal{A}_{\phi_2}$) contains any $(c_1 + 2)$-crossing (resp. $(c_2 + 2)$-crossing), then no graph of $\mathcal{A}_{\phi_1 \wedge \phi_2}$ (i.e. any $\wedge$-node of C) contains any $(c_1 + c_2 + 3)$-crossing thus any $(n + 3)$-crossing. Now ϕ is obtained from the nodes of C without using the operator $\wedge$, since no ancestor of a node of C is a $\wedge$-node, and by adaptating the proof of lemma 2 one can show that we cannot get a new crossing this way. Thus $\mathcal{A}_\phi$ doesn't contain any $(n + 3)$-crossing and this ends the proof.

Let us define the sequence of *TREs* $(\phi_n)_{n>0}$ where $\phi_n = \displaystyle\bigwedge_{i=0}^{n} a^i \cdot \langle a^{n+1} \rangle_1 \cdot a^{n-i}$.

Clearly ϕ_n is n-$\wedge$. We now give a generalization of Proposition 1, namely that ϕ_n is not $(n - 1)$-$\wedge$, even if we allow renaming. Let σ be any renaming:

Proposition 3. *There is no $(n - 1)$-$\wedge$ TRE ψ such that $\sigma(\llbracket \psi \rrbracket) = \llbracket \phi_n \rrbracket$.*

Proof. Let us proceed by contradiction: suppose there exists a $(n-1)$-$\wedge$ *TRE* ψ and a renaming σ satisfying $\sigma(\llbracket\psi\rrbracket) = \llbracket\phi_n\rrbracket$. Note that the alphabet of ψ can be any finite set of actions, but that σ must map every action to a. We check $(a^{2n+1}, (\delta_i)_{1\leq u\leq 2n+1})$ belongs to $\llbracket\phi_n\rrbracket$ if $\delta_i = \frac{1}{n+1}$. Thus there exists an untimed word $x = x_1\cdots x_{2n+1}$ such that $w = (x, (\delta_i)_{1\leq u\leq 2n+1}) \in \llbracket\psi\rrbracket$, and in the TA $\mathcal{A}_\psi$ this timed word w is recognized through a path π of associated graph $G(\pi)$. Since ψ is $(n-1)$-$\wedge$ and by lemma 7, we know there is no $(n+1)$-crossing in this graph. Hence there exists $0 \leq k \leq n$ such that the edge $k \xrightarrow{I} k+n+1$ doesn't appear in $G(\pi)$: it is easy to check by using lemma 1 that the timed word $w' = (x, (\delta_i')_{1\leq i\leq 2n+1})$ with $\delta_i' = \delta_i$ for $i \neq k+n+1$ and $i \neq k+n+2$, $\delta_{k+n+1}' = \frac{3}{2(n+1)}$ and $\delta_{k+n+2}' = \frac{1}{2(n+1)}$ is recognized by $\mathcal{A}_\psi$ through π. Therefore $\sigma(w') = (a^{2n+1}, (\delta_i')_{1\leq i\leq 2n+1}) \in \llbracket\phi_n\rrbracket$, which is as contradiction.

Finally here is an easy result to complete the picture:

Proposition 4. *Let ϕ be a $\wedge$-free TRE and σ a renaming. Then there exists a $\wedge$-free TRE ψ such that $\llbracket\psi\rrbracket = \sigma(\llbracket\phi\rrbracket)$.*

Proof. We denote the morphism from Σ^+ to Σ^+ associated with σ by the same symbol. One checks by an easy structural induction that the *TRE* ψ is obtained by replacing every atom a appearing in ϕ by $\sigma(a)$ (obviously this doesn't work when the $\wedge$-operator is present).

Let $\mathcal{T}_n = \{\llbracket\phi\rrbracket \mid \phi \text{ is n-}\wedge\}$ and $\mathcal{T}_n' = \{\sigma(\llbracket\phi\rrbracket) \mid \phi \text{ is n-}\wedge \text{ and } \sigma \text{ is a renaming}\}$. Clearly $\mathcal{T}_n \subseteq \mathcal{T}_{n+1} \cap \mathcal{T}_n'$ and $\mathcal{T}_n' \subseteq \mathcal{T}_{n+1}'$ for $n \geq 0$. If we sum up the results we obtained, we get the following infinite hierarchy between these sets:

$$
\begin{array}{ccccccccc}
\mathcal{T}_0' & \xrightarrow{\subsetneq} & \mathcal{T}_1' & \cdots & \mathcal{T}_n' & \xrightarrow{\subsetneq} & \mathcal{T}_{n+1}' & \cdots \\
\Big\uparrow{\scriptstyle =} & & \Big\uparrow{\scriptstyle \subseteq} & & \Big\uparrow{\scriptstyle \subseteq} & & \Big\uparrow{\scriptstyle \subseteq} & \\
\mathcal{T}_0 & \xrightarrow{\subsetneq} & \mathcal{T}_1 & \cdots & \mathcal{T}_n & \xrightarrow{\subsetneq} & \mathcal{T}_{n+1} & \cdots
\end{array}
$$

Moreover $\mathcal{T}_i$ and $\mathcal{T}_j'$ are incomparable for $0 < j < i$ (no inclusion holds).

Proof. The equality between $\mathcal{T}_0$ and $\mathcal{T}_0'$ has just been proven in Proposition 4. Now for $n \geq 0$, Proposition 3 means that $\mathcal{T}_{n+1} \not\subseteq \mathcal{T}_n'$ (therefore $\mathcal{T}_{n+1} \not\subseteq \mathcal{T}_n$ and $\mathcal{T}_{n+1}' \not\subseteq \mathcal{T}_n'$ while $\mathcal{T}_i \not\subseteq \mathcal{T}_j'$ for $0 < j < i$) while Proposition 2 tells us $\mathcal{T}_1' \not\subseteq \mathcal{T}_n$ (which implies $\mathcal{T}_{n+1}' \not\subseteq \mathcal{T}_{n+1}$ and $\mathcal{T}_j' \not\subseteq \mathcal{T}_i$ for $0 < j < i$).

9   Discussion

We already mentioned the fact that in the original paper [3], the authors used a signal-based semantics for *TREs*, associating with any *TRE* what is actually a *TA* with *silent* transitions (while we use timed words and plain *TAs*). Thus one may wonder whether or not our results still hold if we allow silent actions ε as atoms in our definition of *TREs*. We believe the core of the proofs of this paper would remain the same if we used those extended *TREs*, although quite a few technical difficulties would be added (the graphs now have 'invisible nodes' scattered among the nodes we considered).

Acknowledgements

Many thanks to Paul Gastin for his remarks that improved the paper a lot. The author also wishes to thank Eugène Asarin and the anonymous referees for their valuable comments.

References

1. R. Alur and D.L. Dill: *Automata for Modelling Real-Time Systems.* Proceedings of ICALP'90, LNCS 443, pages 322-335, 1990.
2. R. Alur and D.L. Dill: *A Theory of Timed Automata,* Theoretical Computer Science 126, pages 183-235, 1994.
3. E. Asarin, P. Caspi and O. Maler: *A Kleene Theorem for Timed Automata.* Proceedings of LICS'97, pages 160-170, 1997.
4. B. Bérard, V. Diekert, P. Gastin and A. Petit: *Characterization of the Expressive Power of Silent Transitions in Timed Automata.* Fundamenta Informaticæ 36, pages 145-182, 1998.
5. V. Diekert, P. Gastin and A. Petit: *Removing ε-Transitions in Timed Automata.* Proceedings of STACS'97, LNCS 1200, pages 583-594, 1997.
6. P. Herrmann: *Timed Automata and Recognizability.* Information Processing Letters 65, pages 313-318, 1998.
7. S.C. Kleene, *Representations of Events in Nerve Nets and Finite Automata.* Automata Studies, pages 3-42, 1956.

Product Interval Automata: A Subclass of Timed Automata

Deepak D'Souza* and P. S. Thiagarajan**

Chennai Mathematical Institute,
92 G. N. Chetty Road, Chennai 600 017, India

Abstract. We identify a subclass of timed automata and develop its theory. These automata, called product interval automata, consist of a network of timed agents. The key restriction is that there is just one clock for each agent and the way the clocks are read and reset is determined by the distribution of shared actions across the agents. We show that the resulting automata admit a clean theory in both logical and language-theoretic terms. It turns out that the study of these timed automata can exploit the rich theory of partial orders known as Mazurkiewicz traces. An important consequence is that the partial order reduction techniques being developed for timed automata [4,10] can be readily applied to the verification tasks associated with our automata. Indeed we expect this to be the case even for the extension of product interval automata called distributed interval automata.

1 Introduction

Timed automata as formulated by Alur and Dill [1] have become a standard model for describing timed behaviours. These automata are very powerful in language-theoretic terms. Their languages are not closed under complementation. Further, their language inclusion problem is undecidable and hence cannot be reduced to the emptiness problem which is decidable. Hence in order to solve verification problems posed as language inclusion problems, one must use deterministic timed automata for specifications (which can be easily complemented) or one must work with a restricted class of timed automata that possess the desired closure properties.

Here we follow the second route and propose a subclass of timed automata called product interval automata (PI automata). Roughly speaking, such an automaton will consist of a network of timed agents $\|_{i=1}^{K} \mathcal{A}_i$ where each $\mathcal{A}_i$ will operate over an alphabet Σ_i of events. Further, there will be a *single* clock c_i associated with each agent i. The agents communicate by synchronising on the timed executions of common events. Suppose a is an event in which the agents $\{1, 3, 4\}$ participate. Then the timing constraint governing each a-execution will

---

* A part of this work has been supported by BRICS, Computer Science Dept., Aarhus University.

** A part of this work has been supported by the IFCPAR Project 1502-1.

C. Pandu Rangan, V. Raman, R. Ramanujam (Eds.): FSTTCS'99, LNCS 1738, pp. 60–71, 1999.

only involve the clocks $\{c_1, c_3, c_4\}$. Moreover, the set of clocks that is reset at the end of each a-execution will be $\{c_1, c_3, c_4\}$. Thus the distribution $\widetilde{\Sigma} = \{\Sigma_i\}_{i=1}^{K}$ of events over the agents will canonically determine the usage of clocks; so much so, we can avoid mentioning the clocks altogether once we fix $\widetilde{\Sigma}$.

This method of structuring timed automata has a number of advantages. In particular, one can provide naturally decomposed and succinct presentations of timed automata with large (control) state spaces. Admittedly, the technique of presenting a global timed automaton as a product of component timed automata has been used by many authors starting from [1]. What is new here, as explained above, is that our decomposed presentation places a corresponding restriction on the manner in which clocks are read and reset. It is worth pointing out that the model considered by Yi and Jonsson [24] in the framework of timed CSP can be easily represented as PI automata. Their main result, in our terms, is that language inclusion problem for PI automata is decidable. We establish a variety of results concerning PI automata which subsume the decidability of the language inclusion problem.

In principle, one could view our automata as a restricted kind of labelled timed Petri net. However the semantics we attach to our automata is the standard one for timed automata whereas the semantics one uses for timed Petri nets – with earliest and latest firing times for the transitions – is somewhat different.

A key feature of our automata is that their timed behaviour can be symbolically represented *without any loss of information* by conventional words. As a consequence, it turns out that their theory can be developed with the help of powerful results available in the theory of Mazurkiewicz traces [6]. We wish to emphasise, our automata will however have a conventional timed semantics with the non-negative reals serving as the time frame. A final aspect of PI automata that we wish to mention is that partial order reduction techniques that are under development [4,10] can be readily applied to our automata.

In pragmatic terms, it is not clear how much modelling power is lost through the restrictions we place on the usage of clocks. In many multi-agent timed systems it seems sufficient to have just one clock for each agent. For instance, a network of timed automata that communicate through shared variables is used to model and analyse the timed behaviour of asynchronous circuits by Maler and Pnueli [17]. It turns out that product interval automata suffice to represent the same class of timed behaviours.

From a theoretical standpoint, PI automata are strictly less expressive than event clock automata due to Alur, Fix, and Henzinger [2] and their state-based version [19] which in turn are strictly less powerful than general timed automata. As a result, the logics we develop here will also be strictly less expressive than the corresponding logics presented in [14] for a generalisation of event recording automata called recursive event recording automata. Nevertheless we feel that PI automata are of independent interest due to the reasons sketched earlier. For basic information about timed automata and their logics we cite the surveys [3,13] and their references. An interesting early instance of timed languages which have nice closure properties and which admit a clean logical characterisation can be found in [22].

In the next section we define PI automata. In section 3 we show that these automata (more precisely their languages) are closed under boolean operations. In section 4 we first present a monadic second order logic denoted TMSO$^\otimes$ to capture the timed regular languages recognised by PI automata. We then formulate a linear time temporal logic denoted TLTL$^\otimes$ and sketch automata-theoretic solutions to the satisfiability and model checking problems for TLTL$^\otimes$ in terms of PI automata.

As we point out in the final section, all our ideas can be extended smoothly to a larger setting in which the underlying "symbolic" automata are asynchronous Büchi automata [11]. The resulting timed automata are called distributed interval automata and they can also be studied using techniques taken from trace theory. It is the case that PI automata are less expressive than distributed interval automata which in turn are less expressive than event recording automata [2]. It turns out that the so called cellular version of distributed interval automata correspond to event recording automata. All these extensions as well detailed proofs of all the results presented below are available in the full paper [7].

2   Product Interval Automata

We fix a finite set of agents $\mathcal{P} = \{1, 2, \ldots, K\}$ and let i, j range over $\mathcal{P}$. A $\mathcal{P}$-distributed alphabet is a family $\widetilde{\Sigma} = \{\Sigma_i\}_{i \in \mathcal{P}}$ where each Σ_i is a finite set of actions. We set $\Sigma = \bigcup_{i \in \mathcal{P}} \Sigma_i$ and call it the global alphabet induced by $\widetilde{\Sigma}$. We let a, b range over Σ. The set of agents that participate in each occurrence of the action a is denoted by $loc(a)$ and is given by: $loc(a) = \{i \mid a \in \Sigma_i\}$. Through the rest of the paper we fix such a $\mathcal{P}$-distributed alphabet $\widetilde{\Sigma}$.

We let $\mathbb{R}^{>0}$ and $\mathbb{R}^{\geq 0}$ denote the set of positive and non-negative reals respectively. Without loss of generality we will use intervals with rational bounds to specify timing constraints (and use ∞ as the upper bound to capture unbounded intervals). If an interval is of the form $[x, y)$ or $[x, y]$ we require x to be a positive rational and if an interval is of the form $(x, y]$ or $[x, y]$ we require y to be a positive rational. An interval defines a subset of reals in the obvious way. Let INT be the set of all intervals of $\mathbb{R}^{>0}$.

A *product interval automaton* over $\widetilde{\Sigma}$ is a structure $(\{\mathcal{A}_i\}_{i \in \mathcal{P}}, Q_{in})$, where for each i, $\mathcal{A}_i$ is a structure $(Q_i, \longrightarrow_i, F_i, G_i)$ such that Q_i is a finite set of states, $\longrightarrow_i$, the transition relation, is a finite subset of $Q_i \times (\Sigma_i \times INT) \times Q_i$, and $F_i, G_i \subseteq Q_i$ are, respectively, finitary and infinitary acceptance state sets. $Q_{in} \subseteq Q = Q_1 \times \cdots \times Q_K$ is a set of global start states.

For convenience we will be interested only in infinite runs of a PI automaton. However in such a run some of the component automata may execute only a finite number of actions. It is for this reason we have two types of accepting states for each component automaton.

In what follows, for an alphabet A we will use A^* and A^ω to denote the set of finite and infinite words over A respectively, and let A^∞ denote the set $A^* \cup A^\omega$.

For a word σ in A^{∞}, we let $prf(\sigma)$ be the set of finite prefixes of σ. A timed word σ over $\widetilde{\Sigma}$ is an element of $(\Sigma \times \mathbb{R}^{>0})^{\infty}$ such that:

(i) σ is non-decreasing: if $\tau(a, t)(b, t')$ is a prefix of σ then $t \leq t'$.
(ii) if $\tau(a, t)\tau'(b, t')$ is a prefix of σ with $t = t'$, then we must have $loc(a) \cap loc(b) = \emptyset$. Thus, simultaneous actions are allowed but only if they are independent – i.e. their locations are disjoint.
(iii) if σ is infinite, then it must be progressive: for each $t \in \mathbb{R}^{>0}$ there exists a prefix $\tau(a, t')$ of σ such that $t' > t$.

We let $T\widetilde{\Sigma}^{\omega}$ denote the set of infinite timed words over $\widetilde{\Sigma}$.

Now suppose $K = 1$ so that $\mathcal{P} = \Sigma_1$. Then the definition above (condition (ii) will not apply) will yield the usual notion of a timed word over the alphabet Σ_i. With this in mind, for an alphabet A, we let TA^* and TA^{ω} be the set of finite and infinite (ordinary) timed words over A, respectively, and set $TA^{\infty} = TA^* \cup TA^{\omega}$.

Let $\sigma \in T\widetilde{\Sigma}^{\omega}$. Then $\sigma \upharpoonright i$ is the i-projection of σ. It is the timed word over Σ_i obtained by erasing from σ all appearances of letters of the form (a, t) with $a \notin \Sigma_i$. It is easy to check that $\sigma \upharpoonright i$ belongs to $T\Sigma_i^{\infty}$.

Finally, let τ be a finite timed word over $\widetilde{\Sigma}$. Then $time_i(\tau)$ is given inductively by: $time_i(\epsilon) = 0$. Further, $time_i(\tau'(a, t)) = t$ if $a \in \Sigma_i$, and equals $time_i(\tau')$ otherwise. Thus $time_i(\tau)$ is the time at which the last i-action took place during the execution of τ. This notion will play an important role in the rest of the paper. We can now define the timed language accepted by a PI automaton.

Let $\mathcal{A} = (\{\mathcal{A}_i\}_{i \in \mathcal{P}}, Q_{in})$ be a PI automaton over $\widetilde{\Sigma}$ and let $\sigma \in T\widetilde{\Sigma}^{\omega}$. Then a run of $\mathcal{A}$ over σ is a map $\rho : prf(\sigma) \to Q$ such that

1. $\rho(\epsilon) \in Q_{in}$
2. Suppose $\tau(a, t)$ is a prefix of σ. Then for each $i \in loc(a)$, there exists a transition $\rho(\tau)[i] \xrightarrow{(a, I)}_i \rho(\tau(a, t))[i]$ with $(t - time_i(\tau)) \in I$. Further, for each $i \notin loc(a)$ we have $\rho(\tau)[i] = \rho(\tau(a, t))[i]$.

A run ρ of $\mathcal{A}$ on σ is accepting iff for each $i \in \mathcal{P}$:

(i) If $\sigma \upharpoonright i$ is finite then $\rho(\tau)[i] \in F_i$ where τ is a prefix of σ such that $\tau \upharpoonright i = \sigma \upharpoonright i$.
(ii) If $\sigma \upharpoonright i$ is infinite then $\rho(\tau)[i] \in G_i$ for infinitely many $\tau \in prf(\sigma)$.

We set $L(\mathcal{A})$ to be the set of words in $T\widetilde{\Sigma}^{\omega}$ accepted by $\mathcal{A}$ (i.e. those on which $\mathcal{A}$ has an accepting run).

We can solve the emptiness problem for PI automata by reducing it to the emptiness problem for timed Büchi automaton (TBA) as formulated by Alur and Dill [1]. The concerned timed simulation is the obvious one and the details can be found in [7]. In going from a PI automaton to a timed automaton, there will be a $2^{O(K)}$ blow-up in the size of the automaton.

3   Closure under Boolean Operations

Here we wish to show that the class of languages accepted by PI automata (over $\widetilde{\Sigma}$) is closed under boolean operations. Using the notion of proper interval sets, we will transform the relevant problems to the domain of ordinary trace languages. We can then apply known constructions involving trace languages to obtain the desired results.

It will be useful to first study a single component of a PI automaton. To this end, let A be a finite alphabet of actions. Then an interval alphabet based on A is a *finite* subset of $A \times INT$. Let Γ be an interval alphabet based on A. Then an interval automaton over Γ is a structure $\mathcal{B} = (Q, \longrightarrow, Q_{in}, F, G)$ where Q is a finite set of states, $\longrightarrow \subseteq Q \times \Gamma \times Q$ is a transition relation, $Q_{in} \subseteq Q$ is a set of initial states, $F \subseteq Q$ is a set of finitary accepting states and $G \subseteq Q$ is a set of infinitary accepting states. $L(\mathcal{B}) \subseteq TA^{\infty}$, the language of timed words accepted by $\mathcal{B}$ is now defined in the obvious way. (To be pedantic, one can set $\mathcal{P} = \{1\}$ and $\Sigma_1 = A$ and apply the definitions from the previous section.) There is also a symbolic language $L_{sym}(\mathcal{B}) \subseteq \Gamma^{\infty}$ that one can associate with $\mathcal{B}$ in the classical manner. The finite words in $L_{sym}(\mathcal{B})$ are accepted using members of F and the infinite words are accepted by viewing G as a Büchi condition (see [21] for a standard treatment). It is also the case that there is a natural map $tw : \Gamma^{\infty} \to TA^{\infty}$ such that $tw(L_{sym}(\mathcal{B})) = L(\mathcal{B})$. The map tw can be defined as follows:

Let $\hat{\sigma} \in \Gamma^{\infty}$ and $\sigma \in TA^{\infty}$. Then $\sigma \in tw(\hat{\sigma})$ iff $|\sigma| = |\hat{\sigma}|$ and for each prefix $\tau(a, t)$ of σ and each prefix $\hat{\tau}(b, I)$ of $\hat{\sigma}$ with $|\tau| = |\hat{\tau}|$ we have $a = b$ and $t - time(\tau) \in I$. The function $time$ is of course given inductively by $time(\epsilon) = 0$ and $time(\tau'(a', t')) = t'$. As usual, for $L \subseteq \Gamma^{\infty}$ we set $tw(L) = \bigcup_{\hat{\sigma} \in L} tw(\hat{\sigma})$. To sum up, the simple observation we wish to start with is:

Proposition 1. *Let $\mathcal{B}$ be an interval automaton over the interval alphabet Γ (based on A). Then $L(\mathcal{B}) = tw(L_{sym}(\mathcal{B}))$.* $\qquad\square$

Now we turn to proper interval sets. We say that $\mathcal{I} \subseteq INT$ is proper iff $\mathcal{I}$ is a *finite* partition of $(0, \infty)$. An interval alphabet Γ based on A is said to be proper iff for each a in A, the set of intervals Γ_a is proper where $\Gamma_a = \{I \mid (a, I) \in \Gamma\}$. The next observation which is easy to verify is crucial for our purposes.

Proposition 2. *Let Γ be a proper interval alphabet based on A and let $\sigma \in TA^{\infty}$. Then there exists a unique word $\hat{\sigma} \in \Gamma^{\infty}$ such that $\sigma \in tw(\hat{\sigma})$.* $\qquad\square$

The uniqueness of $\hat{\sigma}$ as stated in the above result is the key to reducing the study of timed languages to the study of trace languages in the present setting. This can be achieved by embedding interval alphabets into proper interval alphabets. To formalise this idea, let $\mathcal{I}, \mathcal{I}' \subseteq INT$. Then $\mathcal{I}'$ is said to *cover* $\mathcal{I}$ iff each interval in $\mathcal{I}$ can be expressed as the union of a set of intervals taken from $\mathcal{I}'$. Clearly the choice of such a set of intervals (for a given member of $\mathcal{I}$) will be unique in case $\mathcal{I}'$ is a proper interval set.

Let Γ and Γ' be two interval alphabets based on the alphabet A. Then Γ' covers Γ iff Γ'_a covers Γ_a for each $a \in A$.

We will say that $L \subseteq TA^{\infty}$ is a regular interval language over A iff there exists an interval alphabet Γ based on A and an interval automaton $\mathcal{B}$ over Γ such that $L = L(\mathcal{B})$. For convenience we will often say "interval language" to mean "regular interval language."

Proposition 3. *Let A be a finite alphabet of actions.*

(i) *Suppose Γ is an interval alphabet based on A. Then one can effectively construct a* proper *interval alphabet Γ' based on A such that Γ' covers Γ.*

(ii) *The class of interval languages over A is closed under boolean operations.*

Proof sketch. (i) follows from the observation that given an interval set $\mathcal{I} = \{I_1, \ldots, I_m\}$ we can form the set $X \cup \{0, \infty\} = X'$ where X is the set of rational numbers that appear as bounds of the intervals in $\mathcal{I}$. Let $X' = \{0, x_1, \ldots, x_n, \infty\}$ with $0 < x_1 < x_2 < \cdots < x_n < \infty$. Then $\mathcal{I}' = \{(0, x_1), [x_1, x_1], (x_1, x_2), [x_2, x_2], \ldots, (x_n, \infty)\}$ is a proper interval set which covers $\mathcal{I}$.

To prove (ii), one first observes that closure under union is obvious. So let $\mathcal{B}$ be an interval automaton over the interval alphabet Γ. Then we can construct Γ' such that Γ' is proper and covers Γ. Let $\mathcal{B} = (Q, \longrightarrow, Q_{in}, F, G)$. Then we construct the interval automaton $\mathcal{B}' = (Q, \Longrightarrow, Q_{in}, F, G)$ over Γ' where $\Longrightarrow\, \subseteq Q \times \Gamma' \times Q$ is given by: $q \stackrel{(a, I')}{\Longrightarrow} q'$ iff there exists $q \stackrel{(a, I)}{\longrightarrow} q'$ in $\mathcal{B}$ such that $I' \subseteq I$. It is easy to verify that $L(\mathcal{B}) = L(\mathcal{B}')$. Using classical techniques we can now find an interval automaton $\mathcal{B}''$ over Γ' such that $L_{sym}(\mathcal{B}'') = (\Gamma')^\infty - L_{sym}(\mathcal{B}')$. Since Γ' is proper we can now use proposition 2 to conclude that $L(\mathcal{B}'') = TA^\infty - L(\mathcal{B}')$. □

The technique used in the second part of the proof is deployed in different contexts to establish many of our results. It is for this reason we have presented some of the key details in a very simple setting.

We can now address the main concerns of this section. To show closure under boolean operations of the languages accepted by PI automata, we will first characterise these languages using interval languages. Let $\{L_i\}$ be a family of languages such that $L_i \subseteq T\Sigma_i^\infty$ for each i. (Recall $\widetilde{\Sigma} = \{\Sigma_i\}_{i \in \mathcal{P}}$.) Then $\otimes(L_1, \ldots, L_K)$ denotes the language $L \subseteq T\widetilde{\Sigma}^\omega$ defined as:

$$\sigma \in L \text{ iff } \sigma\lceil i \in L_i \text{ for each } i.$$

We will say that $L \subseteq T\widetilde{\Sigma}^\omega$ is a regular direct product interval language iff there exists a family $\{L_i\}$ such that each L_i is a regular interval language over Σ_i and $L = \otimes(L_1, \ldots, L_K)$. A regular product interval language over $\widetilde{\Sigma}$ is a finite union of regular direct product interval languages over $\widetilde{\Sigma}$. Again, for convenience, we will often say "product interval language" to mean "regular product interval language" etc..

Theorem 1. *Let $L \subseteq T\widetilde{\Sigma}^\omega$. Then L is a product interval language iff there exists a PI automaton $\mathcal{B}$ over $\widetilde{\Sigma}$ such that $L = L(\mathcal{B})$.* □

Using the theory of product languages [15], which is a natural subclass of Mazurkiewicz trace languages and the technique used in the proof of Proposition 3 we can now show:

Theorem 2. *The class of product interval languages over $\widetilde{\Sigma}$ is closed under boolean operations.*

4   A Logical Characterisation of Product Interval Languages

Our goal here is to first develop a logical characterisation of product interval languages in terms of a monadic second order logic. We shall then present a related linear time temporal logic for reasoning about timed behaviours captured by PI automata. As done in the previous section it will be convenient to first concentrate on the one component case.

Let A be a finite alphabet of actions. Here and in the logics to follow, we assume a supply of individual variables $x, y, \ldots$ and set variables $X, Y, \ldots$. For each $a \in A$ we have a unary predicate Q_a. There are also predicates of the form $\Delta(x, I)$ in which x is an individual variable, and I is a member of INT.

The syntax of the logic TMSO(A) is given by:

$$\text{TMSO}(A) ::= x \in X \mid x < y \mid Q_a(x) \mid \Delta(x, I) \mid \varphi \vee \varphi' \mid \sim \varphi \mid$$
$$(\exists x)\varphi \mid (\exists X)\varphi.$$

A structure for this logic is a pair $(\sigma, \mathbb{I})$ where $\sigma \in TA^\infty$ and $\mathbb{I}$ is an interpretation which assigns to each individual and set variable, an element and subset, respectively, of the set $pos(\sigma)$. We define $pos(\sigma)$ to be the set $\{1, 2, \ldots, |\sigma|\}$ in case σ is finite, and $\{1, 2, \ldots\}$ if σ is infinite. As is customary, σ can be also viewed as a map from $pos(\sigma)$ into A with $\sigma(n)$ for $n \in pos(\sigma)$ having the obvious meaning. We will use this view of σ whenever convenient. Turning now to the semantics, let $(\sigma, \mathbb{I})$ be a structure. Then $\sigma \models_{\mathbb{I}} (x < y)$ iff $\mathbb{I}(x) <' \mathbb{I}(y)$. (Here $<'$ is the usual ordering over the integers). Further, $\sigma \models_{\mathbb{I}} Q_a(x)$ iff $\sigma(\mathbb{I}(x)) = (a, t)$ for some t. As for the predicate $\Delta(x, I)$, we have

$$\sigma \models_{\mathbb{I}} \Delta(x, I) \quad \text{iff} \quad t - time(\tau) \in I, \text{ where } \tau(a, t) \text{ is the prefix of } \sigma \text{ s.t.}$$
$$|\tau(a, t)| = \mathbb{I}(x).$$

The semantics of the logical connectives $\sim$, and $\vee$, and the existential quantifiers $\exists x$ and $\exists X$ are given in the expected manner. For a sentence $\varphi \in \text{TMSO}(A)$ we set $L(\varphi) = \{\sigma \in TA^\infty \mid \sigma \models \varphi\}$.

We can now use the classical logical characterisation of ω-regular languages and the notion of proper interval alphabets to establish:

Theorem 3. *Let $L \subseteq TA^\infty$. Then L is an interval language iff there exists a sentence $\varphi \in \text{TMSO}(A)$ such that $L = L(\varphi)$.*

Next we wish to present the logic $\text{TMSO}^\otimes(\widetilde{\Sigma})$ which will characterise product interval languages. The syntax of the logic is given by:

$$\text{TMSO}^\otimes(\widetilde{\Sigma}) ::= \varphi(i) \mid \sim \alpha \mid \alpha \vee \beta \mid \alpha \wedge \beta$$

where for each formula $\varphi(i)$ we require $\varphi \in \text{TMSO}(\Sigma_i)$.

A structure for the logic is a pair $(\sigma, \{\mathbb{I}_i\}_{i \in \mathcal{P}})$, where $\sigma \in T\widetilde{\Sigma}^\omega$ and for each $i \in \mathcal{P}$, $\mathbb{I}_i$ is an interpretation for individual and set variables over the set of positions $pos(\sigma \upharpoonright i)$. As for the semantics,

$$\sigma \models_{\{\mathbb{I}_i\}_{i \in \mathcal{P}}} \varphi(i) \quad \text{iff} \quad \sigma \upharpoonright i \models_{\mathbb{I}_i} \varphi.$$

The boolean operators $\sim$, $\vee$ and $\wedge$ are interpreted in the usual manner. Once again, for a sentence $\varphi \in \text{TMSO}^{\otimes}(\widetilde{\Sigma})$, we set $L(\varphi) = \{\sigma \in T\widetilde{\Sigma}^{\omega} \mid \sigma \models \varphi\}$.

Theorem 4. *Let $L \subseteq T\widetilde{\Sigma}^{\omega}$. Then L is a product interval language iff $L = L(\varphi)$ for some sentence $\varphi \in \text{TMSO}^{\otimes}(\widetilde{\Sigma})$.*

This result is established by exploiting Theorem 1 and yet again the notion of proper interval alphabets.

As a natural temporal logic counterpart of $\text{TMSO}^{\otimes}(\widetilde{\Sigma})$, we present here the logic $\text{TLTL}^{\otimes}(\widetilde{\Sigma})$. Again, we will build up the formulas in two steps. For ease of presentation, we will not deal with atomic propositions here. Where necessary, they can be introduced component-wise and dealt with easily.

$$\text{TLTL}(\Sigma_i) ::= \top \mid \sim \varphi \mid \varphi \vee \varphi' \mid \langle a, I \rangle \varphi \mid O\varphi \mid \varphi U \varphi' \,.$$

For each formula of the form $\langle a, I \rangle \varphi$, we require that $a \in \Sigma_i$.

The formulas of $\text{TLTL}^{\otimes}(\widetilde{\Sigma})$ are simply boolean combinations of $\text{TLTL}(\Sigma_i)$-formulas:

$$\text{TLTL}^{\otimes}(\widetilde{\Sigma}) ::= \varphi(i) \mid \sim \alpha \mid \alpha \vee \beta \,.$$

Here we require from each formula of the form $\varphi(i)$ that $\varphi \in \text{TLTL}(\Sigma_i)$.

The semantics of this logic is given by defining the relation $\sigma \models \alpha$ with $\sigma \in T\widetilde{\Sigma}^{\omega}$. To spell out the details, $\sigma \models \varphi(i)$ iff $\sigma \restriction i, \epsilon \models_i \varphi$. The relation $\models_i$ is defined in the obvious way. For instance,

$$\sigma, \tau \models_i \langle a, I \rangle \varphi \text{ iff there exists } t \text{ such that } \tau(a, t) \in prf(\sigma) \text{ and} \\ \sigma, \tau(a, t) \models_i \varphi \text{ and } t - time(\tau) \in I.$$

On the other hand $\sigma, \tau \models_i O\varphi$ iff there exists $\tau(a, t) \in prf(\sigma)$ such that $\sigma, \tau(a, t) \models_i \varphi$.

For each α in $\text{TLTL}^{\otimes}(\widetilde{\Sigma})$ we define $L(\alpha) = \{\sigma \in T\widetilde{\Sigma}^{\omega} \mid \sigma \models \alpha\}$. Clearly, α is satisfiable iff $L(\alpha)$ is non-empty.

Theorem 5. *For each α in $\text{TLTL}^{\otimes}(\widetilde{\Sigma})$ we can effectively construct a PI automaton $\mathcal{A}_{\alpha}$ such that $L(\alpha) = L(\mathcal{A}_{\alpha})$ and the number of global states of $\mathcal{A}_{\alpha}$ is $2^{O(|\alpha|)}$. Further, the largest constant mentioned in the ith component of $\mathcal{A}_{\alpha}$ is at most the largest i-type constant mentioned in α.*

Proof sketch. The proof follows the usual automata-theoretic techniques. To account for negation correctly, the local transition relations of $\mathcal{A}_{\alpha}$ must be over proper alphabets. The obvious idea of converting α into a formula which mentions only intervals taken from a proper interval set will cause an exponential blow-up. To avoid this, we define the (Fisher-Ladner) closure of α to be smallest set of formulas containing the subformulas of α and satisfying: If $(\varphi U \varphi')(i)$ is in the closure then so is $O((\varphi U \varphi')(i))$. Let CL be the closure of α and $\text{CL}_i = \text{CL} \cap \text{TLTL}(\Sigma_i)$ for each i. We now define an i-type atom to be a propositionally consistent subset of CL_i which satisfies: If $\langle a, I \rangle \varphi$ and $\langle b, I' \rangle \varphi'$

are in the atom then $a = b$ and $I \cap I' \neq \emptyset$. We can now use i-type atoms to manufacture the i-local states. The i-local transition relation can then be chosen to be proper. The remaining details are guided by the way the logic $\text{LTL}^{\otimes}(\Sigma)$ is dealt with in [20]. $\qquad\Box$

Using theorem 5, and the emptiness check for PI automata in section 2, we can decide satisfiability of α in time $2^{O(|\alpha|+K \log K)}$. In fact, we can show:

Corollary 1. *The satisfiability problem for* $\text{TLTL}^{\otimes}(\widetilde{\Sigma})$ *is* PSPACE-*complete.*

Given a formula α in $\text{TLTL}^{\otimes}(\widetilde{\Sigma})$ we can check the emptiness of the automaton $\mathcal{A}_\alpha$ of theorem 5 non-deterministically while using space polynomial in the size of α. The argument is similar to the one in [1]. The fact that the problem is PSPACE-hard follows from the observation that the satisfiability problem for LTL can be reduced to the satisfiability problem for the one-agent case of $\text{TLTL}^{\otimes}(\widetilde{\Sigma})$.

Next suppose we consider a real-time program Pr modelled by a PI automaton $\mathcal{A}_{Pr}$, and a formula α of $\text{TLTL}^{\otimes}(\widetilde{\Sigma})$. Then Pr is said to meet the specification α iff $L(\mathcal{A}_{Pr}) \subseteq L(\alpha)$. The model checking problem for $\text{TLTL}^{\otimes}(\widetilde{\Sigma})$ is to determine whether Pr meets the specification α. It is not difficult to show:

Theorem 6. *The model checking problem for* $\text{TLTL}^{\otimes}(\widetilde{\Sigma})$ *is* PSPACE *complete.*

We have not studied in detail the power and limitations of $\text{TLTL}^{\otimes}(\widetilde{\Sigma})$ from a pragmatic standpoint. We have explored $\text{TLTL}^{\otimes}(\widetilde{\Sigma})$ here mainly because it is the natural temporal logic suggested by $\text{TMSO}^{\otimes}(\widetilde{\Sigma})$, the characteristic logic for PI automata. In fact, it is not difficult to show that $\text{TLTL}^{\otimes}(\widetilde{\Sigma})$ is expressively equivalent to the first-order fragment of $\text{TMSO}^{\otimes}(\widetilde{\Sigma})$.

In order to provide some example properties that can be stated in this timed temporal logic, we first sketch briefly how product interval automata can be used to model asynchronous digital circuits. (The full details will appear elsewhere.) Our starting point is the circuit model developed in [17] using inertial delays. The behaviour of a k-gate circuit χ, denoted $L(\chi)$, can be defined as a subset of $T\Sigma^\infty$, where Σ is the alphabet $\{0,1\}^k \times \{0,1\}^k$. An element of $\{0,1\}^k$ will represent the visible values of the gate outputs. The occurrence of an action (s, s') will indicate the visible values of the gate outputs going from s to s' due to the simultaneous switching of an appropriate set of gates. We can associate with χ, a $(k+1)$-component product interval automaton $\mathcal{A}_\chi$ such that $L(\mathcal{A}_\chi) = L(\chi)$. We use one component for each gate in the circuit. This component will keep track of the switching status of the gate (whether excited or quiescent); and, in case it is in an excited state, the time elapsed since its excitement was initiated last. The $(k+1)$-th component keeps track of the current vector of visible values. Suppose its state is s and it permits an action (t, t') at s, then it will be the case that $s = t$. It will also be the case that the circuit allows the transition from t to t' through the switching of a subset of gates that are excited at t. Furthermore, the distribution of actions across the components is such that the action (t, t') will involve – apart from the component $k+1$ – all those components whose switching status is affected by the switches. As for timing constraints, the component i will have timing constraints that enforce the inertial delay corresponding to the output wire of gate-i, while the $(k + 1)$-th component will be free of any timing constraints.

We can use the logic $TLTL^{\otimes}(\widetilde{\Sigma})$ to express (and hence model-check) a variety of properties of χ. Two examples of such properties are:

- If the input signals to χ stabilise then the behaviour of χ eventually stabilises.
- gate i is d-persistent; whenever it becomes excited it turns quiescent only through its switching and further more this switching occurs within d time units since it got excited.

5   Distributed Interval Automata

Here we consider a more expressive class of distributed timed automata but having the same flavour as PI automata. The extension we have in mind will parallel the extension of product automata to asynchronous automata in the setting of traces as detailed in [15].

We shall continue to work with the $\mathcal{P}$-distributed alphabet $\widetilde{\Sigma}$. An a-interval for us will be a map $J : loc(a) \to INT$. Let $INT^{(a)}$ be the set of a-intervals. An a-interval, viewed as a rectangular region, will constitute the guard constraining the timed occurrence of an a-action.

In the definition of our automata and elsewhere we will often write $\{X_i\}$ to denote the family $\{X_i\}_{i\in\mathcal{P}}$. A distributed interval automaton over $\widetilde{\Sigma}$ is a structure

$$\mathcal{A} = (\{Q_i\}, \{\longrightarrow_a\}_{a\in\Sigma}, Q_{in}, \{F_i, G_i\})$$

where the various components of $\mathcal{A}$ are defined as follows. While doing so we will also develop some notations.

(i) Each Q_i is a finite non-empty set called the set of i-local states. Let P be a non-empty subset of $\mathcal{P}$. Then a P-state is a map $q : P \to \bigcup_{i\in\mathcal{P}} Q_i$ such that $q(i) \in Q_i$ for each $i \in P$. An a-state is just a $loc(a)$-state and a $\mathcal{P}$-state will be called a global state. We let Q_a denote the set of a-states and $Q_{\mathcal{P}}$ the set of global states.

(ii) $\longrightarrow_a$ is a finite subset of $Q_a \times INT^{(a)} \times Q_a$ for each $a \in \Sigma$.

(iii) $Q_{in} \subseteq Q_{\mathcal{P}}$ is the set of global initial states.

(iv) $F_i, G_i \subseteq Q_i$ for each i.

Thus a distributed interval automaton is a timed version of an asynchronous Büchi automaton as formulated in [18] which in turn is a minor variant of the original formulation due to Gastin and Petit [11]. If one ignores the acceptance conditions then an asynchronous automaton can be viewed as a labelled 1-safe Petri net. Consequently a distributed interval automaton can also be viewed as a timed 1-safe Petri net. However, as pointed out in the introduction, the timed semantics we assign to our automata is *not* the semantics one encounters in the literature concerning timed Petri nets [5].

We can now systematically extend the theory of PI automata – as sketched in the previous sections – to distributed interval automata. The details can be found in [7].

It turns out that the so called cellular version of distributed interval automata, has the same expressive power as the event recording automata [2]. In this version of asynchronous automata there will be one component for each letter in Σ and one clock associated with each component. The interesting fact is that in the untimed setting both versions of asynchronous automata have the same expressive power. Once again we refer the reader to [7] for the details.

References

1. R. Alur, D. L. Dill: A theory of timed automata, *Theoretical Computer Science* 126: 183–235 (1994).
2. R. Alur, L. Fix, T. A. Henzinger: Event-clock automata: a determinizable class of timed automata, *Proc. 6th International Conference on Computer-aided Verification, LNCS 818*, 1–13, Springer-Verlag (1994).
3. R. Alur, T. A. Henzinger: Logics and Models of Real Time: A Survey, in *Real-Time: Theory in Practice*, J. W. de Bakker, H. Huizing, W. -P. de Roever, G. Rozenberg (Eds.), LNCS 600, 74–106, (1992).
4. J. Bengtsson, B. Jonsson, J. Lilius, W Yi: Partial Order Reductions for Timed Systems, *Proc. CONCUR '98, LNCS 1466* (1998).
5. B. Berthomieu, M. Diaz: Modelling and Verification of Time Dependent Systems Using Time Petri Nets, *IEEE trans. on Soft. Engg.* Vol 17, No. 3, March 1991.
6. V. Diekert, G. Rozenberg: *The Book of Traces*, World Scientific, Singapore (1995).
7. D. D'Souza, P. S. Thiagarajan: Distributed Interval Automata, Internal Report TCS-98-3, Chennai Mathematical Institute (1998). (available at http://www.smi.ernet.in/techreps/)
8. W. Ebinger, A. Muscholl: Logical definability on infinite traces, *Theoretical Computer Science* 154: 67–84 (1996).
9. T. A. Henzinger, P. W. Kopke, A. Puri, P. Varaiya: What's decidable about hybrid automata?, *Proc. 27th Annual Symposium on Theory of Computing*, 373–382, ACM Press (1995).
10. M. Minea: Partial Order Reduction for Model Checking of Timed Automata. To appear in the Proceedings of CONCUR'99.
11. P. Gastin, A. Petit: Asynchronous cellular automaton for infinite traces, *Proceedings of ICALP '92, LNCS 623*, 583–594 (1992).
12. R. Gerth, D. Peled, M. Vardi, P. Wolper: Simple On-the-fly Automatic Verification of Linear Temporal Logic. *Proc. 15th IFIP WG 6.1 Int. Workshop on Protocol Specification, Testing, and Verification.* North-Holland Publ. (1995).
13. T. A. Henzinger: It's About Time: Real-Time Logics Reviewed, *Proc. CONCUR '98, LNCS 1466*, 366–372 (1998).
14. T. A. Henzinger, J.-F. Raskin, and P.-Y. Schobbens: The regular real-time languages, *Proc. 25th International Colloquium on Automata, Languages, and Programming 1998, LNCS 1443*, 580–591 (1998).
15. J. G. Henriksen, P. S. Thiagarajan: A Product Version of Dynamic Linear Time Temporal Logic, *Proc. CONCUR '97, LNCS 1243*, 45–58, (1997).

16. N. Klarlund, M. Mukund, M. Sohoni: Determinizing Büchi Asynchronous Automata, *Proceedings of FSTTCS 15, LNCS 1026*, 456–470 (1995).
17. O. Maler, A. Pnueli: Timing Analysis of Asynchronous Circuits using Timed Automata, in *Proc. CHARME '95, LNCS 987*, 189–205 (1995).
18. M. Mukund, P. S. Thiagarajan: Linear Time Temporal Logics over Mazurkiewicz Traces, *Proc. MFCS 96, LNCS 1113*, 62–92 (1996).
19. J. -F. Raskin, P. -Y. Schobbens: State-clock Logic: A Decidable Real-Time Logic, *Proc. HART '97: Hybrid and Real-Time Systems, LNCS 1201*, 33–47 (1997).
20. P. S. Thiagarajan: A Trace Consistent Subset of PTL, *Proc. CONCUR '95*, LNCS 962, 438–452 (1995).
21. W. Thomas: Automata on Infinite Objects, in J. V. Leeuwen (Ed.), *Handbook of Theoretical Computer Science*, Vol. B, 133–191, Elsevier Science Publ., Amsterdam (1990).
22. Th. Wilke: Specifying Timed State Sequences in Powerful Decidable Logics and Timed Automata, in *Formal Techniques in Real-Time and Fault-Tolerant Systems, LNCS 863*, 694–715 (1994).
23. G. Winskel, M. Nielsen: Models for Concurrency, in S. Abramsky, D. Gabbay (Eds.) *Handbook of Logic in Computer Sc., Vol. 3*, Oxford Univ Press (1994).
24. W. Yi, B. Jonsson: Decidability of Timed Language-Inclusion for Networks of Real-Time Communicating Sequential Processes, in *Proc. FST&TCS 94, LNCS 880* (1994).

The Complexity of Rebalancing
a Binary Search Tree

Rolf Fagerberg*

BRICS, Department of Computer Science, Aarhus University,
DK-8000 Århus C, Denmark
rolf@brics.dk

Abstract. For any function f, we give a rebalancing scheme for binary search trees which uses amortized $O(f(n))$ work per update while maintaining a height bounded by $\lceil \log(n + 1) + 1/f(n) \rceil$. This improves on previous algorithms for maintaining binary search trees of very small height, and matches an existing lower bound. The main implication is the exact characterization of the amortized cost of rebalancing binary search trees, seen as a function of the height bound maintained. We also show that in the semi-dynamic case, a height of $\lceil \log(n + 1) \rceil$ can be maintained with amortized $O(\log n)$ work per insertion. This implies new results for *TreeSort*, and proves that it is optimal among all comparison based sorting algorithms for online sorting.

1   Introduction

The binary search tree is one of the fundamental data structures of computer science. Its importance lies in its ability to maintain a set in sorted order during insertions and deletions, while supporting a wide range of search operations on the elements. It has a vast number of applications, and most undergraduate textbooks on algorithms devote an entire chapter to it.

Much of the research on binary search trees has been centered around rebalancing schemes, and numerous such schemes exist which keep a height of $c \cdot \log(n)$, for some constant $c > 1$, while supporting updates in $O(\log n)$ time. Some of the more well known examples include AVL-trees [1] with $c = 1.44$, red-black trees [9] with $c = 2$, and $BB(\alpha)$-trees [7,14] with $2 \leq c \leq 3.45$ (depending on α), but there are many more.

The trivial lower bound on the height of a binary tree with n nodes is $\lceil \log(n + 1) \rceil$. While being within a constant factor of this lower bound may be sufficient for most practical purposes, a natural theoretical question is, exactly *how* close to this optimum we can keep the height. Presumably, the answer depends on how much rebalancing we are willing to do after an insertion or deletion of a key – for instance, using $\Theta(n)$ work per update, the trivial lower bound

---

* Supported by the Danish National Science Research Council (grant no. 11-0575) and by the ESPRIT Long Term Research Programme of the EU under project no. 20244 (ALCOM-IT). This research was done at Department of Mathematics and Computer Science, University of Southern Denmark, Odense, Denmark.

C. Pandu Rangan, V. Raman, R. Ramanujam (Eds.): FSTTCS'99, LNCS 1738, pp. 72–83, 1999.

is attainable by simply rebuilding the entire tree after each update. Thus, in a more general form, the question is:

Given a function f, what is the best possible height maintainable with $O(f(n))$ rebalancing work per update?

Partial answers to this question exist. In particular, upper bounds better than those mentioned above have been given. Already in 1976, Maurer et al. presented a rebalancing scheme [13] with a height bound of $c \cdot \log(n)$, where c can be chosen arbitrarily close to one. The rebalancing cost is $O(\log n)$, with the constant depending on c. The gap was further closed around 1990, when Andersson and Lai in their theses [3,11] and in resulting papers [2,4,5,6,12] gave a series of schemes for maintaining height $\lceil \log(n + 1) \rceil + 1$, using amortized $O(\log^2 n)$ rebuilding work for the simplest, improving to $O(1)$ for the most complicated.

These very positive results are strongly contrasted by an observation of Lai [11], which shows that $\Omega(n)$ rebuilding work *per update* is necessary, even in the amortized sense, for keeping optimal height $\lceil \log(n + 1) \rceil$ for all n during insertions and deletions. A generalization of this lower bound has been given in [8], where it is shown that for any function f, the maintenance of height $\lceil \log(n + 1) + 1/f(n) \rceil$ requires $\Omega(f(n))$ amortized rebuilding work per update.

In this paper, we give a matching upper bound by showing how to maintain height $\lceil \log(n + 1) + 1/f(n) \rceil$ for all n, in amortized $O(f(n))$ rebuilding work per update.

Taken together, these results provide an exact answer to the question above in the case of amortized complexity – namely a height of

$$\lceil \log(n + 1) + \Theta(1/f(n)) \rceil.$$

This expression may be seen as describing the intrinsic amortized complexity of rebalancing a binary search tree.

The importance of the new upper bound is of course not of a practical nature – in few situations does it matter whether the height bound is $\lceil \log(n + 1) + 1 \rceil$ or $\lceil \log(n + 1) + o(1) \rceil$. Rather, its main virtue is the final settling of the exact amortized cost of rebalancing a binary search tree.

Other implications exist, though. As a corollary to our new upper bound, we prove that in the semi-dynamic case (i.e. insertions only), optimal height $\lceil \log(n + 1) \rceil$ can be maintained at an amortized cost of $O(\log n)$ per insertion. This improves the previous best upper bound of $O(\log^2 n)$ from [3], and matches a lower bound in [8].

This corollary also implies improved results for *TreeSort* (i.e. sorting by repeated insertions into a binary search tree). The best possible bound,

$$\sum_{i=0}^{n-1} \lceil \log n + 1 \rceil = n \lceil \log n \rceil - 2^{\lceil \log n \rceil} + 1,$$

on the number of comparisons for *TreeSort* can now be achieved in $O(n \log n)$ time, as opposed to $O(n \log^2 n)$ before. For *online sorting*, where elements arrive one at a time, and the set has to be sorted at all times, this is best possible for comparison based sorting algorithms ([10], page 184 and 204). Thus, *TreeSort* is optimal (with respect to the number of comparisons *and* to the actual time) for online sorting in the comparison model.

2   Definitions

In this paper, the subject is rebalancing in standard binary search trees. For technical purposes, we will also be dealing with unary-binary search trees.[1] We define both types here: A binary node has a (possibly empty) left and a (possibly empty) right subtree, and contains one key. A unary node has one (possibly empty) subtree and contains no key. A *binary search tree* is a tree containing only binary nodes. A *unary-binary search trees* is a tree that may contain both types of nodes, but with the restriction that all empty subtrees have the same depth. In both types of trees, keys are distributed in an in-order fashion. For examples of unary-binary trees, see the figures later in this paper.

The *size* of a search tree is its number of binary nodes (i.e. its number of keys). The *height* of a tree is defined as zero for empty trees, and as one plus the height of the tallest subtree of the root, otherwise. The *level* of a node is the height of the subtree in which it is the root. The *rebalancing cost* of an update is the number of pointers which are changed in order to rebalance the tree after the update.[2]

3   New Results

Theorem 1. *Let k be a positive integer, let ϵ be a real number between 0 and $1/2$, and let N denote $2^k - 1$. There exists a rebalancing scheme for doing insertions in a binary search tree while its size grows from $\lfloor (1 - \epsilon)N \rfloor$ to $\lceil (1 - \epsilon/2)N \rceil$, which guarantees optimal height k in the tree after each insertion, and which has an amortized rebalancing cost of $O(1/\epsilon)$ per insertion (including the cost of initializing the scheme).*

Theorem 1 is the main result of this paper. It meets a lower bound from [8] stating that in any binary search tree of size $(1 - \epsilon)(2^k - 1)$ and optimal height k, an insertion can be made such that the rebalancing cost of restoring optimal height is $\Omega(1/\epsilon)$. Essentially, these results describe how the amortized cost of maintaining optimal height $\lceil \log(n + 1) \rceil$ in a binary search tree changes when the free room gets scarce – i.e. when n approaches the next power of two.

Theorem 1 is proven in Sects. 4–6. For the remainder of this section, we show how it can be transformed into the results stated in the introduction.

---

[1] Some subclasses of unary-binary trees were studied in the late 70's under the name *brother trees* (see e.g. [15] and the references in it).

[2] Assuming that keys stay in the same nodes. If keys are interchanged between nodes during the rebalancing, we also count the number of moved keys (or view interchanges of keys as pointer changes – the count is the same up to a constant factor)

In the first corollary below, we restrict ourselves to functions f which are non-decreasing, positive functions, for which there exists an integer m such that $f(2n) \leq 2f(n)$ for all $n \geq m$. We call such functions *smooth*. Clearly, any smooth function must be in $O(n)$, but this is no restriction, as $f \in \Omega(n)$ means that the entire tree can be rebuilt after each update, making Corollary 1 obvious. Basically all standard functions in $O(n)$ are smooth, including $\log^k(n)$ for all k and n^ϵ for all $\epsilon \leq 1$.

Corollary 1. *For any smooth function f, there exists a rebalancing scheme for doing insertions and deletions in binary search trees at an amortized rebalancing cost of $O(f(n))$ per update, while keeping the height bounded by $\lceil \log(n+1) + 1/f(n) \rceil$ at all times.*

Proof. From the rebalancing scheme in Theorem 1, we first construct a scheme fulfilling the statement in the corollary for insertions only. We consider, for any k, how to insert 2^{k-1} elements while n grows from $N_1 = 2^{k-1} - 1$ to $N_2 = 2^k - 1$. Let $\epsilon_0 = 1/f(N_2)$, and let i be the smallest integer such that $1/2^i < \epsilon_0 \ln(2)/2$. As n grows from N_1 to $(1 - 1/2^i)N_2$, we employ the scheme from Theorem 1, changing ϵ when necessary. This keeps a height of k at an amortized rebuilding cost of $O(1/\epsilon_0)$ per insertion. As n grows from $(1 - 1/2^i)N_2$ to N_2, we employ Anderssons and Lais best scheme from [6], which keeps a height of $k + 1$ at an amortized price of $O(1)$. The change of scheme requires an $O(n)$ time global rebuilding, but this is of no concern when dealing with amortized complexity.

Using first order approximation to the logarithm function and its convexity, it can be verified that the height is bounded by $\lceil \log(n+1) + \epsilon_0 \rceil$ at all times. Thus, we maintain height $\lceil \log(n+1) + 1/f(n) \rceil$ in amortized time $O(f(n))$, as f is smooth.

This rebalancing scheme can now be made fully dynamic by simply marking nodes as deleted, employing the above scheme with a function $\tilde{f}(n) = 2f(n)$, and each $\Theta(n/f(n))$ updates rebuilding the entire structure while removing marked nodes. This is still amortized $O(f(n))$ work per update, and using first order approximation to the logarithm function it can be verified that the height is bounded by $\lceil \log(n+1) + 1/f(n) \rceil$ at all times. Alternatively, deletions can be done by a procedure similar to the insertion algorithm described later in this paper (by pushing extra unary nodes upwards, instead of pulling missing unary nodes downwards). □

Corollary 2. *There exists a rebalancing scheme for doing insertions only in a binary tree at an amortized rebalancing cost of $O(\log n)$ per insertion, while keeping the optimal height $\lceil \log(n+1) \rceil$ at all times.*

Proof. For any k, we insert 2^{k-1} elements while increasing the size from $2^{k-1} - 1$ to $2^k - 1$ by using the scheme from Theorem 1 with $\epsilon = 1/2^i$ over the next 2^{k-1-i} insertions, for $i = 1, 2, \ldots, k-1$. Summing over the 2^{k-1} insertions, a total work of $O(k2^k)$ can be calculated. This implies the result. □

4   Initialization

Our goal is to keep optimal height in a binary search tree during insertions. If the next insertion increases the height, some rebalancing must be done. Intuitively, this rebalancing process can be seen as one of moving a "hole" from somewhere else in the tree to the insertion point.

Still on the informal level, a key idea utilized in our rebalancing scheme is the observation that "rows of holes" are cheaper to move than "single holes". The figure below illustrates the point. Only edges incident to the marked nodes need to be changed.

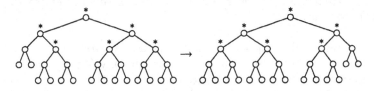

Loosely speaking, a row of $\Theta(r)$ holes can travel past $\Theta(s)$ keys in $\Theta(s/r)$ pointer changes. The central feature of our rebalancing scheme is an initial distribution of holes which allows the observation above to be exploited for an extended sequence of insertions at low amortized cost.

To formalize the ideas above, we use a natural mapping ψ from unary-binary trees of height k to binary trees of height at most k: To each unary-binary tree T we associate a corresponding binary tree $\psi(T)$ by simply contracting all unary nodes. The following figure shows a unary-binary tree T and its corresponding binary tree $\psi(T)$. Unary nodes are square, binary nodes circular.

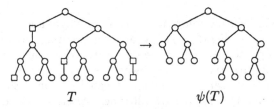

$$T \qquad\qquad \psi(T)$$

We will use unary-binary trees to describe the initial set-up as well as the rebalancing operations. This is purely for ease of notation – the actual trees we are working on are standard binary trees. As for implementation, the unary-binary tree can be used as an *information structure* for rebalancing in the actual binary search tree. It should be clear that the rebalancing cost in the unary-binary tree T is at least as large as the rebalancing cost in the corresponding binary tree $\psi(T)$.

If we define the *weight* of a unary node at level h to be 2^{h-1}, the following holds.

Lemma 1. *In a unary-binary tree T of height k, the number of binary nodes plus the total weight of the unary nodes sum up to $2^k - 1$.*

Proof. Consider building T from a complete binary tree of $2^k - 1$ nodes and height k by changing the appropriate binary nodes into unary nodes, one level at a time in a top-down fashion. When a binary node at level h is changed into a unary node, exactly 2^{h-1} nodes disappear from the tree. From this, the lemma follows.   □

Now let k and ϵ be given, and set $N = 2^k - 1$. We will build a specific initial tree of size $(1 - \epsilon)N$ and height k. By Lemma 1, our task is to distribute a total weight of ϵN in a unary-binary tree of height k.

We partition this total weight into three parts: A *reservoir* of approximately three fourths of the total weight, a set of *layers* having less than one eighth of the total weight, and the *rest*, which is just whatever weight remains when the two first parts have been allotted.

The reservoir is distributed as unary nodes at a certain level $h_{\max}$. The layers consist of unary nodes at odd levels less than $h_{\max}$. Finally, the rest is unary nodes at level one. The underlying idea is that the layers will allow unary nodes to propagate from the reservoir to the insertion point at sufficiently low amortized cost, while the reservoir is big enough to sustain this for the requested number of insertions.

We now describe how to build the initial unary-binary tree in a top down fashion. We define

$$h_{\max} = 2\lfloor (k + \log(\epsilon))/3 \rfloor - 5.$$

Above this level, we place no unary nodes. Hence, the top of the tree consists of a complete binary tree of height $k - h_{\max} + 1$. At level $h_{\max}$, we make $R = \lceil (3\epsilon N/4)/2^{h_{\max}-1} \rceil = \lceil 3\epsilon N/2^{h_{\max}+1} \rceil$ of the nodes unary. Their distribution on the level will not be significant. These unary nodes constitute the *reservoir*.

The unary nodes belonging to the *layers* are distributed on levels $h_{\max} - 2$, $h_{\max}-4$, etc. At each of these levels, we divide the nodes (unary as well as binary) into groups. Each group constitutes a consecutive row of nodes, when going from left to right on that level. We see level $h_{\max}$ as a single group, consisting of all the nodes on that level. The rest of the tree is now constructed in a top-down fashion as follows.

Given a group on level h, we add a level of binary nodes just below it. This determines the number x of nodes on level $h - 2$. These nodes are now divided into *eight* groups as evenly as possible, i.e. into groups of sizes $\lfloor x/8 \rfloor$ and $\lceil x/8 \rceil$. In each of these groups, *four* of the nodes are made unary. Their distribution inside the group is not significant. The figure below illustrates the process for a group of 15 nodes. The first node in each group is indicated by a mark.

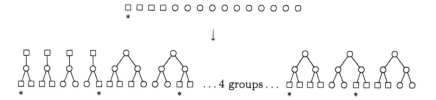

Eight new groups on level $h - 2$ have been created, and the process is now repeated recursively with these. The recursion stops when the groups on level one have been created, i.e. after $\lfloor (k + \log(\epsilon))/3 \rfloor - 3$ steps of the recursion. A sketch of the resulting tree looks like this:

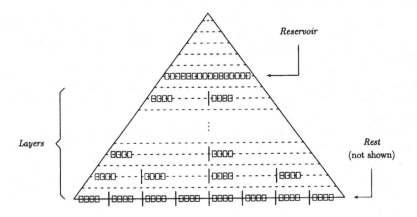

Only unary nodes are shown. The horizontal dashed lines indicates levels in the tree, and the short vertical lines are group borders. The groups above and below the vertical ellipsis are not to scale. For clarity, in the figure each group only spans two groups immediately below it – the actual number is eight.

The weight limit has not been exceeded by the above construction:

Lemma 2. *The total weight of the unary nodes in the reservoir and in the layers is less than ϵN.*

Proof. Summing up this weight is rather straightforward. The details can be found in the full paper. □

Whatever weight remains after having built the tree above, is simply distributed as unary nodes on level one. Their exact distribution will not be significant.

Later, we need the following fact:

Lemma 3. *When the initial tree has been built, the groups on level $1 + 2i$ each contain between $2^{i+6}/\epsilon$ and $2^{i+11}/\epsilon$ nodes (when k is sufficiently large).*

Proof. The proof proceeds by estimating the group sizes at each level in a top-down fashion. The details can be found in the full paper. □

5   Rebalancing

We describe all operations as taking place in unary-binary trees. As usual, an insertion starts by a search down the tree using the in-order ordering of the keys present. Unary nodes contains no keys, and are just passed through. The search stops when an empty tree is encountered. If the father of the empty tree is unary, we simply make it binary and deposit the new key in it. This is illustrated below, where the key 10 is inserted into the tree on the left.

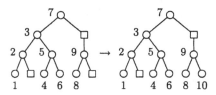

If this is not the case, we rebuild the tree in a way to be described now, with the result that some unary node is moved to the insertion point. Then we proceed as above.

The rebuilding is composed of two basic operations: a horizontal *slide* of a unary node, and a vertical *redistribution* of a unary node on some level into four unary nodes two levels below. The slide moves a unary node within the same level by shifting children between neighboring nodes, as illustrated below. Triangles designate subtrees of the same height. Besides the shifting of children on the level above the triangles, the keys in the marked nodes must be moved around to preserve the in-order ordering of the keys.

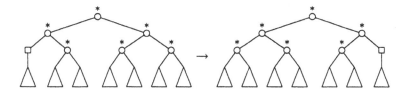

This is essentially the unary-binary formulation of the observation stated in the beginning of section 4. In this way, a unary node on any level can move a horizontal distance of d nodes at a rebuilding cost of $O(d)$. The slide operation in unary-binary trees appears in [13], which has been a starting point for this paper.

The redistribution of a unary node into four unary nodes two levels below proceeds as follows.

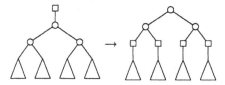

This clearly takes $O(1)$ time. A redistribution makes the lowest of the two groups involved grow by two nodes. To achieve the desired complexity, we need the groups to retain their approximate sizes. Therefore, when a group has doubled in size, we *split* it into two groups of equal size.

To specify where to do a redistribution, we define the *parent node of a group* to be the leftmost node two levels above the group, which has all of its grandchildren inside the group.[3]

Assume now that an insertion is to take place below the binary node v at level one. The process that makes v unary, allowing us to complete the insertion, can be described in pseudo-code as a call REQUEST(v) to the following recursive procedure:

Procedure REQUEST(v)
1.   **if** there is a unary node inside v's group **then**
2.   slide it to v
3.   **else**
4.   REQUEST(parent node of v's group)
5.   redistribute the parent node of v's group
6.   slide a unary node in v's group to v
7.   **if** v's group has grown too big **then**
8.   split the group into two of equal size

We show in the next section that the unary nodes at level $h_{\max}$ will last for the number of insertions stated in Theorem 1. We also show that the amortized complexity of an insertion is $O(1/\epsilon)$.

6   Analysis

The reasoning behind the good amortized behavior of our rebalancing scheme is as follows: To ensure the desired complexity, unary nodes on level one are forced to be $O(1/\epsilon)$ nodes apart. This in turn forces the weight stored at each level to decrease exponentially when going upwards in the tree, as the total weight must be less than ϵN. Correspondingly, the group sizes – and hence the cost of a slide – increase exponentially, in our set-up by a factor of two for each new level in the *layers*. To counteract this increasing cost, the unary nodes in the *layers* are distributed on every *second* level, as this means that on average $2^2 = 4$ requests for a unary node at level $1 + 2(i-1)$ are made for every request for a unary node at level $1 + 2i$. Thus, on average, an insertion requires $(2/4)^i/\epsilon$ work to be done on level $1 + 2i$, which summed over all i is $O(1/\epsilon)$. This calculation is slightly perturbed by the need to split groups, but it turns out that by charging each insertion $(3/4)^i/\epsilon$ work on level $1 + 2i$, more than adequate is charged to cover the work done in group splittings.

The potential function ϕ below provides a formalization of this argument.

Lemma 4. *The amortized rebalancing cost for an insertion is $O(1/\epsilon)$.*

---

[3] It is possible for a group borderline to divide the grandchildren of a node

Proof. We define a suitable potential function. By the *original size* of a group, we mean the size, after the initialization described in section 4, of the original group from which it has descended by a series of group splittings. Let G be a group on level $1 + 2i$ containing j nodes, of which u are unary, and let j_0 be its initial size. We define the potential $\phi(G)$ of the group G as

$$\max\{0, 4 - u\} \cdot 3^{i+2}/\epsilon + (j - j_0)(3/2)^i,$$

and the potential $\phi(T)$ of the entire tree T as the sum of $\phi(G)$ over all groups G in T. Initially, all groups have (at least) four unary nodes, so $\phi(T)$ is zero. Clearly, $\phi(T)$ is never negative. Thus (see [16]), the amortized rebalancing cost of any operation on T is the actual rebalancing cost plus the net increase in $\phi(T)$. Below, we assume that the unit of work used in ϕ is larger than all multiplicative constants in the O-expressions mentioned.

The rebalancing consists of an initial instance of REQUEST, and possibly a series of recursive instances (i.e. instances invoked by other instances of RE-QUEST). Each instance of REQUEST performs some operations, namely one slide (line 2 or line 6), zero or one redistribution (line 5), and zero or one splitting of a group (line 8). The amortized rebalancing costs of these operations depend on the level of the node given as an argument to the particular instance of REQUEST.

The actual rebalancing cost of a *slide* at level $1 + 2i$ is in $O(2^i/\epsilon)$, by the bound on the the group size from Lemma 3 (as groups never become larger than twice their original size). This is also the amortized cost, as $\phi(T)$ is not changed.

A group *splitting* has an actual rebalancing cost of zero.[4] Before the splitting, we have one group of size[5] $2j_0$, containing four nodes. After the splitting, we have two groups of size j_0, containing four nodes in total. Hence, a splitting on level $1 + 2i$ incurs an increase in $\phi(T)$ of $4 \cdot 3^{i+2}/\epsilon - j_0 \cdot (3/2)^i$. As $j_0 \geq 2^{i+6}/\epsilon$ by Lemma 3, the amortized rebalancing cost of a group splitting is actually negative.

A *redistribution* from level $1 + 2i$ to level $1 + 2(i - 1)$ removes one unary node from the upper group involved, and hence increases its potential by $3^{i+2}/\epsilon$. The lower group involved contains no unary nodes and receives four. Also, its size grows by two. Hence, its potential decreases by $4 \cdot 3^{i+1}/\epsilon - 2(3/2)^{i-1}$. The actual rebalancing cost of a redistribution is $O(1)$, making its total amortized cost $1 + 2(3/2)^{i-1} - 3^{i+1}/\epsilon$.

All recursive calls to REQUEST (line 4) are immediately followed by a redistribution (line 5). Hence, the combined amortized cost of the slide in a recursive instance of REQUEST and the redistribution immediately following the call invoking this instance is $2^i/\epsilon + 1 + 2(3/2)^{i-1} - 3^{i+1}/\epsilon$. This is negative for all $i \geq 1$, as $\epsilon \leq 1/2$.

---

[4] Of course, it takes Θ(size of the group) *work* to decide where to split. This work may be included in the count for the slide preceding the split.

[5] Actually, if j_0 is odd, the size is $2j_0 + 1$ before the splitting and j_0 and $j_0 + 1$ after the splitting, since groups always grow by *two* nodes. We leave it to the reader to verify that the amortized cost of a slide is the same in this case.

By this telescoping effect, the total sum of the amortized rebalancing costs for *all* the operations performed during the rebalancing for an insertion, *except* the slide operation in the initial instance of REQUEST, is negative. This slide operation takes place at level one, and hence, as seen above, has an amortized cost of $O(2^0/\epsilon)$. Apart from a call to REQUEST, an insertion also converts a unary node on level one to a binary node. This increases $\phi(T)$ by $3^2/\epsilon$. In total, the amortized rebalancing cost of an insertion is $O(1/\epsilon)$.   □

The amortized cost just calculated does not include the initialization of the rebalancing scheme. This initialization consists of a global rebuilding of the tree, and hence take $O(N)$ time. As stated in the lemma below, the rebalancing scheme will last for more than $\epsilon N/2$ insertions. Thus, amortized over these operations, the initialization cost is $O(1/\epsilon)$ per insertion.

Lemma 5. *If all unary nodes at level h_{max} have been removed by redistributions, at least $5\epsilon N/8$ insertions have taken place.*

Proof. The lemma can be proved using the following facts: The total weight of the initial R unary nodes in the *reservoir* is at least $3\epsilon N/4$. The total weight in the *layers* can grow (as groups get larger) from its initial value, but never to more than $\epsilon N/8$, as can be proved using Lemma 3. Weight can only disappear during insertions, not during rebalancing, and each insertion removes a weight of exactly one.

The details appear in the full paper.   □

7   Conclusion

In this paper, we have given a rebalancing scheme for binary search trees which allows an optimal trade-off between the height bound and the rebalancing cost.

We have, however, only dealt with amortized complexity. For worst case complexity, the amortized lower bound in [8] of course still applies. The best existing upper bound is height $\lceil \log(n+1) + \min\{1/\sqrt{f(n)}, \log(n)/f(n)\} \rceil$ maintained in $O(f(n))$ worst case time, by a combination of results in [3] and [8], and it remains an open problem to close this gap. We conjecture the lower bound to be tight.

References

1. G. M. Adel'son-Vel'skiĭ and E. M. Landis. An Algorithm for the Organisation of Information. Dokl. Akad. Nauk SSSR, 146:263-266, 1962. In Russian. English translation in Soviet Math. Dokl., 3:1259-1263, 1962.
2. A. Andersson. Optimal bounds on the dictionary problem. In Proc. Symp. on Optimal Algorithms, Varna, volume 401 of LNCS, pages 106-114. Springer-Verlag, 1989.
3. A. Andersson. Effcient Search Trees. PhD thesis, Department of Computer Science, Lund University, Sweden, 1990.

4. A. Andersson, C. Icking, R. Klein, and T. Ottmann. Binary search trees of almost optimal height. Acta Informatica, 28:165-178, 1990.
5. A. Andersson and T. W. Lai. Fast updating of well-balanced trees. In SWAT'90, volume 447 of LNCS, pages 111-121. Springer-Verlag, 1990.
6. A. Andersson and T. W. Lai. Comparison-effcient and write-optimal searching and sorting. In ISA'91, volume 557 of LNCS, pages 273-282. Springer-Verlag, 1991.
7. N. Blum and K. Mehlhorn. On the average number of rebalancing operations in weight-balanced trees. Theoretical Computer Science, 11:303-320, 1980.
8. R. Fagerberg. Binary search trees: How low can you go? In SWAT '96, volume 1097 of LNCS, pages 428-439. Springer-Verlag, 1996.
9. L. J. Guibas and R. Sedgewick. A Dichromatic Framework for Balanced Trees. In 19th FOCS, pages 8-21, 1978.
10. D. E. Knuth. Sorting and Searching, volume 3 of The Art of Computer Programming. Addison-Wesley, 1973.
11. T. Lai. Effcient Maintenance of Binary Search Trees. PhD thesis, Department of Computer Science, University of Waterloo, Canada., 1990.
12. T. Lai and D. Wood. Updating almost complete trees or one level makes all the difference. In STACS'90, volume 415 of LNCS, pages 188-194. Springer-Verlag, 1990.
13. H. A. Maurer, T. Ottmann, and H.-W. Six. Implementing dictionaries using binary trees of very small height. Inf. Proc. Letters, 5:11-14, 1976.
14. J. Nievergelt and E. M. Reingold. Binary search trees of bounded balance. SIAM J. on Computing, 2(1):33-43, 1973.
15. T. Ottmann, D. S. Parker, A. L. Rosenberg, H. W. Six, and D. Wood. Minimalcost brother trees. SIAM J. Computing, 13(1):197-217, 1984.
16. R. E. Tarjan. Amortized computational complexity. SIAM J. on Algebraic and Discrete Methods, 6:306-318, 1985.

Fast Allocation and Deallocation with an Improved Buddy System[*]

Erik D. Demaine and Ian J. Munro

Department of Computer Science, University of Waterloo
Waterloo, Ontario N2L 3G1, Canada
{eddemaine,imunro}@uwaterloo.ca

Abstract. We propose several modifications to the binary buddy system for managing dynamic allocation of memory blocks whose sizes are powers of two. The standard buddy system allocates and deallocates blocks in $\Theta(\lg n)$ time in the worst case (and on an amortized basis), where n is the size of the memory. We present two schemes that improve the running time to $O(1)$ time, where the time bound for deallocation is amortized. The first scheme uses one word of extra storage compared to the standard buddy system, but may fragment memory more than necessary. The second scheme has essentially the same fragmentation as the standard buddy system, and uses $O(2^{(1+\sqrt{\lg n})\lg\lg n})$ bits of auxiliary storage, which is $\omega(\lg^k n)$ but $o(n^\varepsilon)$ for all $k \geq 1$ and $\varepsilon > 0$. Finally, we present simulation results estimating the effect of the excess fragmentation in the first scheme.

1 Introduction

The *binary buddy system* [13] is a well-known system for maintaining a dynamic collection of memory blocks. Its main feature is the use of suitably aligned blocks whose sizes are powers of two. This makes it easy to check whether a newly deallocated block can be merged with an adjacent (unused) block, using bit operations on the block addresses. See Section 1.1 for a more detailed description of the method.

While the buddy system is generally recognized as fast, we argue that it is much slower than it has to be. Specifically, the time to allocate or deallocate a block of size 2^k is $\Theta(1 - k + \lg n)$ in the worst case, where n is the size of the memory in bytes. Not only is this a worst-case lower bound, but this much time can also be necessary on an amortized basis. Once we encounter a block whose allocation requires $\Theta(1 - k + \lg n)$ time, we can repeatedly deallocate and reallocate that block, for a total cost of $\Theta(m(1 - k + \lg n))$ over m operations. Such allocations and deallocations are also not rare; for example, if the memory is completely free and we allocate a constant-size block, then the buddy system uses $\Theta(\lg n)$ time. Throughout this paper we assume standard operations on a word of size $1 + \lg n$ or so bits can be performed in constant time.

---

[*] This work was supported by the Natural Science and Engineering Research Council of Canada (NSERC).

C. Pandu Rangan, V. Raman, R. Ramanujam (Eds.): FSTTCS'99, LNCS 1738, pp. 84–96, 1999.
© Springer-Verlag Berlin Heidelberg 1999

1.1   Buddy System

The (binary) buddy system was originally described by Knowlton [11,12]. It is much faster than other heuristics for dynamic memory allocation, such as first-fit and best-fit. Its only disadvantage being that blocks must be powers of two in size, the buddy system is used in many modern operating systems, in particular most versions of UNIX, for small block sizes. For example, BSD [18] uses the buddy system for blocks smaller than a page, i.e., 4 kilobytes.

The classic description of the buddy system is Knuth's [13]. Because our work is based on the standard buddy system, we review the basic ideas now.

At any point in time, the memory consists of a collection of *blocks* of consecutive memory, each of which is a power of two in size. Each block is marked either *occupied* or *free*, depending on whether it is allocated to the user. For each block we also know its size (or the logarithm of its size). The system provides two operations for supporting dynamic memory allocation:

1. Allocate (2^k): Finds a free block of size 2^k, marks it as occupied, and returns a pointer to it.
2. Deallocate (B): Marks the previously allocated block B as free and may merge it with others to form a larger free block.

The buddy system maintains a list of the free blocks of each size (called a *free list*), so that it is easy to find a block of the desired size, if one is available. If no block of the requested size is available, Allocate searches for the first nonempty list for blocks of at least the size requested. In either case, a block is removed from the free list. This process of *finding a large enough free block* will indeed be the most difficult operation for us to perform quickly.

If the found block is larger than the requested size, say 2^k instead of the desired 2^i, then the block is split in half, making two blocks of size 2^{k-1}. If this is still too large ($k - 1 > i$), then one of the blocks of size 2^{k-1} is split in half. This process is repeated until we have blocks of size 2^{k-1}, 2^{k-2}, ..., 2^{i+1}, 2^i, and 2^i. Then one of the blocks of size 2^i is marked as occupied and returned to the user. We will modify this *splitting process* as the first step in speeding up the buddy system.

Now when a block is deallocated, the buddy system checks whether the block can be merged with any others, or more precisely whether we can undo any splits that were performed to make this block. This is where the buddy system gets its name. Each block B_1 (except the initial blocks) was created by splitting another block into two halves, call them B_1 (with the same address but half the size) and B_2. The other block, B_2, created from this split is called the *buddy* of B_1, and vice versa. The *merging process* checks whether the buddy of a deallocated block is also free, in which case the two blocks are merged; then it checks whether the buddy of the resulting block is also free, in which case they are merged; and so on.

One of the main features of the buddy system is that buddies are very easy to compute on a binary computer. First note that because of the way we split and merge blocks, blocks stay *aligned*. More precisely, the address of a block of size 2^k (which we always consider to be written in binary) ends with k zeros. As a result, to find the address of the buddy of a block of size 2^k we simply flip the $(k+1)$st bit from the right.

Thus it is crucial for performance purposes to know, given a block address, the size of the block and whether it is occupied. This is usually done by storing a *block header* in the first few bits of the block. More precisely, we use headers in which the first bit is the *occupied bit*, and the remaining bits specify the size of the block. Thus, for example, to determine whether the buddy of a block is free, we compute the buddy's address, look at the first bit at this address, and also check that the two sizes match.

Because block sizes are always powers of two, we can just encode their logarithms in the block headers. This uses only $\lg \lg n$ bits, where n is the number of (smallest) blocks that can be allocated. As a result, the smallest practical header of one byte long is sufficient to address up to $2^{128} \approx 3.4 \cdot 10^{38}$ blocks. Indeed, if we want to use another bit of the header to store some other information, the remaining six bits suffice to encode up to $2^{64} \approx 1.8 \cdot 10^{19}$ blocks, which should be large enough for any practical purposes.

1.2   Related Work

Several other buddy systems have been proposed, which we briefly survey now. Of general interest are the Fibonacci and weighted buddy systems, but none of the proposals theoretically improve the running time of the Allocate and Deallocate operations.

In Exercise 2.5.31 of his book, Knuth [13] proposed the use of Fibonacci numbers as block sizes instead of powers of two, resulting in the *Fibonacci buddy system*. This idea was detailed by Hirschberg [9], and was optimized by Hinds [8] and Cranston and Thomas [7] to locate buddies in time similar to the binary buddy system. Both the binary and Fibonacci buddy systems are special cases of a generalization proposed by Burton [5].

Shen and Peterson [23] proposed the *weighted buddy system* which allows blocks of sizes 2^k and $3 \cdot 2^k$ for all k. All of the above schemes are special cases of the generalization proposed by Peterson and Norman [20] and a further generalization proposed by Russell [22]. Page and Hagins [19] proposed an improvement to the weighted buddy system, called the *dual buddy system*, which reduces the amount of fragmentation to nearly that of the binary buddy system. Another slight modification to the weighted buddy system was described by Bromley [3,4]. Koch [14] proposed another variant of the buddy system that is designed for disk-file layout with high storage utilization.

The fragmentation of these various buddy systems has been studied both experimentally and analytically by several papers [4,6,17,20,21,22].

1.3   Outline

Sections 2 and 3 describe our primary modifications to the Allocate and Deallocate operations of the binary buddy system. Finding an appropriate free block

for Allocate is the hardest part, so our initial description of Allocate assumes
that such a block has been found, and only worries about the splitting part. In
Sections 4 and 5 we present two methods for finding a block to use for allocation.
Finally, Section 6 gives simulation results comparing the fragmentation in the
two methods.

2   Lazy Splitting

Recall that if we allocate a small block out of a large block, say 2^i out of 2^k units,
then the standard buddy system splits the large block $k - i$ times, resulting in
subblocks of sizes 2^{k-1}, 2^{k-2}, ..., 2^{i+1}, 2^i, and 2^i, and then uses one of the
blocks of size 2^i. The problem is that if we immediately deallocate the block of
size 2^i, then all $k - i + 1$ blocks must be remerged into the large block of size 2^k.
This is truly necessary in order to discover that a block of size 2^k is available;
the next allocation request may be for such a block.

   To solve this problem, we do not explicitly perform the first $k - i - 1$ splits,
and instead jump directly to the last split at the bottom level. That is, the large
block of size 2^k is split into two blocks, one of size 2^i and one of size $2^k - 2^i$. Note
that the latter block has size not equal to a power of two. We call it a *superblock*,
and it contains allocatable blocks of sizes 2^i, 2^{i+1}, ..., 2^{k-2}, and 2^{k-1} (which
sum to the total size $2^k - 2^i$). For simplicity of the algorithms, we always remove
the small block of size 2^i from the left side of the large block of size 2^k, and
hence the allocatable blocks contained in a superblock are always in order of
increasing size.

   In general, we maintain the size of each allocated block as a power of two,
while the size of a free block is either a power of two or a difference of two powers
of two. Indeed, we can view a power of two as the difference of two consecutive
powers of two. Thus, every free block can be viewed as a superblock, containing
one or more allocatable blocks each of a power of two in size.

   To see how the free superblocks behave, let us consider what happens when we
allocate a subblock of size 2^j out of a free superblock of size $2^k - 2^i$. The free block
is the union of allocatable blocks of sizes $2^i, 2^{i+1}, \ldots, 2^{k-1}$, and hence $i \leq j \leq$
$k - 1$. Removing the allocatable block of size 2^j leaves two consecutive sequences:
$2^i, 2^{i+1}, \ldots, 2^{j-1}$ and $2^{j+1}, 2^{j+2}, \ldots, 2^{k-1}$. Thus, we split the superblock into the
desired block of size 2^j and two new superblocks of sizes $2^j - 2^i$ and $2^k - 2^{j+1}$.

2.1   Block Headers

As described in Section 1.1, to support fast access to information about a block
given just its address (for example when the user requests that a block be deallocated) it is common to have a header on every block that contains basic information about the block. Recall that block headers in the standard buddy system
store a single bit specifying whether the block is occupied (which is used to test
whether a buddy is occupied), together with a number specifying the logarithm
of the size of the block (which is used to find the buddy).

With our modifications, superblocks are no longer powers of two in size. The obvious encoding that uses $\Theta(\lg n)$ bits causes the header to be quite large– a single byte is insufficient even for the levels at which the buddy system is applied today (for example, smaller than 4,096 bytes in BSD 4.4 UNIX [18]). Fortunately, there are two observations which allow us to leave the header at its current size. The first is that allocated blocks are a power of two in size, and hence the standard header suffices. The second is that free blocks are a difference of two powers of two in size, and hence two bytes suffice; the second byte can be stored in the data area of the block (which is unused because the block is free).

2.2   Split Algorithm

To allocate a block of size 2^j, we first require a free superblock containing a properly aligned block of at least that size, that is, a superblock of size $2^k - 2^i$ where $k > i$ and $k > j$. Finding this superblock will be addressed in Sections 4 and 5. The second half of the Allocate algorithm is to *split* that superblock down to the appropriate size, and works as follows. Assume the superblock B at address a has an appropriate size. First examine the header of the block at address a. By assumption, the occupied bit must be clear, i.e., B must be free. The next two numbers of the header are k and i and specify that B has size $2^k - 2^i$. In other words, B is a superblock containing allocatable blocks of size $2^i, 2^{i+1}, \ldots, 2^{k-1}$, in that order.

There are two cases to consider. The first case is that one of the blocks in B has size 2^j, that is, $i \leq j \leq k - 1$. The address of this block is $a + \sum_{m=i}^{j-1} 2^m = a + 2^j - 2^i$. First we initialize the header of this block, by setting the occupied bit and initializing the logarithm of the size to j. The address of this block is also returned as the result of the Allocate operation. Next B has to be split into the allocated block and, potentially, a superblock on either side of it. If $j < k-1$ (in other words, we did not allocate the last block in the superblock), then we need a superblock on the right side. Thus, we must initialize the block header at address $a + 2^{j+1} - 2^i$ with a clear occupied bit followed by k and $j + 1$. (That is, the block has size $2^k - 2^{j+1}$.) Similarly, if $j > i$ (in other words, we did not allocate the first block in the superblock), then we need a superblock on the left side. Thus, we modify the first number in the header at address a from k to j, thereby specifying that the block now has size $2^j - 2^i$.

The second case is the undesirable case in which $i > j$, in other words we must subdivide one of the blocks in superblock B to make the allocation. The smallest block in B, of size 2^i, is adequate. It is broken off and the remainder of B is initialized as a free superblock of size $2^k - 2^{i+1}$. The block of size 2^i is broken into the allocated block of size 2^j, and a superblock of size $2^i - 2^j$ which is returned to the structure for future use.

An immediate consequence of the above modification to the split operation is the following:

Lemma 1. *The cost of a split is constant.*

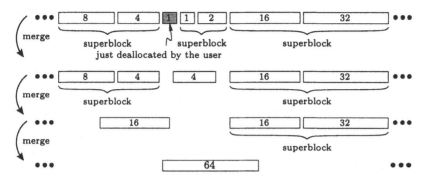

Fig. 1. An example in which blocks have been aggressively merged into superblocks, but a single deallocation causes $\Theta(\lg n)$ merges.

3   Unaggressive Merging

This section describes how merging works in combination with blocks whose sizes are not powers of two. Our goal is for merges to undo already performed splits, because the conditions that caused the split no longer hold. However, we are not too aggressive about merging: we do not merge adjacent superblocks into larger superblocks. Instead, we wait until a collection of superblocks can be merged into a usual block of a power of two in size. This is because we will only use superblocks to speed up splits. An amortized time bound for merges follows immediately, and unfortunately this kind of "aggressive merging" is not enough to obtain a worst-case time bound; see Fig. 1.

Hence, our problem reduces to detecting mergeable buddies in the standard sense, except that buddies may not match in size: the left or right buddy may be a much larger superblock. This can be done as follows. Suppose we have just deallocated a block B and want to merge it with any available buddies. First we clear the occupied bit in the header of B. Next we read the logarithm of the size of the block, call it i, and check whether B can be merged with any adjacent blocks, or in other words whether it can be merged with its buddy, as follows.

Because of the alignment of allocated blocks, the last i bits of the address of B must be zeros. If the $(i+1)$st bit from the right is a zero, then our block is a left buddy of some other block; otherwise, it is a right buddy. In either case, we can obtain the address of B's buddy by flipping the $(i+1)$st bit of B's address, that is, by taking a bitwise exclusive-or applied to 1 shifted left i times.

If the header of B's buddy has the occupied bit clear, we read its size $2^k - 2^j$. If B's size equals the lacking size 2^j (i.e, $i = j$), we merge the buddies and update the header to specify a size of 2^j. In this case, we repeat the process to see whether the buddy of the merged block is also free.

Lemma 2. *The cost of a sequence of merges is constant amortized.*

Proof. The total number of individual merges is at most twice the number of already performed (and not remerged) splits, and hence each sequence of merge operations takes $O(1)$ amortized time.   □

4   Finding a Large Enough Free Block: Fragmentation

This section presents our first approach to the remaining part of the Allocate algorithm, which is to find a block to return (if it is of the correct size) or split (if it is too large). More precisely, we need to find a block that is at least as large as the desired size. The standard buddy system maintains a doubly linked list of free blocks of each size for this purpose. Indeed, the free list is usually stored as a doubly linked list whose nodes are the free blocks themselves (since they have free space to use). The list must be doubly linked to support removal of a block in the middle of the list as the result of a merge.

We do the same, where a superblock of size $2^k - 2^i$ is placed on the free list for blocks of size 2^{k-1}, corresponding to the largest allocatable block contained within it. This will give us the smallest superblock that is large enough to handle the request. However, it may result in splitting a block when unnecessary; we shall readdress this issue in the next section.

The difficulty in finding the smallest large-enough superblock is that when (for example) there is a single, large block and we request the smallest possible block, it takes $\Theta(\lg n)$ time to do a linear scan for the appropriate free list. To find the appropriate list in $O(1)$ worst-case time, we maintain a bitvector of length $\lfloor \lg n \rfloor$, whose $(i+1)$st bit from the right is set precisely if the list for blocks of size 2^i is nonempty. Then the next nonempty list after or at a particular size 2^k can be found by first shifting the bitvector right by k, and then computing the least-significant set bit.

The latter operation is included as an instruction in many modern machines. Newer Pentium chips do it as quickly as an integer addition. It can also be computed in constant time using boolean and basic arithmetic operations [2]. Another very simple method is to store a lookup table of the solutions for all bitstrings of length $\Theta(n^\varepsilon)$ for some constant $\varepsilon > 0$, using $\Theta(n^\varepsilon)$ words of space; cut the bitvector into $\lceil 1/\varepsilon \rceil$ chunks; and check for a set bit in each chunk from right to left. This $\Theta(n^\varepsilon)$ extra space is justified because many data structures require this operation, and it is perfectly reasonable for the operating system to provide a common static table for all processes to access.

Theorem 1. *The described modifications to the buddy system cause Allocate [Deallocate] to run in constant worst-case [amortized] time.*

5   Finding a Large Enough Free Block: Extra Storage

One unfortunate property of the method described above is that even if a block of the desired size is available as part of a superblock, it may not be used because preference is given to a larger block. The reason is that our method prefers a superblock whose largest allocatable block is minimal. Unfortunately, such a superblock may not contain an allocatable block of exactly (or even close to) the desired size, whereas a superblock containing a larger largest block might. Furthermore, even if there is no block of exactly the desired size, our method will not find the smallest one to split. As a result, unnecessary splits may be performed, slightly increasing memory fragmentation. We have not performed a statistical analysis of the effect in fragmentation as a result of this property, but simulation results are presented in Section 6.

In this section, we present a further modification that solves this problem and leaves the fragmentation in essentially the same state as does the standard buddy system. Specifically, we abstract the important properties of the standard buddy system's procedure for finding a large enough free block into the following *minimum-splitting requirement*: the free block chosen must be the smallest block that is at least the desired size. In particular, if there is a block of exactly the desired size, it will be chosen. This requirement is achieved by the standard buddy system, and the amount of block splitting is locally minimized by any method achieving it.

Of course, there may be ties in "the smallest block that is at least the desired size," so different "minimum-splitting" methods may result in different pairs of blocks becoming available for remerging, and indeed do a different amount of splitting on the same input sequence. However, we observe that even different implementations of the "standard buddy system" will make different choices. Furthermore, if we view all blocks of the same size as being equally likely to be deallocated at any given time, then all minimum-splitting systems will have identical distributions of fragmentation.

In the context of the method described so far, the minimum-splitting requirement specifies that we must find a superblock containing a block of the appropriate size if one exists, and if none exists, it must find the superblock containing the smallest block that is large enough. This section describes how to solve the following more difficult problem in constant time: find the superblock whose smallest contained block is smallest, over all superblocks whose largest contained block is large enough to serve the query.

Recall that superblocks have size $2^k - 2^j$ for $1 \leq k \leq \lceil \lg n \rceil$ and $0 \leq j \leq k-1$. For each of the possible (k, j) pairs, we maintain a doubly linked list of all free superblocks of size $2^k - 2^j$ (where "superblock" includes the special case of "block"). By storing the linked list in the free superblocks themselves, the auxiliary storage required is only $\Theta(\lg^2 n)$ pointers or $\Theta(\lg^3 n)$ bits.

For each value of k, we also maintain a bitvector V_k of length $\lfloor \lg n \rfloor$, whose jth bit indicates whether there is at least one superblock of size $2^k - 2^j$. This vector can clearly be maintained in constant time subject to superblock allocations and deallocations. By finding the least-significant set bit, we can also maintain the minimum set j in V_k for each k, in constant time per update.

The remaining problem is to find some superblock of size $2^k - 2^j$ for which, subject to the constraint $k > i$, j is minimized. This way, if $j \leq i$, this superblock contains a block of exactly the desired size; and otherwise, it contains the smallest block adequate for our needs. The problem is now abstracted into having a vector $V_{\min}$, whose kth element ($1 \leq k \leq \lfloor \lg n \rfloor$) is the minimum j for which there is a free superblock of size $2^k - 2^j$; in other words, $V_{\min}[k] = \min\{j \mid V_k[j] > 0\}$. Given a value i, we are to find the smallest value of j in the last $\lfloor \lg n \rfloor - i$ positions of $V_{\min}$; in other words, we must find $\min\{V_{\min}[k] \mid k > i\}$.

The basic idea is that, because each element of $V_{\min}$ value takes only $\lg \lg n$ bits to represent, "many" can be packed into a single $(1 + \lg n)$-bit word. Indeed, we will maintain a dynamic multiway search tree of height 2. The $\lfloor \lg n \rfloor$ elements of $V_{\min}$ are split into roughly $\sqrt{\lg n}$ groups of roughly $\sqrt{\lg n}$ elements each. The pth child of the root stores the elements in the pth group. The root contains roughly $\sqrt{\lg n}$ elements, the pth of which is the minimum element in the pth group. As a consequence, each node occupies $\sqrt{\lg n} \lg \lg n$ bits.

A query for a given i is answered in two parts. First we find the minimum of the first $\lfloor i/\sqrt{\lg n} \rfloor$ elements of the root node, by setting the remaining elements to infinities, and using a table of answers for all possible root nodes. The second part in determining the answer comes from inspecting the $\lceil i/\sqrt{\lg n} \rceil$th branch of the tree, which in general will contain some superblocks (or j values) that are valid for our query and some that are not. We must, then, make a similar query there, and take the smallest j value of the two. The extra space required is dominated by the table that gives the value and position of the smallest element, for all possible $\sqrt{\lg n}$ tuples of $\lg \lg n$ bits each. There are $2^{\sqrt{\lg n} \lg \lg n}$ entries in this table, and each entry requires $2 \lg n$ bits, for a total of $2^{(1+\sqrt{\lg n}) \lg \lg n + 1}$ bits. As a consequence, the total space required beyond the storage we are managing is $o(n^\varepsilon)$ but $\omega(\lg^k n)$. Updating the structure to keep track of the minimum j for each k, in constant time after each memory allocation or deallocation, is straightforward.

Theorem 2.   *The described modifications to the buddy system satisfy the minimum-splitting requirement, and cause* Allocate *[*Deallocate*] to run in constant worst-case [amortized] time.*

6   Simulation

To help understand the effect of the excess block splitting in the first method (Section 4), we simulated it together with the standard buddy system for various memory sizes and "load factors." Our simulation attempts to capture the spirit of the classic study by Knuth [13] that compares various dynamic allocation schemes. In each time unit of his and our simulations, a new block is allocated with randomly chosen size and lifetime according to various distributions, and old blocks are checked for lifetime expiry which causes deallocations. If there is ever insufficient memory to perform an allocation, the simulation halts.

While Knuth periodically examines memory snapshots by hand to gain qualitative insight, we compute various statistics to help quantify the difference between the two buddy systems. To reduce the effect of the choice of random numbers, we run the two buddy schemes on exactly the same input sequence, repeatedly for various sequences. We also simulate an "optimal" memory allocation scheme, which continually compacts all allocated blocks into the left fraction of memory, in order to measure the "difficulty" of the input sequence.

Few experimental results seem to be available on typical block-size and lifetime distributions, so any choice is unfortunately guesswork. Knuth's block sizes are either uniformly distributed, exponentially distributed, or distributed according to a hand-selection of probabilities. We used the second distribution (choosing size 2^i with probability $1/(1 + \lfloor \lg n \rfloor)$), and what we guessed to be a generalization of the third distribution (choosing size 2^i with probability 2^{-i-1}, roughly). We dropped the first distribution because we believe it weights large blocks too heavily—blocks are typically quite small. Note that because the two main memory-allocation methods we simulate are buddy systems, we assume that all allocations ask for sizes that are powers of two. Also, to avoid rapid overflow in the second distribution, we only allow block sizes up to $n^{3/4}$, i.e., logarithms of block sizes up to $\frac{3}{4} \lg n$.

Knuth's lifetimes are uniformly distributed according to one of three ranges. We also use uniform distribution but choose our range based on a given parameter called the *load factor*. The load factor L represents the fraction of memory that tends to be used by the system. Given one of the distributions above on block size, we can compute the expected block size E, and therefore compute a lifetime Ln/E that will on average keep the amount of memory used equal to Ln (where n is the size of memory). To randomize the situation, we choose a lifetime uniformly between 1 and $2Ln/E - 1$, which has the same expected value Ln/E.

The next issue is what to measure. To address this it is useful to define a notion of the system reaching an "equilibrium." Because the simulation starts with an empty memory, it will start by mostly allocating blocks until it reaches the expected memory occupancy, Ln. Suppose it takes t time steps to reach that occupancy. After t more steps (a total of $2t$), we say that the system has reached an *equilibrium*; at that point, it is likely to stay in a similar configuration. (Of course, it is possible for the simulation to halt before reaching an equilibrium, in which case we discard that run.)

One obvious candidate for a quantity to measure is the amount of fragmentation (i.e., the number of free blocks) for each method, once every method has reached an equilibrium. However, this is not really of interest to the user: the user wants to know whether her/his block can be allocated, or whether the system will *fail* by being unable to service the allocation. This suggests a more useful metric, the *time to failure*, frequently used in the area of fault tolerance.

A related metric is to wait until all systems reach an equilibrium (ignoring the results if the system halts before that), and then measure the largest free allocatable block in each system. For the standard buddy system, this is the largest block of size a power of two; for our modified buddy system, it is the largest block in any superblock; and for the optimal system, it is simply the amount of free memory. This measures, at the more-or-less arbitrary time of all systems reaching equilibrium, the maximum-size block that could be allocated.

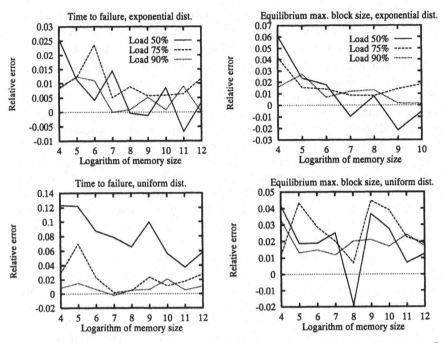

Fig. 2. Simulation results. The distributions refer to the distributions of the logarithms of the block sizes, and "load" refers to the load factor.

We feel that these two metrics capture some notion of what users of a memory-allocation system are interested in. By evaluating them for all three systems under the same inputs, we can measure the difference between the two buddy systems, relative to the optimal system. This kind of "relative error" was measured for 100 runs and then averaged, for each case. Memory size ranges between 2^4 (the smallest power-of-two size for which a difference between the two buddy systems is noticeable) and 2^{12} (the size used by BSD 4.4 UNIX [18]). The tested load factors are 50%, 75%, and 90%.

The results are shown in Fig. 2. The relative errors are for the most part quite small (typically under 5%). Indeed, our first method occasionally does somewhat better than the standard buddy system, because its different choices of blocks to split cause some fortunate mergings. Further evidence is that, for the exponential distribution, less than 10% of the runs showed any difference in time-to-failure between the two systems. (However, the number of differences is greater for the uniform distribution.)

Thus, the difference in distributions of fragmentation between the two buddy systems seems reasonably small. The simplicity of our first method may make it attractive for implementation.

7 Conclusion

We have presented two enhancements to the buddy system that improve the running time of Allocate to constant worst-case time, and Deallocate to constant amortized time. The more complex method keeps the distribution of fragmentation essentially the same as the standard method, while the simpler approach leads to a different and slightly worse distribution. It would be of interest to specify this difference mathematically.

We note that it is crucial for Allocate to execute as quickly as possible (and in particular fast in the worst case), because the executing process cannot proceed until the block allocation is complete. In contrast, it is reasonable for the Deallocate time bound to be amortized, because the result of the operation is not important and the actual work can be delayed until the CPU is idle (or the memory becomes full). Indeed, this delay idea has been used to improve the cost of the standard buddy system's Deallocate [1,10,15,16]. On the other hand, for the purposes of theoretical results, it would be of interest to obtain a constant worst-case time bound for both Allocate and Deallocate.

References

1. R. E. Barkley and T. Paul Lee. A lazy buddy system bounded by two coalescing delays per class. *Operating Systems Review*, pages 167–176, Dec. 1989.
2. Andrej Brodnik. Computation of the least significant set bit. In *Proceedings of the 2nd Electrotechnical and Computer Science Conference*, Portoroz, Slovenia, 1993.
3. Allan G. Bromley. An improved buddy method for dynamic storage allocation. In *Proceedings of the 7th Australian Computer Conference*, pages 708–715, 1976.
4. Allan G. Bromley. Memory fragmentation in buddy methods for dynamic storage allocation. *Acta Informatica*, 14:107–117, 1980.
5. Warren Burton. A buddy system variation for disk storage allocation. *Communications of the ACM*, 19(7):416–417, July 1976.
6. Shyamal K. Chowdhury and Pradip K. Srimani. Worst case performance of weighted buddy systems. *Acta Informatica*, 24(5):555–564, 1987.
7. Ben Cranston and Rick Thomas. A simplified recombination scheme for the Fibonacci buddy system. *Communications of the ACM*, 18(6):331–332, June 1975.
8. James A. Hinds. Algorithm for locating adjacent storage blocks in the buddy system. *Communications of the ACM*, 18(4):221–222, Apr. 1975.
9. Daniel S. Hirschberg. A class of dynamic memory allocation algorithms. *Communications of the ACM*, 16(10):615–618, Oct. 1973.
10. Arie Kaufman. Tailored-list and recombination-delaying buddy systems. *ACM Transactions on Programming Languages and Systems*, 6(1):118–125, Jan. 1984.
11. Kenneth C. Knowlton. A fast storage allocator. *Communications of the ACM*, 8(10):623–625, Oct. 1965.
12. Kenneth C. Knowlton. A programmer's description of L6. *Communications of the ACM*, 9(8):616–625, Aug. 1966.
13. Donald E. Knuth. Dynamic storage allocation. In *The Art of Computer Programming*, volume 1, section 2.5, pages 435–455. Addison-Wesley, 1968.
14. Philip D. L. Koch. Disk file allocation based on the buddy system. *ACM Transactions on Computer Systems*, 5(4):352–370, Nov. 1987.

15. T. Paul Lee and R. E. Barkley. Design and evaluation of a watermark-based lazy buddy system. *Performance Evaluation Review*, 17(1):230, May 1989.
16. T. Paul Lee and R. E. Barkley. A watermark-based lazy buddy system for kernel memory allocation. In *Proceedings of the 1989 Summer USENIX Conference*, pages 1–13, June 1989.
17. Errol L. Lloyd and Michael C. Loui. On the worst case performance of buddy systems. *Acta Informatica*, 22(4):451–473, 1985.
18. Marshall Kirk McKusick, Keith Bostic, Michael J. Karels, and John S. Quarterman. *The Design and Implementation of the 4.4 BSD Operating System*. Addison-Wesley, 1996.
19. Ivor P. Page and Jeff Hagins. Improving the performance of buddy systems. *IEEE Transactions on Computers*, C-35(5):441–447, May 1986.
20. James L. Peterson and Theodore A. Norman. Buddy systems. *Communications of the ACM*, 20(6):421–431, June 1977.
21. Paul W. Purdom, Jr. and Stephen M. Stigler. Statistical properties of the buddy system. *Journal of the ACM*, 17(4):683–697, Oct. 1970.
22. David L. Russell. Internal fragmentation in a class of buddy systems. *SIAM Journal on Computing*, 6(4):607–621, Dec. 1977.
23. Kenneth K. Shen and James L. Peterson. A weighted buddy method for dynamic storage allocation. *Communications of the ACM*, 17(10):558–562, Oct. 1974. See also the corrigendum in 18(4):202, Apr. 1975.

Optimal Bounds for Transformations of ω-Automata

Christof Löding

Lehrstuhl für Informatik VII,
RWTH Aachen, D-52056 Aachen
loeding@informatik.rwth-aachen.de

Abstract. In this paper we settle the complexity of some basic constructions of ω-automata theory, concerning transformations of automata characterizing the set of ω-regular languages. In particular we consider Safra's construction (for the conversion of nondeterministic Büchi automata into deterministic Rabin automata) and the appearance record constructions (for the transformation between different models of deterministic automata with various acceptance conditions). Extending results of Michel (1988) and Dziembowski, Jurdziński, and Walukiewicz (1997), we obtain sharp lower bounds on the size of the constructed automata.

1 Introduction

The theory of ω-automata offers interesting transformation constructions, allowing to pass from nondeterministic to deterministic automata and from one acceptance condition to another. The automaton models considered in this paper are nondeterministic Büchi automata [1], and deterministic automata of acceptance types Muller [12], Rabin [13], Streett [17], and parity [11]. There are two fundamental constructions to achieve transformations between these models. The first is based on the data structure of *Safra trees* [14] (for the transformation from nondeterministic to deterministic automata), and the second on the data structure of *appearance records* [2,3,5] (for the transformations between deterministic Muller, Rabin, Streett, and parity automata). In this paper, we show that for most of the transformations, these constructions are optimal, sharpening previous results from the literature. This requires an analysis and extension of examples as proposed by Michel [10] and Dziembowski, Jurdziński, and Walukiewicz [4].

The first construction of deterministic Rabin automata from nondeterministic Büchi automata is due to McNaughton [9]. Safra's construction [14] generalizes the classical subset construction by introducing trees of states (Safra trees) instead of sets of states, yielding a complexity of $n^{\mathcal{O}(n)}$ (where n is the number of states of the Büchi automaton). Using an example of Michel [10] one obtains the optimality of Safra's construction in the sense that there is no conversion of nondeterministic Büchi automata with n states into deterministic Rabin automata with $2^{\mathcal{O}(n)}$ states and $\mathcal{O}(n)$ pairs in the acceptance condition (see the

C. Pandu Rangan, V. Raman, R. Ramanujam (Eds.): FSTTCS'99, LNCS 1738, pp. 97–109, 1999.
© Springer-Verlag Berlin Heidelberg 1999

survey [18]). A drawback is the restriction to Rabin automata with $\mathcal{O}(n)$ pairs. In the present paper we eliminate this restriction. This is shown in Section 3. Also, using results of Section 4, we obtain an optimal bound for the transformation of nondeterministic Büchi automata into deterministic Streett automata.

For the transformations between deterministic models, the construction of appearance records, introduced by Büchi [2,3] and Gurevich, Harrington [5] in the context of infinite games, is useful. To transform Muller automata into other deterministic automata one uses *state appearance records* (SAR). The main component of an SAR is a permutation of states, representing the order of the last visits of the states in a run. This leads to a size of $(\mathcal{O}(n))!$ of the resulting automaton, where n is the number of states of the original automaton.

For the transformation of Rabin or Streett automata into other deterministic models one uses *index appearance records* (IAR). The main component of an IAR is a permutation of the indices of the pairs (of state sets) in the acceptance condition, representing the order of the last visits of the first components of these pairs. This leads to a size of $(\mathcal{O}(r))!$ of the resulting automaton, where r is the number of pairs of the original automaton.

Dziembowski, Jurdziński, and Walukiewicz [4] have studied the state appearance records as memory entries in automata which execute winning strategies in infinite games. They presented an example of an infinite game over a graph with $2n$ vertices where each winning strategy requires a memory of size $n!$. Starting from this example we introduce families of languages which yield optimal lower bounds for all automata transformations which involve appearance record constructions (either in the form of SAR or IAR).

Table 1 at the end of the paper lists the different transformations considered. In this paper we show that all these transformations involve a factorial blow up.

The lower bounds as exposed in this paper show that in ω-automata theory the single exponential lower bound $2^{\Omega(n)}$ as known from the subset construction from the classical theory of automata on finite words has been extended to $2^{\Omega(n \cdot \log n)}$. So we see in which sense it is necessary to pass from sets of states (classical case) to sequences or even trees of states.

We leave open the question of an optimal lower bound for the transformation of nondeterministic Büchi automata into deterministic Muller automata.

The present results are from the author's diploma thesis [8]. Thanks to Wolfgang Thomas for his advice in this research.

2   Notations and Definitions

For an arbitrary set X we denote the set of infinite sequences (or infinite words) over X by X^ω and the set of finite words over X by X^*. For a sequence $\sigma \in X^\omega$ and for an $i \in \mathbb{N}$ the element on the ith position in σ is denoted by $\sigma(i)$, i.e., $\sigma = \sigma(0)\sigma(1)\sigma(2)\cdots$. The infix of σ from position i to position j is denoted by $\sigma[i,j]$. We define $In(\sigma)$, the *infinity set* of σ, to be the set of elements from X that appear infinitely often in σ. The length of a word $w \in X^*$ is denoted by $|w|$.

An ω-*automaton* $\mathcal{A}$ is a tuple $(Q, \Sigma, q_0, \delta, Acc)$. The tuple (Q, Σ, q_0, δ) is called the *transition structure* of $\mathcal{A}$, where $Q \neq \emptyset$ is a finite set of states, Σ is a finite alphabet, and q_0 is the initial state. The transition function δ is a function $\delta : Q \times \Sigma \to Q$ for deterministic automata and $\delta : Q \times \Sigma \to 2^Q$ for nondeterministic automata. The last component Acc is the *acceptance condition*.

A *run* of $\mathcal{A}$ on a word $\alpha \in \Sigma^\omega$ is an infinite state sequence $\sigma \in Q^\omega$ such that $\sigma(0) = q_0$ and for all $i \in \mathbb{N}$ one has $\sigma(i + 1) = \delta(\sigma(i), \alpha(i))$ for deterministic automata, and $\sigma(i + 1) \in \delta(\sigma(i), \alpha(i))$ for nondeterministic automata.

A run is called *accepting* iff it satisfies the acceptance condition. We will specify this below for the different forms of acceptance conditions. The language $L(\mathcal{A})$ that is *accepted* or *recognized* by the automaton is defined as $L(\mathcal{A}) = \{\alpha \in \Sigma^\omega \mid$ there is an accepting run of $\mathcal{A}$ on $\alpha\}$.

In this paper we consider acceptance conditions of type Büchi, Muller, Rabin, Streett, and parity.

Let $\mathcal{A} = (Q, \Sigma, q_0, \delta, Acc)$ be an ω-automaton. For the different acceptance types mentioned above the acceptance condition Acc is given in different forms.

A Büchi condition [1] refers to a set $F \subset Q$, and a run σ of a Büchi automaton is defined to be accepting iff $In(\sigma) \cap F \neq \emptyset$.

A Muller condition [12] refers to a system of sets $\mathcal{F} \subset 2^Q$; a run σ of a Muller automaton is called accepting iff $In(\sigma) \in \mathcal{F}$.

A Rabin condition [13] refers to a list of pairs $\Omega = \{(E_1, F_1), \ldots, (E_r, F_r)\}$ with $E_i, F_i \subseteq Q$ for $i = 1, \ldots r$. A run σ of a Rabin automaton is called accepting iff there exists an $i \in \{1, \ldots, r\}$ such that $In(\sigma) \cap F_i \neq \emptyset$ and $In(\sigma) \cap E_i = \emptyset$.

A Streett condition [17] also refers to such a list Ω but it is used in the dual way. A run σ of a Streett automaton is called accepting iff for every $i \in \{1, \ldots, r\}$ one has $In(\sigma) \cap F_i = \emptyset$ or $In(\sigma) \cap E_i \neq \emptyset$.

A parity condition [11] refers to a mapping $c : Q \to \{0, \ldots, k\}$ with $k \in \mathbb{N}$. A run σ of a parity automaton is called accepting iff $\min\{c(q) \mid q \in In(\sigma)\}$ is even. The numbers $0, \ldots, k$ are called *colors* in this context.

Obviously the Muller condition is the most general form of acceptance condition. This means every automaton $\mathcal{A}$ of the form above can be transformed into an equivalent Muller automaton just by collecting the sets of states that satisfy the acceptance condition of $\mathcal{A}$.

Let us note that Parity conditions can also be represented as Rabin and as Streett conditions:

Proposition 1. Let $\mathcal{A} = (Q, \Sigma, \delta, q_0, c)$ with $c : Q \to \{0, \ldots, k\}$ be a parity automaton and let $r = \lfloor \frac{k}{2} \rfloor$. Let $\Omega = \{(E_0, F_0), \ldots, (E_r, F_r)\}$ with $E_i = \{q \in Q \mid c(q) < 2i\}$ and $F_i = \{q \in Q \mid c(q) = 2i\}$ for $i = 0, \ldots, r$. Furthermore let $\Omega' = \{(E_0', F_0'), \ldots, (E_r', F_r')\}$ with $E_i' = \{q \in Q \mid c(q) < 2i + 1\}$ and $F_i' = \{q \in Q \mid c(q) = 2i + 1\}$ for $i = 0, \ldots, r$. Then the Rabin automaton $\mathcal{A}_1 = (Q, \Sigma, \delta, q_0, \Omega)$ and the Streett automaton $\mathcal{A}_2 = (Q, \Sigma, \delta, q_0, \Omega')$ are equivalent to $\mathcal{A}$.

For a deterministic automaton $\mathcal{A}$ with a Muller, Rabin, Streett, or parity condition we give a deterministic automaton recognizing the complementary language. This automaton is called the dual of $\mathcal{A}$. For a Muller or a parity automaton the dual automaton is of the same type. For a Rabin automaton the dual automaton is a Streett automaton and vice versa.

Proposition 2. (i) Let $\mathcal{A} = (Q, \Sigma, \delta, q_0, \mathcal{F})$ be a deterministic Muller automaton. The deterministic Muller automaton $\mathcal{A}' = (Q, \Sigma, \delta, q_0, 2^Q \setminus \mathcal{F})$ recognizes $\Sigma^\omega \setminus L(\mathcal{A})$.

(ii) Let $\mathcal{A} = (Q, \Sigma, \delta, q_0, \Omega)$ be a deterministic Rabin (Streett) automaton. The deterministic Streett (Rabin) automaton $\mathcal{A}' = (Q, \Sigma, \delta, q_0, \Omega)$ recognizes $\Sigma^\omega \setminus L(\mathcal{A})$.

(iii) Let $\mathcal{A} = (Q, \Sigma, \delta, q_0, c)$ be a deterministic parity automaton. The deterministic parity automaton $\mathcal{A}' = (Q, \Sigma, \delta, q_0, c')$ with $c'(q) = c(q) + 1$ for every $q \in Q$ recognizes $\Sigma^\omega \setminus L(\mathcal{A})$.

Because of the special structure of Rabin conditions on the one hand and Streett conditions of the onther hand, we can state the following about the union of infinity sets.

Proposition 3. (i) Let $\mathcal{A} = (Q, \Sigma, \delta, q_0, \Omega)$ be a Streett automaton and let $R, S \subseteq Q$ be two infinity sets of possible runs satisfying the acceptance condition of $\mathcal{A}$. Then a run with infinity set $R \cup S$ also satisfies the acceptance condition of $\mathcal{A}$.

(ii) Let $\mathcal{A} = (Q, \Sigma, \delta, q_0, \Omega)$ be a Rabin automaton and let $R, S \subseteq Q$ be two infinity sets of possible runs not satisfying the acceptance condition of $\mathcal{A}$. Then a run with infinity set $R \cup S$ also does not satisfy the acceptance condition of $\mathcal{A}$.

3   Optimality of Safra's Construction

In this section we show the optimality of Safra's construction ([14]) which transforms a nondeterministic Büchi automaton with n states into a deterministic Rabin automaton with $2^{\mathcal{O}(n \cdot \log n)}$ states. The main part of the proof consists of Lemma 5, which states that there exists a family $(L_n)_{n \geq 2}$ of languages, such that L_n can be recognized by a nondeterministic Büchi automaton with $\mathcal{O}(n)$ states, but the complement of L_n can not be recognized by a nondeterministic Streett automaton with less than $n!$ states.

This lemma is essentially due to Michel, who proved that there is a family $(L_n)_{n \geq 2}$ of languages, such that L_n can be recognized by a nondeterministic Büchi automaton with $\mathcal{O}(n)$ states, but the complement of L_n can not be recognized by a nondeterministic Büchi automaton with less than $n!$ states. Here we use the same family of languages as Michel but show the stronger result that there is no nondeterministic Streett automaton with less than $n!$ states recognizing the complement of L_n.

We define the languages L_n via Büchi automata $\mathcal{A}_n$ over the alphabet $\Sigma_n = \{1, \ldots, n, \#\}$. Later we adapt the idea for languages over a constant alphabet. For a technical reason we use a set of initial states instead of one initial state, but recall that we can reduce the automata to the usual format by adding one extra state.

Define the automaton $\mathcal{A}_n = (Q_n, \Sigma_n, Q_0^n, \delta_n, F_n)$ as follows.

- $Q_n = \{q_0, q_1, \ldots, q_n\}$, $Q_0^n = \{q_1, \ldots, q_n\}$, and $F_n = \{q_0\}$.
- The transition function δ_n is defined by

$$\begin{aligned}
\delta_n(q_0, a) &= \{q_a\} && \text{for } a \in \{1, \ldots, n\}, \\
\delta_n(q_0, \#) &= \emptyset, \\
\delta_n(q_i, a) &= \{q_i\} && \text{for } a \in \Sigma_n, i \in \{1, \ldots, n\}, a \neq i, \\
\delta_n(q_i, i) &= \{q_i, q_0\} && \text{for } i \in \{1, \ldots, n\}.
\end{aligned}$$

The automaton is shown in Figure 1. The idea can be adjusted to automata with the constant alphabet $\{a, b, \#\}$ by coding $i \in \{1, \ldots, n-1\}$ with $a^i b$, n with $a^n a^* b$ and $\#$ with $\#$. The resulting automaton is shown on the right hand side of Figure 1 and still has $\mathcal{O}(n)$ states.

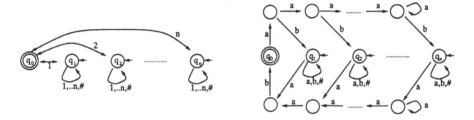

Fig. 1. The transition structure of the Büchi automaton $\mathcal{A}_n$. On the left hand over the alphabet $\{1, \ldots, n, \#\}$ and on the right hand over the alphabet $\{a, b, \#\}$. A nondeterministic Streett automaton for the complementary language needs at least $n!$ states.

As an abbreviation we define $L_n = L(\mathcal{A}_n)$. Before we prove the main lemma, we first give a characterization of the languages L_n, which is not difficult to prove.

Lemma 4. *Let $n \in \mathbb{N}$ and $\alpha \in \Sigma_n^\omega$. Then the following two statements are equivalent.*

(i) $\alpha \in L_n$.
(ii) *There exist $i_1, \ldots, i_k \in \{1, \ldots, n\}$ such that each pair $i_1 i_2, \ldots, i_{k-1} i_k, i_k i_1$ appears infinitely often in α.*

(Our definition corrects an inaccuracy of [18], where the automata are defined in the same way but with q_0 as only initial state. For the languages defined in this way condition (ii) has to be sharpened.)

Lemma 5. *Let $n \geq 2$. The complement $\Sigma_n^\omega \setminus L_n$ of L_n can not be recognized by a nondeterministic Streett automaton with less than $n!$ states.*

Proof. Let $n \in \mathbb{N}$ and let $\mathcal{A}' = (Q', \Sigma_n, q_0', \delta', \Omega)$ be a Streett automaton with $L' := L(\mathcal{A}') = \Sigma_n^\omega \setminus L_n$ and let $(i_1 \ldots i_n)$, $(j_1 \ldots j_n)$ be different permutations of $(1 \ldots n)$. Define $\alpha = (i_1 \ldots i_n \#)^\omega$ and $\beta = (j_1 \ldots j_n \#)^\omega$. From Lemma 4 we obtain $\alpha, \beta \notin L'$. Thus we have two successful runs $r_\alpha, r_\beta \in Q'^\omega$ of $\mathcal{A}'$ on α, β. Define $R = In(r_\alpha)$ and $S = In(r_\beta)$. If we can show $R \cap S = \emptyset$, then we are done because there are $n!$ permutations of $(1 \ldots n)$.

Assume $R \cap S \neq \emptyset$. Under this assumption we can construct $\gamma \in \Sigma_n^\omega$ with $\gamma \in L_n \cap L'$. This is a contradiction, since L' is the complement of L_n.

Let $q \in R \cap S$. The we can choose a prefix u of α leading from q_0 to q, an infix v of α containing the word $i_1 \ldots i_n$, leading from q to q through all states of R, and an infix w of β containing the word $j_1 \ldots j_n$, leading from q to q through all states of S.

Now let $\gamma = u(vw)^\omega$. A run r_γ of $\mathcal{A}'$ on γ first moves from q_0 to q (while reading u), and then cycles alternatingly through R (while reading v) and S (while reading w). Therefore r_γ has the infinity set $R \cup S$. Because R and S satisfy the Streett condition, $R \cup S$ also does (Proposition 3) and we have $\gamma \in L'$.

To show $\gamma \in L_n$ we first note that, if k is the lowest index with $i_k \neq j_k$, then there exist $l, m > k$ with $j_k = i_l$ and $i_k = j_m$. By the choice of the words v and w one can see that γ contains infinitely often the segments $i_1 \ldots i_n$ and $j_1 \ldots j_n$. Thus γ also contains the segments $i_k i_{k+1}, \ldots, i_{l-1} i_l, j_k j_{k+1}, \ldots, j_{m-1} j_m$. Now, using Lemma 4, we can conclude $\gamma \in L_n$.

Theorem 6. *There exists a family $(L_n)_{n \geq 2}$ of languages such that for every n the language L_n can be recognized by a nondeterministic Büchi automaton with $\mathcal{O}(n)$ states but can not be recognized by a deterministic Rabin automaton with less than $n!$ states.*

Proof. Consider the family of languages from Lemma 5. Let $n \in \mathbb{N}$. Assume there exists a deterministic Rabin automaton with less than $n!$ states recognizing L_n. The dual of this automaton is a deterministic Streett automaton with less than $n!$ states recognizing $\Sigma_n^\omega \setminus L_n$. This contradicts Lemma 5.

Since parity conditions are special cases of the Rabin conditions, Theorem 6 also holds for parity automata instead of Rabin automata.

The theorem sharpens previous results of literature (see the survey [18]), where it is shown that there is no conversion of Büchi automata with $\mathcal{O}(n)$ states into deterministic Rabin automata with $2^{\mathcal{O}(n)}$ states and $\mathcal{O}(n)$ pairs. We point out that our proof is almost the same as in [18]. Only a few changes where needed to get this slightly stronger result.

The example demonstrates the optimality of Safra's construction for the transformation of nondeterministic Büchi automata into deterministic Rabin automata. For the transformation of nondeterministic Büchi automata into deterministic Muller automata this question is open. The known lower bound for this transformation is $2^{\Omega(n)}$ ([16]). In the following we will see that the example from above can not be used to show an optimal lower bound for Muller automata: We construct Muller automata $\mathcal{M}_n$, $n \in \mathbb{N}$, with $\mathcal{O}(n^2)$ states recognizing the language L_n.

For $n \in \mathbb{N}$ define the Muller automaton $\mathcal{M}_n = (Q'_n, \Sigma_n, q'_0, \delta'_n, \mathcal{F}_n)$ by $Q'_n = \Sigma_n \times \Sigma_n$, $q'_0 = (\#, \#)$, $\delta'_n((i, j), a) = (j, a)$, and $F \in \mathcal{F}_n$ iff there exist $i_1, \dots, i_k \in \{1, \dots, n\}$ such that $(i_1, i_2), \dots, (i_{k-1}, i_k)$, $(i_k, i_1) \in F$.

The automaton just collects all pairs of letters occurring in the input word and then decides, using the Muller acceptance condition, if the property from Lemma 4 is satisfied. Thus we have $L(\mathcal{M}_n) = L_n$.

In the Muller automata $\mathcal{M}_n$ every superset of an accepting set is accepting too. Therefore the languages considered may even be recognized by deterministic Büchi automata ([7]). Thus, we can restrict the domain of the regular ω-languages, and get a sharpened version of Theorem 6: The factorial blow up in the transformation of Büchi automata into deterministic Rabin automata already occurs over the class G_δ of those languages which are acceptable by a deterministic Büchi automaton.

One may also ask for a lower bound for the transformation of nondeterministic Büchi automata into deterministic Streett automata. The result we obtain belongs to this section and therefore we mention it here: There exists a family $(L_n)_{n \geq 2}$ of languages (L_n over an alphabet of size n) such that for every n the language L_n can be recognized by a nondeterministic Büchi automaton with $\mathcal{O}(n)$ states but can not be recognized by a deterministic Streett automaton with less than $n!$ states.

The proof uses results from the next section and thus the claim will be restated there (Theorem 8).

4   Optimality of the Appearance Record Constructions

Appearance records [2,3,5], abbreviated AR, serve to transform Muller, Rabin, and Streett automata into parity automata. For these constructions two different forms of AR's are used, namely state appearance records (SAR) and index appearance records (IAR).

The SAR construction (see e.g. [18]) serves to transform deterministic Muller automata with n states into an equivalent deterministic parity automata with $(\mathcal{O}(n))!$ states and $\mathcal{O}(n)$ colors. Since parity automata are special kinds of Rabin and Streett automata (Proposition 1), this construction also transforms Muller automata into Rabin or Streett automata.

The IAR construction (see e.g. [15]) transforms a deterministic Streett automaton with n states and r pairs into an equivalent parity automaton with $n \cdot (\mathcal{O}(r))!$ states and $\mathcal{O}(r)$ colors. Because of the duality of Rabin and Streett conditions and the self duality of parity conditions (Proposition 2), the IAR construction can be used for all nontrivial transformations between Rabin, Streett, and parity automata.

In this section we will show that all the AR constructions are of optimal complexity. The idea for the proof originates in [4], where the optimality of the SAR as memory for winning strategies in Muller games was shown. Just to avoid confusion we would like to point out that the family of automata we use in our proof is not just an adaption of the games from [4]. The winning condition of the games and the acceptance condition of the automata are not related. Our proof also does not generalize the one from [4], because the used family of automata can not be adapted to games requiring memory $n!$.

We first give a theorem showing the optimality of the IAR construction for the transformation of Streett into Rabin automata and then explain how to apply the theorem to get the optimality of all other AR transformations.

Theorem 7. *There exists a family $(L_n)_{n \geq 2}$ of languages such that for every n the language L_n can be recognized by a deterministic Streett automaton with $\mathcal{O}(n)$ states and $\mathcal{O}(n)$ pairs but can not be recognized by a deterministic Rabin automaton with less than $n!$ states.*

Proof. We define the languages L_n via deterministic Streett automata $\mathcal{A}_n$ over the alphabet $\{1, \ldots, n\}$. Later we will explain how we can adapt the proof for an alphabet of constant size. The transition structure of $\mathcal{A}_n$ is shown schematically in Figure 2. Formally, for $n \geq 2$, we define the Streett automaton $\mathcal{A}_n = (Q_n, \Sigma_n, q_0^n, \delta_n, \Omega_n)$ as follows.

- $Q_n = \{-n, \ldots, -1, 1, \ldots, n\}$ and $q_0^n = -1$.
- For $i, j \in \{1, \ldots, n\}$ let $\delta_n(i, j) = -j$ and $\delta_n(-i, j) = j$.
- $\Omega_n = \{(E_1, F_1), \ldots, (E_n, F_n)\}$ with $E_i = \{i\}$ and $F_i = \{-i\}$.

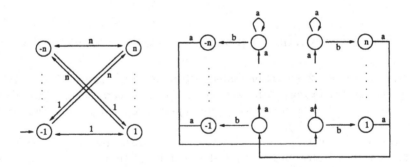

Fig. 2. The transition structure of the Streett automaton $\mathcal{A}_n$. On the left hand over the alphabet $\{1, \ldots, n\}$ and on the right hand over the alphabet $\{a, b\}$. An equivalent deterministic Rabin automaton needs at least $n!$ states.

To characterize the words in L_n we use the following notation. For a word $\alpha \in \Sigma_n^\omega$ let $even(\alpha)$ be the set containing the letters that infinitely often occur on an even position in α and let $odd(\alpha)$ be the set containing the letters that infinitely often occur on an odd position in α. This means $even(\alpha) = In(\alpha(0)\alpha(2)\alpha(4)\cdots)$ and $odd(\alpha) = In(\alpha(1)\alpha(3)\alpha(5)\cdots)$. From the definition of $\mathcal{A}_n$ follows that a word $\alpha \in \Sigma_n^\omega$ is in L_n if and only if $odd(\alpha) \subseteq even(\alpha)$. As a consequence of this, for $\alpha \in \Sigma_n^\omega$ and $u \in \Sigma_n^*$ with $|u|$ even, the word $u\alpha$ is in L_n if and only if α is in L_n. Therefore, in a deterministic automaton recognizing L_n, every state that can be reached by reading a prefix of even length can be used as initial state without changing the accepted language.

We will prove by induction that every deterministic Rabin automaton recognizing L_n needs at least $n!$ states.

We show the base case of the induction for $n = 2$. An automaton recognizing a nonempty proper subset of Σ_2^ω needs at least 2 states. Therefore the base case of the induction holds.

Now let $n > 2$ and let $\mathcal{B} = (Q, \Sigma_n, q_0, \delta, \Omega)$ be a deterministic Rabin automaton with $L(\mathcal{B}) = L_n$. Let Q_{even} be the states that can be reached from q_0 by reading a prefix of even length.

For every $i \in \{1, \ldots, n\}$ and every $q \in Q_{\text{even}}$ we construct a deterministic Rabin automaton $\mathcal{B}_i^q$ over $\Sigma_n \setminus \{i\}$ by removing all i-transitions from $\mathcal{B}$. Furthermore q is the initial state of $\mathcal{B}_i^q$. Since q can be reached in $\mathcal{B}$ after having read a prefix of even length, the language recognized by $\mathcal{B}_i^q$ is L_{n-1} (if $i \neq n$ then the names of the letters are different but the language essentially equals L_{n-1}). Thus, by the induction hypothesis, $\mathcal{B}_i^q$ has at least $(n-1)!$ states.

We can strengthen this statement as follows. In every $\mathcal{B}_i^q$ ($i \in \{1, \ldots, n\}$ and $q \in Q_{\text{even}}$) is a strongly connected component with at least $(n-1)!$ states. Just take a strongly connected component S in $\mathcal{B}_i^q$ such that there is no other strongly connected component reachable from S in $\mathcal{B}_i^q$. Let $p \in S$ be a state that is reachable from q in $\mathcal{B}_i^q$ by reading a prefix of even length. As we have seen above, we can use p as initial state in $\mathcal{B}_i^q$ without changing the accepted language. Therefore, by the induction hypothesis, S must contain at least $(n-1)!$ states.

Now, for $i \in \{1, \ldots, n\}$, we construct words $\alpha_i \in \Sigma_n^\omega$ with runs σ_i of $\mathcal{B}$ such that $|In(\sigma_i)| \geq (n-1)!$ and $In(\sigma_i) \cap In(\sigma_j) = \emptyset$ for $i \neq j$. Then we are done because $|Q'| \geq \sum_{i=1}^n |In(\sigma_i)| \geq n \cdot (n-1)! = n!$.

For $i \in \{1, \ldots, n\}$ construct the word α_i as follows. First take a $u_0 \in (\Sigma_n \setminus \{i\})^*$ such that u_0 has even length and contains every letter from $\Sigma_n \setminus \{i\}$ on an even and on an odd position. Furthermore $\mathcal{B}_i^{q_0}$ should visit at least $(n-1)!$ states while reading u_0. This is possible since $\mathcal{B}_i^{q_0}$ contains a strongly connected component with $\geq (n-1)!$ states. Let q_1 be the state reached by $\mathcal{B}_i^{q_0}$ after having read the word u_0ik, where $k \in \{1, \ldots, n\} \setminus \{i\}$. Then we choose a word $u_1 \in (\Sigma_n \setminus \{i\})^*$ with the same properties as u_0, using $\mathcal{B}_i^{q_1}$ instead of $\mathcal{B}_i^{q_0}$. This means u_1 has even length, contains every letter from $\Sigma_n \setminus \{i\}$ on an even and on an odd position, and $\mathcal{B}_i^{q_1}$ visits at least $(n-1)!$ states while reading u_1.

Repeating this procedure we get a word $\alpha_i = u_0iku_1iku_2ik\cdots$ with $even(\alpha_i) = \{1, \ldots, n\} \setminus \{i\}$, $odd(\alpha_i) = \{1, \ldots, n\}$, and therefore $\alpha_i \notin L_n$. For the run σ_i of $\mathcal{A}'$ on α_i we have $|In(\sigma_i)| \geq (n-1)!$. Hence it remains to show $In(\sigma_i) \cap In(\sigma_j) = \emptyset$ for $i \neq j$.

Assume by contradiction that there exist $i \neq j$ with $In(\sigma_i) \cap In(\sigma_j) \neq \emptyset$. Then we can construct a word α with run σ such that $even(\alpha) = even(\alpha_i) \cup even(\alpha_j) = \{1, \ldots, n\}$, $odd(\alpha) = odd(\alpha_i) \cup odd(\alpha_j) = \{1, \ldots, n\}$ and $In(\sigma) = In(\sigma_i) \cup In(\sigma_j)$, by cycling alternatingly through the infinity sets of σ_i and σ_j (as in the proof of Lemma 5). This is a contradiction since in Rabin automata the union of rejecting cycles is rejecting (Proposition 3), but α is in L_n.

To adapt the proof for an alphabet of constant size we can code every letter $i \in \{1, \ldots, n-1\}$ with $a^i b$ and n with $a^n a^* b$. The resulting automaton looks like shown on the right hand side of Figure 2 and still has $\mathcal{O}(n)$ states.

Theorem 7 shows the optimality of the IAR construction for the transformation of deterministic Streett automata into deterministic Rabin automata. The duality of these two types of automata (Prop. 2) gives us an analogue theorem, with the roles of Rabin automata and Streett automata exchanged. Parity automata are special cases of Rabin automata and of Streett automata. Therefore we also get analogue theorems for the transformation of Rabin automata into parity automata and for the transformation of Streett automata into parity automata. Furthermore the property of the example automata to have $\mathcal{O}(n)$ states and $\mathcal{O}(n)$ pairs also gives us analogue theorems, when starting with Muller automata instead of Rabin or Streett automata. Thus, Theorem 7 shows the optimality of all AR constructions listed in Table 1.

A different construction for the conversion between Rabin and Streett automata is given in [6]. It converts a deterministic Streett automaton with n states and r pairs into a deterministic Rabin automaton with $\mathcal{O}(n \cdot r^k)$ states and l pairs, where k is the Streett index of the language and l is the Rabin index of the language. The Rabin (Streett) index of a language is the number of pairs needed in the acceptance condition to describe the language with a deterministic Rabin (Streett) automaton. The languages from the family $(L_n)_{n \geq 2}$ have Rabin and Streett index $\mathcal{O}(n)$ and therefore the complexity of the construction is of order $n^{\mathcal{O}(n)}$ for our example automata. Hence, as a result of this section, the transformation from [6] is also optimal.

At the end of Section 3 we stated a lower bound for the transformation of nondeterministic Büchi automata into deterministic Streett automata. For that aim we show that the languages $(\Sigma_n^\omega \setminus L_n)_{n \geq 2}$ of the present section can be recognized by Büchi automata with $\mathcal{O}(n)$ states. Then we are done because every deterministic Streett automaton recognizing $\Sigma_n^\omega \setminus L_n$ needs at least $n!$ states (Theorem 7 and Prop. 2).

Theorem 8. *There exists a family $(L_n)_{n \geq 2}$ of languages (L_n over an alphabet of n letters) such that for every n the language L_n can be recognized by a non-deterministic Büchi automaton with $\mathcal{O}(n)$ states but can not be recognized by a deterministic Streett automaton with less than $n!$ states.*

Proof. As mentioned above it suffices to show that there is a family $(\mathcal{B}_n)_{n \geq 2}$ of Büchi automata such that $\mathcal{B}_n$ has $\mathcal{O}(n)$ states and recognizes $\Sigma_n^\omega \setminus L_n$. From the characterization of L_n in the proof of Theorem 7 we know that $\alpha \in L_n$ iff $odd(\alpha) \subseteq even(\alpha)$ and therefore $\alpha \notin L_n$ iff there exists an $i \in \{1, \ldots, n\}$ with

$i \in odd(\alpha)$ and $i \notin even(\alpha)$. Intuitively the Büchi automaton guesses the i and then verifies that it appears infinitely often on an odd position and from some point on never on an even position. Formally $\mathcal{B}_n = (Q_n, \Sigma_n, q_0^n, \delta_n, F_n)$ is defined as follows.

- $Q_n = \{q_o, q_e, q_o^1, q_e^1, q_f^1, \ldots q_o^n, q_e^n, q_f^n\}$, $q_0^n = q_o$, and $F_n = \{q_f^1, \ldots, q_f^n\}$.
- For $i \in \Sigma_n$ and $j \in \{1, \ldots, n\}$ let

$$\delta_n(q_o, i) = \{q_e\}, \qquad \delta_n(q_e, i) = \{q_o, q_o^1, \ldots, q_o^n\},$$
$$\delta_n(q_o^j, i) = \begin{cases} \{q_e^j\} & \text{if } i \neq j \\ \emptyset & \text{if } i = j, \end{cases} \quad \delta_n(q_e^j, i) = \begin{cases} \{q_o^j\} & \text{if } i \neq j \\ \{q_f^j\} & \text{if } i = j, \end{cases}$$
$$\delta_n(q_f^j, i) = \begin{cases} \{q_e^j\} & \text{if } i \neq j \\ \emptyset & \text{if } i = j. \end{cases}$$

The automaton is built in such a way that it is in one of the states from $\{q_e, q_e^1, \ldots, q_e^n\}$ iff the last letter was on an even position.

Let $\alpha \in \Sigma_n^\omega \setminus L_n$ and let $j \in odd(\alpha) \setminus even(\alpha)$. A successful run of $\mathcal{B}_n$ stays in the states q_o and q_e up to the point where j does not appear on an even position anymore. Then it moves to q_o^j. Always when a j appears on an odd position in α, the automaton is in q_e^j and then moves to q_f^j. Since there does not appear a j on an even position anymore, the automaton can continue to infinity and accepts α because it visits q_f^j infinitely often. Therefore we have $\Sigma_n^\omega \setminus L_n \subseteq L(\mathcal{B}_n)$.

Now let $\alpha \in L(\mathcal{B}_n)$. There exists a j such that in an accepting run $\mathcal{B}_n$ from some point on only visits states from $\{q_o^j, q_e^j, q_f^j\}$ and infinitely often visits q_f^j. If the last read letter was on an odd position, then $\mathcal{B}_n$ is in q_o^j or in q_f^j and therefore j may only appear on an even position before $\mathcal{B}_n$ moves to the states $\{q_o^j, q_e^j, q_f^j\}$. But since $\mathcal{B}_n$ infinitely often visits q_f^j, there must be a j on an odd position infinitely often and therefore we have $L(\mathcal{B}_n) \subseteq \Sigma_n^\omega \setminus L_n$.

Table 1. Synopsis of automaton transformations and pointers to optimality results. (The transformation $\star$ is the only one that is not known to be optimal.)

To From	Muller det.	Rabin det.	Streett det.	Parity det.
Büchi ndet.	Safra trees	Safra trees	1.Safra trees	1.Safra trees
			2.IAR	2.IAR
	$\star$	Thm 6	Thm 8	Thm 6
Muller det.		SAR	SAR	SAR
		Thm 7	Thm 7	Thm 7
Rabin det.	trivial		IAR	IAR
			Thm 7	Thm 7
Streett det.	trivial	IAR		IAR
		Thm 7		Thm 7

5   Conclusion

For several different transformations of ω-automata we have seen that the lower bound is $2^{\Omega(n \cdot \log n)}$. The two basic constructions considered in this paper (Safra trees and appearance records) meet these lower bounds and therefore are of optimal complexity. In comparison to the theory of $*$-automata, where determinization is exponential too, but with a linear exponent, we get an additional factor of $\log n$ in the exponent for transformations of ω-automata.

An unsolved problem is the lower bound for the transformation of nondeterministic Büchi automata into deterministic Muller automata. The known lower bound is $2^{\Omega(n)}$, which can be proven by a simple pumping argument as for $*$-automata.

References

1. J.R. Büchi. On a decision method in restricted second order arithmetic. In *Proc. International Congress on Logic, Method and Philos. Sci. 1960*, pages 1–11, 1962.
2. J.R. Büchi. Winning state-strategies for boolean-F_σ games. manuscript, 1981.
3. J.R. Büchi. State-strategies for games in $F_{\sigma\delta} \cap G_{\delta\sigma}$. *Journal of Symbolic Logic*, 48(4):1171–1198, December 1983.
4. S. Dziembowski, M. Jurdziński, and I. Walukiewicz. How much memory is needed to win infinite games? In *Proc. IEEE, LICS*, 1997.
5. Y. Gurevich and L. Harrington. Trees, automata and games. In *Proc. 14th ACM Symp. on the Theory of Computing*, pages 60–65, 1982.
6. S. Krishnan, A. Puri, and R. Brayton. Structural complexity of ω-automata. In *12th Annual Symposium on Theoretical Aspects of Computer Science*, volume 900 of *Lecture Notes in Computer Science*, pages 143–156. Springer, 1995.
7. L.H. Landweber. Decision problems for ω-automata. *Math. System Theory*, 3:376–384, 1969.
8. C. Löding. Methods for the transformation of ω-automata: Complexity and connection to second order logic. Master's thesis, Christian-Albrechts-University of Kiel, 1998.
9. R. McNaughton. Testing and generating infinite sequences by a finite automaton. *Information and Control*, 9:521–530, 1966.
10. M. Michel. Complementation is much more difficult with automata on infinite words. Manuscript,CNET,Paris, 1988.
11. A.W. Mostowski. Regular expressions for infinite trees and a standard form of automata. *Lecture Notes in Computer Science*, 208:157–168, 1984.
12. D.E. Muller. Infinite sequences and finite machines. In *Proc. 4th IEEE Symposium on Switching Circuit Theory and Logical design*, pages 3–16, 1963.
13. M.O. Rabin. Decidability of second order theories and automata on infinite trees. *Transaction of the AMS*, 141:1–35, 1969.
14. S. Safra. On the complexity of ω-automata. In *Proc. 29th IEEE Symp. on Foundations of Computer Science*, pages 319–327, 1988.
15. S. Safra. Exponential determinization for ω-automata with strong-fairness acceptance condition. In *Proc. 24th ACM Symp. on the Theory of Computing*, pages 275–282, 1992.

16. S. Safra and M. Y. Vardi. On ω-automata and temporal logic. In *Proc. 21th ACM Symp. on the Theory of Computing*, 1989.
17. R.S. Streett. Propositional dynamic logic of looping and converse is elementary decidable. *Information and Control*, 54:121–141, 1982.
18. W. Thomas. Languages, automata, and logic. In G. Rozenberg and A. Salomaa, editors, *Handbook of Formal Language Theory*, volume III, pages 385–455. Springer-Verlag, 1997.

CTL$^+$ Is Exponentially More Succinct than CTL

Thomas Wilke

Lehrstuhl für Informatik VII,
RWTH Aachen, 52056 Aachen, Germany
wilke@informatik.rwth-aachen.de

Abstract. It is proved that CTL$^+$ is exponentially more succinct than CTL. More precisely, it is shown that every CTL formula (and every modal μ-calculus formula) equivalent to the CTL$^+$ formula

$$E(Fp_0 \wedge \cdots \wedge Fp_{n-1})$$

is of length at least $\binom{n}{\lceil n/2 \rceil}$, which is $\Omega(2^n/\sqrt{n})$. This matches almost the upper bound provided by Emerson and Halpern, which says that for every CTL$^+$ formula of length n there exists an equivalent CTL formula of length at most $2^{n \log n}$.

It follows that the exponential blow-up as incurred in known conversions of nondeterministic Büchi word automata into alternation-free μ-calculus formulas is unavoidable. This answers a question posed by Kupferman and Vardi.

The proof of the above lower bound exploits the fact that for every CTL (μ-calculus) formula there exists an equivalent alternating tree automaton of linear size. The core of this proof is an involved cut-and-paste argument for alternating tree automata.

1 Introduction

Expressiveness and *succinctness* are two important aspects to consider when one investigates a (specification) logic. When studying the expressiveness of a logic one is interested in characterizing *what* properties can be expressed, whereas when studying the succinctness one is interested in *how short* a formula can be found to express a given property. Succinctness is especially of importance in a situation where one has characterized the expressive power of a logic by a different but equally expressive logic. In such a situation, succinctness is the foremost quantitative measure to distinguish the logics. For instance, linear-time temporal logic (LTL) is known to be exactly as expressive as first-order logic (FO), [9], but FO is much more succinct than LTL: from work by Stockmeyer's, [11], it follows that there exists a sequence of FO formulas of linear length such that the length of shortest equivalent LTL formulas cannot be bounded by an elementary recursive function.

In this paper, the succinctness of computation tree logic (CTL) is compared to the succinctness of CTL$^+$, an extension of CTL, which is known to have exactly the same expressive power as CTL, [4,5]. I present a sequence of CTL$^+$

C. Pandu Rangan, V. Raman, R. Ramanujam (Eds.): FSTTCS'99, LNCS 1738, pp. 110–121, 1999.

formulas of length $\mathcal{O}(n)$ such that the length of shortest equivalent CTL formulas is $\Omega(2^n/\sqrt{n})$. More precisely, I prove that every CTL formula equivalent to the CTL$^+$ formula

$$E(Fp_0 \wedge \cdots \wedge Fp_{n-1})$$

is of length at least $\binom{n}{\lceil n/2 \rceil}$, which shows that CTL$^+$ is exponentially more succinct than CTL. This lower bound is almost tight, because a result by Emerson and Halpern's, [4,5], says that for every CTL$^+$ formula of length n there exists an equivalent CTL formula of length at most $2^{n \log n}$.

It is important to note that this exponential lower bound is not based on any complexity-theoretic assumption, and it does not follow from the fact that model checking for CTL is known to be P-complete whereas model checking for CTL$^+$ is NP- and co-NP-hard (and in Δ_2^p), [3,4,5].

The proof of the lower bound presented in this paper makes use of automata-theoretic arguments, following other approaches to similar questions. The main idea is based on the following fact. For every CTL formula (and for every μ-calculus formula) φ there exists an alternating tree automaton A_φ of size linear in the length of φ that accepts exactly the models of φ, [6,1,2]. So in order to obtain a lower bound on the length of the CTL (or μ-calculus) formulas defining a given class of Kripke structures,[1] it is enough to establish a lower bound on the number of states of the alternating tree automata recognizing the given class of structures.

As mentioned above, automata-theoretic arguments have been used in this way in different places, for instance by Etessami, Vardi, and myself in [8] or Kupferman and Vardi in [10]. The difference, however, is that in this paper the automaton model (alternating automata on trees) is rather intricate compared to the automaton models used in [8] and [10] (nondeterministic automata on words and nondeterministic automata on trees, respectively).

The more elaborate argument that is needed here also answers a question raised in the paper by Kupferman and Vardi. A particular problem they consider is constructing for a given nondeterministic Büchi word automaton an alternation free μ-calculus (AFMC) formula that denotes in every Kripke structure the set of all worlds where all infinite paths originating in this world are accepted by the automaton. They show that if such a formula exists, then there is a formula of size at most exponential in the number of states of the given Büchi automaton, but they cannot give a matching lower bound. This is provided in this paper.

Outline. In Section 2, the syntax and semantics of CTL and CTL$^+$ are briefly reviewed and the main result of the paper is presented. In Section 3, alternating tree automata are briefly reviewed and subsequently, in Section 4, the succinctness problem is reduced to an automata-theoretic problem. Section 5 describes

---

[1] Strictly speaking, a CTL formula defines a class of pointed Kripke structures, see Section 2.

the latter in a more general setting and in Section 6 a sketch is given of the solution of this more general problem. Section 7 presents consequences, and Section 8 gives a conclusion.

This paper is an extended abstract; for details of the proofs, see the technical report [12].

Acknowledgment. I would like to thank Kousha Etessami, Martin Grohe, Neil Immerman, Christof Löding, Philippe Schnoebelen, and Moshe Y. Vardi for having discussed with me the problem addressed in this paper.

Trees and tree arithmetic. In this paper, a tree is a triple (V, E, λ) where (V, E) is a directed tree in the graph-theoretic sense and λ is a labeling function with domain V. By convention, when T denotes a tree, then V, E, and λ always denote the set of nodes, set of edges, and labeling function of T. The same applies to decorations such as T', T^*, T_i, etc.

Let T be an arbitrary tree. A node $v' \in V$ is a *successor* of a node $v \in V$ in T if $(v, v') \in E$. The set of all successors of a node v in T is denoted by $Scs(T, v)$. The set of *leaves* of a tree T, that is, the set of nodes without successors, is denoted by $Lvs(T)$. The set of *inner nodes* is denoted by $In(T)$.

Given a tree T and a vertex v of T, the *ancestors path*, denoted $T{\uparrow}v$, is the unique path from the root of T to v (inclusively). The *descendants tree*, denoted $T{\downarrow}v$, is the subgraph of T induced by all nodes reachable from v (v itself included).

I will use two kinds of concatenations for trees. When T and T' are trees and v is a node of T, then $T \cdot (v, T')$ denotes the tree that results from T by first making an isomorphic copy of T' whose node set is disjoint from the set of nodes of T and then adding an edge from v to the root of T'. Similarly, $T \odot (v, T')$ denotes the tree that results from T by first making an isomorphic copy of T' whose node set is disjoint from the set of nodes of T and then identifying the root of the isomorphic copy of T' with v. By convention, the node v is retained in the resulting tree (rather than the root of T') and the label of v is kept.— These two concatenation operations are extended in a straightforward way: when T is a tree and M a set of pairs (v, T'), with $v \in V$ and T' an arbitrary tree, I might write $T \cdot M$ and $T \odot M$ to denote the result of concatenating (in the respective way) all trees from M to T.

For ease in notation, when π is a finite path (a finite tree with exactly one leaf) with leaf v and T is a tree, I simply write $\pi \cdot T$ for the tree $\pi \cdot (v, T)$ as defined above. To make things even simpler, I view strings as finite paths and vice versa. So when u is a string and T a tree, I might write $u \cdot T$ to denote the tree which is obtained by viewing u as a path and concatenating T to it.

2   CTL, CTL⁺, and Main Result

I start with recalling the syntax and the semantics of CTL and CTL⁺. For technical reasons, I only define formulas in positive normal form. This is not an essential restriction, because every CTL formula is equivalent to a CTL formula in positive normal form of the same length, and the same applies to CTL$^+$.

Let Prop $= \{p_0, p_1, p_2, \dots\}$ be an infinite supply of distinct propositional variables. The set of all *CTL$^+$ formulas* and the set of all *path formulas* are defined simultaneously as follows.

1. 0 and 1 are CTL$^+$ formulas.
2. For $p \in$ Prop, p and $\neg p$ are CTL$^+$ formulas.
3. If φ and ψ are CTL$^+$ formulas, then so are $\varphi \vee \psi$ and $\varphi \wedge \psi$.
4. Every CTL$^+$ formula is a path formula.
5. If φ and ψ are CTL$^+$ formulas, then $\mathsf{X}\varphi$, $\mathsf{U}(\varphi, \psi)$, and $\mathsf{R}(\varphi, \psi)$ are path formulas.
6. If φ and ψ are path formulas, then so are $\varphi \vee \psi$ and $\varphi \wedge \psi$.
7. If φ is a path formula, then $\mathsf{E}\varphi$ and $\mathsf{A}\varphi$ are CTL$^+$ formulas.

A CTL$^+$ formula is a *CTL formula* when it can be constructed without using rule 6. That is, in CTL formulas every path quantifier (E or A) is followed immediately by a temporal modality (X, U, or R). As usual, I use $\mathsf{F}\varphi$ (eventually φ) as an abbreviation for $\mathsf{U}(1, \varphi)$.

CTL and CTL$^+$ formulas are interpreted in Kripke structures, which are directed graphs with specific labeling functions for their nodes. Formally, a *Kripke structure* is a tuple (W, R, α) where W is a set of *worlds*, $R \subseteq W \times W$ is an *accessibility relation*, and $\alpha \colon W \to 2^{\mathrm{Prop}}$ is a *labeling function*, which assigns to each world the set of propositional variables that hold true in it. By convention, Kripke structures are denoted by $\boldsymbol{K}$ or decorated versions of $\boldsymbol{K}$ such as $\boldsymbol{K}'$ or $\boldsymbol{K}^*$, and their components are referred to as W, R, and α, respectively decorated versions of these letters.

Given a world w of a Kripke structure $\boldsymbol{K}$ as above, a world w' is called a *successor* of w in $\boldsymbol{K}$ if $(w, w') \in R$. Just as with trees, the set of all successors of a world w is denoted by $Scs(\boldsymbol{K}, w)$. A *path* through a Kripke structure $\boldsymbol{K}$ as above is a nonempty sequence $w_0, w_1, \dots$ such that $(w_0, w_1) \in R$, $(w_1, w_2) \in R$, $\dots$ A *maximal path* is a path that is either infinite or finite and ends in a world without successors.

A *pointed Kripke structure* is a pair $(\boldsymbol{K}, w)$ of a Kripke structure and a *distinguished* world of it. A *path* through a pointed Kripke structure $(\boldsymbol{K}, w)$ is a path through $\boldsymbol{K}$ starting in w. A *path-equipped Kripke structure* is a pair $(\boldsymbol{K}, \pi)$ of a Kripke structure and a maximal path through it.

For every CTL$^+$ and path formula φ, one defines in a straightforward way what it means for φ to hold in a pointed Kripke structure $(\boldsymbol{K}, w)$ respectively path-equipped Kripke structure $(\boldsymbol{K}, \pi)$ and denotes this by $(\boldsymbol{K}, w) \models \varphi$ respectively $(\boldsymbol{K}, \pi) \models \varphi$. For instance, when φ is a path formula, then $(\boldsymbol{K}, w) \models \mathsf{E}\varphi$ if there exists a maximal path π through $(\boldsymbol{K}, w)$ such that $(\boldsymbol{K}, \pi) \models \varphi$. For details, the reader is referred to [5].

Given a CTL$^+$ formula φ, I write $\mathrm{Mod}(\varphi)$ for the class of all pointed Kripke structures that are *models* of φ, i.e., $\mathrm{Mod}(\varphi) = \{(\boldsymbol{K}, w) \mid (\boldsymbol{K}, w) \models \varphi\}$. CTL$^+$ formulas φ and ψ are *equivalent* if they have the same models, i.e., if $\mathrm{Mod}(\varphi) = \mathrm{Mod}(\psi)$.

The main result of this paper is:

Theorem 1. *Every CTL formula equivalent to the CTL$^+$ formula φ_n defined by*

$$\varphi_n = E(Fp_0 \wedge \cdots \wedge Fp_{n-1}) \tag{1}$$

has length at least $\binom{n}{\lceil n/2 \rceil}$, which is $\Omega(2^n/\sqrt{n})$.

In other words, CTL$^+$ is exponentially more succinct than CTL.

Note that it is easy to come up with a formula of length $\mathcal{O}(n!)$ equivalent to φ_n, namely as a disjunction with $n!$ many disjuncts, each taking care of one possible order in which the p_i's may occur on a path.

3   Alternating Tree Automata

As indicated in the abstract and the introduction, I use an automata-theoretic argument to prove Theorem 1. In this section, the respective automaton model, which differs from other models used in the literature, is introduced.

First, it can handle trees with arbitrary degree of branching in a simple way. Second, the class of objects accepted by an automaton as defined here is a class of pointed Kripke structures rather than just a set of trees. Both facts make it much easier to phrase theorems such as Theorem 2 below and also simplify the presentation of a combinatorial (lower-bound) argument like the one given in Section 6.

An *alternating tree automaton (ATA)* is a tuple $A = (Q, P, q_I, \delta, \Omega)$ where Q is a finite set of *states*, P is a finite subset of Prop, $q_I \in Q$ is an *initial state*, δ is a *transition function* as specified below, and Ω is an *acceptance condition* for ω-automata such as a Büchi or Muller condition. The same notational conventions as with Kripke structures apply.

The transition function δ is a function $Q \times 2^P \to \mathrm{TC}(Q)$, where $\mathrm{TC}(Q)$ is the set of *transition conditions* over Q, which are defined by the following rules.

1. 0 and 1 are transition conditions over Q.
2. For every $q \in Q$, q is a transition condition over Q.
3. For every $q \in Q$, $\Box q$ and $\Diamond q$ are transition conditions over Q.
4. If φ and ψ are transition conditions over Q, then $\varphi \wedge \psi$ and $\varphi \vee \psi$ are transition conditions over Q.

A transition condition is said to be ϵ-*free* if rule 2 is not needed to build it. An ATA is ϵ-free if every condition $\delta(q, a)$ for $q \in Q$ and $a \in 2^P$ is ϵ-free; it is in *normal form* if it is ϵ-free, the conditions $\delta(q, a)$ are in disjunctive normal form, and neither 0 nor 1 occur in these conditions.

ATA's work on pointed Kripke structures. Their computational behavior is explained using the notion of a run. Assume A is an ATA as above and (K, w_I) a pointed Kripke structure as above. A *run* of A on (K, w_I) is a $(W \times Q)$-labeled tree $R = (V, E, \lambda)$ satisfying the conditions described further below. To explain these conditions, some more definitions are needed. For simplicity in notation, I will write $w_R(v)$ and $q_R(v)$ for the first and second component of $\lambda(v)$, respectively.

For every node v of R, I define what it means for a transition condition τ over Q to hold in v, denoted $K, R, v \models \tau$. This definition is by induction on the structure of τ, where the boolean constants 0 and 1 and the boolean connectives are dealt with in the usual way; besides:

— $K, R, v \models q$ if there exists $v' \in Scs(R, v)$ such that $\lambda(v') = (w_R(v), q)$,
— $K, R, v \models \Diamond q$ if there exists $v' \in Scs(R, v)$ and $w \in Scs(K, w_R(v))$ such that $\lambda(v') = (w, q)$, and
— $K, R, v \models \Box q$ if for every $w \in Scs(K, w_R(v))$ there exists $v' \in Scs(R, v)$ such that $\lambda(v') = (w, q)$.

The two additional conditions that are required of a run are the following.

1. *Initial condition.* Let v_0 be the root of (V, E). Then $\lambda(v_0) = (w_I, q_I)$.
2. *Local consistency.* For every $v \in V$,

$$K, R, v \models \tau_v \tag{2}$$

where

$$\tau_v = \delta(q_R(v), \alpha(w_R(v)) \cap P) \ . \tag{3}$$

Note that the intersection with P allows us to deal easily with the fact that in the definition of Kripke structure an infinite number of propositional variables is always present.

A run R is said to be *accepting* if the state labeling of every infinite path through R satisfies the given *acceptance condition* Ω. For instance, if $\Omega \subseteq 2^Q$ is a Muller condition, then every infinite path $v_0, v_1, \ldots$ through R must have the property that the set formed by the states occurring infinitely often in $q_R(v_0), q_R(v_1), \ldots$ is a member of Ω.

A pointed Kripke structure is *accepted* by A if there exists an accepting run of A on the Kripke structure. The class of pointed Kripke structures accepted by A is denoted by $\mathcal{K}(A)$; it is said A *recognizes* $\mathcal{K}(A)$.

Throughout this paper, the same notational conventions as with Kripke structures and alternating tree automata apply to runs.

4   Reduction to Automata-Theoretic Problem

In order to reduce the lower bound claim for the translation from CTL$^+$ to CTL to a claim on alternating tree automata, I describe the models of a CTL formula by an alternating tree automaton, following the ideas of Kupferman, Vardi, and Wolper, [2], but using the more general model of automaton.

Let φ be an arbitrary CTL formula and P the set of propositional variables occurring in φ. The ATA A_φ is defined by $A_\varphi = (Q, P, \varphi, \delta, \Omega)$ where Q is the set of all CTL subformulas of φ including φ itself, Ω is the Muller acceptance condition that contains all sets of subformulas of φ that do not contain formulas starting with EU or AU, and δ is defined by induction, where, for instance, the inductive step for EU is given by

$$\delta(\mathsf{EU}(\psi, \chi), a) = \chi \vee (\psi \wedge \Diamond\mathsf{EU}(\psi, \chi)) \ . \tag{4}$$

The other cases are similar and follow the ideas of [2]. Note that on the right-hand side of (4) the boolean connectives $\vee$ and $\wedge$ are part of the syntax of transition conditions.

Similar to [2], one can prove by a straightforward induction on the structure of φ:

Theorem 2. *Let φ be an arbitrary CTL formula of length l. Then $\mathbf{A}_\varphi$ is an ATA with at most l states such that $Mod(\varphi) = \mathcal{K}(\mathbf{A}_\varphi)$.*

It is quite easy to see that for every ATA there exists an equivalent ATA in normal form with the same number of states. So in order to prove Theorem 1 we only need to show:

Theorem 3. *Every ATA in normal form recognizing $Mod(\varphi_n)$ has at least $\binom{n}{\lceil n/2 \rceil}$ states.*

5   The General Setting

The method I use to prove Theorem 3 (a cut-and-paste argument) does not only apply to the specific properties defined by the φ_n's but to a large class of "path properties." As with many other situations, the method is best understood when presented in its full generality. In this section, I explain the general setting and present the extended version of Theorem 3, namely Theorem 4.

In the following, *word* stands for nonempty string or ω-word. The set of all words over a given alphabet A is denoted by A^∞. A *language* is a subset of $(2^P)^\infty$ where P is some finite subset of Prop. Given a language L over some alphabet 2^P, $\mathsf{E}L$ denotes the class of pointed Kripke structures $(\mathbf{K}, w)$ where there exists a maximal path through $(\mathbf{K}, w)$ whose labeling (restricted to the propositional variables in P) is an element of L. (Remember that a path through a pointed Kripke structure always starts in its distinguished world.)

Observe that for every n, we clearly have $Mod(\varphi_n) = \mathsf{E}L_n$ where

$$L_n = \{a_0 a_1 \cdots \in (2^{P_n})^\infty \mid \forall i(i < n \rightarrow \exists j(p_i \in a_j))\}$$

and $P_n = \{p_0, \ldots, p_{n-1}\}$.

Let L be a regular language. We say a family $\{(u_i, u_i')\}_{i<m}$ is a *discriminating family* for L if $u_i u_i' \in L$ and $u_i u_j' \notin L$ for all $i < m$ and all $j < m$ with $j \neq i$. Obviously, the number of classes of the Nerode congruence[2] associated with L is an upper bound for m. The maximum number such that there exists a discriminating family of that size for L is denoted $\iota(L)$.

The generalized version of Theorem 3 now reads:

---

[2] The Nerode congruence of a language L is the congruence that considers strings u and v equivalent if for every word x (including the empty word), $ux \in L$ iff $vx \in L$.

Theorem 4. *Let L be a regular language. Then every ATA recognizing EL has at least $\iota(L)$ states.*

Before we turn to the proof of this theorem in the next section, let's apply it to the languages L_n (as defined above) to obtain the desired lower bounds.

Fix an arbitrary positive natural number $n > 1$ and let $m = \lceil n/2 \rceil$ and $t = \binom{n}{m}$. Write N for the set $\{0, \ldots, n-1\}$ and $^-$ for set-theoretic complementation with respect to N. For every $M \subseteq N$, let $u(M)$ be a string over 2^{P_n} of length $|M|$ such that for every $p_i \in M$, the letter $\{p_i\}$ occurs in $u(M)$. (In other words, $u(M)$ should be a sequence of singletons where for each $i \in M$ the singleton $\{p_i\}$ occurs exactly once and no other singleton occurs.) Let $M_0, \ldots, M_{t-1}$ be an enumeration of all m-subsets of N and let $u_i = u(M_i)$ and $u_i' = u(\bar{M}_i)$. Then $\{(u_i, u_i')\}_{i < t}$ is a discriminating family for L_n, which means $\iota(L_n) \geq \binom{n}{\lceil n/2 \rceil}$.

This together with Theorem 4 implies Theorem 3 and thus also Theorem 1. (Observe that for $n = 1$ the claims of Theorems 3 and 1 are trivial.)

6   Saturation

In this section, I will introduce the key concepts used in the proof Theorem 4, state the main lemmas, provide as much intuition as is possible within the page limit, and give a rough outline of the proof of Theorem 4.

We will see trees in two different roles. On the one hand, we will look at runs of ATA's, and runs of ATA's are trees by definition. On the other hand, we will consider Kripke structures that are trees. In order to not get confused, I will strictly follow the notational conventions introduced earlier, for instance, that the labeling function of a run R' is referred to by λ'. As we will only work with Kripke structures that are trees, I will use the term *Kripke tree*. A Kripke tree will also be viewed as a pointed Kripke structure where the root of the tree is the distinguished node.

For the rest of this section, fix a language L over some alphabet 2^P, and an ATA A. For each state q, write A_q for the ATA that results from A by changing its initial state to q and $\mathcal{K}_q$ for the class $\mathcal{K}(A_q)$, the class of pointed Kripke structures recognized by A_q.

Let u be a string. A state q is *preventable* for u if there exists a Kripke tree K such that $u \cdot K \notin EL$ and $K \notin \mathcal{K}_q$. We write $pvt(u)$ for set of all states preventable for u, and for every $q \in pvt(u)$, we pick, once and for all, a Kripke tree K as above and denote it by K_q^u. The important observation here is that if K is a Kripke tree, $w \in W$, and $q \in pvt(K{\uparrow}w)$, then K' defined by $K' = K \cdot (w, K_q^u)$ with $u = K{\uparrow}w$ has the following two properties. First, if $K \notin EL$, then $K' \notin EL$. Second, there is no run R of A on K' that has a node v with $w_R(v) = w$ and $K', R, v \models \Box q$. In a certain sense, by adding K_u^q to K, the condition $\Box q$ is "prevented" from being used at w.

A state q is *always successful* for u if there exists a state $q' \in pvt(u)$ such that $K_{q'}^u \in \mathcal{K}_q$. We write $scf(u)$ for the set of states always successful for u, and for every $q \in scf(u)$, we pick, once and for all, a state q' as above and denote it by q^u. (Note that whether or not a state is always successful for a string depends on the particular choices for the K_q^u's.) The important observation here is the following. Choose K, w, u, and K' as in the previous paragraph. If $q \in scf(u)$ and if we want to construct a run R of A on K', then we can always make sure that $K, R, v \models \Diamond q$ holds for a node v with $w_R(v) = w$, because we only need to add to R a successful run of A on $K_{q'}^u$ with $q' = q^u$. Formally, if $R_{q'}^u$ is such a run, we only need to consider $R \cdot (v, R_{q'}^u)$ instead of R.

A world w of a Kripke tree K is said to be *saturated* if for every $q \in pvt(K{\uparrow}w)$, there exists $w' \in Scs(K, w)$ such that $K{\downarrow}w'$ is isomorphic to K_q^u with $u = K{\uparrow}w$.

Let K be an arbitrary Kripke tree. The Kripke tree K^s is defined by

$$K^s = K \cdot \{(w, K_q^u) \mid w \in In(K),\ u = K{\uparrow}w,\ \text{and}\ q \in pvt(u)\}\ , \tag{5}$$

that is, in K^s, every inner world from K is saturated.

Remark 1. Let K be an arbitrary Kripke tree. If $K \in \mathsf{EL}$, then $K^s \in \mathsf{EL}$.

This is because every maximal path through K is also present in K^s; no successors are added to leaves.

Let τ be an arbitrary transition condition over Q and $X, Y \subseteq Q$. The *X-Y-reduct* of τ, denoted $\tau^{X,Y}$, is obtained from τ by replacing

— every atomic subformula of the form $\Box q$ with $q \in X$ by 0,
— every atomic subformula of the form $\Box q$ with $q \in Q \setminus X$ by 1, and
— every atomic subformula of the form $\Diamond q$ with $q \in Y$ by 1.

Let K be an arbitrary Kripke tree. A *partial run* of A on K is defined just as an ordinary accepting run with the following modification of local consistency as defined in (2). For every $v \in V$ such that $w_R(v) \in In(K)$, it is required that

$$K, R, v \models \tau_v^{X_v, Y_v} \tag{6}$$

holds where τ_v is as defined in (3) and

$$X_v = pvt(K{\uparrow}w_R(v))\ , \qquad\qquad Y_v = scf(K{\uparrow}w_R(v))\ .$$

Note that in general neither τ implies $\tau^{X,Y}$ nor $\tau^{X,Y}$ implies τ. So there is no a priori relation between the existence of runs and partial runs. But using Remark 1 and the right notion of restriction of a run one can prove the following.

Lemma 1. *Let K be an arbitrary Kripke tree. Assume $\mathcal{K}(A) = \mathsf{EL}$ and $K \in \mathcal{K}(A)$. Then there exists a partial run of A on K.*

Let R be a partial run of A on a Kripke tree K. The run R is *distributed* if for every $w \in W$ there exists at most one $v \in V$ with $w_R(v) = w$.

The set of all *frontier pairs* of R, denoted by $FrtPrs(R)$, is defined by $FrtPrs(R) = \{\lambda(v) \mid v \in V \text{ and } w_R(v) \in Lvs(K)\}$. Similarly, the set of all *frontier states* of R, denoted $FrtSts(R)$, is defined by $FrtSts(R) = \{q_R(v) \mid v \in V \text{ and } w_R(v) \in Lvs(K)\}$.

The crucial lemma connecting Kripke trees with saturated inner worlds and partial runs is as follows.

Lemma 2. *Let K be a Kripke tree and R a distributed partial run of A on K. Assume that for every $q \in FrtSts(R)$ there exists a Kripke tree $K_q \in \mathcal{K}_q$ such that the tree K^* defined by*

$$K^* = K \odot \{(w, K_q) \mid q \in FrtSts(R)\}$$

does not belong to EL.
Then there exists an accepting run of A on the Kripke tree K^{**} defined by

$$K^{**} = K^s \odot \{(w, K_q) \mid (w, q) \in FrtPrs(R)\} \ ,$$

which does not belong to EL.

Note that because R is supposed to be distributed, the trees K^* and K^{**} are obtained from K and K^s, respectively, by adding to each leaf at most one of the trees K_q.

The proof of this lemma is technically involved and makes extensive use of the aforementioned properties of preventable and always successful states.

I will conclude this section with a rough sketch of the proof of Theorem 4.

Sketch of the Proof of Theorem 4. Let $\{(u_i, u'_i)\}_{i < m}$ be a discriminating family for L of size $\iota(L)$ and A an ATA with $\mathcal{K}(A) = \mathsf{E}L$. I claim that for every $i < m$, there exists a state q such that $u'_i \in \mathcal{K}_q$, but $u'_j \notin \mathcal{K}_q$ for $j < m$ and $j \neq i$. This clearly implies the claim of the theorem.

By way of contradiction, assume this is not the case. Then there exists $i < m$ such that for every $q \in Q$ with $u'_i \in \mathcal{K}_q$ there exists $j \neq i$ such that $u'_j \in \mathcal{K}_q$. For every such q let j_q be an appropriate index j.

Let K be a $|Q|$-branching Kripke tree[3] such that every maximal path starting with the root is labeled $u_i a_i$ where a_i is the first letter of u'_i. Consider the Kripke tree K' defined by $K' = K \odot \{(w, u'_i) \mid w \in Lvs(K)\}$.

Clearly, $K' \in \mathsf{E}L$ (because every maximal path through K' is labeled $u_i u'_i$). Thus, by Lemma 1, there exists a partial run of A on K'. By restricting this run to the worlds in K, we obtain a partial run of A on K. This run has the obvious property that for every $q \in FrtSts(R)$ there exists an accepting run of A_q on u'_{j_q}. By manipulating this run adequately, using the fact that K is $|Q|$-branching, one can transform it into a distributed partial run with the same property. This run together with the u'_{j_q}'s replacing the K_q's satisfies the assumptions of Lemma 2. We can thus conclude the Kripke tree K^{**} as defined in Lemma 2, which does not belong to $\mathsf{E}L$, is accepted by A—a contradiction.

7 Connection with Büchi Automata and μ-Calculus

One can show that Theorem 2 also holds for the modal μ-calculus (see, for instance, [2]). So we also obtain: every modal μ-calculus formula equivalent to

---

[3] A Kripke tree K is *m-branching* if for every world $w \in W$ the following is true. For every successor w_0 of w there exist at least $m - 1$ other successors $w_1, \ldots, w_{m-1}$ of w such that all subtrees $K{\downarrow}w_0, \ldots, K{\downarrow}w_{m-1}$ are isomorphic.

the CTL$^+$ formula φ_n has length at least $\binom{n}{\lceil n/2 \rceil}$. This is interesting because of the following.

As the modal μ-calculus is closed under syntactic negation, the above also says that every modal μ-calculus formula equivalent to the CTL$^+$ formula

$$A(G\neg p_0 \vee \cdots \vee G\neg p_{n-1})$$

has length at least $\binom{n}{\lceil n/2 \rceil}$. And, clearly, this property can easily be expressed by an alternation-free μ-calculus (AFMC) formula (according to the definition of alternation-freeness as introduced by Emerson and Lei in [7]), because it can be expressed in CTL. On the other hand, the set of all ω-words over 2^{P_n} satisfying the linear-time temporal property $G\neg p_0 \vee \cdots \vee G\neg p_{n-1}$ is recognized by a nondeterministic Büchi word automaton (NBW) with $n + 1$ states. We therefore have:

Corollary 1. *There is an exponential lower bound for the translation NBW $\mapsto$ AFMC in the sense of [10].*

This answers a question left open by Kupferman and Vardi in [10].

8   Conclusion

We have seen that there is an exponential gap between the succinctness of CTL$^+$ and CTL, as well as an exponential gap between nondeterministic Büchi word automata and alternation-free μ-calculus. Just as in many other situations, the automata-theoretic approach to understanding the expressive power of (specification) logics has proved to be useful.

References

1. O. Bernholtz [Kupferman] and O. Grumberg. Branching temporal logic and amorphous tree automata. In E. Best, ed., *CONCUR'93*, vol. 715 of *LNCS*, 262–277.
2. O. Bernholtz [Kupferman], M. Y. Vardi, and P. Wolper. An automata-theoretic approach to branching-time model checking. In D. L. Dill, ed., *CAV '94*, vol. 818 of *LNCS*, 142–155.
3. E. M. Clarke, E. A. Emerson, and A. P. Sistla. Automatic verification of finite-state concurrent systems using temporal logic specifications: A practical approach. In *PoPL '83*, 117–126.
4. E. A. Emerson and J. Y. Halpern. Decision procedures and expressiveness in the temporal logic of branching time. In *STOC '82*, 169–181.
5. E. A. Emerson and J. Y. Halpern. Decision procedures and expressiveness in the temporal logic of branching time. *J. Comput. System Sci.*, 30(1):1–24, 1985.
6. E. A. Emerson, C. S. Jutla, and A. P. Sistla. On model-checking for fragments of μ-calculus. In C. Courcoubetis, ed., *CAV '93*, vol. 697 of *LNCS*, 385–396.
7. E. A. Emerson and C.-L. Lei. Efficient model checking in fragments of the propositional mu-calculus (extended abstract). In *LICS '86*, 267–278.
8. K. Etessami, M. Y. Vardi, and Th. Wilke. First-order logic with two variables and unary temporal logic. In *LICS '97*, 228–235.

9. J. A. W. Kamp. *Tense Logic and the Theory of Linear Order.* PhD thesis, University of California, Los Angeles, Calif., 1968.

10. O. Kupferman and M. Y. Vardi. Freedom, weakness, and determinism: From linear-time to branching-time. In *LICS '98*, 81–92.

11. L. J. Stockmeyer. *The Complexity of Decision Problems in Automata Theory and Logic.* PhD thesis, Dept. of Electrical Engineering, MIT, Boston, Mass., 1974.

12. Th. Wilke. *CTL$^+$ is exponentially more succinct than CTL.* Technical Report 99-7, RWTH Aachen, Fachgruppe Informatik, 1999. Available online via `ftp://ftp.informatik.rwth-aachen.de/pub/reports/1999/index.html`.

A Top-Down Look at a Secure Message

Martín Abadi[1], Cédric Fournet[2], and Georges Gonthier[3]*

[1] Bell Labs Research, Lucent Technologies
[2] Microsoft Research
[3] INRIA Rocquencourt

Abstract. In ongoing work, we are investigating the design of secure distributed implementations of high-level process calculi (in particular, of the join-calculus). We formulate implementations as translations to lower-level languages with cryptographic primitives. Cryptographic protocols are essential components of those translations. In this paper we discuss basic cryptographic protocols for transmitting a single datum from one site to another. We explain some sufficient correctness conditions for these protocols. As an example, we present a simple protocol and a proof of its correctness.

1 Introduction

In the last few years, the scope of security protocols has grown, and so has their complexity. In addition to basic functions such as authentication and key establishment, recent protocols sometimes support elaborate transactions. They may comprise preliminary negotiations, where the parties discuss their preferences and expectations, and layers for application records and for error messages (e.g., [15,17]). Correspondingly, research on the analysis of security protocols has started to address the challenges of those sophisticated protocols. Examples of this line of work include the recent analyses of the SSL, TLS, IKE, and SET protocols (e.g., [32,29,28,23,22,8]).

These trends notwithstanding, in this paper we consider only basic protocols with an elementary goal. This goal is to transmit a single datum from one site to another. The protocols consist of one or more lower-level messages. They employ encryption in order to guarantee the integrity and secrecy of the datum, and nonces or other tags in order to protect against replay attacks.

Despite their simplicity, these protocols serve as building blocks for complex systems. Relying on these protocols, we can add cryptographic protection to an arbitrary program, much as is done in systems with remote invocation facilities [7]. More precisely, we can translate from a process calculus with primitive secure channels (the join-calculus [11]) to a lower-level process calculus where communication across sites may take place on public channels and may use cryptography for security. We have studied such translations in recent papers [3,4]. The main purpose of this paper is to show a close-up of an essential part of those translations.

---

* Partly supported by ESPRIT CONFER-2 WG-21836.

C. Pandu Rangan, V. Raman, R. Ramanujam (Eds.): FSTTCS'99, LNCS 1738, pp. 122–141, 1999.

In addition, these protocols provide an example of a helpful top-down approach to the specification and verification of security protocols. Following this approach, we reduce the problem of implementing the join-calculus to that of writing protocols for transmitting a single datum. We isolate a handful of crisp correctness conditions on the protocols. These conditions are sufficient for the overall correctness of the implementation of the join-calculus. They are not strictly necessary, but each of them corresponds to a sensible requirement on protocols, they are hard to weaken, and they can all be met. We also consider the application of these conditions to specific protocols. In particular, as a didactic example, we present a new, simple protocol and a proof of its correctness.

The next section discusses some of the problems that the top-down approach is intended to address. Section 3 presents the join-calculus and some extensions. Section 4 describes our correctness conditions for protocols. The correctness conditions appear in [3]; the aim of this section is to review them and to explain their implications. Section 5 shows our new protocol; section 6 contains the corresponding proofs. Section 7 concludes. Thus, this paper is partly a review; its main novelties are in informal explanations and in sections 5 and 6.

2   A Top-Down Approach

Work on protocols is seldom purely top-down. Protocol designers often proceed bottom-up, building systems from cryptographic algorithms and network services. Indeed, some protocols arise as applications of cryptographic primitives. For example, Ylönen designed version 1 of the popular protocol SSH [33] as an exercise in the use of RSA; the design of version 2 was "more or less top-down" [34]. Therefore, a top-down approach, on its own, is probably unrealistic.

With this caveat, a top-down approach can serve as a guide. In particular, it helps in addressing common confusions about security protocols and their goals (e.g., [16]). These confusions often enable "attacks" (scenarios that violate some of the expected security properties of a protocol). Although some attacks reveal serious flaws in protocols, many alleged attacks are merely the annoying result of poor protocol specifications, or of poor understandings of those specifications. A top-down perspective helps in distinguishing the dangerous attacks from the unimportant ones, and in avoiding the former.

Those confusions arise even in the analysis of elementary protocols for key exchange. For example, suppose that, after running a certain protocol, two parties A and B are supposed to share a session key. Suppose further that an attacker can arrange that A and B end up with two different keys, but does not know those keys. Surely, this scenario violates the key-establishment goal of the protocol. On the other hand, the scenario may be harmless. When A and B attempt to communicate after running the protocol, they are most likely to discover the discrepancy if their subsequent encrypted messages contain some checkable redundancy. Then A and B are likely to rerun the protocol, and eventually they may agree on a key. At worst, the attack will result in a loss of liveness, but not a loss of safety. (Lowe has considered other attacks with similar properties [19].)

Those confusions become more delicate for current, complex protocols. For example, suppose that A and B start out by discussing which cryptosystem to adopt, using some cleartext messages. Suppose further that an attacker can tamper with this negotiation, convincing A and B to use a cryptosystem of its choice. If A and B adjust the contents of the subsequent conversation to the outcome of the negotiation, then this attack may entail only a loss of liveness: A and B may not send certain sensitive messages. However, this attack could also cause the inappropriate use of a poor cryptosystem, and failures of authenticity and secrecy (e.g., [32]).

Characteristically, these problematic scenarios are vague on the context in which a protocol operates, and on the use of the protocol—for example, how retries happen, or what sorts of application data are sent after a key of a certain type is established. The difficulties diminish or disappear if the protocol is seen as part of a larger system. An attack on the protocol is significant only if it has an effect on the behavior of the larger system.

For our purposes, the larger system is an arbitrary join-calculus program (or a lower-level translation of this program). Top-down, we go from the semantics of the join-calculus to the design of particular protocols; if a protocol is correct according to our conditions then it can serve as a building block for the implementation of the join-calculus. Every potential attack on a protocol can be assessed against the conditions, unambiguously.

In a more restricted view, the larger system could be one that serves a specific purpose, for example a file system, rather than an arbitrary join-calculus program. This specialization may allow more efficient protocols; it also gives rise to the danger that the protocols will be used in unintended, inappropriate ways.

A top-down perspective is not unique to our work. In particular, recent papers by Bellare, Canetti, and Krawczyk [6] and Lynch [21] concern modular methods for designing and analyzing protocols. Although those works are largely disjoint from ours in techniques and results, they seem to have similar conceptual basis and potential benefits. A salient characteristic of our work is that it treats cryptographic operations as black boxes (cf. [6,18]). A further refinement would be to replace the black boxes with particular cryptographic algorithms. While this refinement is a natural continuation of our approach, it may be quite hard, and we have yet to study it.

3   The Join-Calculus and Its Extensions

Next we describe the join-calculus and some additions for representing cryptographic operations and a public network. This review is brief; we refer the reader to previous papers for details on the join-calculus, its theory, and its applications [11,12,14,13,10,3,4].

3.1   The Join-Calculus

In the join-calculus, processes communicate through named, one-way channels. Intuitively, the channels of the join-calculus have built-in security properties:

- As in the pi-calculus [26,27], the name of a channel is a transferrable but unforgeable capability.
- A process that sends a message on a channel must have the name of the channel.
- Only the process that creates a channel can receive messages on the channel.

We use lowercase identifiers x, y, *foo*, *bar* ... for names. In addition to a category of names, the syntax of the join-calculus includes categories of values, processes, definitions, and join-patterns. These are defined in the following grammar, where we write $\tilde{v}$ for a tuple of values $v_1, v_2, \ldots, v_n$.

$v ::=$		values
	x	name
$P ::=$		processes
	$x\langle\tilde{v}\rangle$	message
	$\mid$ def D in P	local definition
	$\mid$ if $v = v'$ then P else P'	comparison
	$\mid P \mid P'$	parallel composition
	$\mid 0$	null process
$D ::=$		definitions
	$J \triangleright P$	reaction rule
	$\mid D \wedge D'$	conjunction of definitions
$J ::=$		join-patterns
	$x\langle\tilde{y}\rangle$	message pattern
	$\mid J \mid J'$	join of patterns

Processes have the following informal semantics:

- $x\langle\tilde{v}\rangle$ sends the tuple of values $\tilde{v}$ on x, asynchronously.
- def D in P is the process P with the local definitions given in D. In def D in P, the channel names defined in D are recursively bound in the whole of def D in P, with lexical scoping.
- if $v = v'$ then P else P' tests whether $v = v'$, then runs the process P or the process P' depending on the result of the test.
- $P \mid P'$ is the parallel composition of the processes P and P'.
- 0 is the null process, which does nothing.

A definition $J \triangleright P$ says that the process P may run when there are messages that match the join-pattern J. For example, let D be the definition:

$$(x_1\langle y_1\rangle \mid x_2\langle y_2, y_3\rangle) \triangleright (x_1\langle y_1\rangle \mid z\langle y_1, y_2, y_3\rangle)$$

This definition introduces two channels, with names x_1 and x_2. The names y_1, y_2, and y_3 are also bound; they are formal parameters that may be instantiated to actual values received on x_1 and x_2. The name z is free. When a message $x_1\langle v_1\rangle$ appears on x_1 and a message $x_2\langle v_2, v_3\rangle$ appears on x_2, this definition may fire, consuming both messages, reproducing $x_1\langle v_1\rangle$, and producing $z\langle v_1, v_2, v_3\rangle$. Thus, def D in $x_1\langle v_1\rangle \mid x_2\langle v_2, v_3\rangle$ yields def D in $x_1\langle v_1\rangle \mid z\langle v_1, v_2, v_3\rangle$.

3.2   Syntactic Extensions

It is convenient to introduce some syntactic extensions that do not affect the expressiveness of the calculus. We write repl P for the replication of P, which can be defined as def $x\langle\rangle \triangleright (P \mid x\langle\rangle)$ in $x\langle\rangle$ for a fresh x. In addition, we have notations for data structures that contain unique identifiers. We write the declaration uids t as the definition of an initially empty set t; and write if not tset $t(c)$ then P for a process that atomically tests whether c is in t, and if not adds c to t and then triggers the execution of P. (The test-and-set must be atomic, but the execution of P need not be.)

3.3   Cryptographic Primitives

We define the sjoin-calculus, which is analogous to the spi-calculus [5], by enriching the join-calculus with a few constructs:

- fresh x is a definition that introduces the fresh name x; this name may for example be included as a unique identifier within a message.
- keys x^+, x^- is a definition that introduces a pair of keys, for public-key encryption [24]; x^+ is an encryption key (the "public key") and x^- is the inverse decryption key (the "secret key").
- $\{\widetilde{v}\}_v$ is a value that represents the result of encrypting $\widetilde{v}$ with key v. The inverse of v should be used for decryption.
- decrypt v using v' to $\widetilde{x}$ in P else P' is a process that attempts to decrypt v using v' as key. If the decryption succeeds, then P runs, with the results of the decryption substituted for $\widetilde{x}$; otherwise, P' runs. We may omit P' when it is 0.

With these additions, we can represent systems where low-level processes use cryptography for security.

3.4   A Public Network

In addition, we need a model of a public, asynchronous network over which low-level processes communicate. This network should allow an attacker to intercept, modify, duplicate, and inject messages. Sometimes we may also wish to assume that the attacker has access to certain public keys without knowing the corresponding secret keys. Therefore, we define the contexts:

$$\mathcal{E}[\,\cdot\,] \stackrel{\text{def}}{=} \text{def } recv\langle\kappa\rangle \mid emit\langle m\rangle \triangleright \kappa\langle m\rangle \text{ in } [\,\cdot\,]$$

$$\mathcal{E}nv[\,\cdot\,] \stackrel{\text{def}}{=} \mathcal{E}[\,plug\langle emit, recv\rangle \mid W \mid \cdot\,]$$

$$\mathcal{E}nv_{x^\pm}[\,\cdot\,] \stackrel{\text{def}}{=} \text{def keys } x^+, x^- \text{ in } \mathcal{E}nv\,[\,plug'\langle x^+\rangle \mid \cdot\,]$$

where:

- $emit$ and $recv$ represent the network interface. For output, a process sends its message on $emit$. For input, it sends a continuation channel κ on $recv$, and the network may return a message on κ.

- *plug* and *plug'* are auxiliary channels whose sole purpose is to make certain public names (*emit*, *recv*, x^+) available to outside processes.
- W is the process repl (def fresh m in repl $emit\langle m \rangle$), which repeatedly puts fresh messages on the network, as background noise, helping protect against traffic-analysis attacks.

Intuitively, $\mathcal{E}nv\,[\,\cdot\,]$ represents the network; $\mathcal{E}nv\,[P]$ describes a situation where both a process P and any process running in parallel may use *emit* and *recv*. The other contexts play auxiliary roles.

3.5 Operational Semantics and Types (Notation)

The operational semantics of the join-calculus and its extensions define the following notations and concepts [2].

- $P \downarrow_x$ holds if P may output on x immediately. For example, $x\langle v \rangle \downarrow_x$.
- $P \Downarrow_x$ holds if P may output on x either immediately or after some internal reductions. For example, (def $y\langle\rangle \triangleright x\langle\rangle$ in $y\langle\rangle) \Downarrow_x$.
- $P \equiv Q$ holds if P and Q are structurally equivalent, that is, if P and Q differ only by certain simple syntactic rearrangements. For example, for all P_1 and P_2, $P_1 \mid P_2 \equiv P_2 \mid P_1$. We also take repl $P \equiv (P \mid$ repl $P)$.
- $P \rightarrow Q$ holds if P may reduce to Q, that is, if P may perform one step of internal computation and then behave as Q. For example,

$$\text{def } y\langle\rangle \triangleright x\langle\rangle \text{ in } y\langle\rangle \;\; \rightarrow \;\; \text{def } y\langle\rangle \triangleright x\langle\rangle \text{ in } x\langle\rangle$$

The relations $\rightarrow^=$ and $\rightarrow^*$ are the reflexive closure and the reflexive-transitive closure of $\rightarrow$, respectively.
- An evaluation context $C[\cdot]$ is a context in which computation may immediately take place: if $P \rightarrow Q$ then $C[P] \rightarrow C[Q]$. For example, $P_1 \mid \;\cdot\;$ is an evaluation context, while decrypt v using v' to $\widetilde{x}$ in $\cdot$ else P_2 is not.

We rely on simple, monomorphic type systems for the calculi. In these type systems, each channel has an associated arity (an integer size for the tuples of values transmitted on the channel). We write $\langle \tau_1, \ldots, \tau_n \rangle$ for the type of channels that carry tuples with n values of respective types $\tau_1, \ldots, \tau_n$. We allow types to be recursively defined (formally, using a fixpoint operator), so we may have for example $\tau = \langle \tau, \tau \rangle$.

In addition to channel types, we have a basic type BitString. This is the type of keys, ciphertexts and their contents, and of the fresh names introduced with the construct fresh $\cdot$.

We assume that each name is associated with a type (although we usually keep this type implicit), and that there are infinitely many names for each type. Throughout, we consider only well-typed processes.

4 Correctness

In this section we arrive, more or less top-down, at the problem of transmitting a datum securely. We present correctness conditions for protocols that solve the problem.

4.1   Goals

Now we can write both pure join-calculus processes, where security is based on the scoping of channel names, and processes that communicate using cryptography over a public network (as represented by $\mathcal{E}nv\,[\,\cdot\,]$). Our objective is to see the latter as implementations of the former. Moreover, we wish to do this systematically: we would like to have compilers that map pure join-calculus processes to lower-level code that can execute securely over a public network.

In order to formalize the security requirement for such compilers, we resort to process equivalences. We say that two processes P_1 and P_2 are equivalent, and write $P_1 \approx P_2$, when no context can distinguish one from the other [25,9]. Intuitively, we may view the context as an attacker; then $P_1 \approx P_2$ entails an integrity property (limiting the effect of the attacker on the behavior of P_1 and P_2) and a secrecy property (limiting the observations of the attacker). Formally, we define $\approx$ as the largest symmetric relation $\mathcal{R}$ on processes such that:

1. if $P\,\mathcal{R}\,Q$ and $P \Downarrow_x$ then $Q \Downarrow_x$;
2. $\mathcal{R}$ is a congruence for all evaluation contexts, that is, for all evaluation contexts $C[\,\cdot\,]$, if $P\,\mathcal{R}\,Q$ then $C[P]\,\mathcal{R}\,C[Q]$;
3. $\mathcal{R}$ is a weak bisimulation, that is, if $P\,\mathcal{R}\,Q$ and $P \to^* P'$ then, for some Q', $P'\,\mathcal{R}\,Q'$ and $Q \to^* Q'$.

When we devise implementations of the join-calculus, we wish to preserve equivalences, since equivalences can express security properties [5,1]. More precisely, if P_1 and P_2 are equivalent join-calculus processes, we wish to compile them to processes P_1' and P_2' such that $\mathcal{E}nv\,[\,P_1'\,]$ and $\mathcal{E}nv\,[\,P_2'\,]$ are equivalent. The mention of $\mathcal{E}nv\,[\,\cdot\,]$ accounts for the use of the public network by P_1' and P_2'.

4.2   Protocols

A crucial part of implementing the join-calculus is mapping join-calculus messages to cryptographic protocols. Following an obvious strategy and postponing optimizations, we associate a pair of keys x^+, x^- with each join-calculus channel x.

- The encryption key x^+ corresponds to the capability of sending messages on x, which may be transferred.
- The decryption key x^- corresponds to the capability of receiving messages on x, which only the creator of x has.

In order to simulate communication on a join-calculus channel x, we employ protocols that consist of two processes $E_x[\tilde{v}]$ and R_x, for sending and receiving, respectively.

- Using the key x^+ for encryption, $E_x[\tilde{v}]$ sends $\tilde{v}$. Within $\tilde{v}$, an encryption key y^+ (rather than a pair y^+, y^-) represents the corresponding channel y.
- Using the key x^- for decryption, R_x receives messages, then forwards their cleartext contents on an auxiliary, internal channel x°.

For example, a protocol might be:

$$E_x[\tilde{v}] \stackrel{\text{def}}{=} emit\langle\{\tilde{v}\}_{x^+}\rangle$$

$$R_x \stackrel{\text{def}}{=} \text{def } \kappa\langle m\rangle \triangleright \text{decrypt } m \text{ using } x^- \text{ to } \tilde{y} \text{ in } x^\circ\langle\tilde{y}\rangle$$
$$\text{in } recv\langle\kappa\rangle$$

where the length of $\tilde{y}$ is deduced from the type of x. For instance, using this protocol, the join-calculus process $x\langle y\rangle$ may be mapped to $emit\langle\{y^+\}_{x^+}\rangle$. This naive protocol is subject to message replays and other obvious attacks. The protocol of section 5 thwarts those attacks.

In general, such protocols should guarantee the integrity and secrecy of $\tilde{v}$. These properties are informally appealing, and they are formally necessary for the desired preservation of equivalences. Section 4.3 gives a more precise and complete list of properties.

4.3   Correctness Conditions for Protocols

In the following definition, a protocol is a pair $(R_x, E_x[\cdot])$ consisting of a process for receiving and one for sending, parameterized by a channel name x. The definition relies on a set $\mathbf{R}$ of derivatives of the receiving process R_x. This set represents the different states of a receiver after interaction with its context.

The definition also relies on an expansion relation on processes ($\succeq$) [31], which is similar to $\approx$ but stronger and asymmetric. We define $\succeq$ as the largest relation $\mathcal{R}$ such that $\mathcal{R}$ and its converse meet requirements (1), (2), and (3) of the definition of $\approx$, and such that if $P \mathcal{R} Q$ and $P \to P'$ then, for some Q', $P' \mathcal{R} Q'$ and $Q \to^= Q'$.

Definition 1. *The protocol $(R_x, E_x[\cdot])$ is correct if there is a set of processes $\mathbf{R}$ that satisfies the following conditions.*

1. *$R_x \in \mathbf{R}$.*
2. *The free names of $E_x[\cdot]$ are at most emit, recv, and names of type BitString. For every $R \in \mathbf{R}$, the free names of R are at most emit, recv, x°, and names of type BitString.*
3. *For every $R \in \mathbf{R}$, it is not the case that $R \downarrow_{emit}$ or that $R \downarrow_{x^\circ}$.*
4. *For every $R \in \mathbf{R}$, for every tuple $\tilde{v}$ of values of type BitString whose length matches the arity of x, if x^- does not occur in $\tilde{v}$, then*

$$\mathcal{E}nv_{x^\pm}[R \mid E_x[\tilde{v}]] \succeq \mathcal{E}nv_{x^\pm}[R \mid x^\circ\langle\tilde{v}\rangle]$$

5. *For every value v, if x^+ and x^- do not occur in v and*

$$\mathcal{E}nv_{x^\pm}[R_x \mid emit\langle v\rangle] \to P$$

   then

$$P \succeq \mathcal{E}nv_{x^\pm}[R_x \mid emit\langle v\rangle]$$

6. *For every $R \in \mathbf{R}$, if x^- does not occur in v and*

$$\mathcal{E}nv_{x^{\pm}} [\, R \mid emit\langle v \rangle \,] \to P$$

then

$$P \succeq \mathcal{E}nv_{x^{\pm}} [\, R' \mid Q \,]$$

for some $R' \in \mathbf{R}$ and some process Q such that x^- does not occur in Q.

Condition 1 says that the initial receiving process, R_x, is in $\mathbf{R}$.

Condition 2 restricts the free names in use in the protocol. The sending process $E_x[\cdot]$ may have access to the network-interface channels (*emit* and *recv*). The receiving process R_x and all other processes in $\mathbf{R}$ may have access to those channels and also to x°. The requirement that x° does not occur free in $E_x[\cdot]$ rules out degenerate protocols where the sending process does the work of the receiving process, like the protocol $(0, x^\circ\langle\cdot\rangle)$. In addition to *emit*, *recv*, and x°, the protocol may rely on names of type BitString. Intuitively, condition 2 implies that communication from $E_x[\cdot]$ to R_x is limited to messages of types BitString and $\langle$BitString$\rangle$ on the channels *emit* and *recv*. Therefore, the protocol can be directly implemented over an ordinary network like that represented by the channels *emit* and *recv*, without any additional assumptions about physical security or out-of-band communication.

Condition 3 says that every process $R \in \mathbf{R}$ is passive, in the sense that it does not send messages on *emit* or x° spontaneously. This condition still allows R to send messages on *recv*.

Condition 4 says that the protocol transmits messages reliably and secretly when an instance of the sending process $E_x[\cdot]$ is put in parallel with the receiving process R_x or any other process $R \in \mathbf{R}$. Using the expansion relation, this condition compares $R \mid E_x[\widetilde{v}]$ with $R \mid x^\circ\langle\widetilde{v}\rangle$. The former process has $E_x[\widetilde{v}]$ as a component, while in the latter $E_x[\widetilde{v}]$ is replaced with its intended outcome, namely $x^\circ\langle\widetilde{v}\rangle$ (with no other visible effect). In both processes we have the same component R. Thus, this condition implies that the state of the receiving process remains essentially unchanged as long as it interacts with regular sending processes; any state change in the receiving process, such as the addition of entries in internal tables, should not be observable. (Condition 4 is slightly stronger in [3], where the assumption that x^- does not occur in $\widetilde{v}$ is missing.)

Condition 4 rules out insecure protocols that leak information in the course of transmitting a message; many obviously insecure protocols fall into this category. In particular, the naive protocol of section 4.2 violates condition 4 on several counts. For example, a context that intercepts the message from $E_x[\widetilde{v}]$ and listens on x° can differentiate $\mathcal{E}nv_{x^{\pm}} [\, R \mid E_x[\widetilde{v}] \,]$ and $\mathcal{E}nv_{x^{\pm}} [\, R \mid x^\circ\langle\widetilde{v}\rangle \,]$. Furthermore, a context that interacts with $\mathcal{E}nv_{x^{\pm}} [\, R \mid E_x[\widetilde{v}] \,]$ may be able to guess $\widetilde{v}$, then confirm the guess by computing $\{\widetilde{v}\}_{x^+}$ and comparing $\{\widetilde{v}\}_{x^+}$ with the message from $E_x[\widetilde{v}]$. In contrast, the context cannot obtain the same information in interaction with $\mathcal{E}nv_{x^{\pm}} [\, R \mid x^\circ\langle\widetilde{v}\rangle \,]$.

The last two conditions (5 and 6) describe interactions between the receiving process and the context. These conditions are needed in addition to condition 4 because the context might not behave as a regular sending process. Since the context may communicate with the receiving process only through messages of the form $emit\langle v\rangle$, the last two conditions describe the behavior of the receiving process in reaction to such a message. The conditions concern two cases, distinguished by whether the encryption key x^+ occurs in v.

Condition 5 describes the behavior of R_x in a context that does not have access to the keys x^+ and x^-. Intuitively, this behavior is that exhibited when the attacker has not yet been given x^+, so the receiving process should remain essentially invisible. For every reduction $\mathcal{E}nv_{x\pm}\,[\,R\mid emit\langle v\rangle\,]\rightarrow P$, expansion relates the outcome P to the initial state $\mathcal{E}nv_{x\pm}\,[\,R\mid emit\langle v\rangle\,]$. Thus, if R_x takes a message from the network and the message is unrelated to x^+, x^-, then R_x must resend the message. Similarly, if R_x becomes a process R by internal reductions (so P equals $\mathcal{E}nv_{x\pm}\,[\,R\mid emit\langle v\rangle\,]$), then it must be possible to go back from R to R_x obtaining that $\mathcal{E}nv_{x\pm}\,[\,R\mid emit\langle v\rangle\,]\succeq \mathcal{E}nv_{x\pm}\,[\,R_x\mid emit\langle v\rangle\,]$.

For example, condition 5 excludes a protocol where R_x "swallows" all messages that are not encrypted under x^+, thus revealing its presence. It also excludes a protocol where R_x "swallows" messages selectively, possibly revealing sensitive information.

Condition 6 describes the behavior of a receiving process $R\in \mathbf{R}$ in a context that has access to the encryption key x^+ but not to the decryption key x^-. Intuitively, this behavior corresponds to the case where the attacker has been given the encryption key, and can thus cause messages on x°; in this situation, the attacker should still not interfere with messages from other senders. The reduction $\mathcal{E}nv_{x\pm}\,[\,R\mid emit\langle v\rangle\,]\rightarrow P$ may change the state of the receiving process, for example by completing a run of the protocol and relaying a message on x°. The process P cannot however be arbitrary: it must be in the expansion relation with a process that includes a new receiving process $R'\in \mathbf{R}$ in place of R, in parallel with a process Q in place of $emit\langle v\rangle$. The process Q may contain emissions on x° and $emit$; it typically consists of parts of residues of R that do not need the decryption key x^- any more.

For example, condition 6 excludes an insecure variant of a correct protocol where the receiving process is modified as follows. In answer to messages of a form that a regular sending process never creates (for example, $\{c, c, c\}_{x^+}$ for the protocol of section 5), the receiving process emits a fresh value u; the receiving process works correctly for input tuples that do not contain u, but leaks input tuples that contain u. A flaw appears only after the creation of u, which regular sending processes never cause. Thus, all other conditions can be met. Nonetheless, an attacker may obtain u, pass u to some third party, and harvest secrets later on if the third party includes u in messages to the receiving process. This variant violates condition 6: consider the reduction $\mathcal{E}nv_{x\pm}\,[\,R\mid emit\langle\{\widetilde{v}\}_{x^+}\rangle\,]\rightarrow P$ that will cause the creation of u; then condition 6 requires that $P\succeq \mathcal{E}nv_{x\pm}\,[\,R'\mid Q\,]$ for some $R'\in \mathbf{R}$, while no such R' can satisfy all the other conditions.

As mentioned above, these correctness conditions suffice for our results about the implementations of the join-calculus. Compared to common specifications of cryptographic protocols (e.g., [20]), our correctness conditions are rather extensional [30]: they concern the intended effects of a protocol rather than its internal behavior.

5 A Protocol

In this section, we describe a specific protocol for transmitting a single datum. This protocol is an instructive example: it is simple and correct; moreover, its correctness is not too hard to prove. On the other hand, this protocol is not very practical. Two other correct protocols appear in [3]; they are somewhat more complex and realistic. In current work (discussed briefly in section 7), we are making further efficiency improvements.

Our simple protocol is an enhancement of the naive protocol in several respects:

- Each encrypted message contains a fresh component c, which serves as a confounder (making the message unpredictable and different from other messages with the same payload) and as a unique identifier (against replay attacks).
- Each encrypted message is repeated indefinitely, in case some copies are intercepted.
- When a process receives a message for the first time, it reemits the message and records it before further processing. Later, it reemits but ignores duplicates of this message.

The first correct protocol of [3] is similar in these respects, but it does fewer reemissions and uses records of unique identifiers instead of records of complete messages.

Since calls to $recv$ retrieve messages from the network non-deterministically, and since messages may be duplicated by the sender or by an attacker, processes have to filter for messages that are destined for them. This filtering relies first on a table for discarding duplicates, then on a decryption key for accessing message contents. We arrive at the following definitions.

$$E_x[\widetilde{v}] \;\stackrel{\text{def}}{=}\; \textsf{def fresh } c \textsf{ in repl } emit\langle\{c,\widetilde{v}\}_{x+}\rangle$$

$$R_x \;\stackrel{\text{def}}{=}\; R_x\{\}\,[0]$$

$$R_x\{\widetilde{u}\}\,[P] \;\stackrel{\text{def}}{=}\; \textsf{def uids } t_x\{\widetilde{u}\} \textsf{ in}$$
$$\qquad\qquad\textsf{def } \kappa\langle m\rangle \triangleright emit\langle m\rangle \mid F_{m,x} \textsf{ in } P \mid \textsf{repl } recv\langle\kappa\rangle$$

$$F_{m,x} \;\stackrel{\text{def}}{=}\; \textsf{if not tset } t_x(m) \textsf{ then } F'_{m,x}$$

$$F'_{m,x} \;\stackrel{\text{def}}{=}\; \textsf{decrypt } m \textsf{ using } x^- \textsf{ to } c,\widetilde{y} \textsf{ in } x^\circ\langle\widetilde{y}\rangle$$

where the length of the tuple $\widetilde{y}$ in the definition of $F'_{m,x}$ is deduced from the type of x. In the receiving process, the component $F_{m,x}$ serves as a filter for a single message; such components are replicated, and share a set of previously received messages.

Theorem 1. *The protocol $(R_x, E_x[\cdot])$ is correct.*

The proof of this theorem is the subject of the next section

6   A Correctness Proof

The proof of Theorem 1 relies on a series of lemmas that describes the various stages of communications processing (reception, duplicate elimination, decryption). Proofs for more sophisticated protocols have similar structure, though each of the steps becomes harder.

Throughout, we employ several techniques for establishing equivalences and other relations between processes. These techniques are not specific to cryptography and appear in lemmas in [2]; here, we only present their role informally. We also use the equivalence relation $\asymp$, which is stronger than $\approx$ and is obtained from the definition of $\approx$ in section 4.1 by substituting $\rightarrow^=$ for $\rightarrow^*$. The following "up-to" technique is helpful for proving $\asymp$. In order to show that $\mathcal{R} \subseteq \asymp$, it suffices to prove, for all P and Q such that $P \mathcal{R} Q$, that:

1. if $P \downarrow_x$ then $Q \Downarrow_x$; conversely, if $Q \downarrow_x$ then $P \Downarrow_x$;
2. $C[P] \asymp\mathcal{R}\asymp C[Q]$ for every context $C[\cdot]$ of the form def D in $R \mid [\cdot]$ such that the names bound in P and Q do not occur in D and R;
3. if $P \rightarrow P'$, then $Q \rightarrow^= Q'$ and $P' \asymp\mathcal{R}^=\asymp Q'$ for some Q'; conversely, if $Q \rightarrow Q'$, then $P \rightarrow^= P'$ and $P' \asymp\mathcal{R}^=\asymp Q'$ for some P'.

(Here, $\mathcal{R}^=$ is the reflexive closure of $\mathcal{R}$ and $\asymp\mathcal{R}\asymp$ is its composition with $\asymp$ on both sides.)

We focus on a fixed channel x and the associated names x^+, x^-, and x°. In the remainder of this section, we assume that these names and *emit* and *recv* are never alpha-converted. The up-to proof techniques used in this section allow this assumption.

We say that a value m is a well-formed message when it is of the form $\{c, \widetilde{v}\}_{x^+}$ for some values $c, \widetilde{v}$ of type BitString such that the length of $\widetilde{v}$ is the arity of x. A process P is an internal state when it is a parallel composition of processes $\kappa\langle m \rangle$, $F_{m,x}$, or $F'_{m,x}$, for some BitString values m in which x^- does not appear. A net state is a process of the form:

$$\mathcal{E}\left[\,\text{def keys } x^+, x^- \wedge D \text{ in } R_x\{\widetilde{u}\}\,[P] \mid Q\,\right]$$

where D does not define *emit*, *recv*, x^+, or x^-, the tuple $\widetilde{u}$ contains pairwise distinct BitString values, P is an internal state, and x^- does not appear in D, Q, or $\widetilde{u}$. For any net state S, there exist a unique context $C[\cdot]$, tuple $\widetilde{u}$, process Q, and internal state P such that $S = C[R_x\{\widetilde{u}\}\,[P] \mid Q]$. Moreover, $\widetilde{u}$ and P can be derived (uniquely up to alpha-conversion and reordering for $\widetilde{u}$ and $\equiv$ for P) from any $T \equiv S$, because $R_x\{\widetilde{u}\}\,[P]$ must contain all occurrences of x^- and κ in S.

The first lemma says that net states are closed under application of evaluation contexts and under reduction.

Lemma 1 (Closure). *Let S be a net state. For any evaluation context $E[\cdot]$ in which names bound in S do not appear, we have $E[S] \equiv S'$ for some net state S'. For any reduction step $S \to T$, we have $T \equiv S'$ for some net state S'.*

Proof. Let $S = C[R_x\{\widetilde{u}\}[P] \mid Q]$, with $C[\cdot] = \mathcal{E}[\text{def keys } x^+, x^- \wedge D \text{ in } \cdot]$. For the first claim, it is enough to consider contexts $E[\cdot]$ of the forms $(\cdot \mid Q')$ and $(\text{def } D' \text{ in } \cdot)$. We obtain S' from S by replacing Q with $Q \mid Q'$ or D with $D \wedge D'$, respectively. For the second claim, we base our analysis on the location of the processes involved in the reduction step, and further on the type of reduction if the reduction is local to $R_x\{\widetilde{u}\}[P]$:

- If the step involves only processes in Q (possibly with definitions in $C[\cdot]$), then it leaves $R_x\{\widetilde{u}\}[P]$ unchanged, so there must be D', Q', and $C'[\cdot]$ such that $C'[\cdot] = \mathcal{E}[\text{def keys } x^+, x^- \wedge D' \text{ in } \cdot]$, $T \equiv C'[R_x\{\widetilde{u}\}[P] \mid Q']$, and $C[Q] \to C'[Q']$. Therefore, x^- does not appear in D' and Q', so we take $S' = C'[R_x\{\widetilde{u}\}[P] \mid Q']$.
- If the step involves processes in both Q and $R_x\{\widetilde{u}\}[P]$, then it must use the rule in $\mathcal{E}[\cdot]$ to produce a message $\kappa\langle m \rangle$ from a message $emit\langle m \rangle$ in Q and a message $recv\langle \kappa \rangle$ unfolded from $R_x\{\widetilde{u}\}[\cdot]$. We can assume that this message $emit\langle m \rangle$ does not appear under any definitions in Q, since any such definition can be moved to D. Hence we have $Q \equiv emit\langle m \rangle \mid Q'$ for some process Q', so we take $S' = C[R_x\{\widetilde{u}\}[P \mid \kappa\langle m \rangle] \mid Q']$.
- If the step replaces a message $\kappa\langle m \rangle$ in P with $emit\langle m \rangle \mid F_{m,x}$ using the rule in $R_x\{\widetilde{u}\}[\cdot]$, then $P \equiv P' \mid \kappa\langle m \rangle$ for some internal state P', so we take $S' = C[R_x\{\widetilde{u}\}[P' \mid F_{m,x}] \mid (emit\langle m \rangle \mid Q)]$.
- If the step executes the test-and-set in some process $F_{m,x}$ in P, entering m in the table for x, then $P \equiv P' \mid F_{m,x}$ for some internal state P', so we take $S' = C[R_x\{\widetilde{u}, m\}[P' \mid F'_{m,x}] \mid Q]$.
- Otherwise, the step evaluates the decryption in some process $F'_{m,x}$ in P, yielding $x°\langle\widetilde{v}\rangle$ if m is a well-formed message $\{c, \widetilde{v}\}_{x^+}$ and 0 otherwise, and $P \equiv P' \mid F'_{m,x}$ for some internal state P', so if $m = \{c, \widetilde{v}\}_{x^+}$ we take $S' = C[R_x\{\widetilde{u}\}[P'] \mid (x°\langle\widetilde{v}\rangle \mid Q)]$, and otherwise $S' = C[R_x\{\widetilde{u}\}[P'] \mid Q]$. $\square$

Lemma 2 shows how a receiving process $R_x\{\widetilde{u}\}[P]$ incorporates a message from the public network. The message is immediately reemitted, independently of its contents.

Lemma 2 (Reception). *If $C[R_x\{\widetilde{u}\}[P] \mid (emit\langle m \rangle \mid Q)]$ is a net state, then:*

$$C[R_x\{\widetilde{u}\}[P] \mid (emit\langle m \rangle \mid Q)] \to C[R_x\{\widetilde{u}\}[\kappa\langle m \rangle \mid P] \mid Q]$$
$$\succeq C[R_x\{\widetilde{u}\}[F_{m,x} \mid P] \mid (emit\langle m \rangle \mid Q)]$$

Proof. In more detail, we have:

$$C[R_x\{\widetilde{u}\}[P] \mid (emit\langle m \rangle \mid Q)] \equiv C[R_x\{\widetilde{u}\}[(emit\langle m \rangle \mid recv\langle \kappa \rangle) \mid P] \mid Q]$$
$$\to C[R_x\{\widetilde{u}\}[\kappa\langle m \rangle \mid P] \mid Q]$$
$$\to C[R_x\{\widetilde{u}\}[emit\langle m \rangle \mid F_{m,x} \mid P] \mid Q]$$
$$\equiv C[R_x\{\widetilde{u}\}[F_{m,x} \mid P] \mid (emit\langle m \rangle \mid Q)]$$

The initial structural equivalence unfolds a copy of the replicated process $recv\langle\kappa\rangle$ within $R_x\{\widetilde{u}\}\,[\,\cdot\,]$ and groups this process with $emit\langle m\rangle$. The first reduction step replaces the process $emit\langle m\rangle \mid recv\langle\kappa\rangle$ with the process $\kappa\langle m\rangle$ according to the rule defining $emit$ and $recv$ within $\mathcal{E}[\,\cdot\,]$. The second reduction step consumes $\kappa\langle m\rangle$ according to the rule within $R_x\{\widetilde{u}\}\,[\,\cdot\,]$. The final structural equivalence moves $emit\langle m\rangle$ back to its original position.

The second step depends only on the deterministic definition of κ. Using a standard lemma of the join-calculus, we can substitute $\succeq$ for $\to$ in this case. $\square$

Similarly, Lemma 3 deals with a replicated message. In this case, one copy of the message always remains available, so the first reduction of Lemma 2 is also an expansion.

Lemma 3 (Reception of a replicated message). *Let $C'[\,\cdot\,]$ be a context of the form $C\,[R_x\{\widetilde{u}\}\,[\,\cdot\,\mid P]\mid((\mathsf{repl}\;emit\langle m\rangle)\mid Q)]$, such that $C'[0]$ is a net state. Then $C'[0]\succeq C'[F_{m,x}]$.*

Proof. We have:

$$
\begin{aligned}
C'[0] &\equiv C'[emit\langle m\rangle]\\
&\to C'[\kappa\langle m\rangle]\\
&\to C'[emit\langle m\rangle \mid F_{m,x}]\\
&\equiv C'[F_{m,x}]
\end{aligned}
$$

The first structural equivalence unfolds a copy of the replicated message $emit\langle m\rangle$. The two reduction steps apply the rules for $emit, recv$ and κ, respectively, as in Lemma 2. The final structural equivalence folds back the copy of $emit\langle m\rangle$.

Next, we show that the reduction sequence described above is also an expansion. Given some BitString value m, we let $\mathcal{R}$ be the relation that contains all the pairs of processes $C'[0], C'[F_{m,x}]$ related by the lemma. We show that $\mathcal{R}\subseteq\succeq$. By construction, we have $\mathcal{R}\subseteq\to^2$, so we can apply the special case of the expansion-up-to-expansion proof technique (Lemma 3 of [2]) to analyze two processes $C'[0]$ and $C'[F_{m,x}]$ such that $C'[0]\;\mathcal{R}\;C'[F_{m,x}]$. That proof technique requires matching the reduction steps of $C'[0]$ with those of $C'[F_{m,x}]$, and showing that $\mathcal{R}$ is closed under application of an evaluation context $E[\,\cdot\,]$ in which variables bound in $C'[0]$ or $C'[F_{m,x}]$ do not appear. Every step $C'[0]\to T$ can obviously be matched by a step of $C'[F_{m,x}]$, since the inert term 0 cannot take part in the step: there is a context $C''[\,\cdot\,]$ such that $T\equiv C''[0]$ and $C'[F_{m,x}]\to C''[F_{m,x}]$. By Lemma 1 we can choose $C''[\,\cdot\,]$ so that $C''[F_{m,x}]$ is a net state, whence $C''[0]$ is one as well. The term $\mathsf{repl}\;emit\langle m\rangle$ cannot disappear in the step, so it must occur in the 'Q' part of $C''[\,\cdot\,]$. As in the proof of Lemma 1, we can choose $C''[\,\cdot\,]$ so that this part has the form $(\mathsf{repl}\;emit\langle m\rangle)\mid Q'$, so $C''[0]\;\mathcal{R}\;C''[F_{m,x}]$. A similar argument shows that $\mathcal{R}$ is closed under application of evaluation contexts. $\square$

Lemmas 4 and 5 describe the following steps of communications processing. They concern the freshness of a message and its well-formedness, respectively.

Lemma 4 (Duplicates). *If $C\left[R_x\{\widetilde{u}\}\left[P\right]\right]$ is a net state, then:*

$$C\left[R_x\{\widetilde{u}\}\left[P \mid F_{m,x}\right]\right] \succeq C\left[R_x\{\widetilde{u},m\}\left[P \mid F'_{m,x}\right]\right] \qquad (1)$$
$$C\left[R_x\{\widetilde{u},m\}\left[P \mid F_{m,x}\right]\right] \asymp C\left[R_x\{\widetilde{u},m\}\left[P\right]\right] \qquad (2)$$

Proof. This follows from the definition of tables of unique identifiers, standard up-to proof techniques, and Lemma 1. In the first relation, there may be other copies of $F_{m,x}$ in P attempting to enter m in the table for x, but each copy yields the same process $F'_{m,x}$, so the choice of a particular copy is not observable. $\square$

Lemma 5 (Decryption). *If $C[F'_{m,x}]$ is a net state, then $C[F'_{m,x}] \succeq C[x^\circ\langle\widetilde{v}\rangle]$ when m is a well-formed message $\{c,\widetilde{v}\}_{x^+}$ and $C[F'_{m,x}] \asymp C[0]$ otherwise.*

Proof. This follows from the definition of decryption in the sjoin-calculus, standard up-to proof techniques, and Lemma 1. $\square$

Composing these last three lemmas in the case where there is an emitter $E_x[\widetilde{v}]$, we can summarize a successful run of the protocol as follows:

Lemma 6 (Completion). *If $S = C\left[R_x\{\widetilde{u}\}\left[P\right] \mid (E_x[\widetilde{v}] \mid Q)\right]$ is a net state, c is a BitString name that does not appear in S, and $m = \{c,\widetilde{v}\}_{x^+}$ is a well-formed message, then:*

$$S \equiv \mathsf{def\ fresh}\ c\ \mathsf{in}\ C\left[R_x\{\widetilde{u}\}\left[P\right] \mid ((\mathsf{repl}\ emit\langle m\rangle) \mid Q)\right]$$
$$\succeq \mathsf{def\ fresh}\ c\ \mathsf{in}\ C\left[R_x\{\widetilde{u},m\}\left[P\right] \mid (x^\circ\langle\widetilde{v}\rangle \mid (\mathsf{repl}\ emit\langle m\rangle) \mid Q)\right]$$

Proof. We simply unfold the definitions, use structural equivalence, and successively apply Lemmas 3, 4 (first part), and 5 (first case). The first structural equivalence is obtained by extending the scope of the new unique identifier c out of $E_x[\widetilde{v}] \overset{\text{def}}{=} \mathsf{def\ fresh}\ c\ \mathsf{in\ repl}\ emit\langle\{c,\widetilde{v}\}_{x^+}\rangle$; by hypothesis, the name c does not appear elsewhere in the initial problem, so there is no capture as its definition is lifted outside. As we apply Lemma 4, the replicated message is different from any value appearing in the table for x because no such value may contain c. $\square$

Lemmas 7 and 8 state that what is left after a successful run of the protocol is indistinguishable from noise, and can be discarded up to $\asymp$. Lemma 7 says that a well-formed message m that appears in the table for a channel x can be uniformly replaced with a fresh name d. Lemma 8 says that an ill-formed value can be discarded from the table.

Lemma 7 (Noise). *If $S = C\left[R_x\{\widetilde{u},d\}\left[P\right]\right]$ is a net state, d is a BitString name that is not bound in S, $F'_{d,x}$ does not appear unguarded in P, c is a BitString name that does not appear in S, and $m = \{c,\widetilde{v}\}_{x^+}$, then:*

$$\mathsf{def\ fresh}\ c\ \mathsf{in}\ (S\{^m/_d\}) \asymp \mathsf{def\ fresh}\ d\ \mathsf{in}\ S$$

Proof. We let $\mathcal{R}$ be the relation that contains all pairs of processes equated in the lemma, and prove that $\mathcal{R} \subseteq \asymp$. Every reduction commutes with the substitution; in particular:

- the substitution $\{^m/_d\}$ is an injection from values in which c does not appear to values in which d does not appear, so every reduction step that depends on the comparison of two values selects the same branch before and after the substitution;
- S does not attempt to decrypt d with x^-, and $S\{^m/_d\}$ does not attempt to decrypt m with x^-, since $F'_{d,x}$ does not appear unguarded in P and x^- does not appear elsewhere in S.

By Lemma 1, if $S \to S'$ then we can take S' to be a net state. Furthermore, $F'_{d,x}$ cannot appear unguarded in S': the processes $F_{d,x}$ that may appear in P are inert, because d already appears in the table. It follows that $\mathcal{R}$ is closed under reduction. Closure under application of evaluation contexts follows from Lemma 1. Finally, $\{^m/_d\}$ does not operate on channel names, so for every channel y we have $S \downarrow_y$ if and only if $S\{^m/_d\} \downarrow_y$. $\square$

Lemma 8 (Table simplification). *If $C\left[R_x\{\widetilde{u}, v\}[P]\right]$ is a net state where v is not a well-formed message, then $C\left[R_x\{\widetilde{u}, v\}[P]\right] \asymp C\left[R_x\{\widetilde{u}\}[P]\right]$.*

Proof. We let $\mathcal{R}$ be the relation that contains all pairs of processes equated in the lemma, and prove that $\mathcal{R} \subseteq \asymp$. To establish the bisimulation requirement, we compare the reductions of processes that $\mathcal{R}$ relates. In the case where $P \equiv F_{v,x} \mid P'$ for some internal state P', a reduction step is enabled only on the right-hand side:

$$C\left[R_x\{\widetilde{u}\}[F_{v,x} \mid P']\right] \to C\left[R_x\{\widetilde{u}, v\}[F'_{v,x} \mid P']\right]$$

The left-hand side does not need to match this step, as the second part of Lemma 4 and the second case of Lemma 5 give us:

$$C\left[R_x\{\widetilde{u}, v\}[F_{v,x} \mid P']\right] \asymp C\left[R_x\{\widetilde{u}, v\}[P']\right] \asymp C\left[R_x\{\widetilde{u}, v\}[F'_{v,x} \mid P']\right]$$

In all other cases, reductions are the same on both sides, and lead to related net states by Lemma 1. Closure under application of evaluation contexts is likewise a direct consequence of Lemma 1. Finally, immediate outputs are the same on both sides of $\mathcal{R}$. $\square$

We are now ready to prove the correctness requirements for our protocol (Theorem 1):

Proof. We let **R** consist of all terms $R_x\{\widetilde{u}\}[0]$ where $\widetilde{u}$ is a tuple of pairwise distinct values of type BitString in which x^- does not appear. Thus, **R** consists of derivatives of R_x with different contents in the table of unique identifiers.

1: By definition, $R_x = R_x\{\}[0]$.

2, 3: These conditions are syntactically obvious for $E_x[\cdot]$ and for every $R \in \mathbf{R}$.

4: We must prove that

$$\mathcal{E}nv_{x\pm}\left[\,R\mid E_x[\tilde{v}]\,\right]\;\succeq\;\mathcal{E}nv_{x\pm}\left[\,R\mid x^\circ\langle\tilde{v}\rangle\,\right]$$

for every $R\in\mathbf{R}$ and every tuple $\tilde{v}$ of BitString values whose length matches the arity of x and in which x^- does not appear. We derive this expansion relation by composing Lemmas 6, 7, and 8. Let $R=R_x\{\tilde{u}\}\,[0]$, let c and d be fresh BitString names, and let $m=\{c,\tilde{v}\}_{x+}$; we have:

$$\mathcal{E}nv_{x\pm}\left[\,R_x\{\tilde{u}\}\,[0]\mid E_x[\tilde{v}]\,\right]$$

$$\succeq\;\mathsf{def\ fresh}\ c\ \mathsf{in}\ \mathcal{E}nv_{x\pm}\left[\,R_x\{\tilde{u},m\}\,[0]\mid(\mathsf{repl}\ emit\langle m\rangle)\mid x^\circ\langle\tilde{v}\rangle\,\right]\quad(1)$$

$$\asymp\;\mathsf{def\ fresh}\ d\ \mathsf{in}\ \mathcal{E}nv_{x\pm}\left[\,R_x\{\tilde{u},d\}\,[0]\mid(\mathsf{repl}\ emit\langle d\rangle)\mid x^\circ\langle\tilde{v}\rangle\,\right]\quad(2)$$

$$\asymp\;\mathsf{def\ fresh}\ d\ \mathsf{in}\ \mathcal{E}nv_{x\pm}\left[\,R_x\{\tilde{u}\}\,[0]\mid(\mathsf{repl}\ emit\langle d\rangle)\mid x^\circ\langle\tilde{v}\rangle\,\right]\quad(3)$$

$$\equiv\;\mathcal{E}nv_{x\pm}\left[\,R_x\{\tilde{u}\}\,[0]\mid x^\circ\langle\tilde{v}\rangle\,\right]\quad(4)$$

Let $Q_0=W\mid plug\langle emit,recv\rangle\mid plug'\langle x^+\rangle$. The expansion (1) is obtained by applying Lemma 6, taking $C[\cdot]=\mathcal{E}\,[\mathsf{def\ keys}\ x^+,x^-\ \mathsf{in}\ \cdot\,]$ and $Q=Q_0$. The relations (2) and (3) are obtained by applying Lemmas 7 and 8, respectively, taking $C[\cdot]=\mathcal{E}\,[\mathsf{def\ keys}\ x^+,x^-\ \mathsf{in}\ \cdot\mid(x^\circ\langle\tilde{v}\rangle\mid(\mathsf{repl}\ emit\langle d\rangle)\mid Q_0)\,]$. The structural equivalence (4) is obtained by first restricting the scope of d to the process $\mathsf{repl}\ emit\langle d\rangle$—by hypothesis, d does not occur elsewhere—then by using the structural equivalence $W\equiv W\mid\mathsf{def\ fresh}\ d\ \mathsf{in\ repl}\ emit\langle d\rangle$.

5: Suppose that

$$\mathcal{E}nv_{x\pm}\left[\,R_x\mid emit\langle m\rangle\,\right]\to Q$$

and that x^+ and x^- do not occur in m. The only possible reduction of $\mathcal{E}nv_{x\pm}\left[\,R_x\mid emit\langle m\rangle\,\right]$ is the first reduction shown in Lemma 2, where a message is consumed and turned into a message on a continuation channel. The message can be either $emit\langle m\rangle$ or a copy of $emit\langle d\rangle$ unfolded from W. We treat only the first case; the second one is similar. Since x^+ does not occur in m, it follows that m cannot be a well-formed message for x. We thus have:

$$Q\equiv\mathcal{E}nv_{x\pm}\left[\,R_x\{\}\,[\kappa\langle m\rangle]\,\right]$$

$$\succeq\;\mathcal{E}nv_{x\pm}\left[\,R_x\{\}\,[F_{m,x}]\mid emit\langle m\rangle\,\right]$$

$$\succeq\;\mathcal{E}nv_{x\pm}\left[\,R_x\{m\}\,[F'_{m,x}]\mid emit\langle m\rangle\,\right]$$

$$\asymp\;\mathcal{E}nv_{x\pm}\left[\,R_x\{m\}\,[0]\mid emit\langle m\rangle\,\right]$$

$$\asymp\;\mathcal{E}nv_{x\pm}\left[\,R_x\mid emit\langle m\rangle\,\right]$$

by successively applying Lemmas 2, 4 (first part), 5 (second case), and 8.

6: We prove that if $R_x\{\tilde{u}\}\,[0]\in\mathbf{R}$, x^- does not occur in m, and

$$\mathcal{E}nv_{x\pm}\left[\,R_x\{\tilde{u}\}\,[0]\mid emit\langle m\rangle\,\right]\to Q$$

then

$$Q\succeq\mathcal{E}nv_{x\pm}\left[\,R'\mid T\,\right]$$

for some $R' \in \mathbf{R}$ and some process T such that x^- does not occur in T. As in the previous argument, the reduction step can be only the first one of Lemma 2, and the message can be either $emit\langle m \rangle$ or a copy of $emit\langle d \rangle$ unfolded from W. We detail only the case of $emit\langle m \rangle$; that of $emit\langle d \rangle$ is similar. Let $Q' = \mathcal{E}nv_{x^\pm} [R_x\{\widetilde{u}\} [F_{m,x}] \mid emit\langle m \rangle]$. Applying Lemma 2, we obtain $Q \succeq Q'$. If m is in $\widetilde{u}$, then Lemma 4 (second part) yields

$$Q' \succeq \mathcal{E}nv_{x^\pm} [R_x\{\widetilde{u}\} [0] \mid emit\langle m \rangle]$$

so we can take $R' = R_x\{\widetilde{u}\} [0]$ and $T = emit\langle m \rangle$. Otherwise, let $Q'' = \mathcal{E}nv_{x^\pm} [R_x\{\widetilde{u}, m\} [F'_{m,x}] \mid emit\langle m \rangle]$. Applying Lemma 4 (first part), we obtain $Q' \succeq Q''$. In turn, applying Lemma 5, we obtain

$$Q'' \succeq \mathcal{E}nv_{x^\pm} [R_x\{\widetilde{u}, m\} [0] \mid x^\circ\langle \widetilde{v} \rangle \mid emit\langle m \rangle]$$

or

$$Q'' \succeq \mathcal{E}nv_{x^\pm} [R_x\{\widetilde{u}, m\} [0] \mid emit\langle m \rangle]$$

depending on whether m is well-formed. So we can take $R' = R_x\{\widetilde{u}, m\} [0]$ and $T = x^\circ\langle \widetilde{v} \rangle \mid emit\langle m \rangle$ or $T = emit\langle m \rangle$.   □

7 Conclusion

In this paper, we apply a process calculus and its theory to address correctness issues for basic cryptographic protocols. The protocols are intended to convey securely a single datum, and they are described with only an abstract view of networking and cryptographic algorithms. Nevertheless, these protocols are crucial for translations that map arbitrary join-calculus processes to lower-level processes that communicate over a public network. The security of the translations depends on the correctness of the underlying protocols. We present a few sufficient conditions for correctness, and an example of a correct protocol.

This paper is a partial review and a continuation of an ongoing research project. We are currently investigating extensions of the join-calculus with constructs for authentication. The corresponding protocols implement authentication using digital signatures. They address some of the inefficiencies of the protocols discussed in this paper; for example, the presence of identity information removes the mutual anonymity of emitters and receivers, and enables some reuse of keys. So far, in this study of authentication, we have treated protocols on a case by case basis. However, we hope to obtain general correctness criteria analogous to the ones of this paper.

References

1. Martín Abadi. Protection in programming-language translations. In *Proceedings of the 25th International Colloquium on Automata, Languages and Programming*, pages 868–883, July 1998.

2. Martín Abadi, Cédric Fournet, and Georges Gonthier. Secure implementation of channel abstractions. Manuscript, on the Web at http://join.inria.fr/; subsumes [3] and [4].

3. Martín Abadi, Cédric Fournet, and Georges Gonthier. Secure implementation of channel abstractions. In *Proceedings of the Thirteenth Annual IEEE Symposium on Logic in Computer Science*, pages 105–116, June 1998.

4. Martín Abadi, Cédric Fournet, and Georges Gonthier. Secure communications processing for distributed languages. In *Proceedings of the 1999 IEEE Symposium on Security and Privacy*, pages 74–88, May 1999.

5. Martín Abadi and Andrew D. Gordon. A calculus for cryptographic protocols: The spi calculus. *Information and Computation*, 148(1), January 1999. An extended version appeared as Digital Equipment Corporation Systems Research Center report No. 149, January 1998.

6. Mihir Bellare, Ran Canetti, and Hugo Krawczyk. A modular approach to the design and analysis of authentication and key exchange protocols. In *Proceedings of the 30th Annual ACM Symposium on Theory of Computing*, pages 419–428, May 1998.

7. Andrew D. Birrell. Secure communication using remote procedure calls. *ACM Transactions on Computer Systems*, 3(1):1–14, February 1985.

8. Dominique Bolignano. Towards the formal verification of electronic commerce protocols. In *Proceedings of the 10th IEEE Computer Security Foundations Workshop*, pages 133–146, 1997.

9. Rocco De Nicola and Matthew C. B. Hennessy. Testing equivalences for processes. *Theoretical Computer Science*, 34:83–133, 1984.

10. Cédric Fournet. *The Join-Calculus: a Calculus for Distributed Mobile Programming*. PhD thesis, Ecole Polytechnique, Palaiseau, November 1998.

11. Cédric Fournet and Georges Gonthier. The reflexive chemical abstract machine and the join-calculus. In *Proceedings of POPL '96*, pages 372–385. ACM, January 1996.

12. Cédric Fournet, Georges Gonthier, Jean-Jacques Lévy, Luc Maranget, and Didier Rémy. A calculus of mobile agents. In Ugo Montanari and Vladimiro Sassone, editors, *Proceedings of the 7th International Conference on Concurrency Theory*, volume 1119 of *Lecture Notes in Computer Science*, pages 406–421. Springer-Verlag, August 1996.

13. Cédric Fournet, Cosimo Laneve, Luc Maranget, and Didier Rémy. Implicit typing à la ML for the join-calculus. In Antoni Mazurkiewicz and Jòzef Winkowski, editors, *Proceedings of the 8th International Conference on Concurrency Theory*, volume 1243 of *Lecture Notes in Computer Science*, pages 196–212. Springer-Verlag, July 1997.

14. Cédric Fournet and Luc Maranget. The join-calculus language (version 1.03). Source distribution and documentation available from http://join.inria.fr/, June 1997.

15. Alan O. Freier, Philip Karlton, and Paul C. Kocher. The SSL protocol: Version 3.0. Available at http://home.netscape.com/eng/ssl3/draft302.txt, November 1996.

16. Dieter Gollmann. What do we mean by entity authentication? In *Proceedings of the 1996 IEEE Symposium on Security and Privacy*, pages 46–54, May 1996.

17. D. Harkins and D. Carrel. RFC 2409: The Internet Key Exchange (IKE). Available at ftp://ftp.isi.edu/in-notes/rfc2409.txt, November 1998.

18. Pat Lincoln, John Mitchell, Mark Mitchell, and Andre Scedrov. A probabilistic poly-time framework for protocol analysis. In *Proceedings of the Fifth ACM Conference on Computer and Communications Security*, pages 112–121, November 1998.
19. Gavin Lowe. Some new attacks upon security protocols. In *Proceedings of the 10th IEEE Computer Security Foundations Workshop*, 1996.
20. Gavin Lowe. A hierarchy of authentication specifications. In *Proceedings of the 10th IEEE Computer Security Foundations Workshop*, pages 31–43, 1997.
21. Nancy Lynch. I/O automaton models and proofs of shared-key communications systems. In *Proceedings of the 12th IEEE Computer Security Foundations Workshop*, pages 14–29, 1999.
22. Catherine Meadows. Analysis of the Internet Key Exchange protocol using the NRL protocol analyzer. In *Proceedings of the 1999 IEEE Symposium on Security and Privacy*, May 1999.
23. Catherine Meadows and Paul Syverson. A formal specification of requirements for payment transactions in the SET protocol. In *Proceedings of the Financial Cryptography Conference*, 1998.
24. Alfred J. Menezes, Paul C. van Oorschot, and Scott A. Vanstone. *Handbook of Applied Cryptography*. CRC Press, 1996.
25. Robin Milner. *Communication and Concurrency*. Prentice Hall International, 1989.
26. Robin Milner. Functions as processes. *Mathematical Structures in Computer Science*, 2:119–141, 1992.
27. Robin Milner, Joachim Parrow, and David Walker. A calculus of mobile processes, parts I and II. *Information and Computation*, 100:1–40 and 41–77, September 1992.
28. J. C. Mitchell, V. Shmatikov, and U. Stern. Finite-state analysis of SSL 3.0. In *7th USENIX Security Symposium*, pages 201–216, 1998.
29. Lawrence Paulson. Inductive analysis of the Internet Protocol TLS. *ACM Transactions on Information and System Security*, 2(3), August 1999.
30. A. W. Roscoe. Intensional Specifications of Security Protocols. In *Proceedings of the 9th IEEE Computer Security Foundations Workshop*, pages 28–38. IEEE Computer Society Press, 1996.
31. Davide Sangiorgi and Robin Milner. The problem of "weak bisimulation up to". In W. R. Cleaveland, editor, *Proceedings of CONCUR'92*, volume 630 of *Lecture Notes in Computer Science*, pages 32–46. Springer-Verlag, 1992.
32. David Wagner and Bruce Schneier. Analysis of the SSL 3.0 protocol. In *Proceedings of the Second USENIX Workshop on Electronic Commerce Proceedings*, pages 29–40, November 1996. A revised version is available at `http://www.cs.berkeley.edu/~daw/me.html`.
33. Tatu Ylönen. SSH—Secure login connections over the Internet. In *Proceedings of the Sixth USENIX Security Symposium*, pages 37–42, July 1996.
34. Tatu Ylönen. Private communication. 1999.

Explaining Updates by Minimal Sums*

Jürgen Dix[1][**] and Karl Schlechta[2]

[1] University of Maryland, College Park, MD 20752, USA
dix@cs.umd.edu,
http://www.uni-koblenz.de/~dix
[2] Laboratoire d'Informatique de Marseille, CNRS ESA 6077, CMI,
39 rue Joliot Curie, F-13453 Marseille Cedex 13, France
ks@gyptis.univ-mrs.fr
http://protis.univ-mrs.fr/~ks

Abstract. Human reasoning about developments of the world involves always an assumption of *inertia*. We discuss two approaches for formalizing such an assumption, based on the concept of an *explanation*: *(1)* there is a general preference relation $\prec$ given on the set of all explanations, *(2)* there is a notion of a *distance* dist between models and explanations are *preferred* if their sum of distances is minimal. Each *distance* dist naturally induces a preference relation $\prec_{dist}$. We show exactly under which conditions the converse is true as well and therefore both approaches are equivalent modulo these conditions. Our main result is a general representation theorem in the spirit of Kraus, Lehmann and Magidor.

1 Introduction

Reasoning about *developments* or *changing* situations[1] is an important problem in Artificial Intelligence, as has been recognized very early. Much of human reasoning about these problems is based on the assumption that the world is relatively *static*. We will, for instance, hesitate to accept an explanation as *plausible* which involves many and unmotivated changes.

Generally, there is always an assumption of *inertia* formalizing that certain properties tend to *persist over time*. Many nonmonotonic logics have been used to formalize persistency ([BDK97]). E.g. circumscriptive approaches try to minimize change by circumscribing certain predicates (see [ZF93,KL95]). Default logics formalize persistency by stating special *default rules* ([HM87]). In logic programming, various versions of *negation-as-failure* have been defined to specify that fluents persist if other fluents *can not be proved* to hold (see [BD99,GL93]).

In this paper we generalize a particular approach introduced in [Win88,Win89] and [Dal88]. The overall framework is propositional logic with

---

* A full version of this paper, including detailed proofs, can be obtained under http://www.cs.umd.edu/~dix/pub_journals.html.
** Currently on leave from University of Koblenz-Landau.
[1] The term *situation* is not to be confused with the same term in *situation calculus*.

C. Pandu Rangan, V. Raman, R. Ramanujam (Eds.): FSTTCS'99, LNCS 1738, pp. 142–154, 1999.
© Springer-Verlag Berlin Heidelberg 1999

respect to an underlying signature $\mathcal{L}$. We denote by $Mod_{\mathcal{L}}$ the set of all propositional models with respect to $\mathcal{L}$. In this paper, however, we do not make use of the fact that $Mod_{\mathcal{L}}$ is induced by $\mathcal{L}$. We abstract from this and view this set simply as a set of *worlds* denoted by W. The actual world can then be simply represented as an element of W. In most cases, however, we do not know the actual world. All we know is the current *situation* which is a set of worlds.

Definition 1 (Situation S). *A situation S is a set of worlds: $S \subseteq W$. As usual, S can be also viewed as a set Th of $\mathcal{L}$-formulae: via Gödel's completeness theorem, Th induces the set $\{\mathcal{A} : Th \models \mathcal{A}\} \subseteq W$.*

How the world actually evolves (while certain actions occur) will be described by a *sequence of worlds*.

Definition 2 (Sequence σ, Explanation Expl). *A sequence, denoted by σ, $\langle \mathcal{A}_1, \dots, \mathcal{A}_n \rangle$, is a finite list of worlds: $\mathcal{A}_i \in W$. We also denote it by $\sigma := \langle \sigma_1, \dots, \sigma_n \rangle$. We say that the sequence σ explains the change from situation S to situation $S^\star$, if, by definition, $\sigma_1 \in S$ and $\sigma_n \in S^\star$.*

The sequence σ is also called an explanation *(this use was suggested by Daniel Lehmann and the authors adopt it) for the change of S to $S^\star$. We denote by $\mathsf{Expl}(S, S^\star)$ the set of all such explanations. By Expl we mean the set*

$$\mathsf{Expl}(2^W, 2^W) := \bigcup_{S, S^\star \subseteq W} \mathsf{Expl}(S, S^\star)$$

of all explanations of all possible pairs $(S, S^\star)$.

Thus, a sequence σ describes the development of the world. We note that in general, the change from an initial situation S to another situation $S^\star$ may be described by several different developments, even if both S and $S^\star$ consist of just one world. Before going on with the technical definitions, some general comments about our approach are in order:

1. We assume discrete time, and a sequence of observations: At time 1, we observed S_1, at time 2 S_2 etc., where the S_i are (usually not complete) theories corresponding to sets of worlds.
2. An explanation of this sequence is a sequence of worlds σ_1, σ_2, etc., with $\sigma_i \in S_i$.
3. Thus, given a fixed sequence of observations, its explanations all have the same length.
4. We are interested to single out the *best* or *more plausible* explanations, and do this by an assumption of a *distance*. A *distance* between worlds reflects the "cost" or "probability" of a change from one world to the other.
5. Consequently, we consider sequences of worlds with a small sum of distances between the individual worlds of this sequence as more probable or plausible than those whose sum of distances is big. Thus, such "minimal" explanations will be considered best explanations of a given sequence of observations.

6. A classical example of a distance between worlds is the Hamming distance, i.e. the number of propositional variables in which they differ. Other distances are considered e.g. in [Sch95a].

7. Depending on our assumptions about the world, a number of approaches are possible. First, we can assume an abstract, arbitrary order between explanations, this idea was pursued in [Sch95b]. Second, we can assume that explanations *with repetitions* (i.e. the world has not changed at a certain moment) are better than those without repetitions. Thus, the sequence $\langle w, w \rangle$ is considered better than the sequence $\langle w, w' \rangle$—provided both explain a given sequence of observations. This idea was pursued in [BLS98]. In the present paper, we push the idea of inertia further, minimizing the sum of changes involved in a sequence of worlds.

Such sequences from S to S^*, or explanations, may represent different *grades* of plausibility: some sequences are less plausible than others. This leads to the notion of a *plausible* explanation illustrated in the next example.

Example 1 (Plausible Explanations).

Sequences that contain loops of the form $\langle \mathcal{A}_1, \mathcal{A}_2, \mathcal{A}_1 \rangle$ and thus are unnecessary long, should not be considered as *plausible* explanations. A criterion of *inertia* is needed in order to rule out the unmotivated sequences and to define the set of plausible explanations.

Of course, the most general approach is to just assume any preference relation between explanations.

Definition 3 (Preference Relation $\prec$). *A preference relation $\prec$ is any relation on the set of all explanations* Expl

$$\prec \subseteq \text{Expl} \times \text{Expl}.$$

We call an explanation σ $\prec$-preferred, if by definition, σ is minimal with respect to $\prec$, i. e. there is no other explanation $\sigma' \neq \sigma$ with $\sigma \prec \sigma'$.

In [Sch95b,BLS98], the authors state general representation results for preference relations between arbitrary sequences of models.

A more intuitive approach to exclude such examples is due to [Win88,Win89]. The idea is to assume the notion of a distance between arbitrary worlds.

Definition 4 (Distance dist). *A distance* dist *is a function that associates to any two worlds a nonnegative rational number:*

$$\text{dist} : W \times W \longrightarrow \mathbb{Q}; (\mathcal{A}_1, \mathcal{A}_2) \mapsto \text{dist}(\mathcal{A}_1, \mathcal{A}_2)$$

The idea of [Win88,Win89] was to *measure* the sum of all distances in a sequence and to consider those sequences as most plausible that correspond to minimal sums.

Definition 5 (Plausible Explanations Induced by dist: $\prec_{\mathsf{dist}}$). *Let a distance be given on W^2. Let also two situations S, S^* and two explanations for the change from S to S^* $\sigma := \langle \sigma_1, \dots, \sigma_n \rangle$, $\sigma^* := \langle \sigma_1^*, \dots, \sigma_m^* \rangle$ be given (i. e. σ_1 and σ_1^* are both contained in S and σ_n and σ_m^* are both contained in S^*).*
We say that σ is more plausible than σ^*, denoted by $\sigma \prec_{\mathsf{dist}} \sigma^*$ if, by definition

$$\sum_{i=1}^{n-1} \mathsf{dist}(\sigma_i, \sigma_{i+1}) \le \sum_{i=1}^{m-1} \mathsf{dist}(\sigma_i^*, \sigma_{i+1}^*). \tag{1}$$

The most plausible explanations for the change from S to S^* are those whose sum of distances is minimal.
For a sequence σ we denote by $\mathsf{sum\text{-}dist}(\sigma)$ the number $\sum_{i=1}^{n-1} \mathsf{dist}(\sigma_i, \sigma_{i+1})$.

Thus, if the notion of a *distance* is available, we can immediately define an induced preference relation $\prec_{\mathsf{dist}}$. But is the converse also true? I.e. given an arbitrary preference relation $\prec$ between possible explanations, *does there exist a measure of distance* dist *on W such that $\prec = \prec_{\mathsf{dist}}$?*.

The aim of this paper is to completely solve this question by characterizing those preference relations $\prec$ which can be generated by a distance dist. To do this we use (an adaptation of) an old algorithm, going back to [Far02], to determine whether a set of inequalities of sums has a solution.

The plan of the paper is as follows. After introducing some additional terminology in Section 2, we start in Section 3 with Proposition 1, stating that the preferred sequences are completely determined by the endpoints of certain intermediate sequences. We then give a precise formulation of our main theorem and end with a sketch of the proof of this result. This proof depends heavily on a abstract representation result shortly reviewed in the appendix. We conclude with Section 4 by citing related approaches.

2   Terminology

As already mentioned above, we assume discrete time, given by the integers $\mathbb{N}$. We also assume that we we have only *incomplete information* about the state of affairs at times $t_1, \dots, t_n$. This information is given by a sequence of situations (i. e. sets of models)

Definition 6 ($\Pi_{\mathsf{fin}}(2^W)$). *A sequence $\Sigma := \langle \Sigma_1, \dots, \Sigma_n \rangle$ is a finite list of situations: $\Sigma_i \subseteq W$. Equivalently, we can view Σ as the product $\Pi_{i=1}^n \Sigma_i$. A sequence Σ represents our knowledge about the world at times $t_1, \dots, t_n$. We denote by $\Pi_{fin}(2^W)$ the union of all finite products of situations (sets of worlds) in W:*

$$\Pi_{fin}(2^W) := \bigcup_{n \in \mathbb{N}} \{ \Pi_{i=1}^n \Sigma_i^j \mid \emptyset \ne \Sigma_i^j \subseteq W \}$$

We denote by $\Sigma_{|_{i_0}}$ the restriction of Σ to the first i_0 components.

If there is a distance dist we can determine the set of those sequences σ with $\sigma_i \in \Sigma_i$ for which sum-dist(σ) (see Definition 5) is minimal. We call such sequences dist-preferred sequences. Analogously, given a preference relation $\prec$ defined on W, we call sequences $\prec$-preferred if there are no other sequences σ with $\sigma_i \in \Sigma_i$ that are smaller with respect to the relation $\prec$.

Definition 7 (dist- and $\prec$-Preferred Sequences). *Let a sequence of situations $\Sigma := \langle \Sigma_1, \ldots, \Sigma_n \rangle$ be given. We denote by* $\mathsf{Pref_{dist}}(\Sigma)$ *(resp.* $\mathsf{Pref_\prec}(\Sigma)$*) the set of* dist-preferred *(resp.* $\prec$-preferred*) sequences of worlds that are compatible with Σ: those sequences σ satisfying*

1. $\sigma_i \in \Sigma_i$ *(in particular σ and Σ have the same length),*
2. $\mathsf{sum\text{-}dist}(\sigma)$ *is minimal (resp. σ is $\prec$-preferred) among all sequences satisfying (1).*

Note that $\mathsf{Pref_{dist}}(\Sigma)$ *(resp.* $\mathsf{Pref_\prec}(\Sigma)$*) are plausible explanations for the change of situation Σ_1 to Σ_n.*

We now associate to any sequence of situations Σ the set of endpoints of dist- (resp. $\prec$-) preferred sequences compatible with Σ.

Definition 8 ($\mathsf{End_\prec}, \mathsf{End_{dist}}$). *We define the following functions, depending on the underlying preference relation* dist *or* $\prec$.

$$\mathsf{End_{dist}} : \Pi_{fin}(2^W) \to 2^W;$$
$$\Sigma \quad \mapsto \{\mathcal{A} \in W \mid \exists \sigma \in \mathsf{Pref_{dist}}(\Sigma) \text{ s.t.}^a \ \sigma_n = \mathcal{A}\}$$
$$\mathsf{End_\prec} : \Pi_{fin}(2^W) \to 2^W;$$
$$\Sigma \quad \mapsto \{\mathcal{A} \in W \mid \exists \sigma \in \mathsf{Pref_\prec}(\Sigma) \text{ s.t.}^a \ \sigma_n = \mathcal{A}\}$$

---

^{a} Note that Σ and σ have the same length, say n.

The function $\mathsf{End}_{dist}(\cdot)$ for given dist has certain properties, which we will later use to completely characterize it. This means we will prove a theorem of the form

If the function $\mathsf{End_\prec} : \Pi_{fin}(2^W) \to 2^W$ *satisfies certain properties, then there is a distance* dist *on W^2 such that* $\mathsf{End_\prec}(\Sigma) = \mathsf{End}_{dist}(\Sigma)$ *for all $\Sigma \in \Pi_{fin}(2^W)$.*

We would like to emphasize that our approach only assumes knowledge about $\mathsf{End}_{dist}(\Sigma)$, i. e. about the endpoints in Σ_n. We do not assume anything about the *endpoints of intermediate sequences of length less than n*: $\mathsf{End}_{dist}(\Sigma_{|i})$ for $i < n$.

Consequently, from $\mathsf{End}_{dist}(\Sigma)$ the set of all dist-preferred sequences can not be reconstructed. On the other hand we show in Proposition 1 (Section 3) that knowledge of the endpoints of intermediate sequences allows us to completely reconstruct $\mathsf{Pref_{dist}}(\Sigma)$.

Although it is not needed to formulate our problem, the following extension of $\prec$ from a relation between sequences σ to a relation between sequences Σ is very important in the proof of our main result.

Remark 1 (Extending $\prec$, $\prec_{\text{dist}}$ to sequences Σ). The relations $\prec$, (resp. $\prec_{\text{dist}}$) can be straightforwardly extended to relations between sequences Σ:

$$\Sigma \prec \Sigma' \text{ if, by definition, } \sigma \prec \sigma' \text{ for all } \sigma \in \text{Pref}_\prec(\Sigma), \sigma' \in \text{Pref}_\prec(\Sigma'),$$
$$(resp. \ \sigma \in \text{Pref}_{\text{dist}}(\Sigma), \sigma' \in \text{Pref}_{\text{dist}}(\Sigma')).$$

We assume that we have the *information* $\text{End}_{dist}(\Sigma)$ only about products Σ of sets of models, but not about arbitrary sets of sequences. Thus,

$$\text{End}_{dist}(\{a, a'\} \times \{b, b'\})$$

will be given, but, if $a \neq a'$ and $b \neq b'$, then $\text{End}_{dist}(\{\langle a, b\rangle, \langle a', b'\rangle\})$ may not necessarily be defined—the sequences $\langle a, b'\rangle$, $\langle a', b\rangle$ are missing. On the other hand, we assume that we can *reason* about unions of sets of sequences, in particular if a union of products of sets is itself a product of sets, like

$$\{a, a'\} \times \{b, b'\} = (\{a\} \times \{b\}) \cup (\{a\} \times \{b'\}) \cup (\{a'\} \times \{b\}) \cup (\{a'\} \times \{b'\}).$$

Definition 9 (Legal Sets of Sequences). *We call a set of sequences (of situations) legal, if this set is a product of sets.*

Thus, we can reason about arbitrary sets of sequences, but the *world* does not give us information about arbitrary, only about *legal* sets of sequences. It seems a natural hypothesis that the language of reasoning may be stronger than the language of observation.

Obviously, the Σ_n are in a stronger position than the other intermediate Σ_i, by definition of $\text{End}_{dist}(\Sigma)$. This corresponds to the fact that, considering a development into the future, we are probably most interested in the final outcome. Conversely, given a development from the past to the present, we might have most information about the present.

There are, however, other directions of possible interest, and the reader will see how to adapt our conditions and proofs to the case which interests him. We examine in this paper the two extremes—all $\text{End}_{dist}(\Sigma_{|_i})$ are known, and, only one $\text{End}_{dist}(\Sigma_{|_i})$ is known. It should not be too difficult to modify our results and techniques accordingly.

3 Updating by Minimal Sums

Before formulating our main results, we need some additional notation: If σ is a sequence and a a point, σa will be the concatenation of σ with a. Consequently

1. $\sigma \times A$ will denote the set of all sequences σa, $a \in A$.
2. $\Sigma \times A$ will denote the set of all sequences σa, $\sigma \in \Sigma$, $a \in A$. Likewise $\Sigma \times a$ by abuse of notation.

The following lemma illustrates that, if we also know the preferences for suitable intermediate observations, we can totally determine the preferred sequences. The meaning of "suitable" will become clear in the proof of the lemma.

Proposition 1 ($\mathsf{Pref}_{\mathsf{dist}}(\Sigma)$ **Induced By Intermediate** $\mathsf{End}_{dist}(\Sigma_{|_i})$). *Let Σ be a sequence in the sense of Definition 7.*

$\mathsf{Pref}_{\mathsf{dist}}(\Sigma)$ *is reconstructible from* $\mathsf{End}_{dist}(\Sigma'_{|_i})$ *for suitable Σ' with $\Sigma'_i \subseteq \Sigma_i$.*

Proof. Fix i.

Case 1: $\mathsf{End}_{dist}(\Sigma_{|_i}) = \{a_i\}$. Then for all $x \in \mathsf{End}_{dist}(\Sigma_{|_{i-1}})$ there is a preferred sequence containing $\langle x, a_i \rangle$ as a subsequence. Likewise for $y \in \mathsf{End}_{dist}(\Sigma_{|_{i+1}})$.

Case 2: $|\mathsf{End}_{dist}(\Sigma_{|_i})| > 1$. If $|\mathsf{End}_{dist}(\Sigma_{|_{i-1}})| = 1$, we apply Case 1 to $i-1$. Suppose $|\mathsf{End}_{dist}(\Sigma_{|_{i-1}})| > 1$, and, for the same reason, $|\mathsf{End}_{dist}(\Sigma_{|_{i+1}})| > 1$. Fix $a_i \in \mathsf{End}_{dist}(\Sigma_{|_i})$, and consider $\Sigma[i/\{a_i\}]$, where Σ_i has been replaced by $\{a_i\}$, i.e.

$$\Sigma[i/\{a_i\}] =_{\mathsf{def}} \Sigma_1 \times \ldots \times \{a_i\} \times \ldots \times \Sigma_n.$$

If $a_{i-1} \notin \mathsf{End}_{dist}(\Sigma[i/\{a_i\}]_{|_{i-1}})$, then there is no preferred sequence through $\langle a_{i-1}, a_i \rangle$ in Σ : Any such sequence σ' through $\langle a_{i-1}, a_i \rangle$ is already in $\Sigma[i/\{a_i\}] \subseteq \Sigma$, and there is a better one in $\Sigma[i/\{a_i\}] \subseteq \Sigma$.

Suppose $a_{i-1} \in \mathsf{End}_{dist}(\Sigma[i/\{a_i\}]_{|_{i-1}})$. As $a_i \in \mathsf{End}_{dist}(\Sigma_{|_i})$, there is a preferred sequence in Σ through a_i. It is already in $\Sigma[i/\{a_i\}]$. But in $\Sigma[i/\{a_i\}]$, there is one through all $a_{i-1} \in \mathsf{End}_{dist}(\Sigma[i/\{a_i\}]_{|_{i-1}})$. By rankedness, all are preferred in Σ. So there is a preferred sequence in Σ through $\langle a_{i-1}, a_i \rangle$ for all $a_{i-1} \in \mathsf{End}_{dist}(\Sigma[i/\{a_i\}]_{|_{i-1}})$. The same argument applies to $i+1$.

Suppose now $\sigma, \sigma' \in \mathsf{Pref}_{\mathsf{dist}}(\Sigma)$, and $\sigma_i = \sigma'_i$. Let $\sigma = \sigma^{\#} \Sigma \sigma*$, where $\sigma^{\#} = \sigma_1 \ldots \sigma_i$, $\sigma_* = \sigma_{i+1} \ldots \sigma_n$. Likewise let $\sigma' = \sigma'^{\#} \sigma'^*$. Then also $\sigma^{\#} \sigma'^*$ and $\sigma'^{\#} \sigma^* \in \mathsf{Pref}_{\mathsf{dist}}(\Sigma)$. For if not, then e.g. sum-dist($\sigma^{\#}$) > sum-dist($\sigma'^{\#}$), as $\sigma' \in \mathsf{Pref}_{\mathsf{dist}}(\Sigma)$, but then sum-dist($\sigma^{\#}$)+sum-dist($\sigma^*$) > sum-dist($\sigma'^{\#}$)+ sum-dist($\sigma^*$), contradicting $\sigma \in \mathsf{Pref}_{\mathsf{dist}}(\Sigma)$.

Thus, any sequence constructed as follows:

$$a_i \in \mathsf{End}_{dist}(\Sigma_{|_i})$$
$$a_{i-1} \in \mathsf{End}_{dist}(\Sigma[i/\{a_i\}]_{|_{i-1}})$$
$$a_{i+1} \in \mathsf{End}_{dist}(\Sigma[i/\{a_i\}]_{|_{i+1}})$$

belongs to $\mathsf{Pref}_{\mathsf{dist}}(\Sigma)$, and no others. □

Our next theorem is the main result of this paper. In Section 2 we noted that a preference relation $\prec$ between worlds implies the existence of a function

$$\mathsf{End}_{\prec} : \Pi_{fin}(2^W) \to 2^W.$$

In general, the properties of this function depend on the underlying $\prec$ relation. Indeed, if there is a distance dist then the induced function

$$\mathsf{End}_{\mathsf{dist}} : \Pi_{fin}(2^W) \to 2^W$$

has a lot of properties, due to this distance function. In our main theorem we want to completely characterize the general function $\mathsf{End}_\prec(\cdot)$ by suitable such properties.

For the following, let therefore $\mathsf{End}_\prec(\cdot)$ be any function from $\Pi_{fin}(2^U)$ to 2^U.

We are looking for conditions on $\mathsf{End}_\prec(\cdot)$ which guarantee the existence of a distance with suitable order and addition on the values and which singles out $\mathsf{End}_\prec(\Sigma)$ exactly for all *legal* Σ. If the relation $\prec$ is induced from a distance dist then the following holds:

Criterion 10 (Important Conditions)
(C1) $\mathsf{End}_\prec(\Sigma) \subseteq \Sigma_n$, if Σ_n is the last component of Σ,
(C2) $\Sigma \neq \emptyset \to \mathsf{End}_\prec(\Sigma) \neq \emptyset$,
(C3) $\mathsf{End}_\prec((\bigcup \Sigma_i) \times B) \subseteq \bigcup \mathsf{End}_\prec(\Sigma_i \times B)$.

There is one last condition that we need in order to prove our equivalence result: the (Loop) criterion. Before giving the technical details, we give some intuitive explanations:

1. If a choice function can be represented by a distance, then the relation generated by it must be *free of loops*. So it must not be possible to conclude from the given information that $a + b \preceq c + d \prec a + b$, otherwise, there would be no distances a, b, c, d and addition $+$ representing it. Thus the Loop condition constrains the general preference relation $\prec$. As we put sufficiently many operations into the loop condition, the Farkas algorithm used in our proof will terminate and generate the representing distance.
2. Note that the central conditions for representability in [LMS99] (conditions $(|\ S1), (|\ A2), (|\ A3), (*S1), (*A2), (*A3))$ are also essentially loop conditions. This is not surprising, as the problem there is similar to the one posed here: we try to embed a partial order into a total order, and this can only be done if the strict part of the partial order does not contain any loops.

The proof of our main result uses an abstract representation theorem which is given in the appendix (to make the paper self-contaned). One of the important ingredients of this representation theorem is a certain equivalence relation $\equiv$. We define this relation on the set of all sequences of worlds as follows:

$$\sigma \equiv \sigma' \text{ if, by definition } \sigma \text{ and } \sigma' \text{ have the same endpoint.}$$

1. If σ, σ' have the same length, then $[\sigma, \sigma'] := \{\sigma'' : \sigma_i'' \in \{\sigma_i, \sigma_i'\} \text{ for all } i\}$. Note that $[\sigma, \sigma']$ is a (legal) product of sets (of size ≤ 2). Likewise, if Σ is a legal set of sequences, and σ a sequence, both of same length, then $[\Sigma, \sigma] := \{\sigma' : \sigma_1' \in \Sigma_i \cup \{\sigma_i\}\}$.

2. If σ, σ' are two sequences with $\sigma_n = \sigma_1'$, then $\sigma\sigma'$ denotes their concatenation $\langle \sigma_1, \ldots, \sigma_n, \sigma_n', \ldots, \sigma_{n'}' \rangle$. We write this also as $\sigma_1 \times \ldots \times \sigma_n \times \sigma_1' \times \ldots \times \sigma_{n'}'$.

Definition 11 (Hamming Distance). *If Σ is a set of sequences, σ a sequence, both of the same length, then the Hamming distance $\mathrm{hamm}(\Sigma, \sigma)$ will be the minimum of the Hamming distances $\mathrm{hamm}(\sigma', \sigma)$, $\sigma' \in \Sigma$. (The Hamming distance between two sequences σ, σ' of equal length is the number of i's s.t. $\sigma_i \neq \sigma_i'$.)*

We have to state the following definition which contains the important conditions $(R_1) - -(R_6)$ and $(+1) - -(+5)$ used in the **(Loop)**-condition of Criterion 13.

Definition 12 (Constructing $\prec$ and $\preceq$). *Originally, $\prec$ is only a relation between sequences σ, σ'. Here we extend $\prec$ to (1) a relation between arbitrary sums of sequences, and (2) to a relation between sequences Σ.*

$\prec$, $\preceq$ **and Addition:** *Let us consider in an abstract setting arbitrary sums of distances of sequences σ. I.e. we start with a set $\{a_{\sigma'} : a_{\sigma'}$ is a subsequence of σ, σ is a sequence $\}$ and equip it with a binary function $+$. So we consider the set $\{a_\sigma + \ldots + a_{\sigma'}) : \sigma, \sigma'$ sequences $\}$. In the following we will formulate conditions to constrain the interaction between $+$ and $\prec$. (The terms a_σ, a_τ correspond to one sequence. When they are compared, they are of equal length. ($\lessgtr$ stands for $\preceq$ and $\succeq$ simultaneously.)*

$(+1) \quad a_\sigma + a_\tau \lessgtr a_\tau + a_\sigma,$

$(+2) \quad (a_\sigma + a_\tau) + a_\eta \lessgtr a_\sigma + (a_\tau + a_\eta),$

$(+3) \quad a_\sigma \lessgtr a_{\sigma'} \rightarrow ((a_\tau \lessgtr a_{\tau'}) \leftrightarrow (a_\sigma + a_\tau) \lessgtr (a_{\sigma'} + a_{\tau'})),$

$(+4) \quad a_\sigma \lessgtr a_{\sigma'} \rightarrow (a_\tau \prec a_{\tau'} \leftrightarrow (a_\sigma + a_\tau \prec a_{\sigma'} + a_{\tau'}),$

$(+5) \quad (a_\sigma \prec a_{\sigma'} \wedge a_\tau \prec a_{\tau'}) \rightarrow (a_\sigma + a_\tau \prec a_{\sigma'} + a_{\tau'}).$

$\prec$, $\preceq$ **and Comparisons:** *Here we extend $\prec$ to a relation between sequences Σ. This is done by using the function $\mathrm{End}_\prec$. (In (R4), (R5) i ranges over some index set I.)*

(R1) $\Sigma \times B \preceq \Sigma \times B'$ if $\mathrm{End}_\prec(\Sigma \times (B \cup B')) \cap B \neq \emptyset,$

(R2) $\Sigma \times B \prec \Sigma \times B'$ if $\mathrm{End}_\prec(\Sigma \times (B \cup B')) \cap B' = \emptyset.$

For the left hand side:

(R3) $\Sigma \times B \preceq \Sigma' \times B$ if $\Sigma' \subseteq \Sigma,$

(R4) $(\forall i \in I : \Sigma' \times B \preceq \Sigma_i \times B)$ if $\mathrm{End}_\prec((\Sigma' \cup \bigcup \Sigma_i) \times B) \not\subseteq \bigcup \mathrm{End}_\prec(\Sigma_i \times B).$

(R5) $(\forall i \in I : \Sigma' \times B \prec \Sigma_i \times B)$ if $\bigcap \mathrm{End}_\prec(\Sigma_i \times B) \not\subseteq \mathrm{End}_\prec((\Sigma' \cup \bigcup \Sigma_i) \times B).$

(R6) If $\bigcap_{i \in I} \Sigma_i \neq \emptyset$, then $\mathrm{End}_\prec(\Sigma) = \bigcap_{i \in I} \Sigma_i.$

With the help of the notions introduced in the last definition, we define the *(Loop)*-criterion:

Criterion 13 (Loop) *The (smallest) relation defined by* (R1)–(R6), (+1)–(+5) *(see Definition 12) contains no loops involving $\prec$ (i.e. loops involving $\preceq$ are allowed, but no loops with the "strictly less" relation $\prec$). In other words, the transitive closure of this relation is antisymmetric.*

Again, if the relation $\prec$ is induced from a distance dist then the (Loop) criterion is satisfied, as can be easily checked.

Theorem 1 (Representation Theorem). *Let W, the set of explanations* Expl *and a relation $\prec \subseteq$ Expl $\times$ Expl be given. Then the following are equivalent:*

1. *There is a distance* dist *from $W \times W$ into an ordered abelian group, such that* $\mathsf{End}_\prec(\Sigma) = \mathsf{End}_{\mathsf{dist}}(\Sigma)$ *and* $\mathsf{Pref}_{\mathsf{dist}}(\Sigma) = \mathsf{Pref}_\prec(\Sigma)$.
2. *The function* $\mathsf{End}_\prec$ *satisfies the conditions of Criteria 10 and 13.*

Proof (Sketch). Due to lack of space, we can provide the reader only with a small sketch of the proof. For the detailed proof, which is rather involved and uses the abstract representation result given in the appendix, we refer to URL:http://www.cs.umd.edu/~dix/pub_journals.html.

The direction from *(1)* to *(2)* is trivial. It remains to show that *(2)* implies *(1)*.

Assume a function $\mathsf{End}_\prec$ satisfying Criteria 10 and 13. Suppose the required distance function $\mathrm{dist}(i, j)$ between two neighbouring worlds in a sequence is modelled by the variable $x_{i,j}$. Then, for a sequence $\sigma = \langle 1, 2, ..., m \rangle$, the distance function would yield the sum

$$x_{1,2} + ... + x_{m-1,m}.$$

A similar sum is built up for a sequence τ.

If $\sigma \prec \tau$, this leads to an inequality for the two sums. In this way, a system of inequalities is built up. We solve this system by using a modification of an algorithm communicated by S. Koppelberg, Berlin. The original algorithm seems to be due to [Far02]. The crucial loop criterion is used to ensure that a solution exists. It then remains to show that $\mathsf{End}_\prec(\Sigma) = \mathsf{End}_{dist}(\Sigma)$.

4   Conclusions and Acknowledgements

One of the most distinguishing features of classical reasoning as applied in mathematics and *human reasoning* as applied in everyday life, is the treatment of how the world changes over time. Humans use the fact, often induced by context, that certain properties *persist over time*. Frameworks for studying the formalization of this persistence are very important to develop reasoning calculi that can be applied for realistic scenario. The many frameworks for belief revision—as studied in the last 15 years—all treat this problem.

There have been proposed a lot of systems for dealing with this persistence problem. For example, depending on our assumptions about the world, a number of approaches are possible:

1. We can assume an abstract, arbitrary order between explanations, this idea was pursued in [Sch95b] and [Dal88].
2. We can assume that explanations *with repetitions* (i.e. the world has not changed at a certain moment) are better than those without repetitions. Thus, the sequence $\langle w, w \rangle$ is considered better than the sequence $\langle w, w' \rangle$— provided both explain a given sequence of observations. This idea was pursued in [BLS98].
3. In the present paper, we push the idea of inertia further, minimizing the sum of changes involved in a sequence of worlds ([Win88,Win89]).

We have shown in this paper the exact relationship between these approaches. We developed a general representation result in the spirit of [KLM90], Theorem 1, stating under exactly what conditions an arbitrary *preference ordering* is induced by a *distance on the underlying models*.

We note in particular that although the main theorem can be stated without too much technical machinery, its proof requires quite a bit of technical notation. We also note our use of an old result of Farkas: this shows once again that mathematical results considered quite exotic still find their applications in modern computer science.

We owe special thanks to three anonymous referees for their careful reading and their numerous suggestions to improve this paper. In particular, one of the referees read the paper extremely carefully and his suggestions were greatly appreciated.

References

BD99.   Gerhard Brewka and Jürgen Dix. Knowledge representation with extended logic programs. In D. Gabbay and F. Guenthner, editors, *Handbook of Philosophical Logic, 2nd Edition, Volume 6, Methodologies*, chapter 6. Reidel Publ., 1999.

BDK97. Gerd Brewka, Jürgen Dix, and Kurt Konolige. *Nonmonotonic Reasoning: An Overview*. CSLI Lecture Notes 73. CSLI Publications, Stanford, CA, 1997.

BLS98.   Shai Berger, Daniel Lehmann, and Karl Schlechta. Preferred history semantics for iterated updates. Technical Report TR-98-11, Leibniz Center for Research in Computer Science, Hebrew University, Givat Ram, Jerusalem 91904, Israel, July 1998. to appear in *Journal of Logic and Computation*.

Dal88.   M. Dalal. Investigations into a theory of knowledge bases revisions. In *NCAI'88*, pages 475–479. Morgan Kaufmann, 1988.

LMS99. Menachem Magidor, Daniel Lehmann and Karl Schlechta. Distance semantics for belief revision. *Journal of Symbolic Logic*, to appear, 1999.

Far02.   J. Farkas. Theorie der einfachen Ungleichungen. *Crelles Journal für die Reine und Angewandte Mathematik*, 124:1–27, 1902.

GL93.   Michael Gelfond and Vladimir Lifschitz. Representing Actions and Change by Logic Programs. *Journal of Logic Programming*, 17:301–322, 1993.

HM87.   S. Hanks and D. McDermott. Nonmonotonic Logic and Temporal Projection. *Artificial Intelligence*, 33:379–412, 1987.

KL95.   G. N. Kartha and V. Lifschitz. A Simple Formalization of Actions Using Circumscription. In Chris S. Mellish, editor, *14th IJCAI*, volume 2, pages 1970–1975. Morgan Kaufmann, 1995.

KLM90.   Sarit Kraus, Daniel Lehmann, and Menachem Magidor. Nonmonotonic Reasoning, Preferential Models and Cumulative Logics. *Artificial Intelligence*, 44(1):167–207, 1990.

Sch95a. Karl Schlechta. Logic, topology and integration. *Journal of Automated Reasoning*, 14:353–381, 1995.

Sch95b. Karl Schlechta. Preferential choice representation theorems for branching time structures. *Journal of Logic and Computation*, 5:783–800, 1995.

Win88.   Marianne Winslett. Reasoning about action using a possible models approach. In *Proc. 7th AAAI-88*, pages 89–93, 1988.

Win89.   Marianne Winslett. Sometimes updates are circumscription. In *Proc. 11th Int. Joint Conf. on Artificial Intelligence (IJCAI'89)*, pages 859–863, 1989.

ZF93.   Yan Zhang and Norman Y. Foo. Reasoning about persistence: A theory of actions. In Ruzena Bajcsy, editor, *Proc. of the 13th Int. Joint Conf. on Artificial Intelligence (IJCAI'93)*, pages 718–723. Morgan Kaufmann, 1993.

A An Abstract Representation Result

Definition 14 (The Abstract Framework). *Let the following be given:*

1. *a nonempty universe U, an arbitrary set,*
2. *a function $\Omega : \Pi_{fin}(2^U) \to 2^U$,*
3. *an equivalence relation $\equiv$ on U (we write $[\![u]\!]$ for the equivalence class of $u \in U$ under $\equiv$) such that $[\![u]\!]$ is finite for all $u \in U$,*
4. *two relations $\prec$ and $\preceq$ on U with $\prec \subseteq \preceq$. We denote by $\preceq^*$, (resp. $\prec^*$) the transitive closure of $\preceq$ (resp. $\prec$).*

We also assume that the following holds for Ω, $\prec$ and $\preceq$:

(Ω_0) *We assume two conditions: $\Omega(A) \subseteq A$, and "$A \neq \emptyset$ implies $\Omega(A) \neq \emptyset$",*

(Ω_1) *if $a \in A$, $[\![a]\!] \cap \Omega(A) = \emptyset$, $[\![b]\!] \cap \Omega(A) \neq \emptyset$, then there is $b' \in [\![b]\!] \cap A$, $b' \prec^* a$,*

(Ω_2) *if $a \in A$, $[\![a]\!] \cap \Omega(A) \neq \emptyset$, $[\![b]\!] \cap \Omega(A) \neq \emptyset$, then there is $b' \in [\![b]\!] \cap A$, $b' \preceq^* a$.*

In the first part (Proposition 2 (1)), we construct a ranked[2] order $\lhd$ on U by extending the relation $\prec$ (and $\preceq$), and show that $\Omega = \Omega_\lhd$, where $\Omega_\lhd$ is the minimality operation induced by $\lhd$: $\Omega_\lhd(X) := \{x \in X : \neg \exists x' \in X \, x' \lhd x\}$.

Proposition 2 (Constructing Ranked Orders).

(1) If the relation $\preceq$ is free from cycles containing $\prec$, then $\prec$ can be extended to a ranked order $\lhd$ s.t. for all $A \subseteq U$ and $a \in A$:

$$[\![a]\!] \cap \Omega(A) = \emptyset \ \ \textit{if and only if} \ \ [\![a]\!] \cap \Omega_\lhd(A) = \emptyset.$$

---

[2] $\lhd$ on U is called *ranked*, if, by definition, there exists a function $rank : U \to T$ from U to a strict total order $(T, <_T)$ such that $u \lhd u'$ if and only if $rank(u) <_T rank(u')$.

(2) *If, in addition, U is a set of abstract distances $d(\cdot,\cdot)$ over some space W, i.e. $U = \{d(x,y) : x,y \in W\}$ s.t., in addition to the conditions $\Omega_0, \Omega_1, \Omega_2$ the following holds:*

(d_1) $\forall x,y \in W : x \neq y$ *implies* $d(x,x) \prec d(x,y)$,
(d_2) $\forall x,y \in W : d(x,x) \preceq d(x,y)$

and the relation $\preceq$ is free from cycles containing $\prec$, then there is a totally ordered set $(Z, <)$ with a minimal element 0 and a distance function dist $: W \times W \to Z$ *s.t.*

(a) $0 = \text{dist}(x,x)$ *for any $x \in W$,*
(b) $d(u,v) \prec d(x,y) \to \text{dist}(u,v) < \text{dist}(x,y)$,
$\qquad d(u,v) \preceq d(x,y) \to \text{dist}(u,v) \leq \text{dist}(x,y)$,
(c) *for all $A \subseteq U$, $a \in A$:* $[\![a]\!] \cap \Omega(A) = \emptyset$ *if and only if* $[\![a]\!] \cap \Omega_<(A) = \emptyset$.

A Foundation for Hybrid Knowledge Bases*

James J. Lu[1], Neil V. Murray[2], and Erik Rosenthal[3]

[1] Department of Computer Science, Bucknell University,
Lewisburg, PA 17837. USA
jameslu@bucknell.edu

[2] Department of Computer Science, State University of New York,
Albany, NY 12222,
nvm@cs.albany.edu

[3] Department of Mathematics, University of New Haven,
West Haven, CT 06516,
brodsky@charger.newhaven.edu

Abstract. Hybrid knowledge bases (HKB's) [11] were developed to provide formal models for the mediation of data and knowledge bases [14,15]. They are based on *Generalized Annotated Logic Programming* (GAP)[7] and employ an inference mechanism, *HKB-resolution*, that is considerably simpler than those that have been proposed for GAP. The simplicity of HKB-resolution is explained in this paper by showing that it is a special case of *Ʊ-resolution*, which was introduced in [9]. A generalization of Ʊ-resolution to lattices that are not *ordinary* is also explored.

Keywords: inference, hybrid knowledge bases, deduction

1 Introduction

Hybrid knowledge bases (HKB's) were proposed in [11] to model *mediated systems* [14,15,16,17], which are knowledge base systems that reason across databases and knowledge sources with different structures. HKB's combine several forms of automated reasoning: constraint logic programming [6], non-monotonic reasoning, and annotated logic programming. Several implementations of HKB's – for example, the KOMET system [1,2] and the HERMES system [13] – have been realized and applied to mediation tasks.

Hybrid knowledge bases grew out of the generalized annotated logic programming (GAP) of Kifer and Subrahmanian [7], which is a particular type of multiple-valued logic programming [4].[1] HKB's differ from GAP's in one consequential way: The query processing mechanism developed for HKB's, called HKB-resolution in [11], is considerably simpler than those that have been formulated for GAP's (see, for example, [7], [8], and [12]). The disparity can be

---

* This research was supported in part by the National Science Foundation under grants CCR-9731893, CCR-9404338 and CCR-9504349.
[1] A good survey of multiple-valued logics and their applications, including logic programming, can be found in [5].

C. Pandu Rangan, V. Raman, R. Ramanujam (Eds.): FSTTCS'99, LNCS 1738, pp. 155–167, 1999.
© Springer-Verlag Berlin Heidelberg 1999

summarized as follows: HKB-resolution is an efficient top-down query process-ing procedure similar to SLD-resolution for classical logic programming; no such general, efficient top-down procedures have been found for GAP's.

This paper provides an explanation of the mathematical simplicity of HKB-resolution. The key is that HKB assumes a truth value set that is an *ordinary lattice*, so that, for the specific truth domain formulated in [11], HKB-resolution is a special case of $\mho$-resolution.[2] The simplicity of HKB-resolution comes from underlying structure of the truth domain. It also has the search space pruning advantages inherent in $\mho$-resolution.

The basic ideas of HKB's and GAP's are described in Section 2, and $\mho$-resolution is described in Section 3. The relationship between HKB-resolution and $\mho$-resolution is developed in Section 4, and some computational properties of $\mho$-resolution that are inherited by HKB-resolution are investigated in Section 5. Section 5.2 contains an example from the KOMET system. Proofs are often omitted due to space considerations.

2   Hybrid Knowledge Bases and Generalized Annotated Logic Programming

A generalized annotated logic program consists of a first order language Λ and a complete lattice of truth values Δ under some ordering $\preceq$. An annotated atom is an expression of the form $A : \mu$ where A is an atom in Λ, and μ is an *annotation*—a term over Δ. That is, μ is a constant, a variable, or a complex term built out of constants, variables, and function symbols.

An *annotated clause* is an expression of the form
$$A : \mu \leftarrow B_1 : \beta_1, ..., B_n : \beta_n, n \geq 0$$
where $A : \mu$ and each $B_i : \beta_i$ are annotated atoms. The *head* of the annotated clause is $A : \mu$, and $\{B_1 : \beta_1, ..., B_n : \beta_n\}$ is the *body* of the clause. A *generalized annotated logic program* consists of a finite collection of annotated clauses.[3]

A *hybrid knowledge base* is a GAP whose truth domain is $\Delta = [0, 1]^{\mathcal{N}}$;[4] i.e., Δ is the set of all functions from the non-negative integers $\mathcal{N}$ to the closed interval $[0, 1]$. Without loss of generality, we need only consider *l-representable* functions [11]; that is, functions f that are multiples of characteristic functions of finite sets:
$$f(x) = v \text{ for } x \in \{n_1, ..., n_k\}; \quad f(x) = 0 \text{ otherwise.}$$
We write such a function as the pair $(v, \{n_1, ..., n_k\})$. Intuitively, v is a certainty measure for the time points $n_1, ..., n_k$. The function $(0.5, \{0, 1, 2\})$, for example, indicates a certainty of 0.5 for the time points 0,1,2, and a certainty of 0 for all other times.

---

[2] The symbol $\mho$ is pronounced "mho" because it is an upside-down Ω, which is used for the unit ohm; see Section 4 for the definitions.

[3] GAP's also admit constraints, but they need not be considered for the work presented here.

[4] There is also an assumption on the roles played by constraints in an HKB, and HKB's allow non-monotonic negations.

To simplify the discussion, assume for the remainder of the paper that the annotations in GAP's and HKB's are variable free.

An *interpretation* is a mapping from the set of ground atoms in Λ to elements of Δ. Given an interpretation I, the (variable-free) annotated atom $A : \mu$ is said to be satisfied by I, written $I \models A : \mu$, iff $\mu \preceq I(A)$. For a given GAP, the ordering $\preceq$ induces an ordering on the class of all interpretations as follows:

$$I_1 \preceq I_2 \text{ iff } I_1(A) \preceq I_2(A) \text{ for every ground atom } A.$$

The notion of satisfaction extends to annotated clauses in a straightforward way: I satisfies the clause $A : \mu \leftarrow B_1 : \beta_1, ..., B_n : \beta_n$ if whenever $I \models B_i : \beta_i$, $1 \leq i \leq n$, then $I \models A : \mu$. The next lemma is immediate.

Lemma 1. If $\beta \preceq \mu$, then $A : \mu \models A : \beta$. $\qquad\qquad\qquad\qquad\qquad$ □

It follows that if $p : \mu \leftarrow Body$ is an annotated clause in a GAP P, and if $\leftarrow p : \beta$ is a query with $\beta \preceq \mu$, then the clause and the query may be "resolved," producing the query $\leftarrow Body$.

This simple extension of ordinary resolution for classical logic programming [10] is called *annotated resolution* [7]. (By itself, it is incomplete.) To answer a query over an HKB, the inference rule *HKB-resolution* was defined in [11]. With truth domain $\Delta = [0, 1]^{\mathcal{N}}$, suppose that C is the hybrid knowledge clause $A_1 : (u, t) \leftarrow B_1 : (u_1, t_1), \ldots, B_n : (u_n, t_n)$, and that Q is the query $\leftarrow A_1 : (v_1, s_1), \ldots, A_m : (v_m, s_m)$ with $u \geq v_1$. Then the HKB-resolvent of Q and C with respect to A_1 is the query

$$\leftarrow A_1 : (v_1, s_1 - t), B_1 : (u_1, t_1), \ldots, B_n : (u_n, t_n), A_2 : (v_2, s_2), \ldots, A_m : (v_m, s_m),$$

where $s_1 - t$ is the set difference of s_1 and t. If $s_1 - t = \emptyset$, then the resulting query may be simplified by the removal of the atom $A_1 : (v_1, s_1 - t)$.

Consider, for example, the following HKB:

$$1) \quad p : (0.5, \{1, 2\}) \leftarrow Body_1 \qquad\qquad 2) \quad p : (0.7, \{2, 3\}) \leftarrow Body_2$$

The query $\leftarrow p : (0.5, \{1, 2, 3\})$ may be HKB-resolved against the first clause, yielding the query $\leftarrow p : (0.5, \{3\}), Body_1$. The new query may be HKB-resolved against the second clause to produce the next query, $\leftarrow p : (0.5, \emptyset), Body_1, Body_2$, which simplifies to $\leftarrow Body_1, Body_2$. Note that the same result is obtained if the first resolution is performed on the second clause.

3   Extended $\mho$-Resolution

We assume the reader to be familiar with the standard notions of linear and input deductions. In a logic programming setting, all deductions begin with a query, and all linear deductions are input deductions. Suppose P is a GAP and $Q = \leftarrow A_1 : \mu_1, \ldots, A_n : \mu_n$ is a query, where $P \models A_i : \mu_i, 1 \leq i \leq n$. An inference rule (or set of inference rules) $\mathcal{S}$ for GAP's is said to be *compatible with the linear restriction* if, for any such P and Q, there is a sequence $Q_0 \vdash_\mathcal{S} Q_1 \vdash_\mathcal{S} \ldots \vdash_\mathcal{S} Q_m$ such that $Q_0 = Q$, $Q_m = \square$, and for $1 \leq i < n$, Q_{i+1} is the result of applying an inference rule in S to Q_i and a clause in P or to Q_i alone. (It is possible, for example, to apply an inference rule to a constraint in Q_i.)

Inference rules that are compatible with the linear restriction enjoy a number of practical advantages, including ease of implementation and low memory requirements. It turns out that HKB-resolution *is* compatible with the linear restriction, and it is relatively efficient. However, to date, no effective inference mechanism has been found for GAP's that is compatible with the linear restriction. In [11] it was speculated that the underlying cause is the restriction of HKB's to l-representable functions. As we shall see shortly, this is not quite the case. The underlying cause is the truth domain, which, fixed as $[0,1]^N$, is an *ordinary lattice*. As a result, HKB-resolution is a special case of $\mho$-*resolution*, defined later in this section.

To define ordinary, let Δ be any lattice; if $\mu, \rho \in \Delta$, let $\mathcal{M}(\mu, \rho) = \{\gamma \in \Delta \mid \mathbf{Sup}\{\gamma, \rho\} \succeq \mu\}$, and let $\mho(\mu, \rho) = \mathbf{Inf}\,\mathcal{M}(\mu, \rho)$. Then Δ is said to be an *ordinary lattice* if $\mho(\mu, \rho) \in \mathcal{M}(\mu, \rho)$.

Consider first inference mechanisms in the more general GAP setting. A consequence of the definition of satisfaction is that, given an interpretation I and annotated atoms $A : \mu_1$ and $A : \mu_2$, $I \models A : \mu_1$ and $I \models A : \mu_2$ iff $I \models \mathbf{Sup}\{\mu_1, \mu_2\}$. In practical terms, given a query $\leftarrow A : \mu$, the information required to resolve $A : \mu$ may be distributed across the heads of several clauses in the program. These clauses, therefore, need to be combined in some way. This is illustrated by the next example.

Example. Let Δ and P be the lattice SIX and the GAP shown in Figure 1. Let Q be the query $\leftarrow p : \top$.

$$P_0 \ p{:}\mathbf{t} \leftarrow q{:}\mathbf{t}$$
$$P_1 \ p{:}\mathbf{lf} \leftarrow q{:}\mathbf{f}$$
$$P_2 \ q{:}\mathbf{t} \leftarrow$$
$$P_3 \ q{:}\mathbf{f} \leftarrow.$$

Fig. 1. The Lattice SIX and GAP P

It is clear that P entails $p : \top$. Thus the query $\leftarrow p : \top$ should be provable. It is also clear, however, that $\leftarrow p : \top$ cannot be proved via annotated resolution alone.[5] Both P_0 and P_1 must be used, but neither resolves with the query, and soundly combining them (such as with the *reduction* rule of annotated logic programming) is not linear. To maintain linearity, any inference must initially use the query $\leftarrow p : \top$. No single annotated clause has a "sufficiently high" annotation value to resolve $p : \top$ away. One remedy is to break $p : \top$ down to "simpler" queries, the combination of which is equivalent to the original query.

Considering the lattice SIX, note that there are four possible ways to decompose $\leftarrow p : \top$ into a set of simpler but equivalent queries. They are,

---

[5] Hence the incompleteness of annotated resolution alluded to earlier.

1. $\leftarrow p\!:\!\mathbf{f}, p\!:\!\mathbf{t}$ 2. $\leftarrow p\!:\!\mathbf{f}, p\!:\!\mathbf{lt}$ 3. $\leftarrow p\!:\!\mathbf{lf}, p\!:\!\mathbf{t}$ 4. $\leftarrow p\!:\!\mathbf{lf}, p\!:\!\mathbf{lt}$

This kind of example led to the inference rule *decomposition*, introduced in [9]. Let Q be the query $\leftarrow A_1 : \mu_1, ..., A_m : \mu_m$, and suppose that $A_i : \rho_1$ and $A_i : \rho_2$ are annotated atoms such that $\mu_i \preceq \mathbf{Sup}\{\rho_1, \rho_2\}$. Then $A_i : \mu_i$ is said to *decompose* to $(A_i : \rho_1, A_i : \rho_2)$, and the decomposition of Q with respect to $A_i : \mu_i$ is $\leftarrow A_1 : \mu_1, ..., A_{i-1} : \mu_{i-1}, A_i : \rho_1, A_i : \rho_2, A_{i+1} : \mu_{i+1}, ..., A_m : \mu_m$.

Soundness of decomposition was stated in [9]; the proof is presented here.

Theorem 1. Suppose Q is an annotated query. Let Q_D be a decomposition of Q, and let I be an interpretation. If $I(Q) = true$, then $I(Q_D) = true$.

Proof: Let Q be the query $\leftarrow A : \mu$ where $A : \mu$ can be decomposed to $A : \rho_1, A : \rho_2$. Then the decomposed query Q_D is the query $\leftarrow A : \rho_1, A : \rho_2$. Since $I(Q) = true$, $I(A : \mu) = false$. Hence $\mu \not\preceq I(A)$. It suffices to show that $I(Q_D) = true$, i.e. $\rho_1 \not\preceq I(A)$ or $\rho_2 \not\preceq I(A)$. Because $\mu \not\preceq I(A)$ and $\mu \preceq \mathbf{Sup}\{\rho_1, \rho_2\}$, we have $\mathbf{Sup}\{\rho_1, \rho_2\} \not\preceq I(A)$. To show $\rho_1 \not\preceq I(A)$ or $\rho_2 \not\preceq I(A)$, assume to the contrary that $\rho_1 \preceq I(A)$ and that $\rho_2 \preceq I(A)$. Then $I(A)$ is an upper bound of ρ_1 and ρ_2 by definition, so $\mathbf{Sup}\{\rho_1, \rho_2\} \preceq I(A)$, and we obtain the necessary contradiction.   □

The next theorem extends to GAP's the corresponding result in [9] for annotated logic programming.

Theorem 2. The inference system consisting of annotated resolution and decomposition is complete and is compatible with the linear restriction for GAP's.
□

Consider again the previous example. From the query $\leftarrow p\!:\!\top$, decomposition can be used to produce the query $\leftarrow p\!:\!\mathbf{lf}, p\!:\!\mathbf{t}$. This query may be solved with two annotated resolution steps.

It should be clear, however, that the use of decomposition can be woefully inefficient. Even in the simple example above, there are four choices for decomposing the query $\leftarrow p : \top$. Suppose the initial query had been decomposed to the query $\leftarrow p\!:\!\mathbf{f}, p\!:\!\mathbf{lt}$. Then annotated resolution would still not be applicable. In general, without additional information to guide the search process, a lot of backtracking would be necessary with any implementation.

One way to improve decomposition is to modify it to be a binary inference rule in which the selection of one component is based on the existence of an appropriate head literal in a clause. If μ and ρ are the annotations of the query and the selected head literal, respectively, another annotation γ may be guessed such that $\mu \preceq \mathbf{Sup}\{\gamma, \rho\}$. The inference rule extended $\mathcal{U}$-resolution, described below, exploits this observation.

In [9], $\mathcal{U}$-resolution was introduced, which is defined on a certain class of truth domains—the so called ordinary lattices. There are truth domains for GAP's that are not ordinary; six is an example. The rule defined here, tentatively called *extended $\mathcal{U}$-resolution*, applies to all GAP's, and is the subject of ongoing research. As we shall see, in a setting employing an ordinary lattice as the truth domain,

As we shall see, in a setting employing an ordinary lattice as the truth domain, extended $\mho$-resolution becomes $\mho$-resolution. More importantly, soundness and completeness arguments carry through in the general case.

To develop extended $\mho$-resolution, consider P, a GAP over truth domain Δ, and suppose Q is the query $\leftarrow A_1 : \mu_1, ..., A_m : \mu_m$. Let C be the annotated clause $A : \rho \leftarrow Body$, where $A = A_i$. We would like to soundly infer from Q and C a new query

$$\leftarrow (A_1 : \mu_1, ..., A_{i-1} : \mu_{i-1}, \ A : \gamma, Body, \ A_{i+1} : \mu_{i+1}, ..., A_m : \mu_m)$$

where γ is an appropriate guess for the annotation of the new goal. As the above discussion indicates, any guess in which $\mu_i \preceq \mathbf{Sup}\{\gamma, \rho\}$ has the desired properties. So by simply placing that requirement on γ in the inferred query, we have a sound inference.

Of course, there may be many choices for γ, although certainly fewer in general than with decomposition. Furthermore, the desirable properties of decomposition, namely soundness, completeness, and compatibility with the linear restriction, can be shown to carry over. However, a further improvement is possible.

Returning to the previous example, suppose we had chosen to infer a new query from the initial query $\leftarrow p : \top$ and the clause $p : \mathbf{t} \leftarrow q : \mathbf{t}$. The annotations $\top$ and $\mathbf{t}$ correspond to μ_i and ρ, respectively, in the inference discussed above. The choices left for γ are $\mathbf{f}$ and $\mathbf{lf}$. Hence, we have reduced the number of potential backtracking steps from four, when using decomposition, to two, by these observations.

Now, by Lemma 1, $p : \mathbf{f} \models p : \mathbf{lf}$, so the query $\leftarrow p : \mathbf{f}$ represents a more difficult goal to satisfy than $\leftarrow p : \mathbf{lf}$. Hence, the best choice for γ is $\mathbf{lf}$ since the resulting new query

$$\leftarrow p : \mathbf{lf}, q : \mathbf{t}$$

is simpler than the alternative but achieves the same effect, namely solving the original query.

This leads to the formal definition of extended $\mho$-resolution. Let

$$\mathcal{M}(\mu, \rho) = \{\gamma \in \Delta \mid \mathbf{Sup}\{\gamma, \rho\} \succeq \mu\},$$

and suppose P is a GAP over an ordinary lattice Δ, Q is the query, $\leftarrow A_1 : \mu_1, ..., A_m : \mu_m$, and C is the annotated clause $A : \rho \leftarrow Body$, with $A_i = A$. Then the extended $\mho$-resolvent of Q and C with respect to A_i is the query

$$\leftarrow (A_1 : \mu_1, ..., A_{i-1} : \mu_{i-1}, \ A : \gamma, Body, \ A_{i+1} : \mu_{i+1}, ..., A_m : \mu_m)$$

where γ is $\preceq$-minimal in $\mathcal{M}(\mu_i, \rho)$. An extended $\mho$-deduction of a query from a GAP is an *extended $\mho$-proof* if

$$\leftarrow A_1 : \bot, A_2 : \bot, ..., A_n : \bot$$

is the last clause in the $\mho$-deduction.

Soundness, completeness, and compatibility with the linear restriction still hold; the proof for $\mho$-resolution in [9] can be adapted to extended $\mho$-resolution.

Theorem 3. Extended $\mho$-resolution is sound and complete with respect to GAP in general, and is compatible with the linear restriction. $\square$

4   HKB-Resolution and $\mho$-Resolution

In Section 3 we alluded to the fact that the advantages of HKB-resolution arise because $[0,1]^{\mathcal{N}}$ is an *ordinary lattice*. The next section contains an explanation of these advantages, and the next theorem proves that $[0,1]^{\mathcal{N}}$ is ordinary. First recall the definition of ordinary. Let Δ be any lattice; if $\mu, \rho \in \Delta$, let $\mathcal{M}(\mu, \rho) = \{\gamma \in \Delta \mid \mathbf{Sup}\{\gamma, \rho\} \succeq \mu\}$, and let $\mho(\mu, \rho) = \mathbf{Inf}\, \mathcal{M}(\mu, \rho)$. Then Δ is said to be an *ordinary lattice* if $\mho(\mu, \rho) \in \mathcal{M}(\mu, \rho)$. To see why $[0,1]^{\mathcal{N}}$ is ordinary, given the functions $\mu, \rho \in [0,1]^{\mathcal{N}}$, define $\mu * \rho$ as follows: For each $x \in \mathcal{N}$,

$$\mu * \rho(x) = 0 \; if \; \mu(x) \le \rho(x)$$
$$= \mu(x) \; \text{otherwise.}$$

Lemma 2. If $\mu, \rho \in [0,1]^{\mathcal{N}}$, then $\mu * \rho = \mho(\mu, \rho)$.

Proof: It suffices to show that

1. $\mu * \rho \in \mathcal{M}(\mu, \rho)$, and
2. $\mu * \rho \preceq \gamma$ for every $\gamma \in \mathcal{M}(\mu, \rho)$.

To prove (1), let $x \in \mathcal{N}$. If $\rho(x) \ge \mu(x)$, then $\mu * \rho(x) = 0$, so $(\mathbf{Sup}\{\mu * \rho, \rho\})(x) = \rho(x) \ge \mu(x)$. If $\rho(x) < \mu(x)$, $\mu * \rho(x) = \mu(x)$. In either case, $(\mathbf{Sup}\{\mu * \rho, \rho\})(x) \ge \mu(x)$, so $(\mathbf{Sup}\{\mu * \rho, \rho\}) \succeq \mu$; i.e., $\mu * \rho \in \mathcal{M}(\mu, \rho)$.

To prove (2), consider any γ in $\mathcal{M}(\mu, \rho)$. To show that $\mu * \rho \preceq \gamma$, let $x \in \mathcal{N}$. If $\rho(x) \ge \mu(x)$ then $\mu * \rho(x) = 0 \le \gamma(x)$. Otherwise, since $\gamma \in \mathcal{M}(\mu, \rho)$, $(\mathbf{Sup}\{g, \rho\})(x) \ge \mu(x)$. Since $\rho(x) < \mu(x), \gamma(x) \ge \mu(x) = \mu * \rho(x)$. In either case, $\gamma(x) \ge \mu * \rho(x)$, so $\mu * \rho \preceq \gamma$. $\square$

Theorem 4 now follows easily from the lemma since $\mu * \rho \in [0,1]^{\mathcal{N}}$ whenever $\mu, \rho \in [0,1]^{\mathcal{N}}$.

Theorem 4. The lattice $[0,1]^{\mathcal{N}}$ is ordinary. $\square$

It is easy to see that the theorem applies to the l-representable functions in $[0,1]^{\mathcal{N}}$: If (u, s) and (v, t) are l-representable, then $(u, s) * (v, t) = (u, s)$ if $u > v$, and $(u, s) * (v, t) = (u, s - t)$ if $u \le v$, so the set of l-representable functions is closed under the operator $*$. Moreover, the HKB-resolvent of $A_1 : (u, s)$ and $A_1 : (v, t)$ is $A_1 : (u, s) * (v, t)$. In particular, this proves the next theorem.

Theorem 5. HKB-resolution *is* precisely $\mho$-resolution restricted to the truth domain $[0,1]^{\mathcal{N}}$. $\square$

The next corollary is a restatement of the completeness theorem from [11], but it follows directly from Theorems 3, 4, and 5.

Corollary. HKB-resolution is sound and complete for hybrid knowledge bases and is compatible with the linear restriction. $\square$

5   Computational Considerations

There are many ordinary lattices other than $[0,1]^{\mathcal{N}}$; examples include all finite distributive lattices [9], all powerset lattices, and many of the temporal lattices mentioned by Kifer and Subrahmanian [7], among others. The lattice SIX (Figure 1) is *not* ordinary. Consider, for instance, **t** and **lt**:

$$\mho(\mathbf{t},\mathbf{lt}) = \mathbf{Inf}\,\mathcal{M}(\mathbf{t},\mathbf{lt}) = \mathbf{Inf}\{\mathbf{t},\top,\mathbf{f},\mathbf{lf}\} = \bot \notin \mathcal{M}(\mathbf{t},\mathbf{lt}).$$

There are pairs (μ,ρ) within SIX, and in many lattices that are not ordinary, that satisfy $\mho(\mu,\rho) \in \mathcal{M}(\mu,\rho)$. For such pairs, extended $\mho$-resolution is particularly attractive since, as we have seen for HKB's, $\mho(\mu,\rho)$ represents the "best" choice for the γ in the definition of extended $\mho$-resolution. For pairs that do not satisfy $\mho(\mu,\rho) \in \mathcal{M}(\mu,\rho)$, extended $\mho$-resolution is still applicable and still reduces the number of choices compared with decomposition.

In addition to the aforementioned advantages, extended $\mho$-resolution automatically incorporates a form of search space pruning. To see how, suppose we are given a query $\leftarrow A\!:\!\mu$ and a clause $A\!:\!\rho \leftarrow Body$, where $\mu \npreceq \rho$. Application of extended $\mho$-resolution is useful only if $\mu \npreceq \gamma$, where γ is the annotation selected for the $\mho$-resolvent. The reason is that, if $\mu \preceq \gamma$, then the resulting query is subsumed immediately by the initial query.

Now, if γ is $\preceq$-minimal in $\mathcal{M}(\mu,\rho)$, $\mu \prec \gamma$ can never occur, and the only situation that warrants consideration is when $\mu = \gamma$. More generally, we would like to avoid deductions that are "circular" with respect to annotations, as in the following situation.

$$Q_0 \leftarrow A\!:\!\mu_0,\dots$$
$$Q_1 \leftarrow A\!:\!\mu_1,\dots$$
$$\dots$$
$$Q_n \leftarrow A\!:\!\mu_0,\dots$$

Here, Q_0 is some initial query, and each query Q_i in the deduction is the result of applying extended $\mho$-resolution to the atom $A\!:\!\mu_{i-1}$. Note that the annotation associated with A at step Q_n is the same as that of the initial query. In general, loop detection techniques such as tabling (see, for instance, [3]) are required to prevent circular deductions from occurring. However, for ordinary lattices, extended $\mho$-resolution is $\mho$-resolution, and circular deductions with respect to annotations are avoided by not allowing $\mu = \mho(\mu,\rho)$. The following example assumes the lattice SIX. which, although not ordinary, illustrates the point. Let

$$P = \{\underbrace{p\!:\!\mathbf{f} \leftarrow q\!:\!\mathbf{f}}_{C_1},\ \underbrace{p\!:\!\mathbf{t} \leftarrow q\!:\!\mathbf{t}}_{C_2},\ \underbrace{p\!:\!\mathbf{t}}_{C_3},\ \underbrace{q\!:\!\mathbf{f}}_{C_4}\}.$$

Observe that $P \models p\!:\!\mathbf{lt}$. However, the query $\leftarrow p\!:\!\mathbf{lt}$ is automatically prevented from $\mho$-resolving with C_1, since $\mho(\mathbf{lt},\mathbf{f}) = \mathbf{lt}$; i.e., the resolvent is immediately subsumed by the original query. Thus, C_2 and C_3 are acceptable candidates for extended $\mho$-resolution with the query, but C_1 is not.

In this example, C_1 is not required for solving the query. It could be used if subsuming queries were allowed (see the deduction shown in Figure 2), but whether the query is solvable relies completely on whether it can be solved from $P - \{C_1\}$. Avoiding the use of C_1 may represent considerable savings since the body of C_1 could be arbitrarily large and/or could lead to dead-end deductions. Indeed, proof space reduction was a primary motivation for the introduction of $\mho$-resolution in [9]. Below we consider other deduction properties of extended $\mho$-resolution for both ordinary and non-ordinary lattices. The important result is, for ordinary lattices, the pruning of the search space occurs independently of the order in which clauses are selected for extended $\mho$-resolution.

$$
\begin{aligned}
&\leftarrow p\!:\!\mathrm{lt} && \text{original query}\\
&\leftarrow p\!:\!\mathrm{lt}, q\!:\!\mathrm{f} && \mho\text{-resolution with } C_1\\
&\leftarrow q\!:\!\mathrm{t}, q\!:\!\mathrm{f} && \mho\text{-resolution with } C_2\\
&\leftarrow q\!:\!\mathrm{f} && \mho\text{-resolution with } C_3\\
&\square && \mho\text{-resolution with } C_4
\end{aligned}
$$

Fig. 2. A deduction allowing for subsuming queries

5.1   Properties of $\mho$-Proofs

Perhaps the most important computational property of $\mho$-resolution, proved in [9], is that, in comparison with the inference system **AR** proposed in [7], for each proof that can be obtained through **AR**, there is a $\mho$-resolution proof that is no longer. Moreover, **AR** is not compatible with the linear restriction. The same result is valid for extended $\mho$-resolution, which is introduced here.

Theorem 6. Let P be a GAP and Q a query. Suppose $\mathcal{R}_{\mathbf{AR}}$ is a proof of Q from P with the inference system **AR**. Then there is a proof $\mathcal{R}_{\mho}$ of Q from P with extended $\mho$-resolution that is no longer than $\mathcal{R}_{\mathbf{AR}}$. $\qquad\square$

There is one simple sufficient (but not necessary) condition for detecting $\mu = \mho(\mu, \rho)$: Every ν less than μ is also less than ρ. More precisely, the *downset* of $x \in \Delta$ is $\downarrow x = \{y \in \Delta | y \preceq x\}$. Then μ is said to be *prime relative to* ρ if μ and ρ are not comparable and $(\downarrow \mu - \{\mu\}) \subseteq \downarrow \rho$.

Lemma 3. Suppose Δ is ordinary and μ is prime relative to ρ. Then $\mu = \mho(\mu, \rho)$.

Proof: Since $\mu \in \mathcal{M}(\mu, \rho)$ by definition, it suffices to show that $\mu \preceq \gamma$ for each $\gamma \in \mathcal{M}(\mu, \rho)$. Suppose not. Then for some $\gamma \in \mathcal{M}(\mu, \rho)$, either $\gamma \prec \mu$ or μ and γ are incomparable.

If $\gamma \prec \mu$, then $\gamma \prec \rho$, since μ is prime relative to ρ. It follows that $\mathbf{Sup}\{\gamma, \rho\} = \rho$, which is not comparable to μ. But this contradicts the fact that $\gamma \in \mathcal{M}(\mu, \rho)$.

Suppose now that μ and γ are incomparable; let $\delta = \mathbf{Inf}\{\gamma, \mu\}$. Since $\gamma, \mu \in \mathcal{M}(\mu, \rho)$, $\mho(\mu, \rho) \preceq \delta$. Hence, since Δ is ordinary, $\delta \in \mathcal{M}(\mu, \rho)$, so $\mu \preceq \mathbf{Sup}\{\rho, \delta\}$. On the other hand, since $\delta \preceq \mu$ and μ is prime relative to ρ, $\delta \prec \rho$. But then $\mathbf{Sup}\{\delta, \rho\} = \rho$, which is not comparable to μ. Contradiction.

$\qquad\square$

The condition of relative primeness provides a convenient test for avoiding useless applications of $\mho$-resolution. The converse of the lemma does not hold, however. A counterexample is the lattice shown in Figure 3. The element μ is

Fig. 3. The lattice FIVE

not prime relative to ρ and yet $\mho(\mu, \rho) = \mu$. Observe on the other hand that ρ is prime relative to μ and Lemma 3 applies. Finding a necessary and sufficient condition for characterizing when $\mu = \mho(\mu, \rho)$ remains an open problem. Such a condition would be useful in understanding the extent of the pruning that occurs with $\mho$-resolution.

Another issue in the analysis of proof space: How much is the pruning dependent on the order of clause selection? Consider the query $\leftarrow A : \mu$ and these clauses in a GAP

$$C_1 \ A:\beta_1 \leftarrow Body_1, \quad C_2 \ A:\beta_2 \leftarrow Body_2, \quad \ldots, \quad C_n \ A:\beta_n \leftarrow Body_n .$$

Consider C_1, and suppose $\mu = \mho(\mu, \beta_1)$. An application of $\mho$-resolution to C_1 and the query $\leftarrow A : \mu$ produces a subsumed query, so this step can and should be avoided. The question is, will choosing some other clause first change the status of C_1 so that it can usefully participate in the deduction? The next theorem will help to answer this question.

Theorem 7 (Order Independence). Suppose Δ is ordinary, and suppose $\mu, \beta_1, \beta_2 \in \Delta$. Then
$$\mathcal{M}(\mho(\mu, \beta_1), \beta_2) = \mathcal{M}(\mho(\mu, \beta_2), \beta_1).$$

Proof: Let $\alpha_1 = \mho(\mu, \beta_1)$ and $\alpha_2 = \mho(\mu, \beta_2)$; the following equivalences prove the theorem.

$$
\begin{aligned}
x \in \mathcal{M}(\alpha_1, \beta_2) \ & \text{iff } \alpha_1 \preceq \mathbf{Sup}\{x, \beta_2\} \\
& \text{iff } \mathbf{Sup}\{\alpha_1, \beta_1\} \preceq \mathbf{Sup}\{x, \beta_1, \beta_2\} \\
& \text{iff } \mu \preceq \mathbf{Sup}\{x, \beta_1, \beta_2\} \\
& \text{iff } \mathbf{Sup}\{x, \beta_1\} \in \mathcal{M}(\mu, \beta_2) \\
& \text{iff } \alpha_2 \preceq \mathbf{Sup}\{x, \beta_1\} \\
& \text{iff } x \in \mathcal{M}(\alpha_2, \beta_1)
\end{aligned}
$$
$\square$

A simple corollary of the theorem is that the result of applying $\mho$ to a given μ and two elements, β_1 and β_2, is order independent.

Corollary. Suppose Δ is ordinary. Then for any triple μ, β_1 and β_2,
$$\mho(\mho(\mu, \beta_1), \beta_2) = \mho(\mho(\mu, \beta_2), \beta_1) .$$
$\square$

To answer the question raised before the theorem, if C_1 is not useful for the atom $A\!:\!\mu$, then it must be the case that $\mathcal{U}(\mu,\beta_1) = \mu$. Suppose that C_2 instead of C_1 is chosen first as the clause on which to resolve, and that $\mathcal{U}(\mu,\beta_2) = \rho$. Then the corollary tells us that $\mathcal{U}(\rho,\beta_1)$ must be ρ since

$$\mathcal{U}(\rho,\beta_1) = \mathcal{U}(\mathcal{U}(\mu,\beta_2),\beta_1) = \mathcal{U}(\mathcal{U}(\mu,\beta_1),\beta_2) = \mathcal{U}(\mu,\beta_2) = \rho.$$

This implies that C_1 will be pruned from the search whether it is chosen first or second. The argument can be generalized to the entire set of clauses $C_1, ..., C_n$ in the HKB. It follows that during a $\mathcal{U}$-deduction, a clause that is "unnecessary" in a proof will be pruned independently of the order in which clauses are selected.

Theorem 8 (Clause Selection Independence). Suppose $A : \mu$ is an atom contained in a query Q and C is an annotated clause whose head, $A\!:\!\beta$, satisfies $\mathcal{U}(\mu,\beta) = \mu$. Then C will not participate in any $\mathcal{U}$-proof of the atom $A\!:\!\mu$ in Q.

□

Note that these results pertain to program clauses identifiable as unnecessary for a given query. A situation that seems similar but that is really quite different is the following. Given program clauses $C_1, ..., C_n$ and query Q, several distinct subsets of the clauses may be minimally sufficient to solve the query. For example, suppose both $\{C_1, ..., C_{n-2}\}$ and $\{C_3, ..., C_n\}$ are both sufficient to solve Q but neither has a proper subset that can. Initially, C_1 is relevant to a solution of the query, and it will remain so as the clauses $C_3, ..., C_{n-2}$ are selected for $\mathcal{U}$-resolution. But if C_{n-1} is then selected, C_1 may then be irrelevant to solving the query at hand.

Also note that this is unrelated to the issue of how clauses can be intelligently selected to obtain shorter proofs. Consider the following GAP over the lattice SIX.

$$\begin{aligned}
&C_1 \; p\!:\!\mathbf{lt} \leftarrow \mathbf{Body_1} \\
&C_2 \; p\!:\!\mathbf{lf} \leftarrow \mathbf{Body_2} \\
&C_3 \; \; p\!:\!\mathbf{f} \leftarrow Body_3
\end{aligned}$$

Both C_2 and C_3 resolve away the goal in the query $\leftarrow p\!:\!\mathbf{t}$. However, the resulting queries — $Body_2$ and $Body_3$ — may admit very different proofs; indeed, one may not admit a proof at all.

Remark. This kind of analysis of proof and search space properties of inference engines is rarely available. The result here is strong validation that the deductive component of hybrid knowledge bases is the "right" rule of inference. Theorems 4 and 5, on the other hand, bring to focus the narrowness of HKB-resolution. The computational advantage of HKB-resolution is not limited to the domain $[0,1]^{\mathcal{N}}$—it is applicable to the much richer class of ordinary lattices. The generalization of HKB-resolution, $\mathcal{U}$-resolution, therefore provides an attractive basis for the flexible implementation of HKB systems without sacrificing efficiency.

5.2 KOMET

The KOMET system [2,1], which employs the annotated resolution system of Kifer and Subrahmanian [7], has been applied to the mediation of web searches. An example of a simple mediator written in KOMET, including a detailed explanation, can be found at the web site: http://calmet-pc.ira.uka.de/komet . The truth domain is given by

$$\text{WWW} = \text{POWERSET}(\text{AltaVista},\text{Excite},\text{Yahoo},\text{Lycos}),$$

which defines the powerset lattice of a four element set.

It is intersting to note that KOMET does not restrict its truth domain to $[0,1]^{\mathcal{N}}$, not surprising in view of the remark at the end of the last section. More interesting to note is that in all of the examples on that web page, the truth domains are ordinary.

References

1. Calmet, J., Jekutsch, S., Kullmann, P., and Schü, J., A system for the integration of heterogeneous information sources. *Proceedings of the Symposium on Methodologies for Intelligent Systems*, 1997.
2. Calmet, J. and Kullmann, P., Meta web search with KOMET, *Proceedings of the IJCAI-99 Workshop on Intelligent Information Integration*, Stockholm, July, 1999.
3. Chen, W. and Warren. D. S., Tabled Evaluation With Delaying for General Logic Programs. *J.ACM*, 43(1): 20-74, 1996.
4. Fitting, M., Bilattices and the Semantics of Logic Programming, *The Journal of Logic Programming*, Elsevier Science Publishing Co, Inc., 11:91–116, 1991.
5. Hähnle, R. and Escalada-Imaz, G., Deduction in many-valued logics: a survey, *Mathware & Soft Computing*, IV(2), 69–97, 1997.
6. Jaffar, J. and Lassez, J.L., Constraint Logic Programming, *Proceedings of the ACM Principles of Programming Languages*, 111–119, 1987.
7. Kifer, M., and Subrahmanian, V.S., Theory of generalized annotated logic programming and its applications, the *J. of Logic Programming* 12, 335–367, 1992.
8. Leach, S.M., and Lu, J.J., Query Processing in Annotated Logic Programming: Theory and Implementation, *Journal of Intelligent Information Systems*, 6(1):33–58, 1996.
9. Leach, S.M., Lu, J.J., Murray, N.V., and Rosenthal, E., Ʊ-resolution: an inference for regular multiple-valued logics. *Proceedings of the 6th European Workshop on Logics in AI*, Springer, 1998.
10. Lloyd, J.W., *Foundations of Logic Programming*, 2nd ed., Springer, 1988.
11. Lu, J.J., Nerode, A., and Subrahmanian, V.S., Hybrid Knowledge Bases, *IEEE Transactions on Knowledge and Data Engineering*, 8(5):773–785, 1996.
12. Lu, J.J., Murray, N.V., and Rosenthal, E., A Framework for Automated Reasoning in Multiple-Valued Logics, *J. of Automated Reasoning* 21:39–67, 1998.
13. Subrahmanian, V.S., et. al., *HERMES: Heterogeneous Reasoning and Mediator System*, University of Maryland Technical Report. Available at: http://www.cs.umd.edu//projects/hermes/overview/paper/index.html
14. Wiederhold, G., Mediators in the Architecture of Future Information Systems, *IEEE Computer*, 38–49, 1992.

15. Wiederhold, G., Intelligent Integration of Information, *Proceedings of the ACM SIGMOD Conference on Management of Data*, 434–437, 1993.
16. Wiederhold, G., Jajodia, S., and Litwin, W., Dealing with granularity of time in temporal databases, *Proceedings of the Nordic Conference on Advanced Information Systems Engineering* (R. Anderson et al. eds.), Springer, 124–140, 1991.
17. Wiederhold, G., Jajodia, S., and Litwin, W., Integrating temporal data in a heterogeneous environment, *Temporal Databases*, Benjamin Cummings, 1993.

Hoare Logic for Mutual Recursion and Local Variables

David von Oheimb*

Technische Universität München
D-80290 München, Germany.
www.in.tum.de/~oheimb/

Abstract. We present a (the first?) sound and relatively complete Hoare logic for a simple imperative programming language including mutually recursive procedures with call-by-value parameters as well as global and local variables. For such a language we formalize an operational and an axiomatic semantics of partial correctness and prove their equivalence. Global and local variables, including parameters, are handled in a rather straightforward way allowing for both dynamic and simple static scoping. For the completeness proof we employ the powerful MGF (Most General Formula) approach, introducing and comparing three variants for dealing with complications arising from mutual recursion.

All this work is done using the theorem prover Isabelle/HOL, which ensures a rigorous treatment of the subject and thus reliable results. The paper gives some new insights in the nature of Hoare logic, in particular motivates a stronger rule of consequence and a new flexible Call rule.

Keywords: axiomatic semantics, Hoare logic, mutual recursion, soundness, relative completeness, local variables, call-by-value parameters, Isabelle/HOL.

1 Introduction

Designing a good Hoare logic for imperative languages with mutually recursive procedures and local variables still is an active area of research. By 'good' we mean a provably sound and (relatively) complete calculus that is as simple as possible and thus easy to apply. There are several complications and pitfalls concerning the status of auxiliary variables, initialization of variables, scoping, parameter passing, and mutual recursion. As we will explain in the sequel, the work presented here provides theoretically interesting and practically useful solutions to these problems, and thus is good in the above sense.

Classical verification systems dealing with these subjects – see [1] for an overview – typically neglect mutual recursion and have turned out to be unsound, as mentioned e.g. by [4] and [5], or incomplete, or at least require several auxiliary rules with awkward syntactic side-conditions. Recent investigations

---

* research funded by the DFG Project BALI, http://isabelle.in.tum.de/Bali/.

C. Pandu Rangan, V. Raman, R. Ramanujam (Eds.): FSTTCS'99, LNCS 1738, pp. 168–180, 1999.
© Springer-Verlag Berlin Heidelberg 1999

tend to be much more precise, e.g. on the role of auxiliary variables, and even employ mechanical theorem provers to reliably prove soundness and completeness results. Here we emphasize the work of Kleymann[1][10],[5] who suggests a Hoare logic of total correctness and proves it sound and relatively complete with the mechanical theorem prover LEGO.

The work described in the present paper has been conducted in the context of Project BALI formalizing the semantics of Java and proving key properties like type soundness[7] formally within the theorem proving system Isabelle/HOL. Introducing an axiomatic semantics for a large subset of Java, we felt that there were several issues like mutual recursion and parameter passing where we could not resort to already established techniques. It turned out to be very practical and fruitful to perform our investigations in the reduced setting of a simple imperative programming language. In this respect we benefit from the pioneering work of Nipkow[6] that deals with the basic language (without procedures and local variables) within Isabelle/HOL.

One could argue that mutual recursion can be reduced to the already established results on single recursion (e.g. of Kleymann) by program transformation. But this would require non-trivial syntactic manipulations, which would be difficult to handle in a precise proof of soundness and unsuitable for practical program verification. Concerning local variables, the only fully formal treatment we know of, given by Kleymann, is a bit involved, so that one shrinks back from transferring it to procedure parameters. We are not aware of any previous work tackling even either of mutual recursion and procedure parameters whose soundness and (relative) completeness has been mechanically verified.

Just a few words on Isabelle/HOL: This is the instantiation of the generic interactive theorem prover Isabelle[8] with Church's version of Higher-Order Logic. The appearance of formulas on Isabelle/HOL is standard (e.g. '$\Longrightarrow$' is the infix implication symbol associating to the right) except that logical equivalence is expressed with the equality symbol. Predicates are functions with Boolean result, and function application is written in curried style, e.g. $f\,x$. Logical constants are declared by giving their name and type, such as $c :: \tau$. Basic definitions are written $c \equiv t$. Types follow the syntax of ML; type abbreviations are introduced simply as equations. A free datatype is defined by listing its constructors together with their argument types, separated by '|'. Isabelle offers powerful verification tools like natural deduction involving several variants of search, tableaux reasoning, general rewriting, and combinations thereof.

We deliberately let the style of presentation of this paper be influenced by the fully formal treatment caused by using Isabelle/HOL, which should give an impression of its rigor. On the other hand, we abstract from technical details as much as possible in order to present our results in a generic way.

---

[1] formerly Schreiber.

2   The IMP$_P$ Programming Language

Winskel[11] has introduced a simple imperative programming language for educational purposes called IMP. We enriched it with procedures and local variables, calling the result IMP$_P$. The syntax of its statements ("commands") is

com = SKIP | com; com | $vname$:= $aexp$ | LOCAL loc:=$aexp$ IN com
        | IF $bexp$ THEN com ELSE com | WHILE $bexp$ DO com
        | Call $pname$ | $vname$:= CALL $pname$($aexp$)

where the meanings of most of these constructs (Call being just an auxiliary one) is what you expect. The types $aexp = state \rightarrow val$ and $bexp = state \rightarrow bool$ represent arithmetic and Boolean expressions, which we do not further specify since we need only their (black-box) semantics. The type $state$ has two components, namely the function spaces $globs = glb \rightarrow val$ and $locals = loc \rightarrow val$ representing the stores for global and local variables. The two kinds of variable names are combined into a free datatype $vname$ = Glb glb | Loc loc where Glb and Loc act as tags to distinguish them. The types glb and loc as well as the type of values val are left unspecified. The type of procedure names $pname$ is also arbitrary, but is required to be finite, [2] as motivated in §5.

We model the procedure declarations of a given program by a function body :: $pname \rightarrow com$ mapping procedure names to the corresponding procedure bodies. Our meta-theoretic investigations do not require body to be specified further. For simplicity, each procedure has exactly one parameter, which we model by a generic local variable Arg :: loc, and a result variable Res :: loc (where Res $\neq$ Arg) whose value is returned on procedure exit. These are merely syntactic restrictions avoiding immaterial but cumbersome details like explicit parameter declarations and return statements.

2.1   Operational Semantics

We define the semantics of IMP$_P$ straightforwardly by an evaluation-style operational ("natural") semantics. The evaluation ("execution") of a statement c is described as a relation evalc :: ($com \times state \times state$) set between an initial state σ and a final state σ', written $\langle c, \sigma \rangle \longrightarrow \sigma'$. For lack of space and since the other inductive rules defining evalc are standard, we give only the relevant ones here:

$$Local \ \frac{\langle c, \ \sigma_0[a \ \sigma_0/X] \rangle \longrightarrow \sigma_1}{\langle \text{LOCAL } X := a \text{ IN } c, \ \sigma_0 \rangle \longrightarrow \sigma_1[\sigma_0 \langle X \rangle /X]}$$

$$CALL \ \frac{\langle \text{Call } pn, \ (\text{setlocs } \sigma_0 \text{ newlocs})[a \ \sigma_0/\text{Arg}] \rangle \longrightarrow \sigma_1}{\langle X := \text{CALL } pn(a), \sigma_0 \rangle \longrightarrow (\text{setlocs } \sigma_1 \ (\text{getlocs } \sigma_0))[X := \sigma_1 \langle \text{Res} \rangle]}$$

$$Call \ \frac{\langle \text{body } pn, \ \sigma_0 \rangle \longrightarrow \sigma_1}{\langle \text{Call } pn, \ \sigma_0 \rangle \longrightarrow \sigma_1}$$

---

[2] This is not a real restriction but a handy trick that avoids explicit well-formedness constraints implying that in any program there is only a finite number of procedures.

Note that local variables are initialized immediately when being created. The usual notion of procedure call is split into two parts, which will be very useful for the axiomatic semantics. The CALL statement replaces the local variables of the caller by the actual parameter of the called procedure as the only (by virtue of newlocs) local variable – thus implementing trivial static scoping – and restores them (except for assigning the result variable) after return. The Call statement is responsible for unfolding the procedure body only, thus implementing recursion. If it is invoked directly rather than via CALL, it implements dynamic scoping.

The above definition makes use of a few auxiliary values and functions:

newlocs :: *locals*
setlocs :: *state* → *locals* → *state*
getlocs :: *state* → *locals*   shorthand: $\sigma\langle X\rangle$ ≡ getlocs σ X
_[_:=_] :: *state* → *vname* → *val* → *state*   shorthand: $\sigma[v/X]$ ≡ $\sigma[\text{Loc } X := v]$

Our meta theory does not need define them further as it is independent of their meaning. newlocs is intended to yield the empty set of local variables, setlocs sets the local variables component of the state to a given set of variables, and getlocs returns the local variables of the state. The update function _[_:=_] modifies the state at the given point with a new value, i.e. assigns to a (global or local) variable if it already exists, or otherwise allocates and initializes one.

Properties of the evalc relation, for instance determinism, are typically proved via *rule induction*, i.e. induction on the depth of derivations. In contrast, structural induction (on the syntax of statements) is unsuitable in most cases because rules like *Call* yield structural expansion rather than reduction.

3   Axiomatic Semantics for IMP$_P$

Now that we have introduced the language IMP$_P$, we can describe the core of our work, which is its axiomatic semantics ("Hoare logic").

3.1   Assertions and Hoare Triples

Central to any axiomatic semantics is the notion of *assertions*, which describe properties of the program state before and after executing commands. Semantically speaking, assertions are just predicates on the state. We adopt this abstract view (similarly to our semantic view of expressions) and thus avoid talking explicitly on a syntactic level about terms and substitution and their interpretation. In other words, we do a "shallow embedding" of assertions in our (meta-)logic HOL. Thus, the issue of expressiveness of assertions disappears, and our notion of completeness automatically means completeness (basically) in the sense of Cook[2], i.e. completeness relative to the assumptions that all desired assertions can be expressed syntactically and all valid pure HOL formulas can be proved.

Following Kleymann[5], we give the role of auxiliary variables the attention it deserves. *Auxiliary variables*, also known as "logical" variables (as opposed to program variables), are necessary to relate input and output, in particular to express invariance properties. For example, the proposition that a procedure P does not change the contents of a program variable X is formulated as the Hoare triple $\{X=Z\}$ Call P $\{X=Z\}$, which should mean that whenever X has some value Z before calling P, after return it still has the same value. With this interpretation, Z serves as an auxiliary variable that is implicitly universally quantified. Early works on Hoare logic tended to view Z as a free[3] variable, which gives the desired interpretation only if the triple occurs positively, and otherwise gives incorrect results. Viewing Z as an arbitrary (yet fixed) constant preserves correctness, but this approach suffers from incompleteness: having obtained a procedure specification like $\{X=Z\}$ Call $Quad$ $\{Y=Z*Z\}$, it is often necessary to exploit (i.e., specialize) it for different instantiations of Z, which is impossible if Z is essentially a constant. The classical way out is sets of substitution and adaptation rules involving intricate side-conditions on variable occurrences. A real solution would be explicit quantification like $\forall Z. \{P\ Z\}\ c\ \{Q\ Z\}$, but this changes the structure of Hoare triples and makes them more difficult to handle. Instead we prefer implicit quantification at the level of triple validity, given below, making assertions explicitly dependent not only on the state, but also on auxiliary variables.

Which number of auxiliary variables of which types are required of course depends on the application. So we define the type of assertions with a parameter:

$$\alpha\ assn = \alpha \rightarrow state \rightarrow bool$$

where α may be instantiated as required. Thus the (pretty-printed) postcondition $\{Y=Z*Z\}$ mentioned above fully formally reads as $\{\lambda Z\ \sigma.\ \sigma\langle Y\rangle=Z*Z\}$ where $\alpha = int$. In general it is appropriate (and essential) to let α be the whole *state*, such that all program variables can be monitored when constructing an arbitrary relation between initial and final states.

Built on the type $\alpha\ assn$, we model a *Hoare triple* as the (degenerate) datatype $\alpha\ triple = \{\alpha\ assn\}\ com\ \{\alpha\ assn\}$. It is *valid* wrt. partial correctness, written $\models\{P\}c\{Q\}$, iff $\forall Z\ \sigma.\ P\ Z\ \sigma \Longrightarrow \forall \sigma'.\ \langle c,\sigma\rangle \longrightarrow \sigma' \Longrightarrow Q\ Z\ \sigma'$. Note the universal quantification on the auxiliary variable Z motivated above. This preliminary definition will be refined and extended to judgments with assumptions in §4.1.

3.2 Rules not Dealing with Procedures

The remainder of the current section is dedicated to the question of which Hoare-style rules should be given for the axiomatic semantics of IMP$_P$. For the moment, simple derivation judgments with single triples, written $\vdash \{P\}c\{Q\}$, suffice to capture everything but recursive procedures. So we take the usual rules, with two exceptions.

$$Local\ \frac{\vdash\{P\}\ c\ \{\lambda Z\ \sigma.\ Q\ Z\ (\sigma[\sigma'\langle X\rangle/X])\}}{\vdash\{\lambda Z\ \sigma.\ \sigma'=\sigma \wedge P\ Z\ (\sigma[a\ \sigma/X])\}\ \text{LOCAL}\ X\!:=\!a\ \text{IN}\ c\ \{Q\}}$$

The *Local* rule adapts the pre- and postconditions reflecting the operational semantics directly. To facilitate this, it remembers the initial state in σ' and

---

[3] According to standard conventions, such variables are implicitly universally quantified, i.e. $\Gamma \vdash t\ Z$ is read as $\forall Z.\ \Gamma \vdash t\ Z$. Problems arise if Z occurs also in Γ.

extracts the value of X with $\sigma'\langle X\rangle$. (The meta variable σ' could also be put as an auxiliary variable, but this would complicate matters unnecessarily.) As opposed to the rule given in [5], this yields a straightforward handling of local variables. In particular, we do not require explicit mechanisms catering for static scoping because local variables are kept separate from global ones and are reset completely on procedure call (see §3.3 below). Another option, suggested in [1], would be to simply alpha rename X in c, but this would require a syntactic side-condition, namely that the new name does not already occur in P, Q and c, and an unpleasant modification of the program text.

$$conseq \quad \frac{\forall Z\,\sigma.\ P\,Z\,\sigma \Longrightarrow \exists P'\,Q'.\ \vdash\{P'\}\ c\ \{Q'\}\ \wedge \\ \forall\sigma'.\ (\forall Z'.\ P'\,Z'\,\sigma \Longrightarrow Q'\,Z'\,\sigma') \Longrightarrow Q\,Z\,\sigma'}{\vdash\{P\}\ c\ \{Q\}}$$

Our *conseq* rule is a strengthened version of the generalized rule of consequence discovered by Kleymann. As motivated in [10], it allows adapting the values of the auxiliary variables as required, due to the universal quantification in their interpretation discussed above. Additionally here, the triple in the premise only needs to be derivable if the precondition P holds, and both new pre- and postconditions may depend on the auxiliary variables and the initial state. This allows not only other common structural rules to be derived (rather than asserted), like

$$\top\ \frac{}{\vdash\{P\}c\{\lambda Z\,\sigma.\ \mathsf{True}\}} \qquad \vee\ \frac{\vdash\{P\}c\{Q\} \quad G\vdash\{P'\}c\{Q'\}}{\vdash\{\lambda Z\,\sigma.P\,Z\,\sigma\vee P'\,Z\,\sigma\}c\{\lambda Z\,\sigma.Q\,Z\,\sigma\vee Q'\,Z\,\sigma\}}$$

but also new structural rules, e.g. one facilitating the use of the *Local* and *CALL* rules:

$$export\ \frac{\forall\sigma'.\ \vdash\{\lambda Z\,\sigma.\ \sigma'=\sigma\wedge P\,Z\,\sigma\}c\{Q\}}{\vdash\{P\}c\{Q\}}$$

A typical example is the derivation (modulo predicate-logical steps) for the fact that a local variable does not affect outer local variables with the same name:

$$\begin{array}{l} Local\ \dfrac{\top}{\forall\sigma'.\ \Gamma\vdash\{\lambda Z\,\sigma.\ \mathsf{True}\}\ c\ \{\lambda Z\,\sigma.\ \sigma'\langle X\rangle=(\sigma[\sigma'\langle X\rangle/X])\langle X\rangle\}} \\[4pt] conseq\ \dfrac{}{\forall\sigma'.\ \Gamma\vdash\{\lambda Z\,\sigma.\ \sigma'=\sigma\wedge\mathsf{True}\}\ \mathsf{LOCAL}\ X\!:=\!a\ \mathsf{IN}\ c\ \{\lambda Z\,\sigma.\ \sigma'\langle X\rangle=\sigma\langle X\rangle\}} \\[4pt] export\ \dfrac{}{\Gamma\vdash\{\lambda Z\,\sigma.\ Z=\sigma\langle X\rangle\}\ \mathsf{LOCAL}\ X\!:=\!a\ \mathsf{IN}\ c\ \{\lambda Z\,\sigma.\ Z=\sigma\langle X\rangle\}} \end{array}$$

(middle line: $\forall\sigma'.\ \Gamma\vdash\{\lambda Z\,\sigma.\ \sigma'=\sigma\wedge Z=\sigma\langle X\rangle\}\mathsf{LOCAL}\ X\!:=\!a\ \mathsf{IN}\ c\{\lambda Z\,\sigma.\ Z=\sigma\langle X\rangle\}$)

In a similar way, using some properties of getlocs and $\_[\_:=\_]$, a version of the *Local* rule corresponding to the classical rule leading to dynamic scope (cf. RULE 17 in [1]) can be derived:

$$\frac{\forall v.\ \Gamma\vdash\{\lambda Z\,\sigma.\ P\,Z\,(\sigma[v/X])\wedge\sigma\langle X\rangle=a\,(\sigma[v/X])\}\ c\ \{\lambda Z\,\sigma.\ Q\,Z\,(\sigma[v/X])\}}{\Gamma\vdash\{P\}\ \mathsf{LOCAL}\ X\!:=\!a\ \mathsf{IN}\ c\ \{Q\}}$$

3.3   Simple Procedure Rules

When arriving at procedures, one is faced with the problem that in any practical calculus recursion cannot be handled trivially (i.e. by repeated unfolding). As a first step, we adopt the standard solution of introducing Hoare triples as assumptions of judgments, which enables one to cope with recursive calls of an already unfolded procedure by appealing to a suitable assumption. Revising judgments (currently $\vdash_- :: \alpha \; triple \to bool$) to $_- \vdash_- :: \alpha \; triple \; set \to \alpha \; triple \to bool$, we allow putting triples as assumptions into the contexts of the derivation. In order to reflect this revision, we have to add a context Γ to all judgments in the above rules. Next, we add three rules, the first of them being the well-known $CallN$ ('N' stands for 'nested') rule that makes the specification of the currently unfolded procedure available as an assumption when verifying the procedure body. The second rule enables exploiting assumptions.

$$CallN \; \frac{\{\{P\} \; \texttt{Call} \; pn \; \{Q\}\} \cup \Gamma \vdash \{P\} \; \texttt{body} \; pn \; \{Q\}}{\Gamma \vdash \{P\} \; \texttt{Call} \; pn \; \{Q\}} \qquad asm \; \frac{t \in \Gamma}{\Gamma \vdash t}$$

The third rule, $CALL$, is responsible for adapting the local variables, resembling the *Local* rule, though it adapts not only one variable. It resets all local variables and binds the parameter, and in the postcondition restores them (remembering the initial state in σ') except for the one receiving the result:

$$\frac{\Gamma \vdash \{P\} \; \texttt{Call} \; pn \; \{\lambda Z \, \sigma. \; Q \; Z \, ((\text{setlocs} \; \sigma \; (\text{getlocs} \; \sigma'))[X := \sigma \langle \text{Res} \rangle])\}}{\Gamma \vdash \{\lambda Z \, \sigma. \; \sigma' = \sigma \wedge P \; Z \, ((\text{setlocs} \; \sigma \; \text{newlocs})[a \; \sigma / \text{Arg}])\} X := \texttt{CALL} \; pn(a)\{Q\}}$$

This rule demonstrates how easy it is to include (call-by-value) procedure parameters, which have been left out by [5]. It is inspired by a similar rule from [9], but differs in that it does not have to impose any syntactic restrictions on the variables occuring in the pre- and postconditions.

3.4   Extended Procedure Rules

As we will show in §5.2, the calculus as given up to here is already complete. Yet when using it to verify mutually recursive procedures with non-linear invocation structure, it becomes tedious: since the assumptions about recursive invocations can only be collected stepwise, often large parts of the proof have to be repeated for different invocation contexts. Consider the example of three procedures P, Q and R, where P calls Q and R, Q calls R, and R calls P and Q. Verifying them with the $CallN$ rule yields the following, roughly abstracted, proof tree:

$$\frac{\dfrac{\{P,Q,R\} \vdash \texttt{Call} \; P \quad \{P,Q,R\} \vdash \texttt{Call} \; Q}{\vdots \; \{P,Q,R\} \vdash (\text{body of } R) \; \vdots}{\{P,Q\} \vdash \texttt{Call} \; R}}{\dfrac{\vdots \; \{P,Q\} \vdash (\text{body of } Q) \; \vdots}{\{P\} \vdash \texttt{Call} \; Q} \qquad \dfrac{\{P,R\} \vdash \texttt{Call} \; P \quad \dfrac{\{P,Q,R\} \vdash (\text{body of } Q)}{\{P,R\} \vdash \texttt{Call} \; Q}}{\dfrac{\vdots \; \{P,R\} \vdash (\text{body of } R) \; \vdots}{\{P\} \vdash \texttt{Call} \; R}}}{\dfrac{\vdots \; \{P\} \vdash (\text{body of } P) \; \vdots}{\emptyset \vdash \texttt{Call} \; P}}$$

The bodies of Q and R each are verified twice, which may be very redundant. This can be avoided by conducting a *simultaneous* rather than nested verification of all procedures involved. Verification condition generators such as [4] take this idea to the extreme by verifying all procedures contained in a program simultaneously, forcing the user to identify in advance a single specification for each procedure suitable to cover all invocation contexts. Our solution – given next – is more flexible because it permits, each time a call to a cluster of mutually recursive procedures is encountered, to verify simultaneously as many procedures as required (but not more) and to identify the necessary specifications locally.

We extend the judgments further to $\_ \Vdash \_ :: \alpha\ triple\ set \rightarrow \alpha\ triple\ set \rightarrow bool$ ($\Gamma \vdash t$ now becomes an abbreviation of $\Gamma \Vdash \{t\}$) and replace the $CallN$ rule by

$$Call\ \frac{\Gamma \cup \{\{P_i\}\texttt{Call}\ i\{Q_i\} \mid i \in ps\} \Vdash \{\{P_i\}\texttt{body}\ i\{Q_i\} \mid i \in ps\}\quad p \in ps}{\Gamma \vdash \{P_p\}\ \texttt{Call}\ p\ \{Q_p\}}$$

When using this rule to verify a call of p, one can decide to verify simultaneously an arbitrary family of procedures where ps is the set of their names including p. Of course, we now need introduction rules for (finite) conjunctions of triples, whereas elimination rules like $subset$ may be derived from the others.

$$empty\ \frac{}{\Gamma \Vdash \emptyset}\qquad insert\ \frac{\Gamma \vdash t\quad \Gamma \Vdash ts}{\Gamma \Vdash \{t\} \cup ts}\qquad subset\ \frac{\Gamma \Vdash ts'\quad ts \subseteq ts'}{\Gamma \Vdash ts}$$

Exploiting the simultaneous $Call$ rule, the proof tree of the above example collapses to

$$\frac{\dfrac{}{\overline{\{P,Q,R\} \vdash \texttt{Call}\ P}}\quad \dfrac{}{\overline{\{P,Q,R\} \vdash \texttt{Call}\ Q}}\quad \dfrac{}{\overline{\{P,Q,R\} \vdash \texttt{Call}\ R}}}{\underline{\vdots\ \{P,Q,R\} \vdash (\text{bodies of }P,\ Q\text{ and }R)\ \vdots}}$$
$$\emptyset \vdash \texttt{Call}\ P$$

where no redundancy concerning procedure bodies remains.

Though it is – strictly speaking – not necessary, we found the cut rule very useful in applications, as it helps to adapt the premises of judgments. A similar rule, complementing the $subset$ rule, is the well-known $weaken$ rule. It can be derived from all others by rule induction, or obtained as an immediate consequence of cut and a strengthened version of asm.

$$cut\ \frac{\Gamma' \Vdash ts\quad \Gamma \Vdash \Gamma'}{\Gamma \Vdash ts}\qquad weaken\ \frac{\Gamma' \Vdash ts\quad \Gamma' \subseteq \Gamma}{\Gamma \Vdash ts}\qquad asm'\ \frac{ts \subseteq \Gamma}{\Gamma \Vdash ts}$$

4 The Proof of Soundness

This section motivates our actual definition of validity for Hoare triples, which is influenced by the proof of soundness outlined thereafter.

4.1 Validity

Validity involving assumptions, $\_ \models \_ :: \alpha\ triple\ set \rightarrow \alpha\ triple\ set \rightarrow bool$, could be defined as $\Gamma \models ts \equiv (\forall t \in \Gamma.\ \models t) \implies (\forall t \in ts.\ \models t)$. This would be reasonable, but when attempting to prove the $Call$ rule which adds assumptions about recursive procedure calls, an inductive argument on the depth of these calls is

needed. This could be achieved by syntactic manipulations that unfold procedure calls up to a given depth n, as done in [3]. We prefer a semantic approach
instead, which is influenced by [9] and [5]. We define a variant of the operational
semantics that includes a counter for the *recursive depth* of evaluations, rep-
resented by the judgment $\langle \_,\_ \rangle - \_ \rightarrow \_ :: com \rightarrow state \rightarrow nat \rightarrow state \rightarrow bool$.
The inductive rules using this new form are exactly the same as in §2.1, except
for replacing $\longrightarrow$ by $-n\rightarrow$ and replacing the *Call* rule by

$$Call \quad \frac{\langle body\ pn,\ \sigma_0 \rangle - n\rightarrow \sigma_1}{\langle CALL\ pn,\ \sigma_0 \rangle - n{+}1\rightarrow \sigma_1}$$

This refinement does not affect the semantics, i.e. the parameter n is a mere annotation, stating that evaluation needs to be done only up to recursive depth n.
The equivalence $(\langle c,\sigma \rangle \longrightarrow \sigma') = (\exists n.\ \langle c,\sigma \rangle - n\rightarrow \sigma')$ can be shown by rule induction for each direction, where the '$\Longrightarrow$' direction requires the lemma
$\langle c_1,\sigma_1 \rangle - n_1\rightarrow \sigma'_1 \wedge \langle c_2,\sigma_2 \rangle - n_2\rightarrow \sigma'_2 \Longrightarrow \exists n.\ \langle c_1,\sigma_1 \rangle - n\rightarrow \sigma'_1 \wedge \langle c_2,\sigma_2 \rangle - n\rightarrow \sigma'_2$
which in turn requires non-strictness: $\langle c,\sigma \rangle - n\rightarrow \sigma' \wedge n \leq m \Longrightarrow \langle c,\sigma \rangle - m\rightarrow \sigma'$.

According to the refined notion of statement execution, the notion of validity
for single Hoare triples receives the recursive depth as an extra parameter:
$$\models n{:}\{P\}c\{Q\} \equiv \forall Z\ \sigma.\ P\ Z\ \sigma \Longrightarrow \forall \sigma'.\ \langle c,\sigma \rangle - n\rightarrow \sigma' \Longrightarrow Q\ Z\ \sigma'$$
This definition carries over to sets of triples by $\models n{:}ts \equiv \forall t \in ts.\ \models n{:}t$.
Now we can define the final notion of validity including assumptions as
$$\Gamma \models\!\models ts \equiv \forall n.\ \models n{:}\Gamma \Longrightarrow \models n{:}ts$$
This version is strong and detailed enough to perform induction on the recursive
depth. On the other hand, when the set of assumptions is empty, it is equivalent
to the version given above because the chain $(\emptyset \models\!\models ts) = (\forall n.\ \models n{:}\emptyset \Longrightarrow \models n{:}ts) =$
$(\forall n.\ \models n{:}ts) = (\forall t \in ts.\ \models t) = ((\forall t \in \emptyset.\ \models t) \Longrightarrow (\forall t \in ts.\ \models t))$ holds.

4.2   Actual Soundness Proof

With our new definition of validity we can express soundness as $\emptyset \vdash t \Longrightarrow \emptyset \models t$.
This is a direct instance of $\Gamma \Vdash ts \Longrightarrow \Gamma \models\!\models ts$, which can be shown by rule induction on the derivation of the Hoare judgments and an auxiliary rule induction
for the *Loop* rule. The *Call* rule is the only difficult case, where we benefit from
the proof given in [3] suggesting a lemma that in our case reads as
$\Gamma \cup \{\{P_i\}\ \texttt{Call}\ i\ \{Q_i\} \mid i \in ps\} \models\!\models \{\{P_i\}\ body\ i\ \{Q_i\} \mid i \in ps\} \Longrightarrow$
$\models n{:}\Gamma \Longrightarrow \models n{:}\{\{P_i\}\ \texttt{Call}\ i\ \{Q_i\} \mid i \in ps\}$
Here is the point where the bounded recursive depth comes in, as we conduct the
proof by induction on n. Doing this, we exploit the simple facts $\models n{+}1{:}t \Longrightarrow \models n{:}t$,
$\models 0{:}\{P\}\ \texttt{Call}\ i\ \{Q\}$, and $(\models n{+}1{:}\{P\}\ \texttt{Call}\ i\ \{Q\}) = (\models n{:}\{P\}\ body\ i\{Q\})$. The
CallN rule can of course be derived directly from the *Call* rule.

As we can conclude from this section, the only interesting aspect of the proof
of soundness is to find a suitable notion of validity capable of capturing an
inductive argument on the recursive depth of procedure calls. Of course, due to
the number of rules in the operational and axiomatic semantics, in the inductive
proofs there are a lot of cases involving some amount of detail to be considered,
for which the mechanical theorem prover is of great help.

5   Three Proofs of Completeness

Much more challenging than the proof of soundness is the proof of completeness. Here we benefit heavily from the MGF approach promoted by [5] and others. We extend this approach, which was given for only a single recursive procedure, to several mutually recursive procedures. When dealing with mutual recursion some complications arise, which we overcome in three different ways, each with specific advantages and drawbacks. For lack of space we can describe only proof outlines and mention crucial lemmas.

5.1   The MGF Approach

For proving completeness of Hoare logics involving procedures, typically some variant of *Most General Formula, MGF* for short, is used. A MGF is a judgment $\Gamma \vdash \texttt{MGT}\ c$ where $\texttt{MGT}$ takes a command c and returns a *Most General Triple* which describes the most general property of c, namely its operational semantics. The basic variant of a MGT for partial correctness is

$$\texttt{MGT}\ c \equiv \{\lambda Z\ \sigma_0.\ Z = \sigma_0\}\ c\ \{\lambda Z\ \sigma_1.\ \langle c, Z \rangle \longrightarrow \sigma_1\}$$

Its precondition stores the initial state σ_0 in the auxiliary variable Z, which is consequently of type *state* here. Its postcondition claims that if the execution of command c terminates in some state σ_1, this is the same as the outcome of the operational semantics of c, starting also from σ_0.

Common to all variants of MGTs is that once the corresponding MGF has been proved, completeness almost immediately emerges by virtue of the rule of consequence. For instance, $\emptyset \vdash \texttt{MGT}\ c \implies \emptyset \models \{P\}c\{Q\} \implies \emptyset \vdash \{P\}c\{Q\}$ can be proved in a two-line Isabelle script applying the definition of validity.

5.2   Version 1: Nested Structural Induction

The outline, proposed by Martin Hofmann[3], of our first completeness proof employs two inductions (in very similar situations) on the structure of commands and a variant of $\texttt{MGT}$ that is a bit more involved, namely

$$\texttt{MGT'}\ c \equiv \{\lambda Z\ \sigma_0.\ \forall \sigma_1.\ \langle c, \sigma_0 \rangle \longrightarrow \sigma_1 \implies Z = \sigma_1\}\ c\ \{\lambda Z\ \sigma_1.\ Z = \sigma_1\}$$

We refine the outline a little, first by factoring out structural induction into the *MGT-lemma* $(\forall p.\ \Gamma \vdash \texttt{MGT}\ (\texttt{Call}\ p)) \implies \Gamma \vdash \texttt{MGT}\ c$ such that it is performed only once, and second by replacing $\texttt{MGT'}$ by the simpler $\texttt{MGT}$. [4]

The proof of $\emptyset \vdash \texttt{MGT}\ c$ reveals the crux of structural induction: when arriving at unfolding procedure calls, the new subgoal gets structurally larger, such that we cannot appeal directly to any induction hypothesis. Assumptions in the

---

[4] For the case of the WHILE loop, we return to $\texttt{MGT'}$ because there the auxiliary variable σ_0 has to serve as the (invariant) final state of the iteration. Both variants are equivalent, where $\texttt{MGT'}$ entails $\texttt{MGT}$ only if the language is deterministic (which is true for IMP_P) and there are at least two different program states, which we simply assume since empty or singleton state spaces are of no interest anyway.

judgments come to the rescue. Still, there remains a challenge: when using them naively, one is faced with the need to use structural induction nested as deep as the number of procedures in the program. This problem is overcome by resorting to an auxiliary induction on the number of procedures not yet considered, such that we strengthen our proof goal to $\Gamma' = \{\text{MGT (Call } p) \mid \text{True}\} \implies$ $\forall \Gamma.\ \Gamma \subseteq \Gamma' \implies n \leq |\Gamma'| \implies |\Gamma| = |\Gamma'| - n \implies \forall c.\ \Gamma \vdash \text{MGT } c$ where Γ' equals the set of all possible procedure calls. Its proof is by induction on n, exploiting the MGT-lemma twice. It heavily depends on Γ' being finite as otherwise calculations on cardinality like $|\Gamma| = |\Gamma'| - n$ would be meaningless. Now, $\emptyset \vdash \text{MGT } c$ is an immediate consequence (just specialize Γ to $\emptyset$ and n to $|\Gamma'|$).

Note that this version of completeness proof gets by with the $CallN$ version of the $Call$ rule (thus not requiring the rules $empty$ and $insert$), but on the other hand needs to apply it in a nested way.

5.3   Version 2: Simultaneous Structural Induction

Our desire to circumvent the nesting problem of Version 1 has been the motivation for inventing the $Call$ rule as an extension of $CallN$, which allows handling procedures simultaneously. Version 2 is also by structural induction and makes use of the MGT-lemma, but by exploiting the power of the $Call$ rule, it takes only a much simpler lemma, namely $F \subseteq \{\text{MGT (body } p) \mid \text{True}\} \implies$ $\{\text{MGT (Call } p) \mid \text{True}\} \Vdash F$. The latter is proved by induction on the size of F, so finiteness is vital also here. Comparing Version 2 with Version 1, it requires a more advanced $Call$ rule (and the two simple structural rules $empty$ and $insert$), but handles mutual recursion more directly and thus clearly.

5.4   Version 3: Rule Induction

Our third version of completeness proof takes the MGF approach to the extreme. It gave us surprising insights into the nature of Hoare logic, yet is probably of mainly theoretic interest because we could not avoid supporting it with two additional rules. Our intuition when discovering this approach has been that structural induction is not too nice, in particular when handling recursion, as the other versions show. Let us employ a more direct and powerful induction scheme: rule induction on the operational semantics.

The pattern of rule induction requires that the inductively defined relation, evalc here, occurs negatively in the formula to be proved. Unfortunately, neither $\emptyset \models \{P\}\ c\ \{Q\} \implies \emptyset \vdash \{P\}\ c\ \{Q\}$ itself nor $\emptyset \vdash \text{MGT } c$ are of this pattern. Let us resort to $\forall \sigma_0\ \sigma_1.\ \langle c, \sigma_0 \rangle \longrightarrow \sigma_1 \implies \emptyset \vdash \{\lambda Z\ \sigma.\ \sigma = \sigma_0\} c \{\lambda Z\ \sigma.\ \sigma = \sigma_1\}$ which is a kind of MGF property where the evalc relation has been pulled out of the assertion into the meta logic. From this formula we can easily show completeness applying our strong rule of consequence, but we have to require the (clearly admissible, yet non-derivable) extra rule

$$diverg\ \frac{}{G \vdash \{\lambda Z\ \sigma.\ \neg \exists \sigma'.\ \langle c, \sigma \rangle \longrightarrow \sigma'\} c \{Q\}}$$

The above MGF property itself is directly amenable to the desired rule induction, which yields a surprisingly short proof. Unfortunately, it requires an unfolding variant of the *Loop* rule reflecting the operational semantics:

$$LoopT \quad \frac{\Gamma \vdash \{\lambda Z\, \sigma.\ P\, Z\, \sigma \wedge b\, \sigma\} c\{Q\} \quad \Gamma \vdash \{Q\}\ \texttt{WHILE}\ b\ \texttt{DO}\ c\ \{R\}}{\Gamma \vdash \{\lambda Z\, \sigma.\ P\, Z\, \sigma \wedge b\, \sigma\}\ \texttt{WHILE}\ b\ \texttt{DO}\ c\ \{R\}}$$

On the other hand, only a trivial variant of the *Call* rules (namely one without assumptions) and no auxiliary variables are needed here.

Thus we can conclude that, in principle, the issues of assumptions and auxiliary variables can be circumvented! Of course, this is only a theoretical point as in actual program verification one does not want to be faced with the operational semantics again, which was suitable for the meta-level completeness proof only.

6 Conclusion

In this paper we have described new approaches for dealing with mutual recursion, procedure parameters and local variables in a Hoare-style calculus. The calculus is powerful – and also simple and convenient – enough to be used in actual program verification efforts. In particular, we have introduced a relatively simple handling of local variables, a convenient and flexible rule for simultaneously verifying mutually recursive procedures, and a strong rule of consequence.

All results have been achieved using the theorem prover Isabelle/HOL, which not only gives full confidence in their correctness, but also was a great aid in cleanly formalizing the theory and conveniently conducting the proofs.

We have combined several existing techniques with new ideas, resulting in a lucid soundness proof and three variants of completeness proofs. Once discovered, they should be transferable to other logical systems and programming languages with relative ease. The major current application is to an object-oriented language, namely the investigation of Java within Project BALI.

Acknowledgments

I thank Tobias Nipkow and Martin Hofmann for fruitful discussions on handling mutual recursion. The idea how to perform nested structural induction is due to Martin Hofmann. I also thank Manfred Broy, Tobias Nipkow, Leonor Prensa Nieto, Bernhard Reus, Francis Tang, Markus Wenzel and several anonymous referees for their comments on draft versions of this paper.

References

1. K. R. Apt. Ten years of Hoare logic: A survey – part I. *ACM Trans. on Prog. Languages and Systems*, 3:431–483, 1981.
2. Stephen A. Cook. Soundness and completeness of an axiom system for program verification. *SIAM Journal on Computing*, 7(1):70–90, 1978.

3. Martin Hofmann. Semantik und Verifikation. Lecture notes, in German. http://www.dcs.ed.ac.uk/home/mxh/teaching/marburg.ps.gz, 1997.
4. Peter V. Homeier and David F. Martin. Mechanical verification of mutually recursive procedures. In M. A. McRobbie and J. K. Slaney, editors, *Proceedings of CADE-13*, volume 1104 of *LNAI*, pages 201–215. Springer-Verlag, 1996.
5. Thomas Kleymann. Hoare logic and VDM: Machine-checked soundness and completeness proofs. (Phd Thesis), ECS-LFCS-98-392, LFCS, 1998.
6. Tobias Nipkow. Winskel is (almost) right: Towards a mechanized semantics textbook. In V. Chandru and V. Vinay, editors, *FST&TCS*, volume 1180 of *LNCS*, pages 180–192. Springer-Verlag, 1996.
7. David von Oheimb and Tobias Nipkow. Machine-checking the Java specification: Proving type-safety. In Jim Alves-Foss, editor, *Formal Syntax and Semantics of Java*, volume 1523 of *LNCS*. Springer-Verlag, 1999.
8. Lawrence C. Paulson. *Isabelle: A Generic Theorem Prover*, volume 828 of *LNCS*. Springer-Verlag, 1994. Up-to-date description: http://isabelle.in.tum.de/.
9. A. Poetzsch-Heffter and P. Müller. A programming logic for sequential Java. In S. D. Swierstra, editor, *Programming Languages and Systems (ESOP '99)*, volume 1576 of *LNCS*, pages 162–176. Springer-Verlag, 1999.
10. Thomas Schreiber. Auxiliary variables and recursive procedures. In *TAPSOFT'97*, volume 1214 of *LNCS*, pages 697–711. Springer-Verlag, 1997.
11. Glynn Winskel. *Formal Semantics of Programming Languages*. MIT Press, 1993.

Explicit Substitutions and Programming Languages

Jean-Jacques Lévy and Luc Maranget

INRIA - Rocquencourt,
Jean-Jacques.Levy@inria.fr
Luc.Maranget@inria.fr,
http://para.inria.fr/~{levy,maranget}

Abstract. The λ-calculus has been much used to study the theory of substitution in logical systems and programming languages. However, with explicit substitutions, it is possible to get finer properties with respect to gradual implementations of substitutions as effectively done in runtimes of programming languages. But the theory of explicit substitutions has some defects such as non-confluence or the non-termination of the typed case. In this paper, we stress on the sub-theory of weak substitutions, which is sufficient to analyze most of the properties of programming languages, and which preserves many of the nice theorems of the λ-calculus.

1 Introduction

In the past ten years, several calculi of explicit substitutions have been proposed and studied with various motivations. In their original work, Curien [10] and Hardin [17] considered Categorical Combinators, as an algebraic definition of the syntax of the λ-calculus. In [1,15], their calculus is simplified by using a two-sorted language, with terms and substitutions. The goal was there to study fundamental syntatic properties and applications to the design of runtime interpreters or fancy type-checkers. Unfortunately, the calculus in [1] is neither confluent (Church-Rosser property), nor strongly normalizable on the elementary first-order typed subset [28]. But it is confluent on closed terms (of explicit substitutions), which are sufficient to represent all λ-terms. Later several calculi were designed with full confluence [11], or with both properties by suppressing some of the operations of explicit substitutions such as the composition of substitutions[4,23,7]. Until very recently no fully expressive calculus existed with both properties of confluence and strong normalization. The termination problem, which is connected to cut elimination in linear logic[13], seemed more difficult; according to Martin-Löf or Melliès, it is due to an unlimited use of the η-expansion rule in the λ-calculus, which is known as non terminating when coupled with β-conversion. Recently, there has been a proposal for a new calculus of explicit substitutions with both the confluence and strong normalization properties [12], but this calculus seems rather complex. Therefore, one may be skeptic about the usage of these theories.

C. Pandu Rangan, V. Raman, R. Ramanujam (Eds.): FSTTCS'99, LNCS 1738, pp. 181–200, 1999.
© Springer-Verlag Berlin Heidelberg 1999

Explicit substitutions may be used for a refined study of logical systems with bound variables, for instance, when one wants to axiomatize α-conversion, in higher-order theorem provers, or in recent process algebras such as Action Calculi[30]. In some of these systems, renaming of bound variables has to be defined very carefully. This was the case with the axiomatization of the type-checker for Cardelli's Quest language, or with strategies for higher-order term matching[14].

But explicit substitutions were also introduced to have a formal theory of runtimes in programming languages, with implications for the CAM-machine (kernel of the Caml runtime) [9,22] or Krivine's call-by-name machine. Usually, one then restricts attention to the theory of weak explicit substitutions. There is nothing really new with this remark, and much of the work at end of last decade was related to the correspondence between weak λ-calculus and runtimes of functional languages. But to our knowledge, the theory of weak explicit substitutions has not been much studied, mainly because its properties look easy, but this is quite often a "folk" statement.

In this paper, we present several weak theories, which may be consider as various exercises on weak λ-calculus, and we try to look carefully to the fundamental properties of their syntax. The claim is not that a useful theory has to be confluent or to preserve strong normalization in the typed case, but that keeping in mind these two properties could help for studying extra properties such as dependency analysis, shared evaluation, or stack allocation.

In section 2, following Çağman and Hindley in [8] for Combinatory Logic, we define a confluent calculus of weak λ-calculus, which is not a priori obvious since confluence of this weak theory often fails. This is achieved by allowing redexes under λ-abstractions to be contracted if they do not contain occurrences of bound variables. In section 3, we consider a confluent calculus of weak explicit substitutions exactly corresponding to the calculus of previous section. In section 4, we study the corresponding reductions strategies. In section 5, we map these strategies to runtime interpreters, and show how to state properties, such as stack-allocation or graph-based sharing. In section 6, we consider the ministep semantics of weak calculus of explicit substitutions, and show its connection to more traditional presentations of explicit substitutions with de Bruijn's indices. We conclude in section 7.

2   Confluence of the Weak λ-Calculus

As usual, the set of λ-terms is the minimum set of terms M, N defined by

$$M, N ::= x \mid MN \mid \lambda x.M$$

and the β-reduction rule is

$$(\beta) \qquad (\lambda x.M)N \rightarrow M[\![x \backslash N]\!]$$

where $M[\![x \backslash N]\!]$ is recursively defined by

$$x[\![x\backslash P]\!] = N$$
$$y[\![x\backslash P]\!] = y$$
$$(MN)[\![x\backslash P]\!] = M[\![x\backslash P]\!] \, N[\![x\backslash P]\!]$$
$$(\lambda y.M)[\![x\backslash P]\!] = \lambda y.M[\![x\backslash P]\!]$$

In the last case, the substitution must not bind free variables in P, namely y must not be free in P. Usually in the λ-calculus (as in traditional mathematics), equality is defined up to the renaming of bound variables (α-conversion). Hence it is always possible to find y in $(\lambda y.M)$ filling the previous condition. Sometimes the substitution inside λ-abstractions is analogously defined as

$$(\lambda y.M)[\![x\backslash P]\!] = \lambda y'.M[\![y\backslash y']\!][\![x\backslash P]\!]$$

where y' is a variable not free in M and P.

In the (strong) λ-calculus, every context is active, since any sub-term may be reduced at any time. Formally, reduction is defined inductively by the following set of inference rules

$$(\xi) \ \frac{M \to M'}{\lambda x.M \to \lambda x.M'}$$

$$(\nu) \ \frac{M \to M'}{MN \to M'N} \qquad (\mu) \ \frac{N \to N'}{MN \to MN'}$$

In the weak λ-calculus, the ξ-rule is forbidden, and one cannot reduce inside λ-abstractions. This corresponds to the natural behavior in programming languages, since functions bodies cannot be evaluated without the actual values of their arguments. The ξ-rule corresponds more to partial evaluation of a function, and can be considered only as a compiling transformation.

But such a weak λ-calculus trivially looses the Church-Rosser property (confluence), since when $N \to N'$ we have

$$(\lambda x.\lambda y.M)N \longrightarrow (\lambda x.\lambda y.M)N'$$
$$\downarrow \qquad\qquad\qquad \downarrow$$
$$(\lambda y.M[\![x\backslash N]\!]) \qquad (\lambda y.M[\![x\backslash N']\!])$$

Term $(\lambda y.M[\![x\backslash N]\!])$ is in normal form and cannot be reduced, and the previous diagram does not commute. The problem has been known for a long time in combinatory logic [19], although often kept as a "folk theorem". In [8], it is specifically stated, and shown as being relevant when translating the λ-calculus into combinatory logic. One recovers confluence by adding the new inference rule

$$(\sigma) \ \frac{N \to N'}{M[\![x\backslash N]\!] \to M[\![x\backslash N']\!]}$$

However, this rule is not pleasant, since it axiomatizes more parallel reduction than single reduction steps. Let us first say that a variable x is linear in term M

iff there is a unique occurrence of the free variable x in M. Then we prefer the following variant to the σ-rule

$$(\sigma') \ \frac{N \to N' \quad (x \text{ linear in } M)}{M[\![x \backslash N]\!] \to M[\![x \backslash N']\!]}$$

An alternative statement of (σ') is possible with the context notation. Let a context $C[\,]$ be a λ-term with a missing sub-term, and let $C[M]$ be the corresponding term when M is placed into the hole. Say that $C[\,]$ binds M, when a free variable of M is bound in $C[M]$. Clearly when $C[\,]$ does not bind M, one has $C[M] = C[x][\![x \backslash M]\!]$ for any fresh variable x not in $C[M]$. Hence, the previous rule is equivalent to

$$(\sigma'') \ \frac{M \to M' \quad (C[\,] \text{ does not bind } M)}{C[M] \to C[M']}$$

So, inside λ-abstractions, a redex (reductible expression) does not contain free variables bound in the outside context. Namely, a redex R in M is any sub-term $R = (\lambda y.A)B$ such that $M = M_1[\![x \backslash R]\!]$ for some free x linear in M.

Now, by noticing that the σ'-rule encompasses the μ and ν-rules, it is possible to define the theory of weak λ-calculus as the set of λ-terms with the β and σ' rules. As in [3], the transitive closure of $\to$ is written $\twoheadrightarrow$. So $M \twoheadrightarrow N$ iff M can reduce in several steps (maybe none) to N.

Theorem 1. *The weak λ-calculus is confluent.*

Proof: one can follow Tait–Martin-Löf's axiomatic method used to prove the Church-Rosser property. The main remark is that the problematic previous diagram

$$
\begin{array}{ccc}
(\lambda x.\lambda y.M)N & \longrightarrow & (\lambda x.\lambda y.M)N' \\
\downarrow & & \downarrow \\
(\lambda y.M[\![x \backslash N]\!]) & \longrightarrow\!\!\!\twoheadrightarrow & (\lambda y.M[\![x \backslash N']\!])
\end{array}
$$

now commutes since y is not free in N. $\qquad\qquad\qquad\qquad\qquad\qquad\qquad\square$

In fact, the weak λ-calculus enjoys simple syntactic properties. For instance, in the standard λ-calculus, a rather complex theorem is the so-called finite development theorem, stating that the ordering in which a given set of redexes is contracted is not relevant and thus that there is a consistent definition for parallel reductions. This property is not easy to prove, since residuals of disjoint redexes may be nested. Take for instance term $(\lambda x.Ix)(Iy)$ with $I = \lambda x.x$. Then

$$(\lambda x.\underline{Ix})(\underline{Iy}) \to I(\underline{Iy})$$

It is not the case in the weak λ-calculus since Ix in $(\lambda x.Ix)$ contains the bound variable x and is not a redex in the weak λ-calculus. In order to formally define residuals of redexes, we will use two ways. The first one uses named redexes by extending the set of λ-terms as follows

$$M, N ::= x \mid MN \mid \lambda x.M \mid (\lambda x.M)^a N$$

where a is taken in a given alphabet of names. The calculus is defined by the same β and σ' rules with the addition of a new β'-rule for contraction of named redexes

$$(\beta') \qquad (\lambda x.M)^a N \rightarrow M[\![x \backslash N]\!]$$

and substitution is extended by the following equation

$$((\lambda y.M)^a N)[\![x \backslash P]\!] = (\lambda y.M[\![x \backslash P]\!])^a \, N[\![x \backslash P]\!]$$

In order to track redexes along reductions, redexes may be named and the residuals are redexes with the same names in the term after reduction. Notice that (named) redexes must not contain external bound variables as implied by (σ') and it has to be shown that residuals of redexes are still redexes.

The second way is based on the substitution notation. Let R be a redex in M and let $M \rightarrow M'$ by contraction of R. Then $M = M_1[\![x \backslash R]\!]$, $M' = M_1[\![x \backslash R']\!]$ with x linear in M_1 and $R \rightarrow R'$. Let S be another redex in M. Then $M = N_1[\![y \backslash S]\!]$ with y linear in N_1. Then the residuals of S in M' are defined by case analysis:

- If S contains R, then $M_1 = N_1[\![y \backslash S_1]\!]$, $S = S_1[\![x \backslash R]\!]$ where S_1 is a redex and x linear in S_1. (Clearly, S_1 is a redex in M_1 since x not bound in M_1 cannot be bound in S_1). Then $M = N_1[\![y \backslash S_1]\!][\![x \backslash R]\!]$ and $M' = N_1[\![y \backslash S_1]\!][\![x \backslash R']\!]$. So $M' = N_1[\![y \backslash S_1[\![x \backslash R']\!]]\!]$. The residual of S is this unique redex $S' = S_1[\![x \backslash R']\!]$. It is indeed a redex since $M' = N_1[\![x \backslash R']\!]$. Notice too that $S \rightarrow S'$.

- If S does not contain R. Then $N_1 = N_2[\![x \backslash R_1]\!]$ with x linear in N_2, and $R = R_1[\![y \backslash S]\!]$ where y may not appear in R_1. Then $N_1 = N_2[\![x \backslash R_1]\!] \rightarrow N_2[\![x \backslash R_1']\!] = N_1'$ with $R_1 \rightarrow R_1'$. And $M = N_1[\![y \backslash S]\!] \rightarrow N_1'[\![y \backslash S]\!] = M'$. The residuals of S are all the S-redexes appearing at each occurrence of y in N_1'. Notice then that all residuals are equal to S. Again no free variable in S may be bound in M' since $M' = N_1'[\![y \backslash S]\!]$.

- If $M_1 = N_1$ and $x = y$. Then S and R coincide and S has no residual in M'.

This second definition shows that residuals of redexes are still redexes in the weak λ-calculus. We also remark that a residual R' of any redex R by a given many-step reduction is always such that $R \twoheadrightarrow R'$, which is not true in the normal λ-calculus where one may substitute the free variables of a redex and give much more reduction power to residuals. Finally, it remains to show that the two ways of defining redexes give identical definitions. We leave this in exercise to the reader.

Proposition 1. *Residuals of disjoint redexes are disjoint redexes.*

Proof: Let R and S be two disjoint redexes in M. The only way to get a residual S' of S to go through a residual R' of R by the contraction of a redex $(\lambda x.A)B$ in M is that $A = C[R]$ and x is a free variable of R substituted by $B = C'[S]$. But then R is no longer a redex of the weak λ-calculus, since it contains the free variable x bound in the external context.   □

Let $\mathcal{F}$ be a set of redexes in M. A development of $\mathcal{F}$ is any maximal reduction only contracting residuals of redexes of $\mathcal{F}$.

Proposition 2. *Developments of sets of redexes are always finite.*

Proof: To each R in $\mathcal{F}$, we associate its maximal nesting level $n(R)$, where $n(R) = \max\{1 + n(S) \mid R \text{ directly contains } S \in \mathcal{F}\}$. So $n(R) = 0$ if R contains no redex in $\mathcal{F}$. Consider the multiset $\omega(\mathcal{F}) = \{n(R) \mid R \in \mathcal{F}\}$ with the natural multiset ordering. Then each step of the development decreases this multiset, since by previous proposition no new nesting appears in the residuals $\mathcal{F}'$ of $\mathcal{F}$. However, a redex contained in the contracted redex may have several copies as residuals, but then its nesting level is less that the one of the contracted redex which disappears. As the multiset ordering is well-founded, every development is finite.   □

The rest of the finite development theorem (i.e. confluence and consistency of residuals) is proved as in the standard λ-calculus. A simple way is by use of the labeled weak λ-calculus defined above, and by showing that it is confluent and strongly normalizable when contracting only with the β'-rule. (The strong normalization proof follows exactly the previous outline used for finite developments by considering the nesting levels of β'-redexes).

Proposition 3. *Let $M \to M'$ by contraction of redex R, and let S' be a redex of the weak λ-calculus, residual of S in M not inside R. Then S is also a redex in the weak λ-calculus.*

Proof: Let R be the redex contracted in $M \to M'$. Then $M = M_1[\![x \backslash R]\!]$ and $M' = M'_1[\![x \backslash R']\!]$ where $R = (\lambda z A)B$ and $R' = A[\![z \backslash B]\!]$. We work by contradiction, and suppose S is not a redex. Then a free variable y of S is bound in M. We have several cases:

- R' and S' are disjoint. Then $S' = S$ and if y in S is bound in M, clearly it is same in M'.
- S contains R'. Then $S = S_1[\![x \backslash R]\!] \to S_1[\![x \backslash R']\!] = S'$. If the free y-variable in M is bound in M. Then either y is in S_1 and remains in S', which involves that S' is not a redex of the weak λ-calculus. Either y is in R, which contradicts the fact that R is also a redex of the weak λ-calculus.

                          □

The statement of the previous property may seem over-complicated, but some care is needed since for instance when $I = \lambda x.x$, an easy counterexample is $(\lambda z.Iz)a \to Ia$.

This proposition allows now to state another interesting theorem of the weak λ-calculus, namely Curry's standardization theorem. A standard reduction is usually defined as reduction contracting redexes in an outside-in and left-to-right way. Precisely a reduction of the form

$$M = M_0 \to M_1 \to \ldots M_n = N \quad (n \geq 0)$$

is standard when for all i and j such that $i < j$, the R_j-redex contracted at step j in M_{j-1} is not a residual of a redex internal to or to the left of the R_i-redex contracted at step i in M_{i-1}. We write $M \underset{st}{\twoheadrightarrow} N$ for the existence of a standard reduction from M to N. Notice that the leftmost outermost reduction is a standard reduction (in the usual λ-calculus), but standard reductions may be more general.

Theorem 2. *If $M \twoheadrightarrow M'$, then $M \underset{st}{\twoheadrightarrow} M'$.*

Proof: One follows the proof scheme in [20] or checks the axioms of [29]. The basic step of the proof follows from proposition 3. Take $M \to N \to P$ by contracting R in M and S in N. Suppose R and S are not in the standard ordering. Then S is residual of a redex S' in M to the left of or outside R. By proposition 3, we know that S' is a redex of the weak λ-calculus and we may contract it getting N'. By the finite development theorem, we converge to P by a finite development of the residuals of R in N'.   □

3   Weak Explicit Substitutions

Although its language is minimal, some properties of the weak λ-calculus may look non intuitive and require at least a careful analysis. However it is close to a calculus of closures, which is the language of interpreters for functional languages. We now introduce such a calculus of closures, named the calculus of weak explicit substitutions, and study its connection to the weak λ-calculus.

   The language contains terms which are reductible, and programs which are constant. The new terms with respect to the weak λ-calculus are closures represented as a λ-abstraction coupled with a substitution. We use same names for variables in terms and programs, and will precise their kind when necessary. Programs correspond to all λ-terms. Substitutions are functions from variables to terms represented by their (finite) graph. Notice that the domain of a substitution is always finite

$M, N ::=$		term
	x	variable
	$\mid \ MN$	application
	$\mid \ (\lambda x.P)[s]$	closure

$$P, Q ::= x \mid PQ \mid \lambda x.P \qquad \text{programs}$$
$$s ::= (x_1, M_1), (x_2, M_2), \cdots (x_n, M_n) \qquad x_i \text{ distinct } (n \geq 0)$$

with $domain(s) = \{x_1, x_2, \cdots x_n\}$ and $s(x_1) = M_1, s(x_2) = M_2, \ldots s(x_n) = M_n$. Notice that s is explicitely written as a set of pairs representing the graph of function s. Thus the ordering in which this graph is written does not matter. Now substitutions are extended to every program in the usual way.

$$PQ[\![s]\!] = P[\![s]\!]Q[\![s]\!]$$
$$(\lambda x.P)[\![s]\!] = (\lambda x.P)[s]$$
$$x[\![s]\!] = s(x) \text{ if } x \in domain(s)$$
$$x[\![s]\!] = x \text{ otherwise}$$

Thus substitutions are applied to every subexpression of a program, except for lambda-abstraction where it stays in the substitution part of a closure. A substitution is modified by forcing one of its value

$$s[x\backslash N](y) = N \qquad \text{if } y = x$$
$$= s(y) \quad \text{otherwise}$$

The dynamics of weak explicit substitutions can now be defined by the following β-rule and inference rules for active contexts

$$(\beta) \qquad (\lambda x.P)[s]\, N \rightarrow P[\![\, s[x\backslash N]\,]\!]$$

$$(\xi) \ \frac{s \rightarrow s'}{(\lambda x.P)[s] \rightarrow (\lambda x.P)[s']}$$

$$(\nu) \ \frac{M \rightarrow M'}{MN \rightarrow M'N} \qquad (\mu) \ \frac{N \rightarrow N'}{MN \rightarrow MN'}$$

$$(\sigma) \ \frac{s(x) \rightarrow M' \quad s' = s[x\backslash M']}{s \rightarrow s'}$$

The calculus of weak explicit substitutions is nearly a first-order orthogonal term rewriting system. It manipulates sets for substitutions, which is not allowed in a standard rewriting system where only terms in a free algebra are considered. It is also defined with a scheme of axioms (the β-rule) with respect to programs. Anyhow, the calculus has the good properties of orthogonal systems.

Theorem 3. *The calculus of weak explicit substitutions is confluent.*

Proof: One proof uses the axioms in [29], another more direct proof may again follow the Tait–Martin-Löf's axiomatic technique. The only interesting cases are the two commuting diagrams

$$
\begin{array}{ccc}
(\lambda x.P)[s]N & \longrightarrow & (\lambda x.P)[s']N \\
\downarrow & & \downarrow \\
P[\![\, s[x\backslash N]\,]\!] & \longrightarrow\!\!\!\!\rightarrow & P[\![\, s'[x\backslash N]\,]\!]
\end{array}
\qquad
\begin{array}{ccc}
(\lambda x.P)[s]N & \longrightarrow & (\lambda x.P)[s]N' \\
\downarrow & & \downarrow \\
P[\![\, s[x\backslash N]\,]\!] & \longrightarrow\!\!\!\!\rightarrow & P[\![\, s[x\backslash N']\,]\!]
\end{array}
$$

when $s \rightarrow s'$ and $N \rightarrow N'$. We need then three lemmas showing that one has $s[x\backslash N] \twoheadrightarrow s'[x\backslash N]$, $P[\![s]\!] \twoheadrightarrow P[\![s']\!]$ and $s[x\backslash N] \rightarrow s[x\backslash N']$. □

The standardization theorem also holds in weak explicit substitutions. The main difficulty is in its statement since residuals have to be defined, which can be done. The definition can again be done by considering named redexes, and extending the set of terms and reductions by

$$
\begin{array}{lll}
M, N ::= & & \text{term} \\
\quad | \ \cdots & & \text{as previously} \\
\quad | \ (\lambda x.P)^a[s] \ N & & \text{named redex}
\end{array}
$$

$$(\beta') \qquad (\lambda x.P)^a[s] \ N \rightarrow P[\![\, s[x\backslash N]\,]\!]$$

Proposition 1 can also be shown, and residuals of disjoint redexes keep disjoint. Take for instance

$$M = (\lambda x.Ix)[\,](\underline{I[\,]I[\,]}) \rightarrow I[(x, \underline{I[\,]I[\,]})](\underline{I[\,]I[\,]}) = N$$

when $I = \lambda x.x$. The external redex in N is not a residual of any redex in M. This also greatly simplifies the proof of the finite development theorem (see proposition 2).

We now consider translations of weak explicit substitutions into weak λ-calculus and vice-versa. First the former may be easily translated into the latter, since it suffices to expand substitutions through program abstractions. The translation from weak explicit substitutions to λ-calculus is

$$
\begin{aligned}
\{x\} &= x \\
\{MN\} &= \{M\}\{N\} \\
\{(\lambda x.P)[s]\} &= (\lambda x.P)[\![\, \{s\}\,]\!]
\end{aligned}
$$

$$\{(x_1, M_1), (x_2, M_2), \cdots (x_n, M_n)\} = x_1\backslash\{M_1\}, x_2\backslash\{M_2\}, \cdots x_n\backslash\{M_n\}$$

We assume that no variable x_j is free in a term $\{M_i\}$, which can always be achieved by renaming the x_i variables. Thus, none of the $[\![x_i\backslash M_i]\!]$ substitution interferes with another one and we may safely use the "parallel" substitution notation.

Proposition 4. *If $M \rightarrow M'$, then $\{M\} \twoheadrightarrow \{M'\}$.*

Proof: By structural induction on M. The key point is that, given a closure $(\lambda x.C[x_i])[\cdots (x_i, M) \cdots]$, either the context $C[\ ]$ does not bind x_i, or this occurrence of x_i does not refer the binding (x_i, M). □

The converse translation from λ-calculus to explicit substitutions is a bit more involved. Several translations are possible for any given λ-term M. We consider the translation with maximal substitutions. Let P be a λ-term and Q a sub-term of P, so $P = C[Q]$ with the context notation. Say that Q is a free sub-term of P, iff $C[\ ]$ does not bind Q. We will consider maximal free sub-terms of P. For instance, we underlined them in $\lambda x.x(\underline{y}(\lambda z.x\underline{z}))$ or in $\lambda x.x(\underline{y(\lambda z.yz)})$.

Notice that, given any λ-term P maximal free sub-terms are all disjoint. Hence, P can be written by using the natural generalization of the context notation to n holes. We get: $P = C[x_1, x_2, \cdots x_n][\![x_1 \backslash P_1, x_2 \backslash P_2, \cdots x_n \backslash P_n]\!]$, where $P_1, P_2, \ldots P_n$ are the maximal free sub-terms of P and $x_1, x_2, \ldots x_n$ are fresh variables all distinct. The translation from the weak λ-calculus to weak explicit substitutions is as follows

$$\mathcal{I}(x) = x$$
$$\mathcal{I}(PQ) = \mathcal{I}(P)\mathcal{I}(Q)$$
$$\mathcal{I}(\lambda x.P) = (\lambda x.C[x_1, x_2, \cdots x_n])[(x_1, \mathcal{I}(P_1)), (x_2, \mathcal{I}(P_2)), \cdots (x_n, \mathcal{I}(P_n))]$$
where $P_1, P_2, \ldots, P_n$ are the maximal free sub-terms of P.

Proposition 5. *Given a λ term P, we have $\{\mathcal{I}(P)\} = P$.*

Proof: Easy, the choice of fresh variables for the x_i's is obviously crucial. $\square$

It is no surprise that the converse proposition does not hold, since $\mathcal{I}$ depends on which sub-terms are "abstracted out". Consider $M = (\lambda x.I)[\ ]$, then we get $\mathcal{I}(\{M\}) = (\lambda x.y)[(y, I[\ ])]$.

Proposition 6. *If $P \to P'$, then $\mathcal{I}(P) \to M$, with $\{M\} = P'$*

Proof: By structural induction. The key observation is as follows: let R the redex contracted in the reduction $P \to P'$, since R is a redex there exists Q_i, the maximal free sub-term of P that includes R. Then, we can apply the induction hypothesis to Q_i. $\square$

4   Reduction Strategies with Weak Explicit Substitutions

We consider three different evaluation strategies and show how they are naturally connected to executions of λ-interpreters. We start by call-by-value in weak explicit substitutions, which works on the following subset of terms

$M, N ::=$		term
	x	variable
	$\mid MN$	application
	$\mid (\lambda x.P)[s_v]$	closure
$P, Q ::= x \mid PQ \mid \lambda x.P$		programs
$V ::= x \mid (\lambda x.P)[s_v]$		values
$s_v ::= (x_1, V_1), (x_2, V_2), \cdots (x_n, V_n)$	x_i distinct $(n \geq 0)$	

Values are either variables or closures. Notice that we take the convention that $xV_1V_2 \cdots V_n$ is not a value when $n > 0$. We could have taken a different convention, but it would have just complicated our semantics without much interest. We could also have decided that x was not a value, but this seems more speaking, especially when one adds constants to the set of terms. Now functions need values as arguments in the following β_v reduction rule.

$$(\beta_v) \qquad (\lambda x.P)[s_v]\, V \rightarrow P[\![\, s_v[x\backslash V]\,]\!]$$

For active contexts, it is sufficient to consider μ and ν rules, since ξ and σ rules can never be applied since s_v-substitutions are irreducible.

We first notice that redexes in the call-by-value strategy are innermost redexes in the calculus of weak explicit substitutions, but not all of them since we are interesting to reductions leading to values. An alternative way of expressing this strategy can be done with a bigstep SOS semantics and sequents of form $s \vdash P = V$, meaning that the result of evaluating P with substitution s is value V, as follows

$$s, (x, V) \vdash x = V$$

$$s \vdash x = x \quad (x \notin domain(s))$$

$$s \vdash \lambda x.P = (\lambda x.P)[s]$$

$$\frac{s \vdash P = (\lambda x.P')[s'] \quad s \vdash Q = V' \quad s'[x\backslash V'] \vdash P' = V}{s \vdash PQ = V}$$

Proposition 7. $s \vdash P = V$ *iff* $P[\![s]\!] \twoheadrightarrow V$ *in the calculus of call-by-value weak explicit substitutions.*

Proof: The proof is obvious by induction on the pair $(l, \|P\|)$ where l is the length of the reduction and $\|P\|$ is the size of P. $\square$

Obviously there are similar statements with call-by-name. The set of terms is now the full set of terms in the calculus of weak explicit substitutions. Values are variables or closures. The strategy is defined as the normal reduction $\xrightarrow[\text{norm}]{}$ which always contracts the leftmost outermost redex until reaching a value. The corresponding bigstep semantics is

$$\frac{s' \vdash P = V}{s, (x, (P, s')) \vdash x = V}$$

$$s \vdash x = x \quad (x \notin domain(s))$$

$$s \vdash \lambda x.P = (\lambda x.P)[s]$$

$$\frac{s \vdash P = (\lambda x.P')[s'] \quad s'[x\backslash (Q, s)] \vdash P' = V}{s \vdash PQ = V}$$

In fact, to model call-by-name, our bigstep semantics needed a new kind of delayed substitutions. Now substitutions may contained pairs (Q, s) for any program Q (and not only for program abstractions), which somehow correspond to the Algol 60 "thunks". A close treatment of them could be done with the calculus of gradual weak explicit substitutions exposed in section 6. A value of the bigstep semantics can be mapped to a value of the call-by-name calculus of weak explicit substitutions by replacing all sub-terms of form (Q, s) by terms $Q[\![s]\!]$. Let V^+ be the value mapped from V.

Proposition 8. $s \vdash P = V$ *iff* $P[\![s]\!] \xrightarrow[\text{norm}]{} V^+$ *in the calculus of weak explicit substitutions.*

The proof is similar to the one of proposition 7. Notice that by the standardization theorem, it is possible to show that the value computed by call-by-name is minimal, namely if $M \twoheadrightarrow V$, then $M \xrightarrow[\text{norm}]{} V_0 \twoheadrightarrow V$.

Call-by-need is more delicate, since one must represent some sharing of terms. Following techniques in [24,5,25,27,26], we build a confluent theory of shared reductions as follows. Terms and programs are labeled with names, names for programs are single letters a, b, c ... taken in a given alphabet, names for terms are strings α of letters which can be many-level underlined.

$$
\begin{aligned}
M, N &::= x^\alpha \mid (MN)^\alpha \mid (\lambda x.P)^\alpha[s] & \text{labeled term} \\
P, Q &::= x^b \mid (PQ)^b \mid (\lambda x.P)^b & \text{labeled programs} \\
s &::= (x_1, M_1), (x_2, M_2), \cdots (x_n, M_n) & x_i \text{ distinct } (n \geq 0) \\
\alpha, \beta &::= a \mid \alpha\beta \mid \underline{\alpha} & \text{labels}
\end{aligned}
$$

The new labeled reduction rule β_l is defined as

$$(\beta_l) \qquad ((\lambda x.P)^\alpha[s]N)^\beta \to \beta \cdot (\underline{\alpha} \circ P)[\![\, s[x \backslash N]\,]\!]$$

where the labeled substitution $[\![\ ]\!]$ is defined inductively as follows in a labeled program

$$
\begin{aligned}
x^\beta[\![s]\!] &= \beta \cdot s(x) \quad \text{if } x \in domain(s) \\
x^\beta[\![s]\!] &= x^\beta \quad \text{otherwise} \\
(PQ)^\beta[\![s]\!] &= (P[\![s]\!]\, Q[\![s]\!])^\beta \\
(\lambda x.P)^\beta[\![s]\!] &= (\lambda x.P)^\beta[s]
\end{aligned}
$$

$$
\begin{aligned}
\alpha \cdot x^\beta &= x^{\alpha\beta} & \alpha \circ x^b &= x^{\alpha b} \\
\alpha \cdot (MN)^\beta &= (MN)^{\alpha\beta} & \alpha \circ (PQ)^b &= ((\alpha \circ P)(\alpha \circ Q))^{\alpha b} \\
\alpha \cdot (\lambda x.P)^\beta[s] &= (\lambda x.P)^{\alpha\beta}[s] & \alpha \circ (\lambda x.P)^b &= (\lambda x.P)^{\alpha b}
\end{aligned}
$$

Above, we used two external operations with labels and labeled terms to modify the external label of a term or to broadcast a label on a program. Notice that this labeled calculus is different from the labeled λ-calculus as in [3,24,25], which does not contain the broadcast operation. This new operation means that in the weak β-reduction we need fresh copies of application nodes from the body of the function before its application to an argument. However, it does not copy abstractions, and instead builds new closures.

Proposition 9. *In the labeled calculus of weak explicit substitutions, the following three lemmas hold*

$$
\begin{aligned}
&(i) & (\alpha\beta) \cdot M &= \alpha \cdot \beta \cdot M \\
&(ii) & M \to M' &\Rightarrow \alpha \cdot M \to \alpha \cdot M' \\
&(iii) & s \to s' &\Rightarrow M[\![s]\!] \twoheadrightarrow M[\![s']\!]
\end{aligned}
$$

Proof: (i) is obvious by definition. And $(i) \Rightarrow (ii) \Rightarrow (iii)$.   □

Proposition 10. *The labeled calculus of weak explicit substitutions is confluent.*

Proof: As in previous confluence proofs, we only consider local confluence, leaving the remaining part of the proof to the Tait–Martin-Löf's axiomatic method. When $s \to s'$ and $N \to N'$, we have two interesting cases corresponding to the two diagrams

$$
\begin{array}{ccc}
((\lambda x.P)^\alpha[s]\ N)^\beta & \longrightarrow & ((\lambda x.P)^\alpha[s']\ N)^\beta \\
\downarrow & & \downarrow \\
\beta \cdot (\underline{\alpha} \circ P)[\![\,s[x\backslash N]\,]\!] & \twoheadrightarrow & \beta \cdot (\underline{\alpha} \circ P)[\![\,s'[x\backslash N]\,]\!]
\end{array}
$$

and

$$
\begin{array}{ccc}
((\lambda x.P)^\alpha[s]\ N)^\beta & \longrightarrow & ((\lambda x.P)^\alpha[s]\ N')^\beta \\
\downarrow & & \downarrow \\
\beta \cdot (\underline{\alpha} \circ P)[\![\,s[x\backslash N]\,]\!] & \twoheadrightarrow & \beta \cdot (\underline{\alpha} \circ P)[\![\,s[x\backslash N']\,]\!]
\end{array}
$$

which commutes by using lemmas of proposition 9.   □

As in other theories of labeled λ-calculi, labels are useful for naming redexes, which are either residuals of redexes in a given initial term M, or created along reductions. The name of a redex is the string of labels on the path from its application node to its abstraction node, thus naming the interaction between these two nodes. So, the name of $((\lambda x.P)^\alpha[s]\ N)^\beta$ is α. A complete labeled reduction step $M \overset{\alpha}{\Longrightarrow} N$ is the finite development of all redexes with name α in M. We write $\Longrightarrow$ for a complete anonymous step, and $\twoheadrightarrow$ for several steps. Finally $Init(M)$ will be the predicate which is true iff every label in M is a distinct letter. Intuitively, $Init(M)$ means that term M does not contain shared sub-terms.

Proposition 11. *Let $Init(M_0)$ and $M_0 \twoheadrightarrow M$. If $M \overset{\alpha}{\Longrightarrow} N$ and $M \overset{\beta}{\Longrightarrow} M'$, then $N \overset{\beta}{\Longrightarrow} N'$ and $M' \overset{\alpha}{\Longrightarrow} N'$ for some N'.*

Proof: The proof is far too complex to be exposed in this article.   □
This property is nice since its shows that one has a confluent sub-theory of weak complete reductions.

Call-by-need strategies $\underset{\text{norm}}{\Longrightarrow}$ correspond to complete normal reductions in the labeled calculus of weak explicit substitutions when the initial labeled term M checks predicate $Init(M)$. At each step, all redexes with the same name as the one of the leftmost outermost redex are contracted. One can show that the number of steps to get a value with this reduction is always minimal. No other reductions may get quicker any value. In difference with the theory of full λ-calculus, there is always a simple reduction $\twoheadrightarrow$ which can get the value with the same cost.

Now the goal is to define a bigstep semantics for call-by-need. We make sharing explicit by considering *stores* Σ, which are mappings from *locations* ℓ to either thunks (P, s) or values V. Substitutions s now binds variables to locations. A store may appear as a context or as result in judgments all of the form $s \bullet \Sigma \vdash P = V \bullet \Sigma'$. Accessing to the value of a variable is now a bit more complex, since the value of its location can be of two kinds. When a value, it is as before. When a thunk, one has to evaluate it, and to modify the corresponding location in the store before returning the value and the modified store.

$$\frac{s' \bullet \Sigma, (\ell, (P, s')) \vdash P = V \bullet \Sigma', (\ell, (P, s')))}{s, (x, \ell) \bullet \Sigma, (\ell, (P, s')) \vdash x = V \bullet \Sigma', (\ell, V)}$$

$$s, (x, \ell) \bullet \Sigma, (\ell, V) \vdash x = V \bullet (\ell, V)$$

$$s \bullet \Sigma \vdash x = x \bullet \Sigma \quad (x \notin domain(s))$$

$$s \bullet \Sigma \vdash \lambda x.P = (\lambda x.P)[s] \bullet \Sigma$$

$$\frac{s \bullet \Sigma \vdash P = (\lambda x.P')[s'] \bullet \Sigma' \quad s'[x \backslash \ell] \bullet \Sigma', (\ell, (Q, s)) \vdash P' = V \bullet \Sigma''}{s \bullet \Sigma \vdash PQ = V \bullet \Sigma''}$$

In the last rule, we assume that ℓ is a fresh location.

Notice that, by contrast with Launchbury's [21], there is no need for renaming while substituting a variable in the first rule. No capture of variables can occur here. Another difference is that we make a clear distinction between variables and locations. Let now write M^*, P^*, s^* for the unlabeled terms, programs and substitutions obtained by erasing all labels within the labeled terms M, programs P, and substitutions s. Let also use the V^+ notation defined in the call-by-name subsection.

Proposition 12. *Let $Init(P[\![s]\!])$. Then $s^* \bullet \emptyset \vdash P^* = V^* \bullet \Sigma$ iff $P[\![s]\!] \xrightarrow[norm]{}$ V^+ in the labeled calculus of weak explicit substitutions.*

Proof: The proof is again much complex to be presented here.   □

5   Runtime Interpreters

Sets may be represented by lists with two constructors *nil* and cons ::, and substitutions may become association lists. Substitutions may be then called environments. We do not duplicate the corresponding SOS of call-by-value with environments, which leads to the following recursive λ-evaluator

$$eval(x, (x, V) :: s) = V$$
$$eval(x, (y, V) :: s) = eval(x, s)$$
$$eval(x, nil) = x$$
$$eval(\lambda x.P, s) = (\lambda x.P)[s]$$
$$eval(PQ, s) = eval(P', (x, eval(Q, s)) :: s') \quad \text{if } eval(P, s) = (\lambda x.P')[s']$$

Similarly with call-by-name, one gets the functional interpreter

$eval(x, (x, (P, s')) :: s) = eval(P, s')$
$eval(x, (y, (P, s')) :: s) = eval(x, s)$
$eval(x, nil) = x$
$eval(\lambda x.P, s) = (\lambda x.P)[s]$
$eval(PQ, s) = eval(P', (x, (Q, s)) :: s')$ if $eval(P, s) = (\lambda x.P')[s']$

Finally, the third interpreter is for call-by-need

$eval(x, (x, \ell)) = V$ if $!\ell = (P, s')$ and $V = eval(P, s')$ (side effect $\ell \leftarrow V$)
$eval(x, (x, \ell)) = V$ if $!\ell = V$
$eval(x, (y, \ell) :: s) = eval(x, s)$
$eval(x, nil) = x$
$eval(\lambda x.P, s) = (\lambda x.P)[s]$
$eval(PQ, s) = eval(P', (x, \ell) :: s')$ if $eval(P, s) = (\lambda x.P')[s']$
$$\text{and } \ell = ref(Q, s)$$

In the last case, we use mutable variables ℓ with the ML syntax for creation the reference ℓ and access to its content $!\ell$. The three interpreters are easy mappings of the previous SOS rules seen in the previous section, and their soundness with respect to this operational semantics is straightforward.

We now consider two problems on functional interpreters, and try more to state them within weak explicit substitutions than to give solutions, far beyond the scope of this paper.

The first one is stack allocation for closures. It is well-known that functional languages cannot be implemented with the Algol/Pascal stack discipline, since closures often need to be heap-allocated. In these imperative languages, each environment cell has an extra component, a link to the outer environment, also named "static link" in the compiler terminology. The stack is now represented by the environment at the left of each rule in our bigstep operational semantics. In the call-by-value case, the SOS semantics is now as follows

$$s, (n, (x, V)) \vdash x = V$$

$$\frac{s[1..n] \vdash x = V}{s, (n, (y, V')) \vdash x = V}$$

$$s \vdash \lambda x.P = (\lambda x.P)[\, |s| \,]$$

$$\frac{s \vdash P = (\lambda x.P')[n] \quad s \vdash Q = V' \quad s, (n, (x, V')) \vdash P' = V}{s \vdash PQ = V}$$

where $|s|$ is the length of s, and $s[1..n]$ is the list of the first n elements of s. Intuitively, in the Algol/Pascal subset of terms, the values of arguments of functions can only refer to environments already in the stack. This is why closure values are represented as $(\lambda x.P)[n]$ where n refers to an entry in the current environment. Now it remains to show formally that this works. Clearly, it fails for any currified function. Take for instance, $s \vdash (\lambda x.\lambda y.x)Q = (\lambda y.x)[|s| + 1]$ which yields a result escaping from stack s.

In this implementation of environments with stacks, we need to leave the stack unchanged after the evaluation of each application, which makes the evaluation of currified functions failing. There are other techniques with stacks, dynamic ones as in Caml [22] when functions have arities, or with a static escape analysis as in [6]. The trick is then to try to keep the stack unchanged only after the evaluation of function bodies.

The second problem that we consider in this section is graph implementations of functional languages, which are mainly useful for lazy languages. So we are in the case of weak explicit substitutions with call-by-name. The weak labeled calculus of section 4 can be used to characterize these graphs. Clearly in the β_l-rule,

$$(\beta_l) \qquad ((\lambda x.P)^\alpha [s]N)^\beta \to \beta \cdot (\underline{\alpha} \circ P)[\![\, s[x \backslash N] \,]\!]$$

the broadcast operation $\alpha \circ P$ describes the creation of a fresh copy of the function body (except for its abstraction sub-terms). The rest of the term in which the β-reduction is performed remains unchanged, with the same sharing as before the reduction step. This operation was already considered by Wadsworth in his dissertation, but in the context of sharing for the full λ-calculus. In our case, an intuitive graph β-rule would be written

$$(\beta_g) \qquad ((\lambda x.P)[s]N) \to (\text{copy }(P))[\![\, s[x \backslash N] \,]\!]$$

However, it is fascinating how the proof of the correspondence between the labeled calculus and the straightforward graph implementation is complex. The goal is to prove that nodes connected by a same labeled path are identical in the graph implementation. Notice that this proof was already quite involved in the full λ-calculus where it required to build the so-called context semantics [16]. But one could expect a much simpler proof in the weak case.

6   Weak Explicit Substitutions with Ministep Semantics

Usually, the various authors working on explicit substitutions start here. Their goal is to represent not only bindings of variables, but also the way how substitutions are gradually pushed inside programs. Often, bindings are treated with de Bruijn numbers. Notice that we never used them, since names of variables were sufficient. This is because, in our weak calculi, substitution never cross binders, namely λ-abstractions or other substitutions. Therefore, one has not to care with α-renaming of bound variables. The second part of the motivation of the usual work on explicit substitutions is to study the progression of substitutions in terms, which we avoided since substitutions are always pushed to (free) variables or abstractions.

We consider the following ministep semantics for the calculus of weak explicit substitutions. The set of terms now allows closures on any program, and substitutions are represented by association lists.

$$M, N ::= \qquad\qquad \text{term}$$
$$x \qquad\qquad \text{variable}$$
$$\mid\ MN \qquad\qquad \text{application}$$
$$\mid\ P[s] \qquad\qquad \text{extended closure}$$

$$P, Q ::= x \mid PQ \mid \lambda x.P \qquad \text{programs}$$

$$s ::= nil \qquad\qquad \text{empty substitution}$$
$$\mid\ (x, M) :: s \qquad\qquad \text{association list}$$

The reduction rules are defined by the following five non-overlapping left-linear rewriting rules

$$PQ[s] \rightarrow P[s]Q[s]$$
$$x[(x, M) :: s] \rightarrow M$$
$$x[(y, M) :: s] \rightarrow x[s]$$
$$x[nil] \rightarrow x$$
$$(\lambda x.P)[s]N \rightarrow P[(x, N) :: s]$$

Any sub-term may be reduced, as described by the following definition of active contexts

$$\frac{M \rightarrow M'}{MN \rightarrow M'N} \qquad \frac{N \rightarrow N'}{MN \rightarrow MN'}$$

$$\frac{s \rightarrow s'}{P[s] \rightarrow P[s']}$$

$$\frac{M \rightarrow M'}{(x, M) :: s \rightarrow (x, M') :: s} \qquad \frac{s \rightarrow s'}{(x, M) :: s \rightarrow (x, M) :: s'}$$

We do not detail the different proofs in this new calculus. Notice that the variables of this calculus (in the sense of term rewriting systems) are M, N, s. All other operators are constants. Thus this ministep calculus may be considered as an orthogonal system, and therefore is confluent, with the standardization theorem as stated in [29]. The normal strategy corresponds to the leftmost outermost reduction. And some simulation of the weak calculus of section 3 may easily be shown. Therefore this calculus also implements the weak λ-calculus. Finally, one can investigate sharing within this ministep semantics as in [26,27], which leads to a ministep implementation of graph reduction, as considered at end of previous section or in interpreters of functional lazy languages [31]. Remark that then sharing works for all the set of reduction rules and not only for the β-rule.

7   Conclusion

So, before jumping in the full world of explicit substitutions we hope to have shown that the fundamental properties of the syntax of weak theories in the λ-calculus or in the weak calculus of explicit substitutions are still interesting. In this paper, we did not consider preservation of strong normalization in the typed case, but clearly it holds. We also restricted our study to the sole β-reduction, because it contains many of the problems, but δ-rules may be added in each of the three main calculi considered here, which rapidly provides the power of Plotkin's PCF or of a ML kernel. These extensions are rather easy since we have no critical pairs in the various term rewriting systems. Similarly data structures may be added (lists, records, algebraic structures) as done for records in [2].

We showed that our weak calculi are sufficiently expressive to describe the functional part of the programming languages runtimes, and could be used as a basis to model or to derive some program transformations or program analyses within compilers (stack allocation, graph implementation, dependency analysis, slicing) [6,18,2,15]. But it would be very interesting to understand whether the fundamental properties of the underlining calculi are useful. For instance, how much of confluence or of the standardization property is really used? Also we would like to understand which of the three calculi presented here is the most useful.

Finally, many of the results of this paper were considered as folks theorems, rather easy to prove. We hope to have shown that some of the proofs may deserve some attention. In fact, some of them are not easy at all.

Acknowledgments

We thank Bruno Blanchet for stimulating discussions on stack allocation.

References

1. M. Abadi, L. Cardelli, P.-L. Curien, and J.-J. Lévy. Explicit substitutions. *Journal of Functional Programming*, 6(2):pp. 299–327, 1996.
2. M. Abadi, B. Lampson, and J.-J. Lévy. Analysis and caching of dependencies. In *Proc. of the 1996 ACM SIGPLAN International Conference on Functional Programming*, pages pp. 83–91. ACM Press, May 1996.
3. H. P. Barendregt. *The Lambda Calculus, Its Syntax and Semantics.* North-Holland, 1981.
4. Z.-E.-A. Benaissa, D. Briaud, P. Lescanne, and J. Rouyer-Degli. Lambda-upsilon, a calculus of explicit substitution which preserves strong normalisation. Research Report 2477, Inria, 1995.
5. G. Berry and J.-J. Lévy. Minimal and optima l computations of recursive programs. In *Journal of the ACM*, volume 26. ACM Press, 1979.
6. B. Blanchet. Escape analysis : Correctness proof, implementation and experimental results. In *Proc. of 25th ACM Symposium on Principles of Programming Languages*. ACM Press, 1998.
7. R. Bloo and K. H. Rose. Combinatory reduction systems with explict substitutions thatpreserve strong normalization. In *In Proc. of the 1996 confence on Rewriting Techniques and Applications*. Springer, 1996.
8. N. Çağman and J. R. Hindley. Combinatory weak reduction in lambda calculus. *Theoretical Computer Science*, 198:pp. 239–249, 1998.

9. G. Cousineau, P.-L. Curien, and M. Mauny. The categorical abstract machine. In *Proc. of the second international conference on Functional programming languages and computer architecture*. ACM Press, 1985.

10. P.-L. Curien. *Categorical Combinator, Sequential Algorithms and Functional Programming*. Pitman, 1986.

11. P.-L. Curien, T. Hardin, and J.-J. L'evy. Confluence properties of weak and strong calculi of explicit substitutions. *Journal of the ACM*, 43(2):pp. 362–397, 1996.

12. R. David and B. Guillaume. The lambda_I calculus. In *Proc. of the Second International Workshop on Explicit Substitutions: Theory and Applications to Programs and Proofs*, 1999.

13. R. Di Cosmo and D. Kesner. Strong normalization of explicit substitutions via cut elimination in proof nets. In *In Proc. of the 1997 symposium on Logics in Computer Science*, 1997.

14. G. Dowek, T. Hardin, and C. Kirchner. Higher-order unification via explicit substitutions: the case of higher-order patterns. In M. Maher, editor, *In proc. of the joint international conference and symposium on Logic Programming*, 1996.

15. J. Field. On laziness and optimality in lambda interpreters: Tools for specification and analysis. In *Proc. of the Seventeenth conference on Principles of Programming Languages*, volume 6, pages pp. 1–15. ACM Press, 1990.

16. G. Gonthier, M. Abadi, and J.-J. Lévy. The geometry of optimal lambda reduction. In *Proc. of the Nineteenth conference on Principles of Programming Languages*, volume 8. ACM Press, 1992.

17. T. Hardin. Confluence results for the pure strong categorical logic ccl. lambda-calculi as subsystems of ccl. *Journal of Theoretical Computer Science*, 65:291–342, 1989.

18. T. Hardin, L. Maranget, and B. Pagano. Functional runtimes within the lambda-sigma calculus. *Journal of Functional Programming*, 8(2), march 1998.

19. J. R. Hindley. Combinatory reductions and lambda reductions compared. *Zeit. Math. Logik*, 23:pp. 169–180, 1977.

20. J.-W. Klop. *Combinatory Reduction Systems*. PhD thesis, Mathematisch Centrum, Amsterdam, 1980.

21. J. Launchbury. A natural semantics for lazy evaluation. In *Proc. of the 1993 conference on Principles of Programming Languages*. ACM Press, 1993.

22. X. Leroy. The ZINC experiment: an economical implementation of the ML language. Technical report 117, INRIA, 1990.

23. P. Lescanne. From lambda-sigma to lambda-upsilon, a journey through calculi of explicit substitutions. In *Proc. of the Twenty First conference on Principles of Programming Languages*, 1994.

24. J.-J. Lévy. *Réductions correctes et optimales dans le lambda-calcul*. PhD thesis, Univ. of Paris 7, Paris, 1978.

25. J.-J. Lévy. Optimal reductions in the lambda-calculus. In J. Seldin and J. Hindley, editors, *To H.B. Curry: Essays on Combinatory Logic, Lambda-Calculus and Formalism*. Academic Press, 1980. On the occasion of his 80th birthday.

26. L. Maranget. Optimal derivations in orthogonal term rewriting systems and in weak lambda calculi. In *Proc. of the Eighteenth conference on Principles of Programming Languages*. ACM Press, 1991.

27. L. Maranget. *La stratégie paresseuse*. PhD thesis, Univ. of Paris 7, Paris, 1992.

28. P.-A. Melliès. Typed lambda-calculus with explicit substitutions may not terminate. In *Proc. of the Second conference on Typed Lambda-Calculi and Applications*. Springer, 1995. LNCS 902.

29. P.-A. Melliès. *Description Abstraite des Systèmes de Réécriture*. PhD thesis, Univ. of Paris 7, december 1996.
30. R. Milner. Action calculi and the pi-calculus. In *Proc. of the NATO Summer School on Logic and Computation*. Marktoberdorf, 1993.
31. S. L. Peyton-Jones. *The implementation of Functional Programming Languages*. Prentice-Hall, 1987.

Approximation Algorithms for Routing and Call Scheduling in All-Optical Chains and Rings*

Luca Becchetti[1]**, Miriam Di Ianni[3], and Alberto Marchetti-Spaccamela[2]

[1] Technische Universität Graz, Institut für Mathematik B;
luca@opt.math.tu-graz.ac.at
[2] Dipartimento di Informatica e Sistemistica, Università di Roma "La Sapienza";
alberto@dis.uniroma1.it
[3] Dipartimento di Ingegneria Elettronica e dell'Informazione, Università di Perugia;
diianni@diei.unipg.it

Abstract. We study the problem of routing and scheduling requests of limited durations in an all-optical network. The task is servicing the requests, assigning each of them a starting time and a wavelength, with restrictions on the number of available wavelengths. The goal is minimizing the overall time needed to serve all requests. We propose constant approximation algorithms for both ring and chain networks. In doing this, we also propose a polynomial-time approximation scheme for the problem of routing weighted calls on a directed ring with minimum load.

1 Introduction

All-optical networks allow very high transmission rates, widely exceeding those that can be guaranteed by traditional electronic technology. Wavelength Division Multiplexing (WDM) allows the concurrent transmission of multiple data streams on the same optic fiber; different data streams on the same optical link at the same time and in the same direction use different wavelengths (currently 30-40 in experimental settings) [6,13]. Moreover, the high speed achievable with all-optical networks is mainly due to the fact that the signal is kept in optical form throughout its transmission from source to destination.

In this paper, we address the problem of Call Scheduling in all-optical networks, that is the problem of scheduling a set of communication requests (*calls*) each one characterized by a source-destination pair and a duration. Following [13] we assume that optical links allow only one way communication: if there is a link from x to y then the information flow is unidirectional from source to tail. In order to allow bidirectional connection, we assume that if there is a link from x to y there is also a link from y to x [17]. A different model assumes that the optical links are undirected and allow bidirectional communication; we remark that this model is less realistic with respect to current technology [13].

---

* This work was partially supported by ESPRIT project ALCOM-IT and by the START program Y43-MAT of the Austrian Ministry of Science.
** Part of this research was done while the author was at Dipartimento di Informatica e Sistemistica, University of Rome "La Sapienza".

C. Pandu Rangan, V. Raman, R. Ramanujam (Eds.): FSTTCS'99, LNCS 1738, pp. 201–212, 1999.
© Springer-Verlag Berlin Heidelberg 1999

Given a network and a set of calls MINIMUM CALL SCHEDULING requires to assign a directed path, a wavelength and a starting time to each call, subject to the constraint that no pair of calls assigned to the same wavelength use the same (directed) arc at the same time and that the number of assigned wavelength does not exceed some bound k. The objective is to minimize the overall time to accomodate all calls.

Related work. To the best of our knowledge, call scheduling in all-optical networks received so far little attention.

Concerning optical networks, several related problems have been extensively studied. MIN PATH COLORING [3,4,15,17] aims to find routes and a wavelength assignment for a set of calls of infinite duration that minimize the number of used wavelengths. In the MAX CALL admission problem calls are presented in an on-line fashion; when a call arrives it can be either rejected or accepted; in the latter case it must be immediately satisfied using available resources. The objective is maximizing the number of accepted calls [2,3].

In the MIN RING LOADING problem, arising in the project of SONET networks [18], the aim is that of devising a routing of calls in a ring network such that the maximum load of a link (defined as the sum of the durations of all calls that use the link) is minimized. This problem is polynomial-time solvable under the assumption that all calls have unit durations [19,18]. The undirected ring loading problem (i.e., each link allows two ways communication) has been considered in [18], where a constant approximation algorithm is presented. This result has been improved and a polynomial time approximation scheme has been proposed [14]. The case of links that allow one way communication is the MINIMUM WEIGHTED DIRECTED RING ROUTING (MIN-WDRR) problem and will be considered in the present paper.

Scheduling calls with minimum makespan has been considered in non optical networks. Concerning packet switched networks, a seminal result by Leighton [16] proved the existence of a schedule delivering all packets in a number of steps within a constant factor of the lower bound. In [7,8] the call scheduling problem has been studied in the context of ATM networks. In particular, the authors propose approximation algorithms for star and ring networks.

Results of the paper. In section 3 we propose approximation algorithms for call scheduling in chain networks. Namely we propose a 3-approximation algorithm for the call scheduling problem in chains when just one wavelength is available, by a reduction to the problem of MINIMUM DYNAMIC STORAGE ALLOCATION (MIN-DSA) for which a 3-approximation algorithm is known [11]. Successively, we extend the algorithm to the general case of k available wavelengths, obtaining a 5-approximation algorithm. The above results hold also in the case of undirected chains.

In section 4 we first give a polynomial-time approximation scheme for MINIMUM WEIGHTED DIRECTED RING ROUTING. This result, together with those of Section 3, allows to obtain a $(12 + \epsilon)$-approximation algorithm for the call scheduling problem in ring networks. We remark that the problem of routing weighted calls in a ring network with the aim of minimizing the maximum load, has itself a practical relevance in SONET networks [18,14,19].

The results of section 3 hold also in the case of undirected chains. For the sake of brevity, most proofs are omitted.

2 Preliminaries

In the following, $G = (V, E)$ denotes a simple, directed graph on the vertex set $V = \{v_0, v_1, \ldots, v_{m-1}\}$ such that arc (v_i, v_j) exists *if and only if* arc (v_j, v_i) exists. A *network* is a pair $\langle G, k \rangle$ where k is the number of available wavelengths (from now on, colors) on each arc. A *call* $C = [s, d, l]$ is an ordered pair of vertices s, d completed by an integer $l > 0$ representing the call duration. Namely, s is the source vertex originating the data stream to be sent to the destination d.

Given a network $\langle G, k \rangle$ and a set $\mathcal{C} = \{C_h = [s_h, d_h, l_h] : h = 1, \ldots, n\}$ of calls, a *routing* is a function $R : \mathcal{C} \to \mathcal{P}(G)$, where $\mathcal{P}(G)$ is the set of simple paths in G. Given a network $\langle G, k \rangle$, a set $\mathcal{C}$ of calls and a routing R for $\mathcal{C}$, a *schedule* is an assignment of starting times and colors to calls such that at every time no two calls with the same color use the same arc; formally, a schedule is a pair $\langle S, F \rangle$, with $S : \mathcal{C} \to [0, 1, \ldots \sum_{h=1}^{n} l_h]$ a function assigning starting times to calls and $F : \mathcal{C} \to \{1, \ldots k\}$ a function assigning colors to calls. S and F must be such that, for any $(u, v) \in E$ and $C_i, C_j \in \mathcal{C}$, if $(u, v) \in R(C_i)$, $(u, v) \in R(C_j)$ and $F(C_i) = F(C_j)$ then either $S(C_i) \geq S(C_j) + l_j$ or $S(C_j) \geq S(C_i) + l_i$. $T = \max_{1 \leq h \leq n}\{S([s_h, d_h, l_h]) + l_h\}$ is the *makespan* of schedule $\langle S, F \rangle$. $T^*(x)$ or simply T^* denotes the makespan of an optimal solution for an instance x..

In the MINIMUM CALL SCHEDULING (MIN-CS) problem it is required to find a routing and a schedule for an instance $\langle \langle G, k \rangle, \mathcal{C} \rangle$ having minimum makespan. Since the problem is trivial if $k \geq |\mathcal{C}|$, we assume $k < |\mathcal{C}|$. Notice that, if routes are fixed, the problem of scheduling calls in order to minimize the makespan is unapproximable within $O(|\mathcal{C}|^{1-\delta})$, for any $\delta > 0$ [5] in general networks.

Given an instance of MIN-CS and a routing R, the *load* $L_R(e)$ of arc e is defined as the sum of the durations of calls using it. If $k = 1$, then $L^* = \min_R(\max_e L(e))$ is a natural lower bound to the makespan of any schedule.

In the following we show that MIN-CS in chain networks is closely related to MINIMUM DYNAMIC STORAGE ALLOCATION (MIN-DSA), that requires the allocation of contiguous areas in a linear storage device in order to minimize the overall requested storage space. A *block* $B = (f, e, z)$ represents a storage requirement such that f and e are, respectively, the first and last time instants in which block B must be allocated and z is an integer denoting the memory requirement (size) of B. An instance of MIN-DSA consists of a set of blocks $\mathcal{B} = \{B_1 = (f_1, e_1, z_1), \ldots, B_n = (f_n, e_n, z_n)\}$; an *allocation* of $\mathcal{B}$ is a function assigning blocks to storage locations, such that if both B_i and B_j must stay at the same time in the storage device then they must be assigned to non overlapping portions. Formally, an allocation is a function $F : \mathcal{B} \to N$ such that, for any pair B_i, B_j of blocks, if $f_i \leq f_j \leq e_i$ or $f_j \leq f_i \leq e_j$ then either $F(B_i) + z_i - 1 < F(B_j)$ or $F(B_j) + z_j - 1 < F(B_i)$. MIN-DSA is defined as follows: given an infinite array of storage cells and a set $\mathcal{B}$ of blocks, find an allocation F such that the storage size $M = \max_{1 \leq h \leq n}\{F(B_h) + z_h\}$ is minimum. MIN-DSA is NP-hard [9] but approximable within a constant [10,11].

3 Call Scheduling in Chain Networks

In this section we assume that the network is a chain with m vertices, simply denoted as $0, 1, \ldots, m - 1$. The restriction of MIN-CS to chain networks will be denoted as CHAIN-MIN-CS (or CHAIN-MIN-CS$_k$, when k is the number of available colors). Since the network is a chain, there is only one path between each source-destination pair. This implies that routing of calls is fixed and that the set $\mathcal{C}$ of calls consists of two independent subsets $\mathcal{C}_1 = \{[s_{11}, d_{11}, l_{11}], \ldots, [s_{1n_1}, d_{1n_1}, l_{1n_1}]\}$ and $\mathcal{C}_2 = \{[s_{21}, d_{21}, l_{21}], \ldots, [s_{2n_2}, d_{2n_2}, l_{2n_2}]\}$, where $n_1 + n_2 = n$ and, for any $h = 1, \ldots, n_1$, $s_{1h} < d_{1h}$ while, for any $h = 1, \ldots, n_2$, $s_{2h} > d_{2h}$. This implicitly defines two independent subinstances for each instance of the problem. Without loss of generality, in the following we always refer to just the first of them. The main result of this section is a 5-approximation algorithm for MIN-CS. This result exploits a close relationship between MIN-CS$_1$ and MIN-DSA, which is stated in the next theorem:

Theorem 1. *There exists a polynomial-time reduction from CHAIN-MIN-CS$_1$ to MIN-DSA such that an instance of the first problem admits a schedule with makespan T if and only if the corresponding instance of the second problem admits a storage allocation with size T.*

The reverse reduction also holds, but we do not need it here. Our first approximation result is a direct consequence of theorem 1 and the 3-approximation algorithm for MIN-DSA given in [11].

Corollary 1. *There exists a 3-approximation algorithm for chain-MIN-CS$_1$.*

We now turn to the general case. The algorithm we propose is sketched below: the first step rounds up call durations to the closest power of 2; this worsens the approximation of the makespan of the optimum schedule by a factor at most 2.

Algorithm *chain-cs*
input: chain graph G, set $\mathcal{C}$ of calls, integer k;
begin
    1. **for** $h := 1$ to n **do** Round l_h to the closest upper power of 2;
    2. Find a pseudo-schedule assuming only 1 available color;
    3. Assign calls to colors;
    4. Find a schedule separately for each color;
end.

We assume that call durations are powers of 2. Assuming only one color, in the sequel we will use the following interpretation: a call $C = [s, d, l]$ scheduled in interval $[t, t + l]$ can be graphically represented as a rectangle of height l,

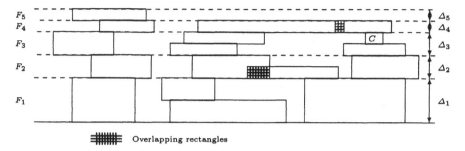

Fig. 1. Example of pseudo-schedule for 1 color

whose x-coordinates vary in $[s, d]$ and whose y-coordinates vary in $[t, t + l]$. We now describe steps 2-4 more in detail.

Find a pseudo-schedule.
We first determine a pseudo-schedule for $\mathcal{C}$, assuming only 1 color is available. A *pseudo-schedule* is an assignment $PS : \mathcal{C} \to \{0, \ldots \sum_{h=1}^{n} l_h\}$ of starting times to calls, such that each arc may be used by more than one call at the same time. The *length* of PS with respect to the set $\mathcal{C}$ of calls is $\max_{h \in \{1, \ldots, n\}} \{PS(C_h) + l_h\}$.

To our purposes, we need a pseudo-schedule in which at most two calls may use the same arc at the same time, based on the following definition: assume that, for some $h \leq n - 1$, we have a pseudo-schedule for calls $C_1, \ldots, C_h$; we say that C_{h+1} is *stable* [10] at time t with respect to the pseudo-schedule of $C_1, \ldots, C_h$ if and only if and only if there exists an arc e such that C_{h+1} uses e and, for every instant $0, \ldots, t-1$, e is used by at least one call in $\{C_1, \ldots, C_h\}$. In order to obtain such a pseudo-schedule, we proceed inductively as follows: i) calls are first ordered according to non-increasing durations (i.e. if $j > i$ then $l_i \geq l_j$); ii) C_1 is assigned starting time 0; iii) assuming $C_1, \ldots, C_h$ have each been assigned a starting time, C_{h+1} is assigned the maximum starting time such that it is stable with respect to the pseudo-schedule of $C_1, \ldots, C_h$.

Under the assumption that call durations are powers of 2 it is possible to show [10] that this choice leads to a pseudo-schedule in which at most two calls use the same arc at the same time. An example of a pseudo-schedule is presented in Figure 1 where, as described above, each call $C_h - [s_h, d_h, l_h]$ corresponds to a rectangle of height l_h whose horizontal coordinates vary between s_h and d_h.

In the following we use PS to denote the particular pseudo-schedule constructed above. Note that the notion of stability implies that the length of PS is a lower bound to the makespan of the optimal schedule for one color.

Assign calls to colors.
We partition the set of calls in the pseudo-schedule into γ subsets $F_1, F_2, \ldots F_\gamma$ called *stripes*, by "slicing" the pseudo-schedule into horizontal stripes (figure 1).

Stripes are defined inductively as follows: i) let C_{h_1} be the longest call starting at time 0 in the pseudo-schedule. Then stripe F_1 has *height* $\Delta_1 = l_{h_1}$ and includes all calls that in PS start not after Δ_1; ii) assume that stripes $F_1, \ldots, F_{i-1}$ have been defined $(i \geq 2)$ and that there are still calls not assigned to a stripe, let t be the time at which the last call in F_{i-1} ends and let C_{h_i} be the longest call starting at t in the pseudo-schedule PS. Then stripe F_i has *height* $\Delta_i = l_{h_i}$ and includes all calls that start not before t and end not after $t + l_{h_i}$. Figure 1 illustrates an example of slicing of a pseudo-schedule into stripes.

Let $\mathcal{F} = \{F_1, F_2, \ldots F_\gamma\}$ be defined by the construction given above. The next lemma states that $\mathcal{F}$ defines a partition of $\mathcal{C}$

Lemma 1. *Every call in $\mathcal{C}$ belongs to one and only one set in $\mathcal{F}$.*

Stripes (and the corresponding calls) are then assigned to colors by defining a proper instance x of MULTIPROCESSOR SCHEDULING with k identical machines with the goal of minimizing the total makespan. Roughly, jobs of instance x correspond to the sets $\{F_1, F_2, \ldots F_\gamma\}$ of the partition $\mathcal{F}$ obtained above. Namely, with each stripe of height Δ we associate a job of the same size.

We use the LPT (*Longest Processing Time* first) rule proposed by Graham [12] to solve the obtained instance of MULTIPROCESSOR SCHEDULING: jobs are first ordered according to decreasing sizes and then greedily assigned to the processors (i.e. a job is assigned to the less heavily loaded processor). The obtained solution yields an assignment of calls to colors, defined as follows: if stripe F_i is assigned to machine j, then all calls in F_i are assigned color j. Now, for each color j, we derive a pseudo-schedule PS_j of the calls that have been assigned color j in the following way: assume machine j has been assigned stripes $F_{j_1}, \ldots, F_{j_r}$ (in this order) and let C be a call assigned to color j and belonging to stripe F_{j_p}, $p \in \{1, \ldots, r\}$. Recall that $PS(C)$ is the starting time of C in the pseudo-schedule obtained in step 2 of *chain-cs*. The starting time $PS_j(C)$ of C in the pseudo-schedule for color j is given by the sum of the heights of stripes $F_{j_1}, \ldots, F_{j_p-1}$ plus the offset of C in its stripe F_{j_p}. Namely, we have that $PS_j(C) = \sum_{i=1}^{p-1} \Delta_{j_i} + PS(C) - \sum_{i=1}^{j_p-1} \Delta_i$.

Figure 2 illustrates the effect of *Assign-Calls-To-Colors* over the pseudo-schedule of figure 1 in the case of 2 colors. In particular, call $C \in F_3$ in figure 1 has been assigned color 2 and $PS_2(C) = \Delta_2 + PS(C) - (\Delta_1 + \Delta_2)$.

We now analyze the makespan of the pseudo-schedule. The assumption that the duration of each call is a power of 2 and the characteristics of the pseudo-schedule obtained in the first step of the algorithm allow a tight analysis of LPT, yielding the next Lemma. For any color $j = 1, \ldots, k$, let T^j be the length of the associated pseudo-schedule after procedure *Assign-Calls-To-Colors* and let $T_{PS} = max_j T^j$ (see fig. 2). Finally, let T^* denote the makespan of an optimal schedule for the instance under consideration.

Lemma 2. $T_{PS} \leq T^*$.

Find a Schedule.

For every color $j = 1, \ldots, k$, let PS_j denote the pseudo-schedule obtained after procedure *Assign-Calls-To-Colors* has been run and let C_j denote the set of calls that are assigned color j. It remains to derive a proper schedule S_j from PS_j, for any color $j = 1, \ldots, k$. The relationship between MIN-DSA and CHAIN-MIN-CS$_1$ stated in theorem 1 implies the following lemma.

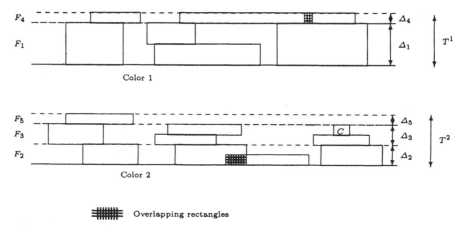

Fig. 2. Example of pseudo-schedule for 2 colors ($T_{PS} = T^1 = T^2$ in this case)

Lemma 3. *[10] If call durations are powers of 2 then there exists an algorithm for scheduling calls in PS_j with makespan at most 5/2 times the length of PS_j.*

Lemmas 1, 2 and 3 imply the following theorem.

Theorem 2. *Algorithm chain-min-cs finds a schedule whose makespan is at most 5 times the optimum.*

Notice that the same result above holds also in the case of undirected chains.

4 Call Scheduling in Ring Networks

In the sequel, Σ_m denotes a directed ring with m vertices, clockwise indexed $0, \ldots, m-1$. The arc set is $\{(v, (v+1) \bmod m), (v, (v-1) \bmod m) : 0 \le i \le m-1\}$, where $(v, (v+1) \bmod m)$ (respectively $(v, (v-1) \bmod m)$) denotes the clockwise (counter clockwise) directed arc between vertices v and $(v+1) \bmod m$ (v and $(v-1) \bmod m$). In the following we perform operations modulo m and we write $v+1$ ($v-1$) for $(v+1) \bmod m$ (respectively $(v\ \ 1) \bmod m$). We define $[s,t]$ to be $\{u | s \le u \le t\}$ if $s \le t$, $[s, m-1] \cup [0, t]$ otherwise.

A call in a ring can be routed clockwise or counter clockwise. We first find a routing of the calls minimizing the load and then we schedule them by using a constant approximation algorithm based on the results of Section 3.

4.1 Routing with Minimum Load

Given a directed ring Σ_m and a set $\mathcal{C}$ of calls, MINIMUM WEIGHTED DIRECTED RING ROUTING (MIN-WDRR) is the problem of routing calls in $\mathcal{C}$ so that the maximum load on the arcs is minimized. A_v (B_v), $v = 0, 1, \ldots, m-1$, denotes the load on arc $(v, v+1)$ (arc $(v+1, v)$). With each call $C_h \in C$ we can associate a binary variable x_h such that $x_h = 1$ if C_h is routed clockwise, $x_h = 0$ otherwise. Under these assumptions, MIN-WDRR can be formally described as follows:

$$\min L = \max\{\max_v A_v, \max_v B_v\}$$

$$A_v = \sum_{h:[v,v+1] \subseteq [s_h, t_h]} l_h x_h, \quad B_v = \sum_{h:[v+1,v] \subseteq [t_h, s_h]} l_h(1 - x_h), \quad x_h \in \{0,1\}, 1 \leq h \leq n.$$

OPT denotes the value of an optimal solution for any given instance of MIN-WDRR. MIN-WDRR is polynomial-time solvable when $l_h = 1$, $h = 1, 2, \ldots, n$ [19].

Theorem 3. MIN-WDRR *is NP-hard.*

Lemma 4. *There is a polynomial-time algorithm that solves* WEIGHTED DIRECTED RING ROUTING *with load at most twice the optimum.*

Sketch of the proof. We solve an instance of Multicommodity Flow [1] obtained by relaxing the integrality constraints in the formulation of MIN-WDRR above. Let $\{x_1^*, \ldots x_n^*\}$ be the fractional optimal solution: the hth component x_h^* is rounded to 0 if its value is $< 1/2$, to 1 otherwise.   □

We now use Lemma 4 to obtain a polynomial-time approximation scheme (PTAS) for WDRR. Let $\overline{L}$ be the load obtained by applying the algorithm described above. In the following, ϵ denotes any fixed positive number. Following [14], a call is routed *long-way* if it uses the longer path to connect its end vertices (ties are broken arbitrarily), it is routed *short-way* otherwise; futhermore, a call C_h is *heavy* if $l_h \geq \epsilon \overline{L}/3$, it is *light* otherwise. The number of heavy calls routed long way is bounded, as stated in the following lemma.

Lemma 5. *In any optimal solution there are at most $12/\epsilon$ heavy, long-way routed calls.*

Let $H \subseteq C$ denote the set of heavy calls. For any set $S \subseteq H$, let LP_S denote the following linear program:

$$\min L(S) = \max\{\max_v A_v, \max_v B_v\}$$

$$A_v = \sum_{h:[v,v+1] \subseteq [s_h, t_h]} l_h x_h + a_v, \quad B_v = \sum_{h:[v+1,v] \subseteq [t_h, s_h]} l_h(1 - x_h) + b_v, \quad x_h \subseteq [0,1], 1 \leq h \leq n.$$

a_v (b_v) denotes the load on clockwise (counter clockwise) directed arc $(v, v+1)$ $((v+1, v))$ resulting from routing long-way calls belonging to S and short-way calls belonging to $H - S$. Finally, $L^*(S)$ denotes the value of the optimal (fractional) solution of $LP^*(S)$. We now give the PTAS for MIN-WDRR:

Algorithm WDRR-PAS(ϵ)

input: $\langle \Sigma_m, \mathcal{C} \rangle$;
output: routing for $\langle \Sigma_m, \mathcal{C} \rangle$;
begin (1)
  for each $S \subseteq H : |S| \leq 12/\epsilon$ **do**
    begin (2)
      for each call $C \in H$ **do**
        if $C \in S$ **then** route C long-way
        else route C short-way;
      solve LP_S
    end(2);
  Let y be the solution corresponding to S' such that $L^*(S') = \min_S L^*(S)$;
  Apply a rounding procedure to y to a feasible routing $\tilde{x}$ with load $\tilde{L}(S')$;
  Output $\tilde{x}$
end(1).

Lemma 6 below shows that, for every $S \subseteq H$, $|S| \leq 12/\epsilon$, we can round the solution of LP_S in such a way that the value $\tilde{L}(S)$ of the integer solution obtained differs at most $\epsilon\overline{L}/2$ from $L^*(S)$. In order to prove the lemma, following [19] we first give the following definition: two calls $C_h = [s_h, t_h, l_h]$ and $C_j = [s_j, t_j, l_j]$ are *parallel* if the intervals $[s_h, t_h]$ and $[t_j, s_j]$ or the intervals $[t_h, s_h]$ and $[s_j, t_j]$ intersect at most at their endpoints. It may eventually happen that $s_h = s_j$ and $t_h = t_j$. Viewing an arc also as a chord, a demand is parallel to an arc when it can be routed through that arc, otherwise it is parallel to that arc's reverse. Finally, for any optimal solution $x^* = \{x_1^*, \ldots, x_n^*\}$ to $LP(S)$, call C_h is said *split* with respect to x^* if $0 < x_h^* < 1$.

Lemma 6. *For any $S \subseteq H$, $|S| \leq 12/\epsilon$, there is a polynomial time rounding procedure of the solution of LP_S, such that $\tilde{L}(S) - L^*(S) \leq \epsilon L/2$.*

Proof. Given $S \subseteq H$, $|S| \leq 12/\epsilon$, let x^* be a solution to LP_S with value $L^*(S)$. Following [19] we first obtain a new fractional solution $\hat{x} = \{\hat{x}_1, \ldots, \hat{x}_n\}$ such that no pair of parallel calls are both split and whose value $\hat{L}(S)$ is not larger than $L^*(S)$. Let us assume that there is a pair $C_i = [s_h, t_h, l_h]$ and $C_j = [s_j, t_j, l_j]$ of parallel calls, with $0 < x_h^*, x_j^* < 1$. We now reroute them in such a way that only one of them remains split. Two cases may arise:

1. $x_h^* + l_j x_j^*/l_h \leq 1$. In this case we set $\hat{x}_h = x_h^* + l_j x_j^*/l_h$ and $\hat{x}_j = 0$. Consider now a clockwise directed arc $(v, v+1)$ belonging to the interval $[s_h, t_h] \cap [s_j, t_j]$ and let $A_v^*, \hat{A}_v$ be its loads respectively before and after rerouting. It is easily verified that $\hat{A}_v = l_h x_h^* + l_j x_j^* = A_v^*$. If instead $(v+1, v)$ is any counter clockwise directed arc in $[t_h, s_h] \cap [t_j, s_j]$, it is easily seen that, if B_v^* and $\hat{B}_v$ respectively denote its loads first and after rerouting, we have $\hat{B}_v = (1 - x_h^*)l_h + (1 - x_j^*)l_j = B_v^*$. If $s_h \neq s_j$ or $t_h \neq t_j$, it also holds that arcs not belonging to $[s_h, t_h] \cap [s_j, t_j]$ or $[t_h, s_h] \cap [t_j, s_j]$ have the same or reduced loads.

2. $x_h^* + l_j x_j^*/l_h > 1$. In this case we set $\hat{x}_h = 1$ and $\hat{x}_j = x_j^* + l_h/l_j(x_h^* - 1)$. The analysis of this case proceeds exactly as in the previous one. Again we have

$\hat{A}_v = A_v^*$ for any clockwise directed arc belonging to interval $[s_h, t_h] \cap [s_j, t_j]$ and for any counter clockwise directed arc belonging to interval $[t_h, s_h] \cap [t_j, s_j]$, while the load on any other arc, if any, doesn't increase or decrease. As before, a similar argument holds for counter clockwise directed arcs.

At the end of this procedure we have a new solution $\hat{x}$ whose load $\hat{L}(S)$ is at most $L^*(S)$ and such that no two split demands are parallel. Without loss of generality we may assume that calls in $\mathcal{C}$ are ordered in such a way that $\{C_1, \ldots, C_q\}$ is the set of split calls for solution $\hat{x}$. We are now left to route split calls. Given any clockwise (counter clockwise) directed arc $(v, v+1)$ $((v+1, v))$, let $F_v \subseteq \{C_1, \ldots, C_q\}$ be the set of all split calls that are parallel to $(v, v+1)$ $((v+1, v))$. Observe that calls in F_v represent an interval in the ordered set $\{C_1, \ldots, C_q\}$; in the following, we shall denote F_v as an interval $[i_v, j_v]$, where i_v and j_v respectively are the first and last index of calls in $\{C_1, \ldots, C_q\}$ that are parallel to $(v, v+1)$ (or $(v+1, v)$). Also observe that F_v is exactly the set of all and only the calls whose rerouting can potentially increase the load on $(v, v+1)$ (or $(v+1, v)$). Let $\tilde{A}_v$ (respectively $\tilde{B}_v$) denote the load resulting on $(v, v+1)$ (respectively $(v+1, v)$) after routing calls that are split in $\hat{x}$ and let $\tilde{x} = \{\tilde{x}_1, \ldots, \tilde{x}_n\}$ be the corresponding integer solution. Finally, let $\tilde{L}(S) = \max\{\max_v \tilde{A}_v, \max_v \tilde{B}_v\}$. We have:

$$\tilde{A}_v = \hat{A}_v + \sum_{h \in [i_v, j_v]} l_h(\tilde{x}_h - \hat{x}_h),$$

$$\tilde{B}_v = \hat{B}_v + \sum_{h \notin [i_v, j_v]} l_h[(1 - \tilde{x}_h) - (1 - \hat{x}_h)] = \hat{B}_v + \sum_{h \notin [i_v, j_v]} l_h(\hat{x}_h - \tilde{x}_h).$$

We are now ready to route calls that are split in $\hat{x}$. For $j = 1, \ldots, q$, C_j is routed as follows:

$$\tilde{x}_j = \begin{cases} 1 \text{ if } -l_j \hat{x}_j + \sum_{h=1}^{j-1} l_i(\tilde{x}_h - \hat{x}_i) < -\frac{l_j}{2} \\ 0 \text{ otherwise} \end{cases}$$

As a consequence, if $\tilde{x}_j = 1$ then $\sum_{h=1}^{j} l_h(\tilde{x}_h - \hat{x}_h) < l_j/2$, while if $\tilde{x}_j = 0$ then $\sum_{h=1}^{j} l_h(\tilde{x}_h - \hat{x}_h) \geq -l_j/2$. In both cases $\sum_{h=1}^{j} l_h(\tilde{x}_h - \hat{x}_h) \in \left[-\frac{l_j}{2}, \frac{l_j}{2}\right)$ for any $j \in [i_v, j_v]$.

We now show that, for each clockwise directed arc $(v, v+1)$, $\tilde{A}_v - \hat{A}_v \leq (3/2)l_{max}$, where $l_{max} = \max_{h \in \{1, \ldots, q\}} l_h$. Since we are considering light calls, we have $l_{max} \leq (\epsilon L)/3$. This implies $\tilde{A}_v - \hat{A}_v \leq (\epsilon L)/2$. In particular, given any clockwise directed arc $(v, v+1)$, two cases may arise:

1. $i_v \leq j_v$. In this case we bound $\tilde{A}_v - \hat{A}_v = \sum_{h=i_v}^{j_v} l_h(\tilde{x}_h - \hat{x}_h)$ as follows:

$$\tilde{A}_v - \hat{A}_v = \sum_{h=1}^{j_v} l_h(\tilde{x}_h - \hat{x}_h) - \sum_{h=1}^{i_v - 1} l_h(\tilde{x}_h - \hat{x}_h) \leq \frac{l_j}{2} - \left(-\frac{l_j}{2}\right) = l_j \leq l_{max}.$$

2. $i_v > j_v$. In this case we have:

$$\tilde{A}_v - \hat{A}_v = \sum_{h=1}^{q} l_h(\tilde{x}_h - \hat{x}_h) - \sum_{h=1}^{i_v - 1} l_h(\tilde{x}_h - \hat{x}_h) + \sum_{h=1}^{j_v} l_h(\tilde{x}_h - \hat{x}_h) < \frac{3}{2}l_j.$$

Since $\frac{3}{2}l_j \leq \frac{3}{2}l_{max}$, it follows that $\tilde{A}_v - \hat{A}_v \leq (3/2)l_{max}$. Since we are only considering light calls (heavy calls have already been routed) and recalling the definition of light call, $\tilde{A}_v - \hat{A}_v \leq (\epsilon L)/2$. In the same way we can prove that $\tilde{B}_v - \hat{B}_v$. Since $\tilde{L}(S) - L^*(S) \leq \max_v\{\max\{A_v - \hat{A}_v, B_v - \hat{B}_v\}\}$ the thesis follows.
$\square$

This is sufficient to prove the main result of this section. In fact, if S_{OPT} denotes the set of heavy calls that are routed long-way in the optimal integer solution, we have $\tilde{L}(S) \leq \epsilon\overline{L}/2 + L^*(S_{OPT})$ where, by lemma 4, $\epsilon\overline{L} \leq 2L^*(S_{OPT})$. This implies that $\tilde{L}(S) \leq (1 + \epsilon)L^*(S_{OPT})$; since $L^*(S_{OPT})$ is a lower bound to the optimal integer solution the following theorem holds

Theorem 4. *The solution provided by algorithm* **WDRR-PAS**(ϵ) *has value at most* $(1 + \epsilon)OPT$. *For any fixed* $\epsilon > 0$, *the time complexity of the algorithm is polynomial.*

4.2   An Approximate Algorithm for RING-MIN-CS

When $\epsilon > 0$, Algorithm WDRR-PAS(ϵ) finds a $(1 + \epsilon)$-approximate algorithm for routing with minimum load. We now turn to the scheduling phase. In the following T^* denotes the makespan of an optimal solution to MIN-CS.
A simple idea is as follows: $i)$ solve MIN-WDRR within $(1 + \epsilon/2)$ the optimum, $ii)$ cut the ring at any link, say e, $iii)$ solve two instances of MIN-CHAIN-CS with calls that do not use link e (in any direction), $iv)$ when the last call scheduled in the previous step is completed schedule calls that use link e using a $(1 + \epsilon/2)$ approximate partitioning algorithm.
It seems reasonable to prove that the above simple algorithm gives an approximation ratio of $6 + \epsilon$. However this is not necessarily true because the optimum scheduling algorithm might use a completely different routing of calls (even one that is not ϵ close to the routing with minimum load). It is possible to prove the following weaker result, whose proof is omitted here. In the full paper we will give a counterexample showing that, using the routing obtained in the previous subsection one is unlikely to prove anything better than $11 + \epsilon$.

Theorem 5. *For any fixed* $\epsilon > 0$, *there exists a polynomial-time algorithm that finds a routing and a schedule for* RING-MIN-CS *having makespan* $\leq (12 + \epsilon)T^*$.

We conclude this section by observing that the same result holds also in the case of undirected rings. In this case we will use the polynomial time approximation scheme proposed in [14] to compute the routing of calls. Details will be given in the full paper.

5   Conclusions

In this paper we have considered the Call Scheduling problem in all-optical, networks. We have proposed approximation algorithms with constant approximation ratio for both chain and ring networks. As a side result, we have also proposed a polynomial-time approximation scheme for the problem of routing weighted calls in a ring network with the aim of minimizing the maximum load, a problem which has itself a practical relevance in SONET networks [18,14,19].

It would be interesting to consider different network topologies and to consider the problem whith release dates and different objective functions. Also the on-line version of the problems deserves future investigation.

References

1. R.K. Ahuja, T.L. Magnanti, and J.B. Orlin. *Network flows*. Prentice-Hall, 1993.
2. B. Awerbuch, Y. Azar, A. Fiat, S. Leonardi, and A. Rosen. On-line competitive algorithms for call admission in optical networks. In *Proc. of the 4th Annual European Symposium on Algorithms*, 1996.
3. Y. Bartal, A. Fiat, and S. Leonardi. Lower bounds for on-line graph problems with application to on-line circuit and optical-routing. In *Proc. of the 28th Annual Symposium on the Theory of Computing*, 1996.
4. Y. Bartal and S. Leonardi. On-line routing in all-optical networks. In *Proc. of the 24th International Colloquium on Automata, Languages and Programming*, 1997.
5. L. Becchetti. *Efficient Resource Management in High Bandwidth Networks*. PhD thesis, Dipartimento di Informatica e Sistemistica, University of Rome "La Sapienza", 1998.
6. C. Brackett. Dense Wavelength Division Multiplexing Networks: Principles and Applications. *IEEE Journal Selected Areas in Comm.*, 8, 1990.
7. T. Erlebach and K. Jansen. Scheduling Virtual Connections in Fast Networks. In *Proc. of the 4th Parallel Systems and Algorithms Workshop PASA '96*, 1996.
8. T. Erlebach and K. Jansen. Call Scheduling in Trees, Rings and Meshes. In Proc. of the 30th Hawaii International Conference on System Sciences, 1997.
9. M. R. Garey and D. S. Johnson. *Computers and Intractability: A Guide to the Theory of NP-Completeness*. W. H. Freeman, San Francisco, 1979.
10. J. Gergov. Approximation algorithms for dynamic storage allocation. In *Proc. of the 4th Annual European Symposium on Algorithms*, 1996.
11. J. Gergov. Algorithms for Compile-Time Memory Optimization. Private communication, 1998.
12. R. L. Graham. Bounds for multiprocessing timing anomalies. *SIAM Journal of Applied Math.* , 17, 1969.
13. P. E. Green. *Fiber-optic Communication Networks*. Prentice-Hall, 1992.
14. S. Khanna. A Polynomial Time Approximation Scheme for the SONET Ring Loading Problem. *Bell Labs Tech. J.*, Spring, 1997.
15. H. A. Kierstead and W. T. Trotter. An extremal problem in recursive combinatorics. *Congressus Numerantium*, 33, 1981.
16. T. Leighton, B. Maggs, and S.Rao. Packet routing and jobshop scheduling in O(congestion+dilation) steps. *Combinatorica*, 14, 1994.
17. P. Raghavan and E. Upfal. Efficient Routing in All-Optical Networks. In *Proc. of the 26th Annual Symposium on the Theory of Computing*, 1994.
18. A. Schrijver, P. Seymour, and P. Winkler. The Ring Loading Problem. *SIAM J. on Discrete Math.*, 11, 1998.
19. G. Wilfong and P. Winkler. Ring Routing and Wavelength Translation. In *Proc. of the European Symposium on Algorithms*, pages 333–341, 1998.

A Randomized Algorithm for Flow Shop Scheduling

Naveen Garg[1], Sachin Jain[2], and Chaitanya Swamy[3]

[1] Department of Computer Science, Indian Institute of Technology,
Hauz Khas, New Delhi - 110016, India
naveen@cse.iitd.ernet.in
[2] Department of Computer Sciences, The University of Texas at Austin,
Austin, TX 78712-1188, USA
sachin@cs.utexas.edu
[3] Department of Computer Science, Cornell University,
Ithaca, NY 14853, USA
swamy@cs.cornell.edu

Abstract. Shop scheduling problems are known to be notoriously intractable, both in theory and practice. In this paper we give a randomized approximation algorithm for flow shop scheduling where the number of machines is part of the input problem. Our algorithm has a multiplicative factor of $2(1 + \delta)$ and an additive term of $O(m \ln(m + n) p_{max})/\delta^2)$.

1 Introduction

Shop scheduling has been studied extensively in many varieties. The basic shop scheduling model consists of *machines* and *jobs* each of which consists of a set of *operations*. Each operation has an associated machine on which it has to be processed for a given length of time. The processing times of operations of a job cannot overlap. Each machine can process at most one operation at a given time. We assume that there are m machines and n jobs. The processing time(of an operation) of job j on machine i is denoted by p_{ij} and $p_{max} = \max p_{ij}$. We will use the terms job(operation) size and processing time interchangeably.

The three well-studied models are the open shop, flow shop and job shop problems. In an open shop problem, the operations of a job can be performed in any order; in a job shop, they must be processed in a specific, job-dependent order. A flow shop is a special case of job shop in which each job has exactly m operations – one per machine, and the order in which they must be processed is same for all the jobs. The problem is to minimize the makespan, ie. the overall length, of the schedule with the above constraints.

All the above problems are strongly **NP**-Hard in their most general form. For job shops, extremely restricted versions are also strongly **NP**-Hard; for example when there are two machines and the all operations have processing times of one or two time units. For the flow shop problem, the case when there are more than two machines is strongly **NP**-Hard, although the two machine version is polynomially solvable [3]. The open shop problem is weakly **NP**-Hard when the

C. Pandu Rangan, V. Raman, R. Ramanujam (Eds.): FSTTCS'99, LNCS 1738, pp. 213–218, 1999.
© Springer-Verlag Berlin Heidelberg 1999

number of machines is fixed(but arbitrary) and its relation to being strongly NP-Hard is open. For two machines there exists a polynomial algorithm for open shops.

As far as approximability of these models is concerned, Williamson et al. [10] proved a lower bound 5/4 for the problems in their most general form. For the general open shop problem a greedy heuristic is a 2-approximation algorithm. In the case of job shop Shmoys, Stein and Wein [8] give a randomized $O(\log^2(m\mu)/\log\log(m\mu))$ approximation algorithm where μ is the maximum number of operations per job. This bound was slightly improved by Goldberg, Paterson, Srinivasan and Sweedyk [1]. Schmidt, Siegel and Srinivasan [4] give a deterministic $\log^2(m\mu)/\log\log(m\mu))$ approximation algorithm. For flow shop, an algorithm by Sevast'janov [5] gives an additive approximation of $m(m-1)p_{max}$. When m is not fixed these are the best results known. For fixed m we have $(1+\epsilon)$ polynomial approximation schemes for all the three problems. An approximation scheme for flow shop was given by Hall [2]. Recently an approximation scheme for the open shop problem was given by Sevast'janov and Woeginger [6] while Sviridenko, Jansen and Solis-Oba [9] give one for job shops.

Our Contribution

In this paper we present a randomized approximation algorithm for flow shop scheduling when the number of machines is not fixed. Our algorithm is based on the rounding of the solution of an LP formulation of the flow shop problem. The LP formulation imposes some additional constraints which makes the rounding scheme possible. The makespan returned by our algorithm is within $2(1+\delta)$ times the optimal makespan and has an additive term of $O(m\ln(m+n)p_{max}/\delta^2)$. This shows a tradeoff between the additive and multiplicative factors; if n is bounded by some polynomial in m the additive factor we obtain is better than the one in the Sevast'janov algorithm, and the multiplicative factor is better than that in the algorithm by Shmoys et al.

The remaining part of this paper is organized as follows. In Section 2, we discuss the new slotting constraints imposed by us. Section 3 gives a integral multicommodity flow formulation of the problem and Section 4 deals with the randomized algorithm and its analysis.

2   Slotting Constraints

It is no loss of generality to assume that all operations have size at least 1. Let p_{max} be the largest operation size. The machines are numbered in the order in which the operations of each job are to be processed.

We divide time into slots of size s, $s \geq 2p_{max}$. For our randomized rounding scheme to work we require that the slots be *independent* of each other. By this we mean that the order in which operations are scheduled on a machine in any time-slot is independent of the order of the operations in other time-slots and on other machines. To ensure this we impose the restriction that no operation straddles a slot boundary and that no job moves from one machine to another in the middle of a slot. The second condition is equivalent to saying that the operations of a job are performed in distinct slots. Thus a job's operation on the ith machine can only start if the operation on the $(i-1)^{th}$ machine has been completed by the end of the previous slot.

Consider a flow shop schedule with minimum makespan, OPT. We now show how to modify the schedule to satisfy the slotting constraints. First divide time into slots of size $s - p_{max}$. Since there could be operations straddling slot-boundaries, we insert gaps of duration p_{max} after each slot. Operations starting in a slot and going over to the next now finish in these gaps. Finally we merge each gap with the slot just before it. This yields slots of size s and each operation now finishes in the slot in which it starts. Since OPT was the makespan of the original schedule the makespan of this modified schedule is $s \cdot OPT/(s - p_{max})$. Next we shift all operations on the second machine by s, on the third machine by $2s, \ldots$, on the mth machine by $(m-1)s$. This increases the makespan by $(m-1)s$ and gives a schedule which satisfies the restriction that all operations of a job are performed in distinct time-slots. Thus we have obtained a schedule which satisfies the slotting constraints and has makespan at most $\frac{s}{s-p_{max}}OPT + (m-1)s$. Note that $\frac{s}{s-p_{max}}$ is at most 2 for $s = 2p_{max}$.

3   An Integral Multicommodity Flow Formulation

In this section we obtain an approximation to the flow shop scheduling problem with the slotting restriction. We begin by "guessing" the number of time-slots required by the schedule. Construct a directed graph $G = (V, E)$ which has a vertex for each (time-slot, machine) pair. There is an edge directed from vertex $u = (a, i)$ to vertex $v = (b, j)$ if $j = i + 1$ and $a < b$. For each job j we have two vertices s_j and t_j. s_j has edges to all vertices corresponding to the first machine and t_j has edges from all vertices corresponding to the last machine.

With this graph we associate a multicommodity flow instance; there is one commodity associated with each job j and s_j, t_j are the source and sink for this commodity. The flow of each commodity should be conserved: the total flow of a commodity entering a vertex (other than the source/sink of that commodity) is equal to the flow of that commodity leaving the vertex. We wish to route one unit of each commodity subject to the following throughput constraints on the vertices. Consider a vertex $v = (a, i)$ and let $x_{v,j}$ be the flow of commodity j through v. Then

$$\sum_j x_{v,j} p_{i,j} \leq s$$

Note that an integral multicommodity flow corresponds to a flow shop schedule satisfying the slotting restrictions. The feasibility of a multicommodity flow instance can be determined in polynomial time by formulating it as a linear program [7]. Infeasibilty of the multicommodity flow instance implies that our guess on the makespan is too small. When this happens we increase our guess on the number of time slots. Let k be the smallest number of time slots for which the multicommodity flow instance is feasible and let F be the corresponding flow. If T denotes the minimum makespan of a flow shop schedule satisfying the slotting constraints then $T \geq (k-1)s$.

4   The Algorithm and Its Analysis

F is a (fractional) multicommodity flow which routes one unit of each commodity while respecting the throughput constraints on the vertices. Flow-theory says that the flow of any commodity can be viewed as a collection of at most $|E|$ paths. With each path we associate a weight which is just the flow along that path. Hence the total weight of all paths corresponding to commodity j is one.

The randomized algorithm picks exactly one path for each commodity with the probability of picking a path equal to its weight. This collection of paths, one for each commodity, gives an integral multicommodity flow which (possibly) violates the throughput constraints. The integral multicommodity flow in turn defines a flow shop schedule which satisfies the slotting constraints but which might be infeasible as the total processing time of operations schedule on a machine in a specific time-slot might exceed s.

Consider vertex $v = (a, i)$. Let X_j be a random variable which is 1 if job j is scheduled on machine i in slot a and 0 otherwise. Let X be a random variable defined as

$$X = \sum_j p_{ij} X_j$$

Claim. $E[X] \leq s$.

Proof. The probability that X_j is 1 equals $x_{v,j}$. Hence, $E[X_j] = x_{v,j}$. By linearity of expectations it follows that $E[X] = \sum_j p_{ij} E[X_j] = \sum_j p_{ij} x_{v,j}$. The throughput constraint on vertex v implies $\sum_j p_{ij} x_{v,j} \leq s$ from which the claim follows.

Since the random variable X is a linear combination of n independent Poisson trials $X_1, X_2, \ldots, X_n$ such that $\mathbf{Pr}[X_j = 1] = x_{v,j}$, using Chernoff bounds we obtain

$$\mathbf{Pr}[X > (1+\delta)\mathbf{E}[X]] \leq \left[\frac{e^\delta}{(1+\delta)^{(1+\delta)}} \right]^{\mathbf{E}[X]/p_{\max}}.$$

Observe that a trivial flow shop schedule satisfying the slotting constraints can be obtained by assigning job j to slot $j + i - 1$ on machine i, $1 \leq j \leq n$ and $1 \leq i \leq m$. This schedule has makespan $(n + m - 1)s$ and hence the number of vertices in the graph (excluding the source and sink vertices) is at most $m(n + m - 1)$.

Let $c = \frac{(1+\delta)^{(1+\delta)}}{e^\delta}$ and choose $s = \max\{2, \log_c[2m(n + m - 1)]\}p_{\max}$. From the above argument it follows that $\mathbf{Pr}[X > (1+\delta)s] \leq \frac{1}{2m(n+m-1)}$. Hence the probability that for some machine and some time-slot the total processing time of operations scheduled on this machine in this time-slot is more than $(1 + \delta)s$

is at most $1/2$. Equivalently, with probability at least $1/2$ the processing time in every slot is less than $(1 + \delta)s$. Expanding each slot of size s to a slot of size $(1 + \delta)s$ then gives us a flow shop schedule of makespan $(1 + \delta)ks$.

Recall that any flow shop schedule satisfying the slotting restriction has makespan at least $(k-1)s$. Further, a flow shop schedule of makespan OPT yields a schedule satisfying the slotting restrictions and having makespan $\frac{s}{s-p_{max}} OPT +$ $(m - 1)s$. This implies that

$$(k - 1)s \leq \frac{s}{s - p_{max}} OPT + (m - 1)s$$

Hence with probability at least $1/2$ the randomized rounding procedure gives us a feasible schedule whose makespan is bounded by

$$(1 + \delta)ks \leq \frac{(1 + \delta)s}{s - p_{max}} OPT + (1 + \delta)ms$$

where $s = \max\{2, \log_c[2m(n + m - 1)]\}p_{max}$, $c = \frac{(1+\delta)^{(1+\delta)}}{e^\delta}$ and δ is a positive constant chosen so that $c > 1$.

Theorem 4.1. *There exists a polynomial time randomized algorithm for flow shop scheduling which with probability at least $1/2$ finds a schedule with makespan at most*

$$2(1 + \delta)OPT + m(1 + \delta)p_{max} \log_c[2m(n + m - 1)]$$

where $c = \frac{(1+\delta)^{(1+\delta)}}{e^\delta}$.

References

1. Leslie A. Goldberg, Mike Paterson, Aravind Srinivasan, and Elizabeth Sweedyk. Better approximation guarantees for job shop scheduling. In *Proc. 8th ACM-SIAM Symp. on Discrete Algorithms(SODA)*, pages 599–608, 1997.
2. Leslie A. Hall. Approximability of flow shop scheduling. In *Proc. 36th IEEE Annual Symp. on Foundations of Computer Science*, pages 82–91, 1995.
3. S.M. Johnson. Optimal two- and three- stage procution schedules with setup times included. *Naval Research Logistics Quarterly*, 1:61–68, 1954.
4. J.P. Schmidt, A. Siegel, and A. Srinivasan. Chernoff-holding bounds for applications with limited independence. In *Proc. 4th ACM-SIAM Symp. on Discrete Algorithms(SODA)*, pages 331–340, 1993.
5. S. Sevast'janov. Bounding algorithm for the routing problem with arbitrary paths and alternative servers. *Cybernetics*, 22:773–780, 1986.
6. S. Sevast'janov and G. Woeginger. Makespan minimization in open shops : A polynomial time approximation scheme. *Mathematical Programming*, 82(1-2), 1998.
7. D.B. Shmoys. *Approximation algorithms for **NP**-Hard problems*, chapter Cut problems and their applications to divide and conquer, pages 192–234. PWS, 1997.
8. D.B. Shmoys, C. Stein, and J. Wein. Improved approximation algorithms for shop scheduling problems. *SIAM Journal of Computing*, 23:617–632, 1994.

9. Maxim Sviridenko, K. Jansen, and R. Solis-Oba. Makespan minimization in job shops : A polynomial time approxmiation scheme. In *Proc. 31st Annual ACM Symp. on Theory of Computing*, pages 394–399, 1999.
10. D.P. Williamson, L.A. Hall, J.A. Hoogeveen, C.A.J. Hurkens, J.K. Lenstra, and S.V. Sevast'janov. Short shop schedules. *Operations Research*, 45:288–294, 1997.

Synthesizing Distributed Transition Systems from Global Specifications*

Ilaria Castellani[1], Madhavan Mukund[2], and P. S. Thiagarajan[2]

[1] INRIA Sophia Antipolis,
2004 route des Lucioles, B.P. 93, 06092 Sophia Antipolis Cedex, France
Ilaria.Castellani@sophia.inria.fr
[2] Chennai Mathematical Institute,
92 G.N. Chetty Road, Chennai 600 017, India
{madhavan,pst}@smi.ernet.in

Abstract. We study the problem of synthesizing distributed implementations from global specifications. In particular, we characterize when a global transition system can be implemented as a synchronized product of local transition systems. Our work extends a number of previous studies in this area which have tended to make strong assumptions about the specification—either in terms of determinacy or in terms of information concerning concurrency.

We also examine the more difficult problem where the correctness of the implementation in relation to the specification is stated in terms of bisimulation rather than isomorphism. As an important first step, we show how the synthesis problem can be solved in this setting when the implementation is required to be deterministic.

1 Introduction

Designing distributed systems has always been a challenging task. Interactions between the processes can introduce subtle errors in the system's overall behaviour which may pass undetected even after rigorous testing. A fruitful approach in recent years has been to specify the behaviour of the overall system in a global manner and then *automatically synthesize* a distributed implementation from the specification.

The question of identifying when a sequential specification has an implementation in terms of a desired distributed architecture was first raised in the context of Petri nets. Ehrenfeucht and Rozenberg [ER90] introduced the concept of *regions* to describe how to associate places of nets with states of a transition system. In [NRT92], Nielsen, Rozenberg and Thiagarajan use regions to characterize the class of transition systems which arise from elementary net systems. Subsequently, several authors have extended this characterization to larger classes of nets (for a sample of the literature, see [BD98,Muk92,WN95]).

---

* This work has been sponsored by IFCPAR Project 1502-1. This work has also been supported in part by BRICS (Basic Research in Computer Science, Centre of the Danish National Research Foundation), Aarhus University, Denmark.

C. Pandu Rangan, V. Raman, R. Ramanujam (Eds.): FSTTCS'99, LNCS 1738, pp. 219–231, 1999.

Here, we focus on *product transition systems*—networks of transition systems which coordinate by synchronizing on common actions [Arn94]. This model comes with a natural notion of component and induced notions of concurrency and causality. It has a well-understood theory, at least in the linear-time setting [Thi95]. This model is also the basis for system descriptions in a number of model-checking tools [Kur94,Hol97].

We establish two main sets of results in this paper. First, we characterize when an *arbitrary* transition system is isomorphic to a product transition system with a specified distribution of actions. Our characterization is effective—for finite-state specifications, we can synthesize a finite-state implementation. We then show how to obtain implementations when concurrency is specified in terms of an abstract independence relation, in the sense of Mazurkiewicz trace theory [Maz89]. We also present realizability relationships between product transition systems in terms of a natural preorder over the distribution of actions across agents. Our result subsumes the work of Morin [Mor98] on synthesizing product systems from *deterministic* specifications.

Our second result deals with the situation when we have global specifications which are behaviourally equivalent to, but not necessarily isomorphic to, product systems. The notion of behavioural equivalence which we use is strong bisimulation [Mil89]. The synthesis problem here is to implement a global transition system TS as a product transition system $\widetilde{TS}$ such that TS and $\widetilde{TS}$ are bisimilar to each other. We show how to solve this problem when the implementation is deterministic. Notice that the specification itself may be nondeterministic. Since distributed systems implemented in hardware, such as digital controllers, are deterministic, the determinacy assumption is a natural one. Solving the synthesis problem modulo bisimulation in the general case where the implementation may be nondeterministic appears to be hard.

The problem of expressing a global transition system as a product of component transition systems modulo bisimilarity has also been investigated in the context of process algebras in [Mol89,MM93]. In [GM92], Groote and Moller examine the use of decomposition techniques for the verification of parallel systems. These results are established in the context of transition systems which are generated using process algebra expressions. Generalizing these results to arbitrary, unstructured, finite-state systems appears hard.

The paper is organized as follows. In the next section we formally introduce product transition systems and formulate the synthesis problem. In Section 3, we characterize the class of transition systems which are isomorphic to product transition systems. The subsequent section extends these results to the context where the distributed implementation is described using an independence relation. Next, we show that deterministic systems admit canonical minimal implementations. In Section 6, we present our second main result, characterizing the class of transition systems which are bisimilar to deterministic product systems.

2   The Synthesis Problem for Product Transition Systems

Labelled transition systems provide a general framework for modelling computing systems. A labelled transition system is defined as follows.

Definition 2.1. *Let Σ be a finite nonempty set of actions. A labelled transition
system over Σ is a structure $TS = (Q, \rightarrow, q_{in})$, where Q is a set of states,
$q_{in} \in Q$ is the initial state and $\rightarrow \subseteq Q \times \Sigma \times Q$ is the transition relation.*

We abbreviate a transition sequence of the form $q_0 \xrightarrow{a_1} q_1 \cdots \xrightarrow{a_n} q_n$ as
$q_0 \xrightarrow{a_1 \cdots a_n} q_n$. In every transition system $TS = (Q, \rightarrow, q_{in})$ which we encounter,
we assume that each state in Q is reachable from the initial state—that is, for
each $q \in Q$ there exists a transition sequence $q_{in} = q_0 \xrightarrow{a_1 \cdots a_n} q_n = q$.

A large class of distributed systems can be fruitfully modelled as networks of
local transition systems whose moves are globally synchronized through common
actions. To formalize this, we begin with the notion of a distributed alphabet.

Definition 2.2. *A distributed alphabet over Σ, or a distribution of Σ, is a tuple
of nonempty sets $\widetilde{\Sigma} = \langle \Sigma_1, \ldots, \Sigma_k \rangle$ such that $\bigcup_{1 \leq i \leq k} \Sigma_i = \Sigma$. For each action
$a \in \Sigma$, the locations of a are given by the set $loc_{\widetilde{\Sigma}}(a) = \{i \mid a \in \Sigma_i\}$. If $\widetilde{\Sigma}$ is
clear from the context, we write just $loc(a)$ to denote $loc_{\widetilde{\Sigma}}(a)$.*

We consider two distributions to be the same if they differ only in the order of
their components.

Henceforth, for any natural number k, $[1..k]$ denotes the set $\{1, 2, \ldots, k\}$.

Definition 2.3. *Let $\langle \Sigma_1, \ldots, \Sigma_k \rangle$ be a distribution of Σ. For each $i \in [1..k]$, let
$TS_i = (Q_i, \rightarrow_i, q_{in}^i)$ be a transition system over Σ_i. The product $(TS_1 \parallel \cdots \parallel
TS_k)$ is the transition system $TS = (Q, \rightarrow, q_{in})$ over $\Sigma = \bigcup_{1 \leq i \leq k} \Sigma_i$, where:*

- $q_{in} = (q_{in}^1, \ldots, q_{in}^k)$.
- $Q \subseteq (Q_1 \times \cdots \times Q_k)$ *and* $\rightarrow \subseteq Q \times \Sigma \times Q$ *are defined inductively by:*
  - $q_{in} \in Q$.
  - *Let* $q \in Q$ *and* $a \in \Sigma$. *For* $i \in [1..k]$, *let* $q[i]$ *denote the* i^{th} *component
    of* q. *If for each* $i \in loc(a)$, TS_i *has a transition* $q[i] \xrightarrow{a}_i q_i'$, *then*
    $q \xrightarrow{a} q'$ *and* $q' \in Q$ *where* $q'[i] = q_i'$ *for* $i \in loc(a)$ *and* $q'[j] = q[j]$ *for*
    $j \notin loc(a)$.

We often abbreviate the product $(TS_1 \parallel \cdots \parallel TS_k)$ by $\parallel_{i \in [1..k]} TS_i$.

The synthesis problem

The synthesis problem can now be formulated as follows. If $TS =
(Q, \rightarrow, q_{in})$ is a transition system over Σ, and $\widetilde{\Sigma} = \langle \Sigma_1, \ldots, \Sigma_k \rangle$ is a distri-
bution of Σ, does there exist a $\widetilde{\Sigma}$-implementation of TS—that is, a tuple of
transition systems $\langle TS_1, \ldots, TS_k \rangle$ such that TS_i is a transition system over Σ_i
and the product $\parallel_{i \in [1..k]} TS_i$ is isomorphic to TS?

Example 2.4. Let $\widetilde{\Sigma} = \langle \{a,c\}, \{b,c\} \rangle$ *be a distribution of* $\{a,b,c\}$. *The first transition system below is* $\widetilde{\Sigma}$*-implementable—using expressions in the style of process algebra, we can write the product as* $(a+c) \parallel (bc+c)$. *Similarly, the second transition system may be implemented as* $(ac + c) \parallel (bc + c)$.

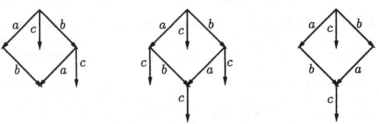

On the other hand, the system on the right is not $\widetilde{\Sigma}$-*implementable. Intuitively, the argument is as follows. If it were implementable in two components with alphabets* $\{a,c\}$ *and* $\{b,c\}$, c *would be enabled at the initial state in both components. But* c *can also occur after both actions* a *and* b *have occurred. So,* c *is possible after* a *in the first component and after* b *in the second component. Thus, there are two* c *transitions in both components and the product should exhibit all their combinations, giving rise to the system in the centre. We can formalize this argument once we have proved the results in the next section.*

3   A Characterization of Implementable Systems

We now characterize $\widetilde{\Sigma}$-implementable systems. In this section, unless otherwise specified, we assume that $\widetilde{\Sigma} = \langle \Sigma_1, \ldots, \Sigma_k \rangle$, with $\Sigma = \bigcup_{i \in [1..k]} \Sigma_i$.

The basic idea is to label each state of the given system by a k-tuple of local states (corresponding to a global state of a product system) such that the labelling function satisfies some consistency conditions. We formulate this labelling function in terms of local equivalence relations on the states of the original system—for each $i \in [1..k]$, if two states q_1 and q_2 of the original system are i-equivalent, the interpretation is that the global states assigned to q_1 and q_2 by the labelling function agree on the ith component. Our technique is similar to the one developed independently by Morin for the more restrictive class of *deterministic* transition system specifications [Mor98].

Theorem 3.1. *A transition system* $TS = (Q, \rightarrow, q_{in})$ *is* $\widetilde{\Sigma}$-*implementable with respect to a distribution* $\widetilde{\Sigma} = \langle \Sigma_1, \ldots, \Sigma_k \rangle$ *if and only if for each* $i \in [1..k]$ *there exists an equivalence relation* $\equiv_i \subseteq (Q \times Q)$ *such that the following conditions are satisfied:*

(i) *If* $q \xrightarrow{a} q'$ *and* $a \notin \Sigma_i$, *then* $q \equiv_i q'$.
(ii) *If* $q \equiv_i q'$ *for every* i, *then* $q = q'$.
(iii) *Let* $q \in Q$ *and* $a \in \Sigma$. *If for each* $i \in loc(a)$, *there exist* $s_i, s_i' \in Q$ *such that* $s_i \equiv_i q$, *and* $s_i \xrightarrow{a} s_i'$, *then for each choice of such* s_i*'s and* s_i'*'s there exists* $q' \in Q$ *such that* $q \xrightarrow{a} q'$ *and for each* $i \in loc(a)$, $q' \equiv_i s_i'$.

Proof. ($\Rightarrow$) : Suppose $\|_{i\in[1..k]} TS_i$ is a $\widetilde{\Sigma}$-implementation of TS. We must exhibit k equivalence relations $\{\equiv_i\}_{i\in[1..k]}$, such that conditions *(i)—(iii)* are satisfied. Assume, without loss of generality, that TS is not just isomorphic to $\|_{i\in[1..k]} TS_i$ but is in fact equal to $\|_{i\in[1..k]} TS_i$.

For $i \in [1..k]$, let $TS_i = (Q_i, \rightarrow_i, q_{\text{in}}^i)$. We then have $Q \subseteq (Q_1 \times \cdots \times Q_k)$ and $q_{\text{in}} = (q_{\text{in}}^1, \ldots, q_{\text{in}}^k)$. Define $\equiv_i \subseteq (Q \times Q)$ as follows: $q \equiv_i q'$ iff $q[i] = q'[i]$.

Since TS is a product transition system, it is clear that conditions *(i)* and *(ii)* are satisfied. To establish condition *(iii)*, fix $q \in Q$ and $a \in \Sigma$. Suppose that for each $i \in loc(a)$ there is a transition $s_i \xrightarrow{a} s_i'$ such that $s_i \equiv_i q$. Clearly, for each $i \in loc(a)$, $s_i \xrightarrow{a} s_i'$ implies $s_i[i] \xrightarrow{a}_i s_i'[i]$. Moreover $s_i[i] = q[i]$ by the definition of $\equiv_i$. Since TS is a product transition system, this implies $q \xrightarrow{a} q'$, where $q'[i] = s_i'[i]$ for $i \in loc(a)$ and $q'[i] = q[i]$ otherwise.

($\Leftarrow$) : Suppose we are given equivalence relations $\{\equiv_i \subseteq (Q \times Q)\}_{i\in[1..k]}$ which satisfy conditions *(i)—(iii)*. For each $q \in Q$ and $i \in [1..k]$, let $[q]_i \overset{\text{def}}{=} \{s \mid s \equiv_i q\}$. For $i \in [1..k]$, define the transition system $TS_i = (Q_i, \rightarrow_i, q_{\text{in}}^i)$ over Σ_i as follows:

- $Q_i = \{[q]_i \mid q \in Q\}$, with $q_{\text{in}}^i = [q_{\text{in}}]_i$.
- $[q]_i \xrightarrow{a}_i [q']_i$ iff $a \in \Sigma_i$ and there exists $s \xrightarrow{a} s'$ with $s \equiv_i q$ and $s' \equiv_i q'$.

We wish to show that TS is isomorphic to $\|_{i\in[1..k]} TS_i$. Let $\|_{i\in[1..k]} TS_i = (\widehat{Q}, \rightsquigarrow, \hat{q}_{\text{in}})$. We claim that the required isomorphism is given by the function $f : Q \rightarrow \widehat{Q}$, where $f(q) = ([q]_1, \ldots, [q]_k)$.

- We can show that f is well-defined—that is $f(q) \in \widehat{Q}$ for each q—by induction on the length of the shortest path from q_{in} to q. We omit the details.
- We next establish that f is a bijection. Clearly condition *(ii)* implies that f is injective. To argue that f is onto, let $([s_1]_1, \ldots, [s_k]_k) \in \widehat{Q}$ be reachable from $\hat{q}_{\text{in}}$ in n steps. We proceed by induction on n.
  - *Basis:* If $n = 0$, $([s_1]_1, \ldots, [s_k]_k) = \hat{q}_{\text{in}} = f(q_{\text{in}})$.
  - *Induction step:*
    Let $([r_1]_1, \ldots, [r_k]_k)$ be reachable from $\hat{q}_{\text{in}}$ in $n-1$ steps. Consider a move $([r_1]_1, \ldots, [r_k]_k) \xrightarrow{a} ([s_1]_1, \ldots, [s_k]_k)$. By the induction hypothesis there exists $q \in Q$ such that $f(q) = ([q]_1, \ldots, [q]_k) = ([r_1]_1, \ldots, [r_k]_k)$. Now, $([r_1]_1, \ldots, [r_k]_k) \xrightarrow{a} ([s_1]_1, \ldots, [s_k]_k)$ implies that $[r_i]_i \xrightarrow{a}_i [s_i]_i$ for each $i \in loc(a)$. Hence, for each $i \in loc(a)$, there exist r_i', s_i' such that $r_i' \equiv_i r_i$, $s_i' \equiv_i s_i$ and $r_i' \xrightarrow{a} s_i'$. By condition *(iii)*, since $q \equiv_i r_i$, for any choice of such r_i''s and s_i''s there exists q' such that $q \xrightarrow{a} q'$ and $q' \equiv_i s_i'$ for each $i \in loc(a)$. We want to show that $f(q') = ([s_1]_1, \ldots, [s_k]_k)$. For $i \in loc(a)$ we already know that $q' \equiv_i s_i' \equiv_i s_i$. So suppose $i \notin loc(a)$. In this case $q \xrightarrow{a} q'$ implies $[q']_i = [q]_i$, by condition *(i)*. From $[q]_i = [r_i]_i$ and $[r_i]_i = [s_i]_i$ it follows that $[q']_i = [s_i]_i$.
- It is now easy to argue that f is an isomorphism—we omit the details.

An effective synthesis procedure

Observe that Theorem 3.1 yields an effective synthesis procedure for finite-state specifications which is exponential in the size of the original transition system and the number of components in the distributed alphabet. The number of ways of partitioning a finite-state space using equivalence relations is bounded and we can exhaustively check each choice to see if it meets criteria *(i)–(iii)* in the statement of the theorem.

4   Synthesis and Independence Relations

An abstract way of enriching a labelled transition system with information about concurrency is to equip the underlying alphabet with an independence relation—intuitively, this relation specifies which pairs of actions in the system can be executed independent of each other.

Definition 4.1. *An* independence relation *over* Σ *is a symmetric, irreflexive relation* $I \subseteq \Sigma \times \Sigma$.

Each distribution $\widetilde{\Sigma} = \langle \Sigma_1, \ldots, \Sigma_k \rangle$ induces a natural independence relation $I_{\widetilde{\Sigma}}$ over Σ—two actions are independent if they are performed at nonoverlapping sets of locations across the system. Formally, for $a, b \in \Sigma$, $a\ I_{\widetilde{\Sigma}}\ b \Leftrightarrow loc(a) \cap loc(b) = \emptyset$. The following example shows that different distributions may yield the same independence relation.

Example 4.2. If $\Sigma = \{a, b, c, d\}$ *and* $I = \{(a, b), (b, a)\}$, *then the distributions* $\widetilde{\Sigma} = \langle \{a, c, d\}, \{b, c, d\} \rangle$, $\widetilde{\Sigma}' = \langle \{a, c, d\}, \{b, c\}, \{b, d\} \rangle$ *and* $\widetilde{\Sigma}'' = \langle \{a, c\}, \{a, d\},$ $\{b, c\}, \{b, d\}, \{c, d\} \rangle$ *all give rise to the independence relation* I.

However for each independence relation I there is a *standard distribution* inducing I, whose components are the maximal cliques of the dependency relation $D = (\Sigma \times \Sigma) - I$. In the example above, the standard distribution is the one denoted $\widetilde{\Sigma}$. Henceforth, we will denote the standard distribution for I by $\widetilde{\Sigma}_I$.

The synthesis problem with respect to an independence relation

We can phrase the synthesis problem in terms of independence relations as follows. Given a transition system $TS = (Q, \rightarrow, q_{\text{in}})$ and a nonempty independence relation I over Σ, does there exist an *I-implementation* of TS, that is a $\widetilde{\Sigma}$-implementation of TS such that $\widetilde{\Sigma}$ induces I?

We show that if a transition system admits a $\widetilde{\Sigma}$-implementation then it also admits a $\widetilde{\Sigma}_{I_{\widetilde{\Sigma}}}$-implementation. Thus the synthesis problem with respect to independence relations reduces to the synthesis problem for standard distributions.

We begin by showing that a system that has an implementation over a distribution $\widetilde{\Sigma}$ also has an implementation over any coarser distribution $\widetilde{\Gamma}$, obtained by merging some components of $\widetilde{\Sigma}$ and possibly adding some new ones.

Definition 4.3. *Let* $\widetilde{\Sigma} = \langle \Sigma_1, \ldots, \Sigma_k \rangle$ *and* $\widetilde{\Gamma} = \langle \Gamma_1, \ldots, \Gamma_\ell \rangle$ *be distributions of* Σ. *Then* $\widetilde{\Sigma} \lesssim \widetilde{\Gamma}$ *if for each* $i \in [1..k]$, *there exists* $j \in [1..\ell]$ *such that* $\Sigma_i \subseteq \Gamma_j$. *If* $\widetilde{\Sigma} \lesssim \widetilde{\Gamma}$ *we say that* $\widetilde{\Sigma}$ *is finer than* $\widetilde{\Gamma}$, *or* $\widetilde{\Gamma}$ *is coarser than* $\widetilde{\Sigma}$.

We then have the following simple observation.

Proposition 4.4. *If* $\widetilde{\Sigma} \lesssim \widetilde{\Gamma}$ *then* $I_{\widetilde{\Gamma}} \subseteq I_{\widetilde{\Sigma}}$.

Note that $\lesssim$ is not a preorder in general. In fact $\widetilde{\Sigma} \lesssim \widetilde{\Gamma}$ means that the maximal elements of $\widetilde{\Sigma}$ are included in those of $\widetilde{\Gamma}$. Let us denote by $\simeq$ the relation $\lesssim \cap \gtrsim$. Then, $\widetilde{\Sigma} \simeq \widetilde{\Gamma}$ just means that $\widetilde{\Sigma}$ and $\widetilde{\Gamma}$ have the same maximal elements—in general, it does not guarantee that they are identical. However, when restricted to distributions "without redundancies", $\lesssim$ becomes a preorder.

Definition 4.5. *A distribution* $\widetilde{\Sigma} = \langle \Sigma_1, \ldots, \Sigma_k \rangle$ *of* Σ *is said to be* simple *if for each* $i, j \in [1..k]$, $i \neq j$ *implies that* $\Sigma_i \not\subseteq \Sigma_j$.

Proposition 4.6. *Let* $\widetilde{\Sigma}$ *and* $\widetilde{\Gamma}$ *be simple distributions of* Σ. *If* $\widetilde{\Sigma} \simeq \widetilde{\Gamma}$ *then* $\widetilde{\Sigma} = \widetilde{\Gamma}$.

For any independence relation I over Σ, the associated standard distribution $\widetilde{\Sigma}_I$ is a simple distribution, and is the coarsest distribution inducing I. At the other end of the spectrum, we can define the *finest* distribution inducing I as follows:

Definition 4.7. *Let* I *be an independence relation over* Σ, *and* $D = (\Sigma \times \Sigma) - I$. *The distribution* $\widetilde{\Delta}_I$ *over* Σ *is defined by:*

$$\widetilde{\Delta}_I = \{\{x, y\} \mid (x, y) \in D, x \neq y\} \cup \{\{x\} \mid x \, I \, y \text{ for each } y \neq x\}$$

Proposition 4.8. *Let* I *be an independence relation over* Σ. *Then the distribution* $\widetilde{\Delta}_I$ *is the finest simple distribution over* Σ *that induces* I.

A finer distribution can be faithfully implemented by a coarser distribution.

Lemma 4.9. *Let* $\widetilde{\Sigma} = \langle \Sigma_1, \ldots, \Sigma_k \rangle$ *and* $\widetilde{\Gamma} = \langle \Gamma_1, \ldots, \Gamma_\ell \rangle$ *be distributions of* Σ *such that* $\widetilde{\Sigma} \lesssim \widetilde{\Gamma}$. *Then, for each product transition system* $\|_{i \in [1..k]} TS_i$ *over* $\widetilde{\Sigma}$, *there exists an isomorphic product transition system* $\|_{i \in [1..\ell]} \widehat{TS}_i$ *over* $\widetilde{\Gamma}$.

Proof. For each $i \in [1..k]$ let $f(i)$ denote the least index j in $[1..\ell]$ such that $\Sigma_i \subseteq \Gamma_j$. For each $j \in [1..\ell]$, define $\widehat{TS}_j = (\widehat{Q}_j, \leadsto_j, \hat{q}_{in}^j)$ as follows.

- If j is not in the range of f, then $\widehat{Q}_j = \{\hat{q}_{in}^j\}$, and $\hat{q}_{in}^j \overset{a}{\leadsto}_j \hat{q}_{in}^j$ for each $a \in \Sigma_j$.
- If j is in the range of f, let $f^{-1}(j) = \{i_1, i_2, \ldots, i_m\}$. Set $\widehat{TS}_j = (TS_{i_1} \| \cdots \| TS_{i_m})$.

It is then straightforward to verify that $\|_{i \in [1..k]} TS_i$ is isomorphic to $\|_{i \in [1..\ell]} \widehat{TS}_i$. We omit the details.

Corollary 4.10. *Let $TS = (Q, \rightarrow, q_{\text{in}})$ be a transition system over Σ, and $\widetilde{\Sigma}$ and $\widetilde{\Gamma}$ be two distributions of Σ such that $\widetilde{\Sigma} \lesssim \widetilde{\Gamma}$. If TS is $\widetilde{\Sigma}$-implementable it is also $\widetilde{\Gamma}$-implementable.*

Let I be an independence relation over Σ. We have already observed that $\widetilde{\Sigma}_I$ is the coarsest distribution of Σ whose induced independence relation is I. Coupling this remark with the preceding corollary, we can now settle the synthesis problem with respect to independence relations.

Corollary 4.11. *Let $TS = (Q, \rightarrow, q_{\text{in}})$ be a transition system over Σ, and $\widetilde{\Sigma}$ be a distribution of Σ inducing the independence relation I. Then if TS is $\widetilde{\Sigma}$-implementable it is also $\widetilde{\Sigma}_I$-implementable. Moreover TS is I-implementable if and only if it is $\widetilde{\Sigma}_I$-implementable.*

We remark that the converse of Lemma 4.9 is not true—if $\widetilde{\Sigma} \lesssim \widetilde{\Gamma}$, it may be the case that TS is $\widetilde{\Gamma}$-implementable but not $\widetilde{\Sigma}$-implementable. Details can be found in the full paper [CMT99].

5   Canonical Implementations and Determinacy

A system may have more than one $\widetilde{\Sigma}$-implementation for a fixed distribution $\widetilde{\Sigma}$. For instance, the system

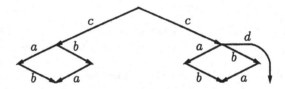

has two implementations with respect to the distributed alphabet $\widetilde{\Sigma} = \langle\{a, c, d\}, \{b, c, d\}\rangle$, namely $ca + c(a + d) \parallel c(b + d)$ and $c(a + d) \parallel cb + c(b + d)$.

One question that naturally arises is whether there exists a unique minimal or maximal family of equivalence relations $\{\equiv_1, \ldots, \equiv_k\}$ on states that makes Theorem 3.1 go through. We say that a family $\{\equiv_1, \ldots, \equiv_k\}$ is *minimal* (respectively, *maximal*) if there is no other family $\{\equiv'_1, \ldots, \equiv'_k\}$ with respect to which TS is $\widetilde{\Sigma}$-implementable with $\equiv'_i \subseteq \equiv_i$ (respectively, $\equiv_i \subseteq \equiv'_i$) for each $i \in [1..k]$.

It turns out that for deterministic systems, we can find unique minimal implementations. A transition system $TS = (Q, \rightarrow, q_{\text{in}})$ is *deterministic* if $q \xrightarrow{a} q'$ and $q \xrightarrow{a} q''$ imply that $q' = q''$.

Theorem 5.1. *Suppose $TS = (Q, \rightarrow, q_{\text{in}})$ is deterministic and $\widetilde{\Sigma}$-implementable. Then there exists a unique minimal family $\{\equiv_1, \ldots, \equiv_k\}$ of equivalence relations on states with respect to which TS is $\widetilde{\Sigma}$-implementable.*

Proof. Suppose that $\{\equiv_1, \ldots, \equiv_k\}$ and $\{\equiv'_1, \ldots, \equiv'_k\}$ are two families which represent $\widetilde{\Sigma}$-implementations of TS. Let $\{\hat{\equiv}_1, \ldots, \hat{\equiv}_k\}$ be the intersection family, given by $q \hat{\equiv}_i q' \Leftrightarrow q \equiv_i q' \wedge q \equiv'_i q'$.

By definition, $\hat{\equiv}_i \subseteq \equiv_i$ and $\hat{\equiv}_i \subseteq \equiv'_i$. Thus, it suffices to show that $\{\hat{\equiv}_1, \ldots, \hat{\equiv}_k\}$ represents a $\widetilde{\Sigma}$-implementation of TS. From the definition of the relations $\{\hat{\equiv}_1, \ldots, \hat{\equiv}_k\}$, it is obvious that both conditions *(i)* and *(ii)* of Theorem 3.1 are satisfied. Now suppose that $q \in Q$ and $a \in \Sigma$. For every $i \in loc(a)$, let $s_i, s'_i \in Q$ be such that $s_i \hat{\equiv}_i q$ and $s_i \overset{a}{\longrightarrow} s'_i$. This means that for every $i \in loc(a)$, both $s_i \equiv_i q$ and $s_i \equiv'_i q$. Hence by condition *(iii)* there exists q' such that $q \overset{a}{\longrightarrow} q'$ and $q' \equiv_i s'_i$ for every $i \in loc(a)$. Similarly, there exists q'' such that $q \overset{a}{\longrightarrow} q''$ and $q'' \equiv_i s'_i$ for every $i \in loc(a)$. Since TS is deterministic, it must be $q' = q''$. Thus, $q' = q''$ is such that $q' \hat{\equiv}_i s'_i$ for each $i \in loc(a)$.

This result leads us to conjecture that the synthesis problem for deterministic systems is much less expensive computationally than the synthesis problem in the general case, since it suffices to look for the unique minimal family of equivalence relations which describe the implementation.

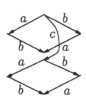

We conclude this section by observing that a deterministic system may have more than one maximal family $\{\equiv_1, \ldots, \equiv_k\}$ for which it is $\widetilde{\Sigma}$-implementable. For instance the system on the left has two distinct maximal implementations with respect to the distribution $\langle\{a, c\}, \{b, c\}\rangle$, namely (fix $X.\ a + cX$) $\parallel b + cb$ and $a + ca \parallel$ (fix $Y.\ b + cY$), whose components are not even language equivalent.

6   Synthesis Modulo Bisimulation

In the course of specifying a system, we may accidentally destroy its inherent product structure. This may happen, for example, if we optimize the design and eliminate redundant states. In such situations, we would like to be able to reconstruct a product transition system from the reduced specification. Since the synthesized system will not, in general, be isomorphic to the specification, we need a criterion for ensuring that the two systems are behaviourally equivalent. We use strong bisimulation [Mil89] for this purpose.

In general, synthesizing a behaviourally equivalent product implementation from a reduced specification appears to be a hard problem. In this section, we show how to solve the problem for reduced specifications which can be implemented as *deterministic* product transition systems—that is, the global transition system generated by the implementation is deterministic. Notice that the specification itself may be nondeterministic. Since many distributed systems implemented in hardware, such as digital controllers, are actually deterministic, our characterization yields a synthesis result for a large class of useful systems.

We begin by recalling the definition of bisimulation.

Definition 6.1. *A* bisimulation *between a pair of transition systems $TS_1 = (Q_1, \to_1, q_{in}^1)$ and $TS_2 = (Q_2, \to_2, q_{in}^2)$ is a relation $R \subseteq (Q_1 \times Q_2)$ such that:*

- $(q_{in}^1, q_{in}^2) \in R$.
- *If $(q_1, q_2) \in R$ and $q_1 \xrightarrow{a}_1 q_1'$, there exists q_2', $q_2 \xrightarrow{a}_2 q_2'$ and $(q_1', q_2') \in R$.*
- *If $(q_1, q_2) \in R$ and $q_2 \xrightarrow{a}_2 q_2'$, there exists q_1', $q_1 \xrightarrow{a}_1 q_1'$ and $(q_1', q_2') \in R$.*

The synthesis problem modulo bisimilarity

The synthesis problem modulo bisimilarity can now be formulated as follows. If $TS = (Q, \to, q_{in})$ is a transition system over Σ, and $\widetilde{\Sigma} = \langle \Sigma_1, \dots, \Sigma_k \rangle$ is a distribution of Σ, does there exist a product system $\|_{i \in [1..k]} TS_i$ over $\widetilde{\Sigma}$ such that $\|_{i \in [1..k]} TS_i$ is bisimilar to TS?

To settle this question for deterministic implementations, we need to consider product languages.

Languages Let $TS = (Q, \to, q_{in})$ be a transition system over Σ. The *language* of TS is the set $L(TS) \subseteq \Sigma^*$ consisting of the labels along all runs of TS. In other words, $L(TS) = \{w \mid q_{in} \xrightarrow{w} q, q \in Q\}$.

Notice that $L(TS)$ is always prefix-closed and always contains the empty word. Moreover, $L(TS)$ is regular whenever TS is finite. For the rest of this section, we assume all transition systems which we encounter are finite.

Product languages Let $L \subseteq \Sigma^*$ and let $\widetilde{\Sigma} = \langle \Sigma_1, \dots, \Sigma_k \rangle$ be a distribution of Σ. For $w \in \Sigma^*$, let $w{\restriction}_{\Sigma_i}$ denote the projection of w onto Σ_i, obtained by erasing all letters in w which do not belong to Σ_i.

The language L is a *product language* over $\widetilde{\Sigma}$ if for each $i \in [1..k]$ there is a language $L_i \subseteq \Sigma_i^*$ such that $L = \{w \mid w{\restriction}_{\Sigma_i} \in L_i, i \in [1..k]\}$.

For deterministic transition systems, bisimilarity coincides with language equivalence. We next show that we can extend this result to get a simple characterization of transition systems which are bisimilar to deterministic product transition systems. We first recall a basic definition.

Bisimulation quotient Let $TS = (Q, \to, q_{in})$ be a transition system and let $\sim_{TS}$ be the *largest* bisimulation relation between TS and itself. The relation $\sim_{TS}$ defines an equivalence relation over Q. For $q \in Q$, let $[q]$ denote the $\sim_{TS}$-equivalence class containing q. The *bisimulation quotient* of TS is the transition system $TS/\sim_{TS} = (\widehat{Q}, \rightsquigarrow, [q_{in}])$ where

- $\widehat{Q} = \{[q] \mid q \in Q\}$.
- $[q] \xrightarrow{a}_{\rightsquigarrow} [q']$ if there exist $q_1 \in [q]$ and $q_1' \in [q']$ such that $q_1 \xrightarrow{a} q_1'$.

The main result of this section is the following.

Theorem 6.2. *Let TS be a transition system over Σ and let $\widetilde{\Sigma}$ be a distribution of Σ. The system TS is bisimilar to a deterministic product transition system over $\widetilde{\Sigma}$ iff TS satisfies the following two conditions.*

- *The bisimulation quotient $TS/_{\sim_{TS}}$ is deterministic.*
- *The language $L(TS)$ is a product language over $\widetilde{\Sigma}$.*

To prove this theorem, we first recall the following basic connection between product languages and product systems [Thi95].

Lemma 6.3. $L(\|_{i\in[1..i]} TS_i) = \{w \mid w\!\upharpoonright_{\Sigma_i} \in L(TS_i), i \in [1..k]\}.$

We also need the useful fact that a product language is always the product of its projections [Thi95].

Lemma 6.4. *Let $L \subseteq \Sigma^*$ and let $\widetilde{\Sigma} = \langle \Sigma_1, \ldots, \Sigma_k \rangle$ be a distribution of Σ. For $i \in [1..k]$, let $L_i = \{w\!\upharpoonright_{\Sigma_i} \mid w \in L\}$. Then, L is a product language iff $L = \{w \mid w\!\upharpoonright_{\Sigma_i} \in L_i, i \in [1..k]\}.$*

Notice that Lemma 6.4 yields an effective procedure for checking if a finite-state transition system accepts a product language over a distribution $\langle \Sigma_1, \ldots, \Sigma_k \rangle$. For $i \in [1..k]$, construct the finite-state system TS_i such that $L(TS_i) = L\!\upharpoonright_{\Sigma_i}$ and then verify that $L(TS) = L(\|_{i\in[1..k]} TS_i)$.

Next, we state without proof some elementary facts about bisimulations.

Lemma 6.5.   *(i) Let TS be a deterministic transition system. Then, $TS/_{\sim_{TS}}$ is also deterministic.*

(ii) Let TS_1 and TS_2 be deterministic transition systems over Σ. If $L(TS_1) = L(TS_2)$ then TS_1 is bisimilar to TS_2.

(iii) Let TS_1 and TS_2 be bisimilar transition systems. Then $TS_1/_{\sim_{TS_1}}$ and $TS_2/_{\sim_{TS_2}}$ are isomorphic. Further $L(TS_1) = L(TS_2)$.

We now prove both parts of Theorem 6.2.

Lemma 6.6. *Suppose that a transition system TS is bisimilar to a deterministic product transition system. Then, $TS/_{\sim_{TS}}$ is deterministic and $L(TS)$ is a product language.*

Proof. Let $\widehat{TS} = \|_{i\in[1..i]} \widehat{TS_i}$ be a deterministic product transition system such that TS is bisimilar to $\widehat{TS}$.

By Lemma 6.5 (iii), $L(TS) = L(\widehat{TS})$. Since $L(\widehat{TS})$ is a product language, it follows that $L(TS)$ is a product language.

To check that $TS/_{\sim_{TS}}$ is deterministic, we first observe that $\widehat{TS}/_{\sim_{\widehat{TS}}}$ is deterministic, by Lemma 6.5 (i). By part (iii) of the same lemma, $TS/_{\sim_{TS}}$ must be isomorphic to $\widehat{TS}/_{\sim_{\widehat{TS}}}$. Hence, $TS/_{\sim_{TS}}$ is also deterministic.

Lemma 6.7. *Let TS be a transition system over Σ and $\widetilde{\Sigma}$ be a distribution of Σ, such that $TS/_{\sim_{TS}}$ is deterministic and $L(TS)$ is a product language over $\widetilde{\Sigma}$. Then, TS is bisimilar to a deterministic product transition system over $\widetilde{\Sigma}$.*

Proof. Let $\widetilde{\Sigma} = \langle \Sigma_1, \ldots, \Sigma_k \rangle$. For $i \in [1..k]$, let $L_i = \{w \restriction_{\Sigma_i} \mid w \in L(TS)\}$. We know that each L_i is a regular prefix-closed language which contains the empty word. Thus, we can construct the minimal deterministic finite-state automaton $A_i = (Q_i, \rightarrow_i, q_{in}^i, F_i)$ recognizing L_i. Since L_i contains the empty word, $q_{in}^i \in F_i$. Consider the restricted transition relation $\rightarrow_i' = \rightarrow_i \cap (F_i \times \Sigma \times F_i)$. It is easy to verify that the transition system $TS_i = (F_i, \rightarrow_i', q_{in}^i)$ is a deterministic transition system such that $L(TS_i) = L_i$.

Consider the product $\widehat{TS} = \|_{i \in [1..k]} TS_i$. Lemma 6.3 tell us that $L(\widehat{TS}) = \{w \mid w \restriction_{\Sigma_i} \in L_i, i \in [1..k]\}$. From Lemma 6.4, it follows that $L(\widehat{TS}) = L(TS)$.

We claim that TS is bisimilar to $\widehat{TS}$. Consider the quotient $TS/_{\sim_{TS}}$. Both $TS/_{\sim_{TS}}$ and $\widehat{TS}$ are deterministic and $L(TS/_{\sim_{TS}}) = L(TS) = L(\widehat{TS})$. Thus, by Lemma 6.5 (ii), it must be the case that $TS/_{\sim_{TS}}$ is bisimilar to $\widehat{TS}$. By transitivity, TS is also bisimilar to $\widehat{TS}$.

Using standard automata theory, we can derive the following result from Theorem 6.2.

Corollary 6.8. *Given a finite-state transition system $TS = (Q, \rightarrow, q_{in})$ and a distributed alphabet $\widetilde{\Sigma}$, we can effectively decide whether TS is bisimilar to a deterministic product system over $\widetilde{\Sigma}$.*

The synthesis problem modulo bisimilarity appears to be quite a bit more difficult when the product implementation is permitted to be nondeterministic. We have a characterization of the class of systems which are bisimilar to nondeterministic product systems [CMT99]. Our characterization is phrased in terms of the structure of the execution tree obtained by unfolding the specification. Unfortunately, the execution tree may be infinite, so this characterization is not effective, even if the initial specification is finite-state. The difficulty lies in bounding the number of transitions in the product implementation which can collapse, via the bisimulation, to a single transition in the specification.

References

Arn94.   A. Arnold: *Finite transition systems and semantics of communicating sytems*, Prentice-Hall (1994).

BD98.   E. Badouel and Ph. Darondeau: Theory of Regions. *Lectures on Petri nets I (Basic Models), LNCS* **1491** (1998) 529–588.

CMT99.   I. Castellani, M. Mukund and P.S. Thiagarajan: Characterizing decomposable transition systems, *Internal Report, Chennai Mathematical Institute* (1999).

ER90.   A. Ehrenfeucht and G. Rozenberg: Partial 2-structures; Part II, State spaces of concurrent systems, *Acta Inf.* **27** (1990) 348–368.

GM92.   J.F. Groote and F. Moller: Verification of Parallel Systems via Decomposition, *Proc. CONCUR'92, LNCS* **630** (1992) 62–76.

Hol97.   G.J. Holzmann: The model checker SPIN, *IEEE Trans. on Software Engineering*, **23**, 5 (1997) 279–295.

Kur94.   R.P. Kurshan: *Computer-Aided Verification of Coordinating Processes: The Automata-Theoretic Approach*, Princeton University Press (1994).

Maz89.   A. Mazurkiewicz: Basic notions of trace theory, in: J.W. de Bakker, W.-P. de Roever, G. Rozenberg (eds.), *Linear time, branching time and partial order in logics and models for concurrency*, LNCS **354** (1989) 285–363.

Mil89.   R. Milner: *Communication and Concurrency*, Prentice-Hall, London (1989).

Mol89.   F. Moller: *Axioms for Concurrency*, Ph.D. Thesis, University of Edinburgh (1989).

MM93.   F. Moller and R. Milner: Unique decomposition of processes, *Theor. Comput. Sci.* **107** (1993) 357–363.

Mor98.   R. Morin: Decompositions of asynchronous systems, *Proc. CONCUR'98*, LNCS **1466** (1998) 549–564.

Muk92.   M. Mukund: Petri Nets and Step Transition Systems, *Int. J. Found. Comput. Sci.* **3**, 4 (1992) 443–478.

NRT92.   M. Nielsen, G. Rozenberg and P.S. Thiagarajan: Elementary transition systems, *Theor. Comput. Sci.* **96** (1992) 3–33.

Thi95.   P.S. Thiagarajan: A trace consistent subset of PTL, *Proc. CONCUR'95*, LNCS **962** (1995) 438-452.

WN95.   G. Winskel and M. Nielsen: Models for concurrency, in S. Abramsky, D. Gabbay and T.S.E. Maibaum, eds, *Handbook of Logic in Computer Science, Vol 4*, Oxford (1995) 1–148.

Beyond Region Graphs:
Symbolic Forward Analysis of Timed Automata

Supratik Mukhopadhyay and Andreas Podelski

Max-Planck-Institut für Informatik
Im Stadtwald, 66123 Saarbrücken, Germany
{supratik,podelski}@mpi-sb.mpg.de

Abstract. Theoretical investigations of infinite-state systems have so far concentrated on decidability results; in the case of timed automata these results are based on region graphs. We investigate the specific procedure that is used practically in order to decide verification problems, namely symbolic forward analysis. This procedure is possibly non-terminating. We present basic concepts and properties that are useful for reasoning about sufficient termination conditions, and then derive some conditions. The central notions here are constraint transformers associated with sequences of automaton edges and *zone trees* labeled with successor constraints.

1 Introduction

A *timed automaton* [1] models a system whose transitions between finitely many control locations depend on the values of clocks. The clocks advance continuously over time; they can individually be reset to the value 0. Since the clocks take values over reals, the state space of a timed automaton is infinite.

The theoretical and the practical investigations on timed automata are recent but already quite extensive (see e.g. [1,7,11,2,4]). Many decidability results are obtained by designing algorithms on the *region graph*, which is a finite quotient of the infinite state transition graph [1]. Practical experiments showing the feasibility of model checking for timed automata, however, employ *symbolic forward analysis*. We do not know of any practical tool that constructs the region graph. Instead, symbolic model checking is extended directly from the finite to the infinite case; logical formulas over reals are used to 'symbolically' represent infinite sets of tuples of clock values and are manipulated by applying the same logical operations that are applied to Boolean formulas in the finite state case.

If model checking is based on *backward* analysis (where one iteratively computes sets of predecessor states), termination is guaranteed [9]. In comparison, symbolic forward analysis for timed automata has the theoretical disadvantage of possible non-termination. Practically, however, it has the advantage that it is amenable to on-the-fly local model checking and to partial-order reduction techniques (see [8] for a discussion of forward vs. backward analysis).

In symbolic forward analysis applied to the timed automata arising in practical applications (see e.g. [11]), the theoretical possibility of non-terminating does

C. Pandu Rangan, V. Raman, R. Ramanujam (Eds.): FSTTCS'99, LNCS 1738, pp. 232–244, 1999.
© Springer-Verlag Berlin Heidelberg 1999

not seem to play a role. Existing versions that exclude this possibility (through built-in runtime checks [4] or through a static preprocessing step [7]) are not used in practice.

This situation leads us to raising the question whether there exist 'interesting' sufficient conditions for the termination of symbolic model checking procedures for timed automata based on forward analysis. Here, 'interesting' means applicable to a large class of cases in practical applications. The existence of a practically relevant class of infinite-state systems for which the practically employed procedure is actually an algorithm would be a theoretically satisfying explanation of the success of the ongoing practice of using this procedure, and it may guide us in designing practically successful verification procedures for other classes of infinite-state systems.

As a first step towards answering the question that we are raising, we build a kind of 'toolbox' consisting of basic concepts and properties that are useful for reasoning about sufficient termination conditions. The central notions here are constraint transformers associated with sequences of automaton edges and *zone trees* labeled with successor constraints. The constraint transformer associated with the sequences of edges $e_1, \ldots, e_n$ of the timed automaton assigns a constraint φ another constraint that 'symbolically' represents the set of the successor states along the edges $e_1, \ldots, e_n$ of the states in the set represented by φ. We prove properties for constraint transformers associated with edge sequences of a certain form; these properties are useful in termination proofs as we then show. The zone tree is a vehicle that can be used to investigate sufficient conditions for termination without having to go into the algorithmic details of symbolic forward analysis procedures. It captures the fact that the constraints enumerated in a symbolic forward analysis must respect a certain tree order.

We show how the zone tree can characterize termination of (various versions of) symbolic forward analysis. A combinatorial reasoning is then used to derive sufficient termination conditions for symbolic forward analysis. We prove that symbolic forward analysis terminates for three classes of timed automata. These classes are not relevant practically; the goal is merely to demonstrate how the presented concepts and properties of the successor constraint function and of the zone tree can be employed to prove termination. Termination proofs can be quite tedious, as the third case shows; the proof here distinguishes many cases.

2   The Constraint Transformer $\varphi \mapsto [\![w]\!](\varphi)$

A *timed automaton* $\mathcal{U}$ can, for the purpose of reachability analysis, be defined as a set $\mathcal{E}$ of guarded commmands e (called edges) of the form below. Here L is a variable ranging over the finite set of *locations*, and $\boldsymbol{x} = \langle x_1, \ldots, x_n \rangle$ are the variables standing for the clocks and ranging over nonnegative real numbers. As usual, the primed version of a variable stands for its value after the transition. The 'time delay' variable z ranges over nonnegative real numbers.

$$e \equiv L = \ell \wedge \gamma_e(\boldsymbol{x}) \parallel L' = \ell' \wedge \alpha_e(\boldsymbol{x}, \boldsymbol{x}', z).$$

The guard formula $\gamma_e(x)$ over the variables x is built up from conjuncts of the form $x_i \sim k$ where x_i is a clock variable, $\sim$ is a comparison operator (i.e., $\sim \in \{=, <, \le, >, \ge\}$) and k is a natural number.

The action formula $\alpha_e(x, x', z)$ of e is defined by a subset Reset_e of $\{1, \ldots, n\}$ (denoting the clocks that are *reset*); it is of the form

$$\alpha_e(x, x', z) \equiv \bigwedge_{i \in \text{Reset}_e} x_i' = z \wedge \bigwedge_{i \notin \text{Reset}_e} x_i' = x_i + z.$$

We write ψ_e for the logical formula corresponding to e (with the free variables x and x'; we replace the guard symbol $\|$ with conjunction).

$$\psi_e(x, x') \equiv L = \ell \wedge \gamma_e(x) \wedge L' = \ell' \wedge \exists z \, \alpha_e(x, x', z)$$

The states of $\mathcal{U}$ (called *positions*) are tuples of the form $\langle \ell, v \rangle$ consisting of values for the location and for each clock. The position $\langle \ell, v \rangle$ can make a *time transition* to any position $\langle \ell, v + \delta \rangle$ where $\delta \ge 0$ is a real number.

The position $\langle \ell, v \rangle$ can make an *edge transition* (followed by a time transition) to the position $\langle \ell', v' \rangle$ using the edge e if the values ℓ for L, v for x, ℓ' for L' and v' for x' define a solution for ψ_e. (An edge transition by itself is defined if we replace the variable z in the formula for α by the constant 0.)

We use *constraints* φ in order to represent certain sets of positions (called *zones*). A constraint is a conjunction of the equality $L = \ell$ with a conjunction of formulas of the form $x_i - x_j \sim c$ or $x_i \sim c$ where c is an integer (i.e. with a *zone constraint* as used in [4]). We identify solutions of constraints with positions $\langle \ell, v \rangle$ of the timed automaton.

We single out the *initial constraint* φ^0 that denotes the time successors of the initial position $\langle \ell^0, \mathbf{0} \rangle$.

$$\varphi^0 \equiv L = \ell^0, x_1 \ge 0, x_2 = x_1, \ldots, x_n = x_1$$

A constraint φ is called *time-closed* if its set of solutions is closed under time transitions. Formally, $\varphi(x)$ is equivalent to $(\exists x \exists z (\varphi \wedge x_1' = x_1 + z \wedge \ldots \wedge x_n' = x_n + z))[x/x']$. For example, the initial constraint is time-closed. In the following, we will be interested only in time-closed constraints.

In the definition below, $\varphi'[x'/x]$ denotes the constraint obtained from φ' by α-renaming (replace each x_i' by x_i).

We write $e_1. \ldots .e_m$ for the word w obtained by concatenating the 'letters' $e_1, \ldots, e_m$; thus, w is a word over the set of edges $\mathcal{E}$, i.e. $w \in \mathcal{E}^*$.

Definition 1 (Constraint Transformer $[\![w]\!]$). *The constraint transformer wrt. to an edge e is the 'successor constraint function' $[\![w]\!]$ that assigns a constraint φ the constraint*

$$[\![e]\!](\varphi) \equiv (\exists x (\varphi \wedge \psi_e))[x'/x].$$

The successor constraint function $[\![w]\!]$ wrt. a string $w = e_1. \ldots .e_m$ of length $m \ge 0$ is the functional composition of the functions wrt. the edges $e_1, \ldots, e_m$, i.e. $[\![w]\!] = [\![e_1]\!] \circ \ldots \circ [\![e_m]\!]$.

Thus, $[\![\varepsilon]\!](\varphi) = \varphi$ and $[\![w.e]\!](\varphi) = [\![e]\!]([\![w]\!](\varphi))$. The solutions of $[\![w]\!](\varphi)$ are exactly the ("edge plus time") successors of a solution of φ by taking the sequence of transitions via the edges $e_1, \ldots, e_m$ (in that order).

We will next consider constraint transformers $[\![w]\!]$ for strings w of a certain form. In the next definition, the terminology 'a clock x_i is queried in the edge e' means that x_i is a variable occurring in the guard formula γ of e; 'x_i is reset in e' means that $i \in \text{Reset}_e$.

Definition 2 (Stratified Strings). *A string $w = e_1 \ldots e_m$ of edges is called stratified if*

- *each clock $x_1, \ldots, x_n$ is reset at least once in w, and*
- *if x_i is reset in e_j then x_i is not queried in $e_1, \ldots, e_j$.*

Proposition 1. *The successor constraint function wrt. a stratified string w is a constant function over satisfiable constraints (i.e. there exists a unique constraint φ_w such that $[\![w]\!](\varphi) = \varphi_w$ for all satisfiable constraints φ).*

Proof. We express the successor constraint of the constraint φ wrt. the stratified string $w = e_1 \ldots e_m$ equivalently by

$$[\![w]\!](\varphi) \equiv (\exists \boldsymbol{x} \exists \boldsymbol{x}^1 \ldots \exists \boldsymbol{x}^{m-1} \exists z^1 \ldots \exists z^m (\varphi \wedge \psi_1 \wedge \ldots \wedge \psi_m))[\boldsymbol{x}/\boldsymbol{x}^m]$$

where ψ_k is the formula that we obtain by applying α-renaming to the (quantifier-free) conjunction of the guard formula $\gamma_{e_k}(\boldsymbol{x})$ and the action formula $\alpha_{e_k}(\boldsymbol{x}, \boldsymbol{x}', z)$ for the edge e_k; i.e.

$$\psi_k \equiv \gamma_{e_k}(\boldsymbol{x}^{k-1}) \wedge \alpha_{e_k}(\boldsymbol{x}^{k-1}, \boldsymbol{x}^k, z^k).$$

Thus, in the formula for e_k, we rename the clock variable x_i to x_i^{k-1}, its primed version x_i' to x_i^k, and the 'time delay' variable z to z^k.

We identify the variables x_i (applying in φ) with their "0-th renaming" x_i^0 (appearing in ψ_1); accordingly we can write $\boldsymbol{x}^0$ for the tuple of variables $\boldsymbol{x}$.

We will transform $\exists \boldsymbol{x}^1 \ldots \exists \boldsymbol{x}^{m-1}(\psi_1 \wedge \ldots \wedge \psi_m)$ equivalently to a constraint ψ containing only conjuncts of the form $x_i^m = z^l + \ldots + z^m$ and of the form $z^l + \ldots + z^m \sim c$ where $l > 0$; i.e. ψ does not contain any of the variables x_i of φ. Thus, we can move the quantifiers $\exists \boldsymbol{x}$ inside; formally, $\exists \boldsymbol{x}(\varphi \wedge \psi)$ is equivalent to $(\exists \boldsymbol{x} \varphi) \wedge \psi$. Since φ is satisfiable, the conjunct $\exists \boldsymbol{x} \varphi$ is equivalent to *true*. Summarizing, $[\![w]\!](\varphi)$ is equivalent to a formula that does not depend on φ, which is the statement to be shown.

The variable x_i^k (the "k-th renaming of the i-th clock variable") occurs in the action formula of ψ_k, either in the form $x_i^k = z^k$ or in the form $x_i^k = x_i^{k-1} + z^k$, and it occurs in the guard formula of ψ_{k+1}, in the form $x_i^k \sim c$.

If the i-th clock is not reset in the edges $e_1, \ldots, e_{k-1}$, then we replace the conjunct $x_i^k = x_i^{k-1} + z^k$ by $x_i^k = x_i + z^1 + \ldots z^k$.

Otherwise, let l be the largest index of an edge e_l with a reset of the i-th clock. Then we replace $x_i^k = x_i^{k-1} + z^k$ by $x_i^k = z^l + \ldots + z^k$.

If $k = m$, the first case cannot arise due to the first condition on stratified strings (the i-th clock must be reset at least once in the edges $e_1, \ldots, e_m$). That is, we replace $x_i^m = x_i^{m-1} + z^k$ always by a conjunct of the form $x_i^k = z^l + \ldots + z^k$.

If the conjunct $x_i^k \sim c$ appears in ψ_{k+1}, then, by assumption on w (the second condition for stratified strings), the i-th clock is reset in an edge e_l where $l \leq k$. Therefore, we can replace the conjunct $x_i^k \sim c$ by $z_l + \ldots + z_k \sim c$.

Now, each variable x_i^k (for $0 < k < m$) has exactly one occurrence, namely in a conjunct C of the form $x_i^k = x_i + z^1 + \ldots z^k$ or $x_i^k = z^l + \ldots z^k$. Hence, the quantifier $\exists x_i^k$ can be moved inside, before the conjunct C; the formula $\exists x_i^k \, C$ can be replaced by *true*.

After the above replacements, all conjuncts are of the form $x_i^m = z^l + \ldots + z^m$ or of the form $z^l + \ldots + z^m \sim c$; as explained above, this is sufficient to show the statement.   □

We say that an edge e is *reset-free* if $\text{Reset}_e = \emptyset$, i.e., its action is of the form $\alpha_e \equiv \bigwedge_{i=1,\ldots,n} x_i' = x_i$. A string w of edges is reset-free if all its edges are.

Proposition 2. *If the string w is reset-free, and the successor constraint of a time-closed constraint of the form $L = \ell \wedge \varphi$ is of the form $L = \ell' \wedge \varphi'$, then φ' entails φ, formally $\varphi' \models \varphi$.*

Proof. It is sufficient to show the statement for w consisting of only one reset-free edge e. Since φ is time-closed, it is equivalent to $(\exists x \exists z (\varphi \wedge x' = x + z))[x/x']$.

Then $[\![w]\!](L = \ell \wedge \varphi)$ is equivalent to $(\exists \ldots (L = \ell' \wedge \varphi \wedge x' = x + z' \wedge \gamma_e(x') \wedge x'' = x' + z')[x/x'']$. This constraint is equivalent to $L = \ell' \wedge \varphi(x) \wedge \gamma(x)$. This shows the statement.   □

3   Zone Trees and Symbolic Forward Analysis

Definition 3 (Zone Tree). *The zone tree of a timed automaton $\mathcal{U}$ is an infinite tree with domain $\mathcal{E}^*$ (i.e., the nodes are the strings over $\mathcal{E}$) that labels the node w by the constraint $[\![w]\!](\varphi^0)$.*

That is, the root ε is labeled by the initial constraint φ^0. For each node w labeled φ, and for each edge $e \in \mathcal{E}$ of the timed automaton, the successor node $w.e$ is labeled by the constraint $[\![e]\!](\varphi)$. Clearly, the (infinite) disjunction of all constraints labeling a node of the zone tree represents all reachable positions of $\mathcal{U}$.

We are interested in the termination of various versions of symbolic forward analysis of a timed automaton $\mathcal{U}$. All versions have in common that they traverse (a finite prefix of) its zone tree, in a particular order. The following definition of a non-deterministic procedure abstracts away from that specific order.

Definition 4 (Symbolic Forward Analysis). *A symbolic forward analysis of a timed automaton $\mathcal{U}$ is a procedure that enumerates constraints φ_i labeling the nodes w_i of the zone tree of $\mathcal{U}$ in a tree order such that the enumerated constraints together represent all reachable positions. Formally,*

- $\varphi_i = [\![w_i]\!](\varphi^0)$ *for* $0 \le i < B$ *where the bound B is a natural number or ω,*
- *if w_i is a prefix of w_j then $i \le j$,*
- *the disjunction $\bigvee_{0 \le i < B} \varphi_i$ is equivalent to the disjunction $\bigvee_{0 \le i < \omega} \varphi_i$.*

We assume that the constraint φ_i is computed by applying any of the known quantifier elimination algorithms (see e.g. [12]) to a conjunction of constraints.

The number i is a *leaf* of a symbolic forward analysis if the node w_i is a leaf of the tree formed by all the nodes w_i where $0 \le i \le B$.

We say that a symbolic forward analysis *terminates* if the bound B is finite (i.e. not ω). We define that symbolic forward analysis terminates *with local subsumption* if for all its leafs i there exists $j < i$ such that the constraint φ_i entails the constraint φ_j. In contrast, it terminates with *global* subsumption if for all its leafs i there the constraint φ_i entails the disjunction of all constraints φ_j where $j < i$. Model checking is more efficient with local subsumption than with global subsumption, both practically and theoretically [5].

A depth-first symbolic forward analysis depends on a chosen order of edges. Symbolic forward analysis terminates if and only if the depth-first symbolic forward analysis of $\mathcal{U}$ terminates for *every* order chosen.

If the symbolic depth-first forward analysis of $\mathcal{U}$ terminates for at least one order of edges, then also the breadth-first version terminates. The converse need not be true, as the counterexample of Figure 1 shows.

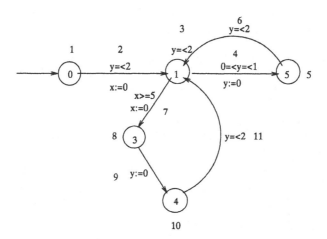

Fig. 1. Example of a timed automaton for which the breadth-first version of symbolic forward analysis terminates but the depth-first version does not, if the edge numbered 4 is followed before the edge numbered 7.

A *path* p in a zone tree is an infinite string over $\mathcal{E}$, i.e., $p \in \mathcal{E}^\omega$; p contains a node w if the string w is a prefix of p, written $w < p$. A node v precedes a node w if v is a prefix of w, written $v < w$.

Definition 5 (Local finiteness). *A path p of a zone tree is locally finite if and only if it contains a node w labeled by a constraint that entails the constraint labeling some node v preceding w (formally, there exist v and w such that $v < w < p$ and $[\![w]\!](\varphi^0) \models [\![v]\!](\varphi^0)$). A zone tree is locally finite if every path is.*

Proposition 3. *Every symbolic forward analysis of a timed automaton $\mathcal{U}$ terminates with local subsumption if and only if the zone tree of $\mathcal{U}$ is locally finite.*

We will next investigate the special class of *strings* (that we call *cycles*) that correspond to cycles in the control graph of the given timed automaton. Each cycle in the graph-theoretic sense corresponds to finitely many cycles in the sense defined here (as strings), depending on the entry location.

We say that an edge e of the form $L = \ell \ldots \parallel L' = \ell' \ldots$ leads from the location ℓ to the location ℓ'. This terminology refects the fact that there exists a directed edge from ℓ to ℓ' labeled by the corresponding guarded command in the control graph of the given timed automaton (we will not formally introduce the control graph). Semantically, all transitions using such an edge go from a position with the location ℓ to a position with the location ℓ'. We canonically extend the terminology 'leads to' from edges e to strings w of edges.

Definition 6 (Cycle). *The string $w = e_1.\ldots.e_m$ of length $m \geq 1$ is a cycle if the sequence of edges $e_1, \ldots, e_m$ lead from a location ℓ to the same location ℓ such that there exists a sequence of edges that leads from the initial location ℓ^0 to ℓ whose last edge is different from e_m.*

The last condition above expresses that ℓ is an entry point to the corresponding cycle in the control graph of the given timed automaton $\mathcal{U}$. The next notion is used in effective sufficient termination conditions.

Definition 7 (Simple Cycle). *A cycle $w = e_1.\ldots.e_m$ is called simple if it does not contain a proper subcycle; formally, no string $e_i.\ldots.e_j$ where $1 \leq i < j \leq m$ is also a cycle.*

Proposition 4. *A locally infinite path $p \in \mathcal{E}^\omega$ in the zone tree of the timed automaton $\mathcal{U}$ contains infinitely many occurrences of a simple cycle w; formally, p is an element of the omega-language $(\mathcal{E}^*.w)^\omega$.*

Proof. Let p be a locally infinite path. Then there exists a location ℓ such that infinitely many nodes on this path are labeled by ℓ (i.e. a constraint of the form $L = \ell \wedge \ldots$. The strings formed by the edges connecting two nodes labeled by ℓ must all contain a simple cycle. Since the number of simple cycles is finite, some simple cycles must be repeated infinitely often.   □

A string is *stratifiable* if contains a stratified substring (a substring of a string $e_1.\ldots.e_m$ is any string of the form $e_i.\ldots.e_j$ where $1 \leq i \leq j \leq m$).

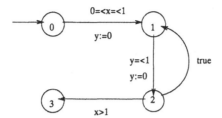

Fig. 2. Example of a timed automaton showing that the property: "Every reachable location is reachable through a simple path" does *not* entail termination of depth-first symbolic forward analysis.

Proposition 5. *If every simple cycle of the timed automaton $\mathcal{U}$ is either reset-free or stratifiable, the zone tree of $\mathcal{U}$ is locally finite.*

Proof. Follows from Propositions 1, 2 and 4.   □

We apply the above results to obtain our first sufficient termination condition.

Theorem 1. *Symbolic depth-first forward analysis of a timed automaton $\mathcal{U}$ terminates if all simple cycles of $\mathcal{U}$ are either reset-free or stratifiable.*

Proof. Follows from Propositions 3 and 5.   □

4 RQ Automata

A timed automaton $\mathcal{U}$ is called RQ [10] if for each clock x, $\mathcal{U}$ contains exactly one edge with a reset of x and exactly one edge with a query of x, and moreover, for every transition sequence of $\widetilde{\mathcal{U}}$ starting from the initial position, the sequence of resets and queries of x is alternating, with a reset before the first query; here, $\widetilde{\mathcal{U}}$ refers to the timed automaton from $\mathcal{U}$ obtained by replacing all conjuncts $x \sim c$ in the guard formulas by the conjunct $x \geq 0$. We may require wlog. that no edge e of a timed automaton $\mathcal{U}$ contains both a reset of a clock and a query of a clock.

RQ automata have the following interesting property: if a location is reachable then it is reachable through a *simple path*, i.e. a sequence of edges that form a string not containing a cycle [10]. So it is possible to derive specialized terminating graph algorithms for reachability for RQ automata. Moreover, a cycle is traversable infinitely often if it is traversable once [10]. We will now investigate how a generic model checker based on symbolic forward analysis behaves on RQ automata. We do not know whether we obtain termination for this special case. We know that the distinguished property of RQ automata (that reachability is equivalent to reachability through a simple path) by itself is not sufficient for termination; Figure 2 gives a counterexample.

We will consider two special classes of RQ automata. The first one is characterized by the cut condition.

A timed automaton $\mathcal{U}$ satisfies the *cut condition* if any two simple cycles w and w' are either identical or their sets of edges are disjoint. Graph-theoretically, every simple cycle in the control graph has exactly one entry point (which is then called the 'cut vertex').

Theorem 2. *Symbolic depth-first forward analysis of an RQ timed automaton $\mathcal{U}$ terminates if it satisfies the cut condition and in every simple cycle, either all or no clock is reset.*

Proof. A simple cycle containing a reset for each clock in an RQ automaton satisfying the cut condition is stratified. Hence, Theorem 1 yields the statement. □

The second class of RQ automata is obtained by restricting the number of clocks to two.

Theorem 3. *Symbolic depth-first forward analysis of an RQ timed automaton with two clocks terminates.*

Proof. We name the two clock variables of the automaton x and y. We note R_x the unique edge of the time automaton where x is reset, and Q_y the one where x is queried; similarly we define R_y and Q_y. By our non-proper restriction, $R_x \neq Q_x$ etc..

A *segment* S of a path p in a zone tree is a sequence of nodes $n_1, \ldots, n_m$ of the zone tree. The string $w = e_1 \ldots e_{m-1}$ *labels* the segment S if n_m is reached from n_1 by following the edges $e_1, \ldots, e_m$ in the zone tree.

For a proof by contradiction, assume that p is an infinite branch of the zone tree. By Proposition 4, there exists a simple cycle w (leading, say, from the location ℓ to ℓ) that repeats infinitely often on p. We write $S_1, S_2, \ldots$ for the segments that are labeled by w (in consecutive order). We write L_i for the segment between S_i and S_{i+1}. We note v^i the string labeling the segment L_i; each string v^i is a cycle (leading also from the location ℓ to ℓ). Below we will use the terminology 'w labels S_i' and 'v^i labels L_i'.

We first distinguish between the cases whether the edge R_x is part of the string w ("$R_x \in w$") or not.
Case 1 $R_x \in w$.
The edge Q_x must then also be an element of w (if the cycle w can be executed once then even infinitely often [10]; if it contained R_x but not Q_x then the RQ condition would be violated).
Case 1.1 $R_y \in w$.
Again, we must have that $Q_y \in w$.
We distinguish between the cases that the edge R_y appears strictly before the edge Q_y in the strings w ("$R_y < Q_y$") or after ("$Q_y < R_y$").
Case 1.1.1 $R_y < Q_y$.
Repeating the above reasoning for x instead of y, we distinguish between the cases "$R_y < Q_y$" and "$Q_y < R_y$".

Case 1.1.1.1 $R_x < Q_x$.

The two assumptions $R_x < Q_x$ and $R_y < Q_y$ mean that the string w is *stratified*. Hence, by Proposition 1, the successor constraint function wrt. w is constant. Hence, the constraint labeling the last node of S_2 entails the constraint labeling the last node of S_1. Thus, the path p is locally finite, which achieves the contradiction.

Case 1.1.1.2 $Q_x < R_x$.

We distinguish the cases whether the edge Q_x appears before the edge R_y or strictly after.

Case 1.1.1.2.1 $Q_x < R_y$.

Combining the assumptions leading to this case, namely $R_x \in w$ (and hence also $Q_x \in w$) and $R_y \in w$ (and hence also $Q_y \in w$) and $R_y < Q_y$ and $Q_x < R_x$ and $Q_x < R_y$, we know that the string w is of the form $w = w_1.Q_x.w_2$ such that w_2 contains R_x and R_y. Hence, the substring w_2 of w *stratified*. By Proposition 1, the successor constraint function wrt. w_2 is constant, and hence also the one wrt. w. As in the case above, we achieve a contradiction.

Case 1.1.1.2.2 $R_y < Q_x$.

Again we combine the assumptions leading to this case: namely $R_x, Q_x, R_y, Q_y \in w$ and $R_y < Q_y$ and $Q_x < R_x$ and $R_y < Q_x$.

Only using that $R_y < R_x$, we know that the string w is of the form $w = w_1.R_y.w_2.R_x.w_3$.

One of the two cases, namely $R_x \notin L_i$ or $R_x \in L_i$, will hold for infinitely many segments L_i's.

Case 1.1.1.2.2.1 $R_x \notin L_i$.

Then also $Q_x \notin L_i$ (because of the RQ-condition and since L_i is a cycle). We then distinguish between the analogue cases for y instead of x.

Case 1.1.1.2.2.1.1 $R_y \notin L_i$.

Again, then $Q_y \notin L_i$.

We are assuming that $R_x, Q_x, R_y, Q_y \notin L_i$ for infinitely many L_i. We take two such segments, calling them L and L'. Let v and v' be the string labeling (the edge linking the nodes in) L and L'. Then, the successor constraint functions wrt. v and v' are the identity.

We form the *stratified* strings $V = R_x.w_3.v.w_1.R_y$ and $V' = R_x.w_3.v'.w_1.R_y$. Since the successor constraint functions wrt. v and v' are the identity, the successor constraint functions wrt. V and V' are the same *constant* function. The same reasoning as above leads to a contradiction.

Case 1.1.1.2.2.1.2 $R_y \in L_i$.

Then also $Q_y \in L_i$. Because of the RQ-condition and since the edge R_y precedes Q_y in S_i, the first occurrence of R_y precedes the first occurrence of Q_y in L_i. Hence, the strings v and v' (defined as above, labeling of some L_i's) is of the form $v = v_1.R_y.v_2$ or $v = v_1'.R_y.v_2'$ where v_1, v_2, v_1' and v_2' do not contain any reset or any query of a clock variable (and hence, yield the identity as the successor constraint function). We form the *stratified* substrings $V = R_x.w_3.v_1.R_y$ and $V' = R_x.w_3.v_1'.R_y$, which yield the same constant successor constraint function for the same reason as above. Again, this leads to a contradiction.

Case 1.1.1.2.2.2 $R_x \in L_i$.

Again, then $Q_x \in L_i$. Now we are assuming that $R_x, Q_x, R_y, Q_y \in L_i$ for infinitely many L_i.

As in Case 1.1.1.2.2.1.2, the first occurrence of R_y must precede the first occurrence of Q_y in L_i.

Assume that there is a reset of x in L_i before the first reset of y. We form the string $R_x.w_2.v_1, R_x$ where $w = w_1.R_x.w_2$ is such that w_2 does not contain any reset (by the assumptions for the cases 1.1.1.2 and 1.1.1.2.2) and $v^i = v_1.R_x.v_2$ (the string labeling L_i) is such that v_1 does not contain any reset. Following the lines of the proof for Proposition 2 one can show that for any constraint φ, $[\![R_x.w_2.v_1.R_x]\!](\varphi)$ entails $[\![R_x]\!](\varphi)$. This is a contradiction (to the fact that the path p is locally infinite).

Assume that there is no reset of x in L_i before the first reset of y. Then the string formed by the edges leading from the reset of x in S_i to the first reset of y in L_i is stratified. We can then apply the same reasoning as in Case 1.1.1.2.1 to derive a contradiction.

Case 1.1.2 $Q_y < R_y$.

Thus now $R_x \in w$ (and hence $Q_x \in w$), $R_y \in w$ (and hence $Q_y \in w$) and $Q_y < R_y$. Now we consider the following subcases of this case.

Case 1.1.2.1 $R_x < Q_x$

This case is symmetric to Case 1.1.1.2.1 where $R_x, R_y \in w$, $Q_x < R_y$ and $R_y < Q_y$.

Case 1.1.2.2 $Q_x < R_x$.

The assumption of the case is that the reset occurs after the query for both clocks. Due to the RQ condition, there cannot be any query between the two resets. Therefore, $R_x.w_1.R_y$ (or, symmetrically, $R_y.w_1.R_x$) forms a *stratified* substring of w. As before, we obtain a contradiction.

We refer to the full version of this paper [13] for the remaining cases. □

5 Future Work

The presented work targets theoretical investigations of timed automata not at the verification problem itself but, instead, at the termination behavior of the procedure solving it in practice, namely symbolic forward analysis. This work is a potential starting point for deriving interesting sufficient termination conditions. There are, however, other open questions along these lines.

Our setup may also be used to derive *necessary* termination conditions. These are useful obviously in the cases when their test is negative. Another question is whether there exist decidable necessary and sufficient conditions.

We may also consider logical equivalence instead of local subsumption for a practically more efficient, but theoretically weaker fixpoint test (used in tools such as Uppaal [11]). We observe that Proposition 1 is still directly applicable in the new context, but Proposition 2 is not. The comparison of the different fixpoint tests (equivalence, local and global subsumption) is an interesting subject of research.

We may be able to derive natural and less restrictive sufficient termination conditions when we consider the enhancement of symbolic forward analysis with techniques from [3] to compute the effect of loops, i.e. essentially the constraint transformer $[\![w^\omega]\!]$ for simple cycles w.

The constraint transformers $[\![w]\!]$ form a 'symbolic version' of the *syntactic monoid* [6] for timed automata. This notion may be of intrinsic interest and deserve further study.

Acknowledgement

We thank Tom Henzinger and Jean-Francois Raskin for discussions.

References

1. R. Alur and D. Dill. A theory of timed automata. *Theoretical Computer Science*, 126(2):183–236, 1994.
2. F. Balarin. Approximate reachability analysis of timed automata. In *Proceedings of 17th IEEE Real-Time Systems Symposium*, pages 52–61. IEEE Computer Society Press, 1996.
3. B. Boigelot. *Symbolic Methods for Exploring Infinite State Spaces*. PhD thesis, Université de Liège, 1998.
4. C. Daws and S. Tripakis. Model checking of real-time reachability properties using abstractions. In B. Steffen, editor, *Proceedings of the 4th International Conference on Tools and Algorithms for the Construction of Systems*, LNCS 1384, pages 313–329. Springer-Verlag, 1998.
5. G. Delzanno and A. Podelski. Model checking in CLP. In Rance Cleaveland, editor, *Proceedings of TACAS'99, the Second International Conference on Tools and Algorithms for the Construction and Analysis of Systems*, volume 1579 of *Springer LNCS*. Springer-Verlag, 1999.
6. S. Eilenberg. *Automata, Languages and Machines*, volume B. Academic Press, 1976.
7. T. A. Henzinger, P. W. Kopke, A. Puri, and P. Varaiya. What's decidable about hybrid automata? In *Proceedings of the 27th Annual Symposium on Theory of Computing*, pages 373–382. ACM Press, 1995.
8. T. A. Henzinger, O. Kupferman, and S. Qadeer. From pre-historic to post-modern symbolic model checking. In *Proceedings of the International Conference on Computer-Aided Verification*, pages 195–206. Springer, 1998.
9. T.A. Henzinger, X. Nicollin, J. Sifakis, and S. Yovine. Symbolic model checking for real-time systems. *Information and Computation*, 111(2):193–244, 1994. Special issue for LICS 92.
10. W. K. C. Lam and R. K. Brayton. Alternating RQ timed automata. In C. Courcoubetis, editor, *Proceedings of the 5th International Conference on Computer-Aided Verification*, LNCS 697, pages 236–252. Springer-Verlag, 1993.
11. K.G. Larsen, P. Pettersson, and W. Yi. Compositional and symbolic model checking of real-time systems. In *Proceedings of the 16th Annual Real-time Systems Symposium*, pages 76–87. IEEE Computer Society Press, 1995.
12. K. Marriott and P. J. Stuckey. *Programming with Constraints: An Introduction*. MIT Press, 1998.

13. S. Mukhopadhyay and A. Podelski. Beyond region graphs: Symbolic forward analysis of timed automata, 1999. Full Version. Available at http://www.mpi-sb.mpg.de/~podelski.

Implicit Temporal Query Languages: Towards Completeness

Nicole Bidoit[1] and Sandra de Amo[2]*

[1] LaBRI–UMR 5800 du CNRS, Université Bordeaux 1, France
`Nicole.Bidoit@labri.u-bordeaux.fr`
[2] Departement of Computer Science, Federal University of Uberlândia, Brazil
`deamo@ufu.br`

Abstract. In the propositional case, it is known that *temporal logic* (TL) and first order logic with timestamp (TS-FO) have the same expressive power. Recent work has proved that, in contrast, there are first order logic queries on timestamp databases that are *not* expressible in first order temporal logic: *TL is not complete*. Specifying a complete implicit temporal query language remains an open problem. We investigate two extensions of TL, namely NTL and RNTL. A strict hierarchy in expressive power among fragments $RNTL^i$ of RNTL is established. On the one hand, it leads to the conjecture that RNTL is complete. On the other hand, it provides a new promising perspective towards proving the well-known conjecture that there is a strict hierarchy among the i time-variable fragments $TS\text{-}FO^i$ of TS-FO.

Keywords: Temporal database, Query languages, Temporal logic, Expressive power, Communication protocol.

1 Introduction

There are two alternative ways [5,6] of extending the relational model in order to represent temporal data. The first approach captures time in an implicit manner: a relationnal temporal database instance is then a finite sequence of relational instances. The second approach relies on augmenting each relation with a "timestamp" column storing the time instants of validity of each tuple. Figure 1 illustrates these two equivalent representations.

In the context of an implicit representation of time, query languages, called implicit temporal query languages, are usually based on first order temporal logic [8]. When time is explicitly represented, queries are specified using the standard relational query languages [3] with built-in linear order on the timestamps. One of these languages, called TS-FO, is the relational calculus (i.e. first order logic) with timestamps. A non trivial question arises then: how implicit

* Work by this author was funded by the University Bordeaux 1 and done during her visit at Labri in 1998–1999.

C. Pandu Rangan, V. Raman, R. Ramanujam (Eds.): FSTTCS'99, LNCS 1738, pp. 245–257, 1999.

$$\mathcal{I}^{est} = R^{est} \begin{array}{|c c|} \hline A & T \\ \hline a & 1 \\ b & 1 \\ c & 1 \\ b & 2 \\ d & 2 \\ a & 3 \\ c & 3 \\ \hline \end{array}$$

$$\mathcal{I} = \quad I_1 \qquad I_2 \qquad I_3$$

R	A		R	A		R	A
	a			b			a
	b			d			c
	c						

Fig. 1. The implicit and explicit representations of time

temporal languages and explicit temporal languages relate to each other with respect to expressive power. It has been studied from various angles in [2] [5] [7] [9] [13] [10].

[2] and [1] provide a hierarchy of temporal languages[1] with respect to expressivity (see also [13]) : (1) it is shown that *future* first order temporal logic (FTL) is strictly weaker than first temporal order logic (TL) and (2) that TL is strictly weaker than TS-FO. The first result (FTL $\subset$ TL) follows from the fact that the query Does there exist an instant (in the future) whose state equals the initial state? cannot be expressed in FTL but is expressible in TL. The second result (TL $\subset$ TS-FO) is derived by showing that the query Does there exist two distinct instants (in the future) whose respective states are equal? cannot be expressed in TL but is expressible in TS-FO. These two results are of major interest and stand in contrast with the propositional case. In [9], the notion of complete temporal language is introduced via equivalence with TS-FO and the authors show that propositional TL is complete (see also [11]). Moreover, it is shown that propositional FTL is equivalent to propositional TL.

In the present paper, we study the following open problem : find an implicit first order temporal language which is complete i.e. equivalent to TS-FO. We enrich the hierarchy described above by investigating two languages NTL and RNTL. Both languages are shown to be more powerful than TL. The language NTL is not complete. The language RNTL is more expressive than NTL. Completeness of RNTL remains a conjecture.

The language NTL is the first order linear version of NCTL*[12]. It extends TL by a temporal modality $\aleph$ ("From Now On"). Intuitively, the modality $\aleph$ allows one to choose a new initial time instant (called relative origin) and forget about all previous instants. In this sense, it can be said that NTL introduces a notion of relative past: the past modalities (Previous and Since) are evaluated with respect to the relative origin. This stands in contrast with TL whose modalities are of course always evaluated with respect to the absolute origin. Relative past increases the expressive power of TL in the first order case even if in the propositional case relative past is redundant with the other temporal operators [12]. However, the language NTL is not complete: we show that the query Does there exist 3 distinct instants whose respective

---

[1] Other languages investigated by these authors are stronger with respect to expressivity than TS-FO and TL.

states (let say S_1, S_2 and S_3) satisfy $S_1 \cap S_2 = \emptyset$ and $S_1 \cup S_2 = S_3$? is expressible in TS-FO but not in NTL[2]. This result is proved by extending the proof technique based on communication protocol developed in [2].

Because NTL fails to be complete, we extend it by investigating a rather simple idea: allowing one to forget the past is coupled together with allowing one to restore the past. The implicit language RNTL is defined by introducing a temporal modality $\Re$ whose task is to restore the segment of the past which has been "removed" by the last "application" of $\aleph$. Once again, the ability to restore the past does not add any expressive power in the propositional case [4]. However, in the first order case, RNTL is strictly more expressive than NTL. A strict hierarchy in expressive power among fragments RNTL[i] of RNTL is established. The fragment RNTL[i] is defined by restriction on the maximal number of $\Re$ operators in formulas. On the one hand, this leads to the conjecture that RNTL is complete. On the other hand, this provides a new promising perspective towards proving the well-known conjecture that there is a strict hierarchy in expressive power among the i time-variable fragments TS-FO[i] of TS-FO. The fragment TS-FO[i] is the subclass of TS-FO formulas built by restricting the number of distinct time-variables to be at most i.

The paper is organized as follows. The next section is devoted to introductory material. Section 3 discusses the expressive power of NTL and introduces the communication protocols which are the key tools for the main results of the paper. The last section introduces the language RNTL and investigates the RNTL[i] hierarchy as well as its relationship with the conjectured TS-FO[i] hierarchy. We also discuss [13] results.

Because of space limitation, some technical proofs are omitted and other proofs are simply sketched.

2   Preliminaries

We assume the reader is familiar with relational databases concepts [3]. A *database schema* is a finite set of relation names with associated arity. An *instance* of a schema assigns to each relation name a *finite* relation of appropriate arity over a fixed countably infinite domain of data elements. The *active domain* of an instance is the set of all data elements appearing in some of its relations. Valuations of variables are always assumed in the active domain. A *temporal instance* over a database schema $\mathcal{R}$ is a non-empty finite sequence $\mathcal{I}=I_1,\ldots,I_l$ ($l \geq 1$) of instances of $\mathcal{R}$.

The query language TL. Temporal logic [8] is an obvious candidate language for querying temporal databases. The syntax of TL over some database schema $\mathcal{R}$ is obtained using the formation rules for standard first order logic over $\mathcal{R}$ together with the additional formation rule: if φ_1 et φ_2 are formulas then φ_1 Until φ_2, φ_1 Since φ_2, Next φ_1 and Prev φ_1 are formulas.

---

[2] One says that the query separates NTL and TS-FO.

The semantics of TL is briefly recalled. Given a temporal instance $\mathcal{I}=I_1,\ldots,I_l$ over $\mathcal{R}$ and a TL formula φ, the *truth* of φ at time $i \in \{1,\ldots,l\}$ given the valuation ν of the free variables of φ, denoted $[\mathcal{I},i,\nu] \models_{tl} \varphi$, is defined as follows:

$[\mathcal{I},i,\nu] \models_{tl} R(t_1,,\ldots,t_k)$ if $(\nu(t_1),\nu(t_2),\ldots,\nu(t_l)) \in I_i(R)$.

If φ is a boolean combination of formulas or a quantification ($\exists$, $\forall$) of a formula then the definition is as usual.

$[\mathcal{I},i,\nu] \models_{tl} \varphi_1$ Until φ_2 iff there exists $j > i$ such that $[\mathcal{I},j,\nu] \models_{tl} \varphi_2$ and for each k such that $i \leq k < j$, $[\mathcal{I},k,\nu] \models_{tl} \varphi_1$.

$[\mathcal{I},i,\nu] \models_{tl} \varphi_1$ Since φ_2 iff there exists $j < i$ such that $[\mathcal{I},j,\nu] \models_{tl} \varphi_2$ and for each k such that $j < k \leq i$, $[\mathcal{I},k,\nu] \models_{tl} \varphi_1$.

$[\mathcal{I},i,\nu] \models_{tl}$ Next φ_1 iff $i < l$ and $[\mathcal{I},i+1,\nu] \models_{tl} \varphi_1$.

$[\mathcal{I},i,\nu] \models_{tl}$ Prev φ_1 iff $i > 1$ and $[\mathcal{I},i-1,\nu] \models_{tl} \varphi_1$.

It is sometimes convenient to use the following derived temporal modalities: $F \varphi_1 \equiv$ true Until φ_1 ("*sometimes in the future φ_1*"), $P \varphi_1 \equiv$ true since φ_1 ("*sometimes in the past φ_1*"), first $\equiv \neg$ Prev True ("*initial state*"), last $\equiv \neg$ Next True ("*final state*").

A query Q in TL is specified by $\{(\overrightarrow{x}) \mid \varphi(\overrightarrow{x})\}$ where $\varphi(\overrightarrow{x})$ is a formula in TL and $\overrightarrow{x}$ its free variables. The answer of Q over a temporal instance $\mathcal{I}$ is the relation $Q(\mathcal{I})=\{\nu(\overrightarrow{x}) \mid [\mathcal{I},1,\nu] \models_{tl} \varphi(\overrightarrow{x}), \nu$ a valuation of $\overrightarrow{x}\}$.

The timestamp representation and the language TS-FO. A timestamp temporal instance $\mathcal{I}^{est}$ over a schema $\mathcal{R}$ is a two sorted relational structure over the timestamp schema $\mathcal{R}^{est}$. The schema $\mathcal{R}^{est}$ contains for each k-ary relation R of $\mathcal{R}$ an extended relation R^{est} of arity $k+1$. The extra column of this relation holds timestamps, the other ones hold data elements. Given an implicit temporal instance $\mathcal{I}=I_1,\ldots,I_l$ its timestamp representation is denoted $\mathcal{I}^{est}$.

TS-FO is a query language over timestamp temporal databases defined in a straightforward way by two-sorted first order formulas. The data variables ranges over data elements and time variables over integers.

A query Q in TS-FO is specified by $\{\overrightarrow{x} \mid \varphi(\overrightarrow{x})\}$ where $\varphi(\overrightarrow{x})$ is a formula in TS-FO and $\overrightarrow{x}$ its free variables, each free variable is of data sort. The answer of Q evaluated on a timestamp temporal instance $\mathcal{I}^{est}$ is the relation $Q(\mathcal{I}^{est})=\{\nu(\overrightarrow{x}) \mid \mathcal{I}^{est} \models \varphi(\overrightarrow{x}), \nu$ a valuation of $\overrightarrow{x}\}$.

Example 1. Let $\mathcal{R}$ and $\mathcal{I}$ be the schema and temporal instance of figure 1. The TL query $\{(x) \mid R(x) \wedge \exists y (R(y)$ Until $(R(x) \wedge$ last)) $\}$ evaluated on $\mathcal{I}$ returns $\{a, c\}$. This query can be equivalently expressed by the TS-FO expression $\{(x) \mid R^{est}(x,1) \wedge \exists y \exists t (t > 1 \wedge R^{est}(x,t) \wedge \neg$ true$(t+1) \wedge \forall s ((1 \leq s \wedge s < t) \rightarrow R^{est}(y,s)))\}$.

3   The Expressive Power of NTL

In this section, we investigate the implicit temporal query language NTL. This language has been initially introduced in [12] as a propositional branched temporal language. Here we consider the linear first order version of this language.

NTL extends TL by a modality ℵ. Intuitively, this modality stands for *from now on* or *henceforth*. It is meant to restrict the scope of past-time operators by forgetting the past instants with respect to an instant which becomes the (relative) origin of time. In the propositional case, [12] shows that this new modality is redundant with pure-future modalities. Here we will show that, in the first order case, NTL is strictly more expressive than TL. However, NTL is not complete since it is not equivalent to TS-FO.

The syntax of NTL is defined by adding the following formation rule to the ones defining TL: if φ is a formula, so is ℵ φ.

Notation: Let $\mathcal{I} = (I_1,\ldots,I_l)$ be a temporal instance. The temporal instance $\mathcal{J} = J_1,\ldots, J_{l-i+1}$ such that for $j = 1,\ldots,l-i+1$, $J_j=I_{j+i-1}$ is denoted by $\mathcal{I}_{|i}$. Intuitively, $\mathcal{I}_{|i}$ is the instance $\mathcal{I}$ where the first $i-1$ states have been removed.

Definition 1. Given a temporal instance $\mathcal{I}=(I_1,\ldots,I_l)$ and a NTL formula φ, the *truth* of φ *at time* p, given the *relative origin* t and the valuation ν of the free variables, denoted $[\mathcal{I},t,p,\nu] \models_{ntl} \varphi$ is defined as follows:

 - if φ is obtained by TL formation rules then $[\mathcal{I},t,p,\nu] \models_{ntl} \varphi$ is defined in a way similar to $[\mathcal{I}_{|t},p-t+1,\nu] \models_{tl} \varphi$.
 - If φ is of the form ℵ φ_1 then $[\mathcal{I},t,p,\nu] \models_{ntl} \varphi$ if $[\mathcal{I},p,p,\nu] \models_{ntl} \varphi_1$.   □

The effect of the temporal operator ℵ is to position the relative origin at p, leaving the value p of the "at" time unchanged.

Example 2. The formula Fℵ$(R(d) \wedge \text{first})$ is satisfied by the temporal instance of Figure 1. Intuitively, in $\mathcal{I}$, there exists an instant (here 2) such that when the previous states are removed (here state 1 only), it is the first instant and its associated state contains $R(d)$.

The answer of the NTL query $\{(\overrightarrow{x}) \mid \phi(\overrightarrow{x})\}$ evaluated on the temporal instance $\mathcal{I}$ is the relation $\{\nu(\overrightarrow{x}) \mid [\mathcal{I},1,1,\nu] \models_{ntl} \phi(\overrightarrow{x}), \nu \text{ a valuation of } \overrightarrow{x} \}$.

NTL versus TL The first result establishes that the modality ℵ increases the expressive power of TL. This is entailed by the query **Does there exist two distinct instants whose respective states are equal?** which is not expressible in TL (see the introduction) but can be expressed in NTL by the boolean formula $\{() \mid F$ℵ$F(\forall x(R(x) \leftrightarrow P(\text{first} \wedge R(x))))\}$. So we have:

Theorem 1. TL $\subset$ NTL

NTL versus TS-FO The increase in expressive power provided by the temporal operator ℵ is not sufficient to express all TS-FO queries. Of course, each NTL query can be expressed as a TS-FO query. This is implied by the following first result:

Lemma 1. Let $\varphi(\overrightarrow{x})$ be a NTL formula. There exists a TS-FO formula $\psi(i,j,\overrightarrow{x})$ such that for each implicit temporal instance $\mathcal{I}$, $[\mathcal{I},t,p,\nu] \models_{ntl} \varphi(\overrightarrow{x})$ iff $[\mathcal{I}^{\text{est}},\mu] \models \psi$ where μ is the valuation ν extended by $[i/t, j/p]$.

The proof does not present any difficuly and it is not presented here. We now show that the inclusion NTL⊆TS-FO is in fact a strict inclusion:

Theorem 2. NTL ⊂ TS-FO

The proof technique is a generalization of that introduced in [2] and is based on the following three steps:
1. We extend the communication protocols proposed in [2]. These protocols allow us to introduce the notion of *polynomial communication complexity* of k-ary predicates on sets of sets of data elements.
2. We exhibit a ternary predicate whose communication complexity is not polynomial.
3. Finally, it is proved that if a ternary predicate can be expressed by a NTL query, then it has a polynomial communication complexity.

Communication protocols revisited Let $\mathcal{D}$ be a finite set of elements. Sets of non-empty subsets of $\mathcal{D}$, called super-sets are denoted by $\mathcal{X}$, $\mathcal{Y}$, $\mathcal{Z}$ ... The communication protocols involve two partners I and II who exchange messages i.e. finite relations of fixed arity over $\mathcal{D}$. The two partners are supposed to collaborate in the process of computing some predicate on super-sets. Given n super-sets, they are distributed over the two partners in such a way that they share $n-2$ super-sets. This implies that each partner hides one super-set to the other. The super-sets shared are called the pivot and the number of super-sets shared the order of the pivot.

Definition 2. A *communication protocol* Pr of *arity k* with *pivot of order v* is specified by two abstract functions f, g. A message of arity k is a finite subset of $\mathcal{D}^k$ or a boolean (true or false).
Given $v+2$ super-sets enumerated by $\mathcal{Y}$, $\mathcal{Z}$, $\mathcal{X}_1$, ..., $\mathcal{X}_v$, the execution of the protocol Pr over $\mathcal{Y}$, $\mathcal{Z}$ with pivot $\mathcal{X}_1$, ..., $\mathcal{X}_v$, is a finite sequence $((a_1, b_1) \ldots (a_r, b_r))$ of pairs of messages of arity k exchanged by the partners I and II as follows:

$a_i = f(\mathcal{D}, \mathcal{X}_1, ..., \mathcal{X}_v, \mathcal{Y}, [b_1, b_2, \ldots, b_{i-1}])$ is the ith message of I to II.
$b_i = g(\mathcal{D}, \mathcal{X}_1, ..., \mathcal{X}_v, \mathcal{Z}, [a_1, a_2, \ldots, a_i])$ is the ith answer of II to I.   □

Next, we omit to specify the arity of a communication protocol and denote by Pr_v a communication protocol whose pivot is of order v. For each $v+2$-tuples of super-sets $\mathcal{Y}$, $\mathcal{Z}$, $\mathcal{X}_1$, ..., $\mathcal{X}_v$, $\mathrm{Pr}_v(\mathcal{X}_1, \ldots, \mathcal{X}_v \| \mathcal{Y}, \mathcal{Z})$ denotes the *last* message a_r of partner I. We say that the protocol is boolean when this last message is a boolean. The notion of polynomial communication complexity is introduced as follows:

Definition 3. The communication protocol Pr_v is *polynomial* if there exists a function κ of polynomial order such that for each $v+2$-tuple of super-sets $\mathcal{Y}$, $\mathcal{Z}$, $\mathcal{X}_1$, ..., $\mathcal{X}_v$ we have: $r = \kappa(\sum_{i=1}^{v} |\mathcal{X}_i|)$ where $a_r = \mathrm{Pr}_v(\mathcal{X}_1, \ldots, \mathcal{X}_v \| \mathcal{Y}, \mathcal{Z})$.   □

The messages exchanged by the two partners during the execution of a communication protocol aims at computing a predicate of arity $v+2$ when the order of the pivot is v.

Definition 4. A boolean communication protocol Pr_v *computes* the predicate $\mathcal{P}$ of arity $v+2$ if there exists a permutation π on $\{1, \ldots, v+2\}$ such that for each tuple of super-sets $\mathcal{X}_1, \ldots, \mathcal{X}_{v+2}$ we have:

$$\mathcal{P}(\mathcal{X}_1, \ldots, \mathcal{X}_{v+2}) = \textit{true} \text{ iff } \text{Pr}_v(\mathcal{X}_{\pi(1)}, \ldots, \mathcal{X}_{\pi(v)} \| \mathcal{X}_{\pi(v+1)}, \mathcal{X}_{\pi(v+2)}) = \textit{true} \quad \square$$

Next, we say that the predicate $\mathcal{P}$ of arity $v+2$ has a *polynomial communication complexity* (for short, we say that $\mathcal{P}$ is v-polynomial) if there exists a polynomial communication protocol Pr_v computing $\mathcal{P}$. Note that the communication protocols defined in [2] correspond exactly to our protocols with pivot of order 0 and that the predicates considered there are binary.

For the purpose of proving that NTL$\subset$TS-FO, we will focus on communication protocol with pivot of order 1 and on ternary predicates. Communication protocols with pivot of higher order will be brought into play in the last section when investigating a hierarchy of subclasses of RNTL.

Example 3. The predicate specified by **there exists** $X \in \mathcal{X}$, $Y \in \mathcal{Y}$ **and** $Z \in \mathcal{Z}$ **such that** $X = Y = Z$ is 1-polynomial. Assuming that the partner I (resp. II) knows the super-sets $\mathcal{X}$ and $\mathcal{Y}$ (resp. $\mathcal{Y}$ and $\mathcal{Z}$), the protocol computing this predicate proceeds as follows: partner I identifies the sets in $\mathcal{X} \cap \mathcal{Y}$ and sends them one by one to its partner who checks if one of these sets is also in $\mathcal{Y} \cap \mathcal{Z}$. Thus the number of exchanges is bounded by $|\mathcal{Y}|$.

Lemma 2. The predicate N-SEP$(\mathcal{X},\mathcal{Y},\mathcal{Z})$ specified by **there exists** $X \in \mathcal{X}$, $Y \in \mathcal{Y}$ **and** $Z \in \mathcal{Z}$ **such that** $X \cap Y = \emptyset$ **and** $X \cup Y = Z$ is not 1-polynomial.

Proof (Sketch): Let us assume that there exists a polynomial communication protocol computing N-SEP with $\mathcal{X}$ as the pivot. Let us fix $\mathcal{X}$ as $\{\{p_0\}\}$. Then the number of exchanges needed to derive N-SEP$(\mathcal{X},\mathcal{Y},\mathcal{Z})$ is constant for any choice of $\mathcal{Y}$ and $\mathcal{Z}$. On the other hand, for any choice of $\mathcal{Y}$ and $\mathcal{Z}$, checking N-SEP$(\mathcal{X},\mathcal{Y},\mathcal{Z})$ leads to check TL-SEP$(\mathcal{U}, \mathcal{V})$[3] where $\mathcal{U} = \{Y \mid Y \in \mathcal{Y}$ and $p_0 \notin Y\}$ and $\mathcal{V} = \{Z - \{p_0\} \mid Z \in \mathcal{Z}$ and $p_0 \in Z\}$. This implies that the property TL-SEP is 0-polynomial, a contradiction with [2]. The proof is similar when the pivot of the protocol assumed to compute N-SEP is $\mathcal{Y}$ (resp. $\mathcal{Z}$). Due to space limitation, these parts of the proof are not developed here. $\quad \square$

Complexity of NTL queries It remains to show that the language NTL is only able to express predicates (of arity 3) which are 1-polynomial. In order to link predicates and NTL queries, we restrict our attention to particular temporal instances meant to encode triples of super-sets.

From now on the database schema is reduced to contain a unique unary relation.

Definition 5. A temporal instance $\mathcal{I} = (I_1, \ldots, I_l)$ is 3-splitable (splitable, for short) if there exists exactly two distincts instants n and m such that (a) $2 < n+1 < m < l-1$, (b) $I_n = I_m = \emptyset$, and (c) $\forall i \neq j \in [1, n-1]$, $I_i \neq I_j$, $\forall i \neq j \in [n+1, m-1]$, $I_i \neq I_j$, $\forall i \neq j \in [m+1, l]$, $I_i \neq I_j$. $\quad \square$

---

[3] TL-SEP$(\mathcal{U}, \mathcal{V})$ is specified by $\mathcal{U} \cap \mathcal{V} \neq \emptyset$.

In the sequel, we say that a splitable temporal instance $\mathcal{I}$ encodes a tuple $\mathcal{X}_1$, $\mathcal{X}_2$, $\mathcal{X}_3$ of super-sets when $\mathcal{X}_1 = \{I_i \mid i = 1..n - 1\}$, $\mathcal{X}_2 = \{I_i \mid i = n+1..m - 1\}$, and $\mathcal{X}_3 = \{I_i \mid i = m + 1..l\}$. Thus when $\mathcal{I}$ encodes $\mathcal{X}_1$, $\mathcal{X}_2$, $\mathcal{X}_3$, its *left part* defined by $(I_1,\ldots,I_m)$ and denoted by $\mathcal{I}^{\text{left}}$ can be viewed as encoding the two super-sets known by the partner I of a communication protocol. Symmetrically, the *right part* of $\mathcal{I}$ defined by $\mathcal{I}_{|n}$ and denoted $\mathcal{I}^{\text{right}}$ can be viewed as encoding the two super-sets known by the partner II of a communication protocol.

Definition 6. A boolean NTL formula φ expresses the predicate $\mathcal{P}$ of arity 3 if there exists a permutation π on $\{1, 2, 3\}$ such that for each tuple of super-sets $(\mathcal{X}_1, \mathcal{X}_2, \mathcal{X}_3)$ we have: $\mathcal{P}(\mathcal{X}_1, \mathcal{X}_2, \mathcal{X}_3)$=true iff $\mathcal{I} \models_{ntl} \varphi$ where $\mathcal{I}$ encodes $(\mathcal{X}_{\pi(1)}, \mathcal{X}_{\pi(2)}, \mathcal{X}_{\pi(3)})$. $\qquad\qquad\square$

In order to prove that the predicates expressible in NTL are always 1-polynomial, we need to define an intermediate language called split-NTL which has the same expressive power as NTL over splitable temporal instances. Syntactically, split-NTL is a version of NTL where each temporal modality has a *left* and a *right* versions. For instance, Next is replaced by Next^{right} and Next^{left}. Informally, the left (resp. right) version of a temporal operator behaves like the operator itself except that it is intended to be evaluated on the left part of the splitable temporal instance. Defining the semantics of split-NTL requires two functions left and right presented below:

- for $i \in [1..m]$, $\text{left}(i) = i$ and for $i \in [m + 1..l]$, $\text{left}(i) = m$.
- for $i \in [1..n - 1]$, $\text{right}(i) = 1$ and for $i \in [n..l]$, $\text{right}(i) = i - n + 1$.

Definition 7. Given a splitable temporal instance $\mathcal{I}= (I_1,\ldots,I_l)$ and a formula φ of split-NTL, the *truth* of φ at time p given the relative origin t and the valuation ν of the free variables, denoted $[\mathcal{I}, t, p, \nu] \models_{\text{split}} \varphi$, is defined[4] as follows:

- Let OP be either Next or Prev or $\aleph$

If φ is OP$^{\text{left}}\varphi_1$ then $[\mathcal{I}, t, p, \nu] \models_{\text{split}} \varphi$ if $[\mathcal{I}^{\text{left}}, \text{left}(t), \text{left}(p), t, \nu] \models_{\text{left}} \text{OP}\varphi_1$.

If φ is OP$^{\text{right}}\varphi_1$ then $[\mathcal{I}, t, p, \nu] \models_{\text{split}} \varphi$ if $[\mathcal{I}^{\text{right}}, \text{right}(t), \text{right}(p), t, \nu] \models_{\text{right}} \text{OP}\varphi_1$.

- The case where OP is a binary temporal operator is treated in a similar manner. $\qquad\qquad\square$

The relations $\models_{\text{left}}$ and $\models_{\text{right}}$ are defined for hybrid formulas whose upper temporal operator is unmarked but whose subformulas are in split-NTL. Unformally, these relations are defined like $\models_{ntl}$ over either the left or right part of a splitable temporal instance. Due to space limitation, these relations are not formally defined here. For instance, $[\mathcal{I}^{\text{left}}, i, j, t, \nu] \models_{\text{left}} \varphi_1$ Since φ_2 if there exists u such that $i \le u < j$ and $[\mathcal{I}, t, u, \nu] \models_{\text{split}} \varphi_2$ and $\forall k \in]u, j]$, $[\mathcal{I}, t, k, \nu] \models_{\text{split}} \varphi_1$. Note that in $[\mathcal{I}^{\text{left}}, i, j, t, \nu] \models_{\text{left}} \ldots$ i is the relative origin , j is the "at" instant of evaluation with respect to the left (or right) part of $\mathcal{I}$ and t stores the relative origin with respect to the entire splitable temporal instance $\mathcal{I}$.

---

[4] The definition of $\models_{\text{split}}$ is only given for the temporal operators.

The next result shows that split-NTL can be used instead of NTL to query splitable temporal instances.

Lemma 3. Given a NTL formula φ, there exits a formula ψ of split-NTL such that for each splitable temporal instance $\mathcal{I}$ we have: $[\mathcal{I}, t, p, \nu] \models_{\mathrm{ntl}} \varphi$ iff $[\mathcal{I}, t, p, \nu] \models_{\mathrm{split}} \psi$.

The proof is technical and does not present any difficulty. It is omitted here. This lemma is however the key to prove that:

Theorem 3. Each ternary predicate $\mathcal{P}$ expressible by a boolean NTL formula is 1-polynomial.

Proof (Sketch): Assume that φ is a boolean NTL formula which expresses the predicate $\mathcal{P}$. By lemma 3, the predicate $\mathcal{P}$ is expressible by a split-NTL closed formula ψ. In order to show that $\mathcal{P}$ is 1-polynomial, we need to exhibit a polynomial communication protocol computing $\mathcal{P}$. This protocol is derived from the split-NTL formula ψ. The arity of the messages is given by the maximal number of free variables of any subformulas of ψ. Of course, the order of the pivot is 1. The protocol computing $\mathcal{P}(\mathcal{X}, \mathcal{Y}, \mathcal{Z})$ is now sketched while assuming that the triple $(\mathcal{X}, \mathcal{Y}, \mathcal{Z})$ is encoded by the splitable temporal instance $\mathcal{I}$ and that the partner I (resp. II) controls $\mathcal{X}$ and $\mathcal{Y}$ (resp. $\mathcal{Y}$ and $\mathcal{Z}$) encoded by $\mathcal{I}^{\mathrm{left}}$ (resp. $\mathcal{I}^{\mathrm{right}}$). This entails that $\mathcal{Y}$ is the pivot. It is assumed that all temporal subformulas of ψ are enumerated by $\psi_1, ..., \psi_s = \psi$ such that each subformula occurs after its own subformulas. The execution protocol can be decomposed into s steps and each step turns out to consist of $\sum_{i=1}^{m-n+1}(m - n + 1 - i + 1) = \sum_{i=1}^{|\mathcal{Y}|+2}(|\mathcal{Y}| + 3 - i)$ exchanges of messages.

Each step i corresponds to "what is necessary" for the evaluation of the split-NTL subformula ψ_i, assuming that all subformulas $\psi_k, k < i$ have been dealt with. Let us consider for instance that the upper temporal modality of ψ_i is marked by *left*. In this case, the partner I who controls the left part of the instance, is going to evaluate the messages $\{\nu(\overrightarrow{x}) \mid [\mathcal{I}^{\mathrm{left}}, t, p, 1, \nu] \models_{\mathrm{left}} \psi\}$ for $t = 1$ to $m - n + 1$ and for $p = t$ to $m - n + 1$. In order to do that, the partner I may have to use the preceeding messages computed for the subformulas $\psi_k, k < i$. $\qquad\square$

The proof of theorem 2 stating that NTL$\subset$TS-FO follows directly from theorem 3, lemma 2 and the fact that the predicate N-SEP can be expressed by the TS-FO closed formula $\exists t \; \exists s \; \exists u \forall x (R(x, t) \leftrightarrow \neg R(x, s)) \wedge \forall x (R(x, u) \leftrightarrow (R(x, t) \vee R(x, s)))$.

4   The Expressive Power of RNTL

The results of the previous section show that NTL is not complete. This motivates the investigation of a new extension of TL. The language RNTL extends NTL by the introduction of a modality $\Re$ allowing one to restore the most recently erased segment of the past. The syntax of RNTL is defined by adding the following formation rule to the ones defining NTL: if φ is a formula then $\Re\varphi$ is a formula.

The ability to restore segments of the past requires that relative origins be memorized. These instants can be considered as checkpoints. Roughly speaking, RNTL adds to TL a stack of relative origins, the semantics of $\aleph$ formalizing *push* and the semantics of $\Re$ formalizing *pop*.

Definition 8. Given a temporal instance $\mathcal{I}=(I_1,\ldots,I_l)$ and a RNTL formula φ, the *truth* of φ at time p, given the sequence of checkpoints $(t_1, \ldots, t_k)$ and the valuation ν, denoted $[\mathcal{I}, (t_1,...,t_k), p, \nu] \models_{rntl} \varphi$, is defined by:
- If φ is obtained by one of the formation rules defining TL then
$[\mathcal{I}, (t_1,...,t_k), p, \nu] \models_{rntl} \varphi$ is defined in the same way as $[\mathcal{I}_{|t_k}, p - t_k + 1, \nu] \models_{tl} \varphi$.
- If φ is $\aleph \varphi_1$ then $[\mathcal{I}, (t_1,...,t_k), p, \nu] \models_{rntl} \varphi$ if $[\mathcal{I}, (t_1,...,t_k,p), p, \nu] \models_{rntl} \varphi_1$.
- If φ is $\Re \varphi_1$ then $[\mathcal{I}, (t_1,...,t_k), p, \nu] \models_{rntl} \varphi$ if $[\mathcal{I}, (t_1,...,t_{k-1}), p, \nu] \models_{rntl} \varphi_1$
when $k > 1$ and $[\mathcal{I}, (t_1), p, \nu] \models_{rntl} \varphi_1$, otherwise. $\square$

Above, it is assumed that the list of checkpoints is nonempty (it always contains the absolute origin) and increasing. Multiple occurrences of an instant is allowed in this list. Note also that the last checkpoint t_k is the "active" relative origin and obviously, it is assumed that the "at" instant p satisfies $t_k \le p \le l$.

NTL versus RNTL and TS-FO Although in the propositional case, the operator $\Re$ adds no expressive power to NTL[4], it is relatively simple to show that:

Theorem 4. NTL $\subset$ RNTL $\subseteq$ TS-FO.

In fact the predicate N-SEP which was used to separate NTL and TS-FO can be expressed by the following RNTL closed formula (pfR(x) denotes the subformula $P(\text{first} \wedge R(x))$):
$$F\aleph F\aleph F(\ \forall x[\Re \text{ pfR}(x) \leftrightarrow \neg \text{ pfR}(x)] \wedge \forall x[(\Re \text{ pfR}(x) \vee \text{ pfR}(x)) \leftrightarrow R(x)]).$$
The proof that RNTL $\subseteq$ TS-FO is technical but rather easy.

A hierarchy for RNTL^i The completeness of RNTL remains a conjecture. The hierarchy of RNTL's subclasses presented next is an encouraging important step. The fragment RNTL^i is the subset of RNTL formulas whose $\Re$-depth is less or equal to i. The $\Re$-depth of a formula is determined from its tree representation and the maximal number of modalities $\Re$ over branches of this tree. For instance the formula expressing N-SEP is in RNTL^1. Note also that NTL matches RNTL^0.

Theorem 5.
- for $i > 1$, RNTL$^{i-1} \subset$ RNTL^i
- for $i \ge 0$ RNTL$^i \subset$ TS-FO

The proof is similar to the one of theorem 2. Of course, it makes use of the general communication protocols introduced in section 3 for proving:

Lemma 4. (1) Each predicate of arity $i + 2$ (over super-sets) expressible by a boolean RNTL^{i-1} formula is i-polynomial.

(2) There exists a predicate SEP-i of arity $i + 2$ which is not i-polynomial.

The predicate SEP-i when $i \geq 1$ is specified by : SEP-i$(\mathcal{X}_1, \ldots, \mathcal{X}_{i+2})$=true iff $\exists X_1 \in \mathcal{X}_1, \ldots \exists X_{i+2} \in \mathcal{X}_{i+2}$ such that $X_j \cap X_k = \emptyset$ for $j, k \in [1..i+1]$ $(j \neq k)$ and $X_1 \cup \ldots \cup X_{i+1} = X_{i+2}$. For example, SEP-1 is the predicate N-SEP introduced in section 3 to separate NTL and TS-FO.

The binary predicate SEP-0 is the predicate TL-SEP introduced in section 3 and defined by SEP-0$(\mathcal{X}_1, \mathcal{X}_2)$=true iff $\exists X_1 \in \mathcal{X}_1, \exists X_2 \in \mathcal{X}_2$ such that $X_1 = X_2$. It is one of the predicate introduced in [2] in order to separate TL and TS-FO.

The predicate SEP-i of arity $i + 2$ is not i-polynomial thus not expressible in RNTL^{i-1}. However it can be expressed in RNTL^i and as a matter of fact this predicate (extended in a trivial way to make it of arity i+3) is $i + 1$-polynomial. Recall that:

SEP-0$\equiv$TL-SEP is not expressible in TL because it is not 0-polynomial but it is expressible in $\mathrm{RNTL}^0 \equiv$NTL. A trivial protocol with pivot $\mathcal{X}_2$ (thus of order 1) which assigns both super-sets $\mathcal{X}_1$ and $\mathcal{X}_2$ to partner I and $\mathcal{X}_2$ together with the empty super-set to partner II computes SEP-0$\equiv$TL-SEP. This protocol has a constant complexity. This shows that the predicate SEP-0 is 1-polynomial.

SEP-1$\equiv$N-SEP is not expressible in NTL because it is not 1-polynomial but it is expressible in RNTL^1. It can be showed that this predicate is 2-polynomial.

Relationship with TS-FO^i We conclude this section by some comments on hightly interesting links between (1) the strict hierarchy in expressive power among the fragments RNTL^i established above, and (2) the conjecture that there is a strict hierarchy in expressive power among the fragments of TS-FO^i. The language TS-FO^i is the subclass of TS-FO formed by formulas built using at most i distinct time-variables. It is known that TS-FO$^1 \subset$ TS-FO$^2 \subset$ TS-FO^3. However, for $i \geq 3$ the strict inclusion TS-FO$^i \subset$ TS-FO^{i+1} remains a conjecture.

It has been already established and it is not difficult to verify that each TL formula is expressible by a formula in TS-FO using at most 3 distinct time-variables. The proof of the separation of TL and TS-FO [2] entails that TL is strictly less expressive than TS-FO^3 because the predicate TL-SEP$\equiv$SEP-0 can be expressed by a formula with 2 distinct (thus at most 3) time-variables. This generalizes when considering RNTL^i:

Lemma 5. $\mathrm{RNTL}^i \subset$ TS-FO^{i+4}

Note that a NTL formula can be translated into a TS-FO^4 formula. The proof of separating RNTL^i and RNTL^{i+1} entails that RNTL^i is stricly less expressive than TS-FO^{i+4}: the predicate SEP-i+1 can be expressed by a formula with i+2 (thus at most i+4) distinct time-variables. Note that one of the possible TS-FO formula translating SEP-i+1 is $\exists t_1 \ldots \exists t_{i+2}[(\forall x R(x, t_1) \leftrightarrow \neg R(x, t_2)) \wedge \ldots \wedge (\forall x R(x, t_{i+1}) \leftrightarrow \neg R(x, t_{i+2})) \wedge (\forall x R(x, t_1) \vee \ldots \vee R(x, t_{i+1}) \leftrightarrow R(x, t_{i+2}))]$.

The main contribution of these results to the conjecture[5] that there exists a strict TS-FO^i hierarchy is that, proving this conjecture reduces to proving that TS-FO$^i \subseteq$ RNTL^k for some $k \geq i - 3$. We are currently investigating techniques in order to prove this result. Completeness of RNTL would then directly follow.

Note that results in [13] are not contradicting this conjecture. [13] proves that, for dense linear order (generalizable to discrete linear order), extending TL by a finite number of temporal connectives always leads to languages less expressive than TS-FO (their temporal connectives are formulas in TS-FO with a fixed finite number of free time variables). NTL and each fragment RNTL^i fall into this class of TL-extensions. However, we claim that it is not the case for RNTL.

Acknowledgements

We thank S. Abiteboul, J-M. Couvreur and D. Niwinski for informal helpful discussions.

References

1. Abiteboul, S., Herr, L. and Van den Bussche J.: Temporal Connectives Versus Explicit Timestamps in Temporal Query Languages, In Recent Advances in Temporal Databases, S. Clifford and A. Tuzhilin, Eds, Springer Verlag (1995) 43–60
2. Abiteboul, S., Herr, L. and Van den Bussche, J.: Temporal Versus First-Order Logic to Query Temporal Databases, Proceedings of PODS'96, (1996) 49–57
3. Abiteboul, S., Hull, R. and Vianu, V.: Foundations of Databases, Addison-Wesley (1995)
4. Bidoit, N. and De Amo, S.: Branching time temporal logic for querying multiversion databases, Technical Report in preparation (1999)
5. Chomicki, J.: Temporal Query Languages: a survey, Temporal Logic, First Int. conf., LNAI 827 (1994) 506–534
6. Chomicki, J. and Toman, D.: Temporal Logic in Information Systems, Logics for Databases and Information Systems (1998) 31-70
7. Clifford, J., Croker, A. and Tuzhilin, A.: On Completeness of Historical Relational Query Languages, ACM Transactions on Database Systems, 19, 1 (1994) 64–116
8. Emerson, E. A.: Temporal and Modal Logic, In Handbook of Theoretical Computer Science, Volume B: Formal Models and Semantics, Jan van Leeuwen, Ed., Elsevier Science Publishers (1990) 995–1072
9. Gabbay, D., M., Pnueli, A., Shelah, S., and Stavi, J.: On the Temporal Basis of Fairness, Symposium on Principles of Programming Languages, (1980) 163–173
10. Hafer, T., and Thomas, W.: Computation Tree Logic CTL* and Path Quantifiers in the Monadic Theory of the Binary Tree, In Automata, Languages and Programming, 14th International Colloquium, LNCS 267 (1987) 269–279
11. H.W. Kamp: Tense Logic and the Theory of Linear Order, PhD thesis, University of California, Los Angeles (1968)

---

[5] A closely related question is whether there is a strict FO^i hierarchy on the class of ordered finite graphs.

12. Laroussinie, L., and Schnoebelen, PH: A hierarchy of temporal logics with past, TCS, 148, 2, (1995) 303–324
13. Toman, D., and Niwinski, D.: First Order Queries over temporal Databases Inexpressible in Temporal Logic, EDBT (1996) 307–324.

On the Undecidability of Some Sub-classical First-Order Logics[*]

Matthias Baaz[1], Agata Ciabattoni[2], Christian Fermüller[1], and Helmut Veith[1]

[1] Technische Universität Wien, Austria
[2] Università di Milano, Italy

Abstract. A general criterion for the undecidabily of sub-classical first-order logics and important fragments thereof is established. It is applied, among others, to Urquart's (original version of) **C** and the closely related logic **C***. In addition, hypersequent systems for (first-order) **C** and **C*** are introduced and shown to enjoy cut-elimination.

1 Introduction

A wide range of non-classical logics can be viewed as sub-classical. By this we mean that they are based on (a subset of) the signature of classical logic and can be extended to classical logic. Intuitionistic logic, fragments of linear logic, Łukasiewicz logic, Gödel/Dummett logic (in fact, all intermediate logics, and most substructural or many-valued logics), are just a few prominent examples.

(Un)decidability issues are a fundamental topic of Logic in the context of Theoretical Computer Science. In the literature there are results on the decision problem for *particular* first-order sub-classical logics. For instance, in [14,11,10] the decision problem for some substructural logics, that is logics that when formulated as a Gentzen system do not have all the structural rules, has been investigated. It was proved that removing the weakening rule from standard Gentzen-style formulations of classical or intuitionistic first-order logic gives a logic that is still undecidable. On the other hand, removing the contraction rule from a sequent calculus for classical or intuitionistic logic results in a decidable logic. However it is clear that the lack of contraction in itself is *not sufficient* for the decidability of a sub-classical logic, as witnessed, e.g., by Łukasiewicz's and other important "fuzzy logics" (see [8]).

In this paper we provide a sufficient and quite general criterion for the undecidability of first-order sub-classical logics, with or without function symbols (in fact: already fragments thereof). Moreover, we continue the proof theoretical investigation—initiated in [3]—of the logics **C** (introduced in by A. Urquhart in [16][1], see also [13]), and the related **C***. These are underlying the most important formalizations of fuzzy logic, namely Gödel, Łukasiewicz and Product logic

---

[*] Partly supported by COST-Action No. 15, FWF grant P-12652-MAT and WTZ Austria-Italy 1999-2000, Project No. 4
[1] Recently A. Urquhart discovered that his axiom system for **C** as given in [16] is incomplete with respect to an intended semantics [17]. Because of its proof theoretic interest we—like others, e.g. [13]—will deal with the original system **C** here.

C. Pandu Rangan, V. Raman, R. Ramanujam (Eds.): FSTTCS'99, LNCS 1738, pp. 258–268, 1999.

(see, e.g., [8]). $\mathbf{C}$ and $\mathbf{C}^*$ lack contraction, i.e., $[A \supset (A \supset B)] \supset (A \supset B)$ does not hold. Moreover, they share the property that their truth values are linearly ordered, that is they satisfy the linearity axiom $(A \supset B) \lor (B \supset A)$.

In [3] we introduced *propositional* cut-free hypersequent calculi for $\mathbf{C}$ and $\mathbf{C}^*$. This allowed us to establish that derivability in these logics is decidable. In this paper we consider the *first-order* version of these logics and define analytic (cut-free) calculi for them. The analyticity of the logics $\mathbf{C}$ and $\mathbf{C}^*$ is contrasted with their undecidability, which is a corollary to the general undecidability result mentioned above. It should also be seen in light of the fact that their purely intuitionistic counterparts $\mathbf{I}^-$ and $\mathbf{I}^{-*}$—obtained by dropping the linearity axiom from $\mathbf{C}$ and $\mathbf{C}^*$, respectively—are decidable.

2   Basic Definitions

We investigate logics that are based on the language of classical first-order logic without equality. More precisely, we use the binary *connectives* $\lor$, $\land$, $\supset$ and the *truth constant* $\bot$. Negation is defined, as usual, by: $\neg A := (A \supset \bot)$.

Remark 1. Treating negation as a defined connective is just a matter of convenience. Explicit axioms and rules for negation could easily be added in all our cases. However—as witnessed by intuitionistic logic—the other connectives and the quantifiers will not be interdefinable in general.

Object variables are denoted by x, y, ... The *quantifiers* $\forall$ (for all) and $\exists$ (exists) refer to these variables. Moreover, there is an infinite supply of n-ary *predicate symbols* for every $n \geq 0$. We consider both, logics with and without *function symbols*. In the former case an infinite supply of n-ary function symbols and constants is assumed for every $n \geq 0$. Constants are considered as 0-ary function symbols. *Terms* and *Formulas* are inductively defined in the usual way. Classical first-order logic with function symbols is denoted as $\mathbf{CL}f$, without function symbols as $\mathbf{CL}$.

Here, we assume logics to be specified by Hilbert-style systems. *Axioms* are always considered as schemata. (I.e., all instances are axioms, too.) *Rules* of derivation are written in the form

$$\vdash A_1, \ldots, \vdash A_n \Longrightarrow \vdash B.$$

A *logic* $\mathcal{L}$ is identified with the set of provable formulas. We write $\vdash_{\mathcal{L}} F$ for $F \in \mathcal{L}$.

Definition 1. *A logic $\mathcal{L}$ (with or without function symbols) is called* sub-classical *if it can be extended to classical logic. More formally: if $\vdash_{\mathcal{L}} F$ implies $\vdash_{\mathbf{CL}f} F$.*

Definition 2. *A rule $\vdash A_1, \ldots, \vdash A_n \Longrightarrow \vdash B$ is* admissible *in logic $\mathcal{L}$ if $\vdash_{\mathcal{L}} B'$ whenever $\vdash_{\mathcal{L}} A'_i$ for all $1 \leq i \leq n$, where B' and A'_i are corresponding instances of the schematic formulas of the rule.*

We write $\alpha[F/p]$ for the formula that arises by instantiating the formula variable p of the schema $\alpha[p]$.

3   Two General Undecidability Results

We establish sufficient conditions that allow to embed (undecidable fragments of) classical logic into sub-classical logics. As a central notion for the "simulation of two-valued-ness" in non-classical contexts we introduce the following:

Definition 3. *A schema of form* $\alpha[p] \vee \overline{\alpha}[p]$ *with exactly one formula variable* p *is called a* duality principle *of a logic* $\mathcal{L}$ *if for all formulas* F: $\vdash_{\mathcal{L}} \alpha[F/p] \vee \overline{\alpha}[F/p]$ *but there exists a formula* G *such that* $\nvdash_{CL} \alpha[G/p]$ *and* $\nvdash_{CL} \overline{\alpha}[G/p]$.

For the undecidabilty proof below we will assume admissibility[2] of the following rules for disjunction:

$\vdash A \Longrightarrow \vdash A \vee B$	(weak)	$\vdash A \vee (B \vee C) \Longrightarrow \vdash (A \vee B) \vee C$	(assoc-l)
$\vdash A \vee B \Longrightarrow \vdash B \vee A$	(comm)	$\vdash (A \vee B) \vee C \Longrightarrow \vdash A \vee (B \vee C)$	(assoc-r)
$\vdash A \vee A \Longrightarrow \vdash A$	(idem)		

The following rule expresses a weak form of distributivity:

$$\vdash A \vee C, \ \vdash B \vee C \Longrightarrow \vdash (A \wedge B) \vee C \qquad \text{(distr)}$$

A "minimal" rule for existential quantification is:

$$\vdash A(t) \Longrightarrow \vdash \exists x \colon A(x) \qquad (\exists\text{-in'})$$

where t is any term.

Definition 4. *We call a rule* $\vdash A_1, \ldots, \vdash A_n \Longrightarrow \vdash B$ $\vee$-normal *in a logic* $\mathcal{L}$ *if its admissibility in* $\mathcal{L}$ *implies the admissibilty of* $\vdash A_1 \vee C, \ldots, \vdash A_n \vee C \Longrightarrow \vdash B \vee C$

Theorem 1. *Any sub-classical logic* $\mathcal{L}$ *with function symbols that has a duality principle and in which the rules* weak, assoc-l, assoc-r, comm, idem, distr, $\exists$-in' *are admissible and* $\vee$-normal *is undecidable.*

Proof. The proof proceeds as follows: we consider any classical formula F in conjunctive normal form and apply Herbrand's theorem. By using the duality principle we can translate the resulting Herbrand instance F_{Her} into a formula $\delta(F_{Her})$ such that $\vdash_{CL_f} F_{Her}$ if and only if $\vdash_{\mathcal{L}} \delta(F_{Her})$. This relation transfers to F and $\delta(F)$. Hence follows that $\mathcal{L}$ is undecidable.

It is well known that validity (provability) of formulas of form

$$F = \exists \boldsymbol{x} \colon \bigvee_{1 \leq i \leq m} \ \bigwedge_{1 \leq j \leq n_i} L_{i,j}(\boldsymbol{x})$$

where the $L_{i,j}$ are literals and $\boldsymbol{x}$ is a vector of variables, is undecidable in classical logic with function symbols.

---

[2] Remember that admissibility of rules is a much weaker condition than the derivability of corresponding formulas. E.g., the admissibility of idem in $\mathcal{L}$, in general, does *not* imply $\vdash_{\mathcal{L}} (A \vee A) \supset A$.

Assume that $\vdash_{\mathbf{CL}f} F$. By Herbrand's theorem there exist vectors of ground terms $t^{(1)}, \ldots t^{(k)}$ such that also the Herbrand instance

$$F_{Her} = \bigvee_{1 \le \ell \le k} \bigvee_{1 \le i \le m} \bigwedge_{1 \le j \le n_i} L_{i,j}(t^{(\ell)})$$

is provable.

A disjunction $\bigvee_{1 \le i \le (km)} L_i$ is called a *path* of F_{Her} if each of the literals L_i occurs in exactly one of the "clauses" $\bigwedge_{1 \le j \le m_i} L_{i,j}(t^{(\ell)})$ of F_{Her}. A pair of literals of the form $(P(t), \neg P(t))$ is called *complementary*. It is easy to check that $\vdash_{\mathbf{CL}f} F_{Her}$ iff every path π of F_{Her} contains a complementary pair $(P(t), \neg P(t))$, i.e. if both $P(t)$ and $\neg P(t)$ occur as disjuncts in π.

Let $\alpha[p] \vee \overline{\alpha}[p]$ be a duality principle of $\mathcal{L}$. Consider the following translation $\delta(.)$ of literals:

$$\delta(P) = \alpha[P/p]$$
$$\delta(\neg P) = \overline{\alpha}[P/p]$$

The translation extends homomorphically to arbitrary formulas.

We first show that the classical provability of F_{Her} implies that $\vdash_{\mathcal{L}} \delta(F_{Her})$. Let $(P(t), \neg P(t))$ be the complementary pair of a path π of F_{Her}. By definition we have $\delta(P(t) \vee \neg P(t)) = \delta(P(t)) \vee \delta(\neg P(t)) = \alpha[P(t)/p] \vee \overline{\alpha}[P(t)/p]$, and therefore $\vdash_{\mathcal{L}} \delta(P(t) \vee \neg P(t))$. Using rule weak we can add all the other literals of π disjunctively to $\delta(P(t) \vee \neg P(t))$ in $\mathcal{L}$. By the admissibility of the rules assoc-l, assoc-r, comm and the fact that they are $\vee$-normal in $\mathcal{L}$ we can reorder the disjuncts as needed. After having derived (in $\mathcal{L}$) the formulas corresponding to every path of F_{Her}, we join them conjunctively using rule distr. We thus have

$$\vdash_{\mathcal{L}} \delta(F_{Her})$$

Now since rule ∃-in' is admissible and $\vee$-normal we can re-introduce the existential quantifiers in front of the inner disjunction. I.e. we obtain

$$\vdash_{\mathcal{L}} \bigvee_{1 \le \ell \le k} \exists x : \bigvee_{1 \le i \le m} \bigwedge_{1 \le j \le n_i} \delta(L_{i,j}(x)).$$

Observe that we used the same variables as in the original formula F. Therefore all k disjuncts are identical. Since rule idem is admissible and $\vee$-normal we can contract this to

$$\vdash_{\mathcal{L}} \delta(F).$$

We thus have proved: $\vdash_{\mathbf{CL}f} F$ implies $\vdash_{\mathcal{L}} \delta(F)$.

For the other direction, observe that $\vdash_{\mathcal{L}} \delta(F)$ implies $\vdash_{\mathbf{CL}f} \delta(F)$ since $\mathcal{L}$ is sub-classical. It remains to check that $\delta(F)$ is equivalent to F in classical logic with respect to provability. For this observe that, since p is the only formula variable of the duality principle, $\alpha[p]$ is classically equivalent either to p or to $\neg p$. (It cannot be constantly equivalent to $\top$ by $\nvdash_{\mathbf{CL}} \alpha[G/p]$, for some G. It also cannot be constantly equivalent to $\bot$ since all instances of $\alpha[p] \vee \overline{\alpha}[p]$ are provable in $\mathcal{L}$ and therefore also in classical logic.) The same holds for $\overline{\alpha}[p]$. Since all instances of $\alpha[p] \vee \overline{\alpha}[p]$ are provable in $\mathcal{L}$ and therefore also in classical logic, we conclude that

either: $\alpha[p] \equiv p$ and $\overline{\alpha}[p] \equiv \neg p$,
   or: $\alpha[p] \equiv \neg p$ and $\overline{\alpha}[p] \equiv p$.

where $\equiv$ denotes classical equivalence. Both ways the duality principle is classically equivalent to $p \vee \neg p$. Summarizing, $\vdash_{\mathcal{L}} \delta(F)$ implies $\vdash_{\mathrm{CL}_f} F$, qed.

From the proof of Theorem 1 we immediately obtain:

Corollary 1. *The existential fragments of logics fullfilling the conditions stated in the previous theorem are undecidable.*

To obtain a useful criterion for undecidability of logics without function symbols we need to consider also the universal quantifier. Consider the standard rule for universal quantification:

$$A(x) \vdash \forall x \colon A(x) \qquad (\forall\text{-in'})$$

Remark 2. We require that, if rule $\forall$-in' is $\vee$-normal, the *eigenvariable condition* is obeyed in its application. I.e., in

$$A(x) \vee C \vdash (\forall x \colon A(x)) \vee C \qquad (\forall\text{-in'})$$

x must not occur in C.

If, in addition to the assumptions of Theorem 1 we require $\forall$-in' to be admissible and $\vee$-normal (including the eigenvariable condition) we obtain undecidablity for logics without function symbols.

Theorem 2. *Any sub-classical logic $\mathcal{L}$ that has a duality principle and in which the rules* weak, assoc-l, assoc-r, comm, idem, distr, $\exists$-in', *as well as* $\forall$-in' *are admissible and $\vee$-normal is undecidable, even without function symbols.*

Proof. The proof is analogous to that of Theorem 1. We start with formulas in prenex conjuntive normal form:

$$F = Qx_1 \ldots Qx_n \colon \bigvee_{1 \leq i \leq m} \bigwedge_{1 \leq j \leq n_i} L_{i,j}(\boldsymbol{x}),$$

where $Q \in \{\exists, \forall\}$ and $\boldsymbol{x} = x_1, \ldots, x_n$.

Since we have no function symbols we apply Herbrand's theorem in its orginal form (i.e., without Skolemizing). In the Herbrand instance

$$F_{Her} = \bigvee_{1 \leq \ell \leq k} \bigvee_{1 \leq i \leq m} \bigwedge_{1 \leq j \leq n_i} L_{i,j}(\boldsymbol{y}^{(\ell)})$$

the variables x_i are replaced by variables $y_i^{(\ell)}$ in such a way that introducing the quantifiers in a proof of F given F_{Her}, can be done without violating the eigenvariable condition. Again the proof can be transferred to the logic $\mathcal{L}$ using the duality principle and the fact that the stated rules are admissible and $\vee$-normal. (Note that it is essential that $\forall$-in' is applied in its $\vee$-normal variant, obeying the eigenvariable condition.) The rest of the proof proceeds exactly as in Theorem 1. I.e., we have $\vdash_{\mathrm{CL}} F$ if only if $\vdash_{\mathcal{L}} \delta(F)$.

4   First-Order Extensions of Urquhart's (Original) C

In [3] we investigated some propositional logics underlying the most important formalizations of fuzzy logic [8]. Here we focus attention to the extensions of these logics to the *first-order level*.

Our logics are defined by their Hilbert-style axiomatizations. We refer to the following list of axioms (cf. [3]):

Ax1: $A \supset (B \supset A)$	*Res1 :* $[(A \wedge B) \supset C] \supset [A \supset (B \supset C)]$
Ax2: $(A \supset B) \supset [(C \supset A) \supset (C \supset B)]$	*Res2 :* $[(A \supset (B \supset C)] \supset [(A \wedge B) \supset C]$
Ax3: $[A \supset (C \supset B)] \supset [(C \supset (A \supset B)]$	*Lin :* $(A \supset B) \vee (B \supset A)$
Ax4: $(A \wedge B) \supset A$	*Com :* $[A \wedge (A \supset B)] \supset [B \wedge (B \supset A)]$
Ax5: $(A \wedge B) \supset B$	*Abs :* $\perp \supset A$
Ax6: $A \supset [C \supset (A \wedge C)]$	*Contr :* $[A \supset (A \supset B)] \supset (A \supset B)$
Ax7: $A \supset (A \vee B)$	*Dneg :* $\neg\neg A \supset A$
Ax8: $B \supset (A \vee B)$	
Ax9: $(A \supset B) \supset [(C \supset B) \supset [(A \vee C) \supset B]]$	

Taking *Modus Ponens* as the only rule of derivation we define the propositional fragments of the logics $\mathbf{I}^-$, $\mathbf{I}^{-*}$, $\mathbf{C}$, and $\mathbf{C}^*$ by the following sets of axioms[3]:

$$\mathbf{I}^- : \{Ax1, \ldots, Ax9, Abs\} \qquad \mathbf{C} : \mathbf{I}^- \cup \{Lin\}$$
$$\mathbf{I}^{-*} : \mathbf{I}^- \cup \{Res1, Res2\} \qquad \mathbf{C}^* : \mathbf{I}^{-*} \cup \{Lin\}$$

The first-order logics are given by adding to the corresponding propositional systems the axioms:

$$Ax\forall : \forall x : A(x) \supset A(t) \qquad Ax\exists : A(t) \supset \exists x : A(x)$$

and the rules

$$\vdash B \supset A(x) \Longrightarrow \vdash B \supset \forall x : A(x) \quad (\forall\text{-in}) \qquad \vdash A(x) \supset B \Longrightarrow \vdash \exists x : A(x) \supset B \quad (\exists\text{-in})$$

The quantifier axioms and rules are subject to the usual variable conditions: t is free for x and x does not ocur free in B.

$\mathbf{C}$ coincides with Urquhart's "basic" many-valued logic as defined in [16] (see also [13]). $\mathbf{I}^{-*}$ extends to intuitionistic logic $\mathbf{I}$ by adding *Contr* (contraction). By adding to $\mathbf{I}^{-*}$ the axiom *Dneg* (involutivity of negation) we obtain the affine multiplicative fragment of linear logic [6] **aMLL** (extended with the additive disjunction). Hájek's Basic Logic **BL** [8] corresponds to the logic obtained by adding axiom *Com* to $\mathbf{C}^*$ (as proved in [3]). If we add to **BL** axiom *Dneg* we get Łukasiewicz logic [12]; while by extending **BL** with the axioms $\neg\neg A \supset ((B \wedge A \supset C \wedge A) \supset (B \supset C))$ and $A \wedge \neg A \supset \perp$ we obtain Product logic [9]. $\mathbf{C}$, $\mathbf{C}^*$, and **BL** all turn into Gödel logic $\mathbf{G}_\infty$ [7,5] if we add axiom *Contr*.

In [3,1] analytic—i.e., cut-free Gentzen-style—calculi for the *propostional versions* of the logics $\mathbf{I}^-$, $\mathbf{I}^{-*}$, $\mathbf{C}$, $\mathbf{C}^*$, and $\mathbf{G}_\infty$ have been introduced. These calculi are called $\mathbf{LJ}^-$, $\mathbf{LJ}^{-*}$, $\mathbf{HC}$, $\mathbf{HC}^*$ and $\mathbf{GLC}$, respectively. As a consequence one

---

[3] In fact, *Res1* is redundant in presence of *Ax1, Ax2, Ax3*.

obtains that the derivability problem for these logics is decidable and is at most in PSPACE. **HC**, **HC*** and **GLC** are based on *hypersequents*—a simple and natural generalization of Gentzen sequents to *mulitsets* of sequents (see [2] for an overview).

The axioms and rules of the calculus **HC** are as follows:

Axioms:

Cut Rule:

$$A \vdash A \qquad \bot \vdash A$$

$$\frac{G \mid \Gamma_1 \vdash A \quad G' \mid A, \Gamma_2 \vdash B}{G \mid G' \mid \Gamma_1, \Gamma_2 \vdash B} \quad \text{(cut)}$$

Structural Rules:

$$\frac{G \mid \Gamma \vdash C}{G \mid \Gamma, A \vdash C} \quad \text{(W)} \qquad\qquad \frac{G \mid \Gamma \vdash \Delta}{G \mid \Gamma \vdash \Delta \mid \Gamma' \vdash \Delta'} \quad \text{(EW)}$$

$$\frac{G \mid \Gamma \vdash \Delta \mid \Gamma \vdash \Delta}{G \mid \Gamma \vdash \Delta} \quad \text{(EC)} \qquad \frac{G \mid \Pi_1, \Gamma_1 \vdash A \quad G' \mid \Pi_2, \Gamma_2 \vdash B}{G \mid G' \mid \Pi_1, \Pi_2 \vdash A \mid \Gamma_1, \Gamma_2 \vdash B} \quad \text{(Comm)}$$

Logical Rules:

$$\frac{G \mid \Gamma, A \vdash B}{G \mid \Gamma \vdash A \supset B} \quad (\supset\text{-right}) \qquad \frac{G \mid \Gamma_1 \vdash A \quad G' \mid B, \Gamma_2 \vdash C}{G \mid G' \mid \Gamma_1, \Gamma_2, A \supset B \vdash C} \quad (\supset\text{-left})$$

$$\frac{G \mid \Gamma_1 \vdash A \quad G' \mid \Gamma_2 \vdash B}{G \mid G' \mid \Gamma_1, \Gamma_2 \vdash A \wedge B} \quad (\wedge\text{-right}) \qquad \frac{G \mid \Gamma, A_i \vdash C}{G \mid \Gamma, A_1 \wedge A_2 \vdash C} \quad (\wedge_i\text{-left})$$

$$\frac{G \mid \Gamma \vdash A_i}{G \mid \Gamma \vdash A_1 \vee A_2} \quad (\vee_i\text{-right}) \qquad \frac{G \mid \Gamma, A \vdash C \quad G' \mid \Gamma, B \vdash C}{G \mid G' \mid \Gamma, A \vee B \vdash C} \quad (\vee\text{-left})$$

The calculus **HC*** is obtained by substituting in the above calculus the $(\wedge_i-\text{left})$ rule with the following one:

$$\frac{G \mid \Gamma, A, B \vdash C}{G \mid \Gamma, A \wedge B \vdash C} \quad (\wedge\text{-left})$$

The calculus **GLC** (see [1]) is obtained by simply adding the contraction rule to **HC** or **HC***. The *first-order* calculi **HC**, **HC***, and **GLC** are obtained by adding to the corresponding calculi, the following rules:

$$\frac{G \mid \Gamma \vdash F(a)}{G \mid \Gamma \vdash \forall x F(x)} \quad (\forall\text{-right}) \qquad \frac{G \mid \Gamma, F(t) \vdash C}{G \mid \Gamma, \forall x F(x) \vdash C} \quad (\forall\text{-left})$$

$$\frac{G \mid \Gamma \vdash F(t)}{G \mid \Gamma \vdash \exists x F(x)} \quad (\exists\text{-right}) \qquad \frac{G \mid \Gamma, F(a) \vdash C}{G \mid \Gamma, \exists x F(x) \vdash C} \quad (\exists\text{-left})$$

where—in the rules ($\forall$-right) and ($\exists$-left)—a does not occur in the lower hypersequent.

The correctness and completeness results for our hypersequent calculi (with respect to the Hilbert style systems) are easily extended from the propositional level (see [3]) to the first-order level.

Theorem 3. *Provability of " ⊢ F" in* **HC, HC*,** *and* **GLC** *coincides with provability of "F" in* **C, C*,** *and* **G$_\infty$,** *respectively.*

Corollary 2. *(1) The existential fragments and of the logics* **C, C*,** *and* **G$_\infty$** *with function symbols are undecidable. (2) The prenex fragments of these logics without function symbols are undecidable.*

Proof. It is easy to check that the rules weak, assoc-l, assoc-r, comm, idem, distr, ∃-in' are both: admissible and ∨-normal, in the mentioned logics.

The schema $(p \supset \neg p) \vee (\neg p \supset p)$, which is an instance of the linearity axiom is a duality principle for **C, C*,** and **G$_\infty$**. Indeed, by using Avron's Communication rule (Comm) it can be derived in the corresponding calculi (see [1,3]) but neither $p \supset \neg p$ nor $\neg p \supset p$ is provable in classical logic. Therefore the statements (1) and (2) follow from Theorems 1 and Theorem 2, respectively.

This should be contrasted with the fact that decision procedures for **I$^-$** and **I^{-*}** can directly be extracted from corresponding cut-free calculi as proved in [14].

Finally, we remark that our general undecidability results also apply to (other) important fuzzy logics:

Corollary 3. *Also for* **BL,** *Łukasiewicz and Product logic the existential fragment with function symbols and the prenex fragments without function symbols are undecidable.*

5   Analyticity and a Decidable Subclass

Proposition 1. *The first-order calculi* **LJ$^-$, LJ^{-*}** *and* **LJ** *admit cut-elimination.*

Proof. (Sketch) For the sequent calculus **LJ** for intuitionistic logic, this is well-known. An inspection of the classical proof shows that the absence of contraction does not affect cut-elimination.

In [3] ([1]) it was proved that the hypersequent calculi **HC** and **HC*** (respectively, **GLC**) admit cut-elimination. In order to extend these results to the first-order level we need the following "eigenvariable" lemma. (The straightforward proof is omitted for space reasons.)

Lemma 1. *Let* $\mathcal{P}(a)$ *be a proof in the first-order calculus* **HC** *(***HC*, GLC***) of the hypersequent S that contains the variable a. If, throughout the proof, we replace a by a term t, containing only variables that do not occur in* $\mathcal{P}(a)$*, we obtain a proof* $\mathcal{P}(t)$ *ending with* $S[t/a]$*.*

Theorem 4. *First-order* **HC, HC*** *and* **GLC** *admit cut-elimination.*

Proof. It is enough to show that if $\mathcal{P}$ is a proof of a hypersequent H containing only one cut rule which occurs as the last inference of $\mathcal{P}$, then H is derivable without the cut rule.

Cut-elimination for hypersequent calculi works essentially in the same way as for the corresponding sequent calculi. As was noticed in [4], a simple way to make the inductive argument work in the presence of the (EC) rule, is to consider the number of applications of this rule in a given derivation as an independent parameter. The proof will proceed by induction on lexicographically-ordered triple (r, c, h), where r is the number of the applications of the (EC) rule in the proofs of the premises of the cut rule, c is the complexity of the cut formula, and h is the sum of the length of the proofs of the premises of the cut rule.

It suffices to consider the following cases according to which inference rule is being applied just before the application of the cut rule:

1. either $G \mid \Gamma \vdash A$ or $G' \mid \Gamma', A \vdash B$ is an initial hypersequent;
2. either $G \mid \Gamma \vdash A$ or $G' \mid \Gamma', A \vdash B$ is obtained by a structural rule;
3. both $G \mid \Gamma \vdash A$ and $G' \mid \Gamma', A \vdash B$ are lower sequents of some logical rules such that the principal formulas of both rules are just the the cut formulas;
4. either $G \mid \Gamma \vdash A$ or $G' \mid \Gamma', A \vdash B$ is a lower sequent of a logical rule whose principal formula is not the cut formula.

We will give here a proof for some relevant cases. Suppose that the last inference in the proof of one premise of the cut is the (EC) rule, and the proof ends as follows:

$$\frac{\dfrac{G \mid \Gamma \vdash A \mid \Gamma \vdash A}{G \mid \Gamma \vdash A} \text{(EC)} \qquad G' \mid A, \Gamma' \vdash B}{G \mid G' \mid \Gamma', \Gamma \vdash B} \text{(cut)}$$

Let r' be the number of applications of the (EC) rule in the above proof. This proof can be replaced by

$$\frac{\dfrac{G \mid \Gamma \vdash A \mid \Gamma \vdash A \qquad G' \mid A, \Gamma' \vdash B}{G \mid G' \mid \Gamma, \Gamma' \vdash B \mid \Gamma \vdash A} \text{(cut)} \qquad G' \mid A, \Gamma' \vdash B}{\dfrac{G \mid G' \mid G' \mid \Gamma', \Gamma \vdash B \mid \Gamma', \Gamma \vdash B}{G \mid G' \mid \Gamma', \Gamma \vdash B} \text{(EC)}} \text{(cut)}$$

which contains two cuts with $r' - 1$ applications of the (EC) rule. Then, they can be eliminated by induction hypothesis.

Cases concerning the remaining structural rules and the logical ones are treated as in the corresponding proofs of the cut-elimination theorem for the hypersequent calculi $\mathbf{HC}, \mathbf{HC^*}$ and $\mathbf{GLC}$ (see [3]). In the following we show how to eliminate a cut involving the rules for quantifiers.

Suppose that the last inference in the proof of one premise of the cut is the ($\exists$-left) rule, and the proof, say $\mathcal{P}$, ends as follows:

Suppose that the last inference in the proof of one premise of the cut is the ($\exists$-left) rule, and the proof, say $\mathcal{P}$, ends as follows:

$$\dfrac{\dfrac{G \mid B(a), \Gamma \vdash A}{G \mid \exists x B(x), \Gamma \vdash A}\text{ ($\exists$-left)} \qquad G' \mid \Sigma, D(a), A \vdash C}{G \mid G' \mid \exists x B(x), \Gamma, \Sigma, D(a) \vdash C}\text{ (cut)}$$

Let $\mathcal{P}(a)$ be the proof ending with $G' \mid \Sigma, D(a), A \vdash C$. By Lemma 1, by replacing a by b throughout $\mathcal{P}(a)$ one gets a proof ending with $G'[b/a] \mid \Sigma[b/a], D(a), A \vdash C$ Thus the proof $\mathcal{P}$ can be replaced by

$$\dfrac{\dfrac{G \mid B(a), \Gamma \vdash A \qquad G'[b/a] \mid \Sigma[b/a], D(b), A \vdash C}{G \mid G'[b/a] \mid B(a), \Gamma, \Sigma[b/a], D(b) \vdash C}\text{ (cut)}}{G \mid G'[b/a] \mid \exists x B(x), \Gamma, \Sigma[b/a], D(b) \vdash C}\text{ ($\exists$-left)}$$

in which the cut is shifted upward, so it can be eliminated by induction hypothesis.

Suppose that both the premises of the cut rule are lower sequents of the rules for the quantifier $\exists$ and that the cut formula is the principal formula of both rules, that is

$$\dfrac{\dfrac{G \mid \Gamma \vdash B(t)}{G \mid \Gamma \vdash \exists x B(x)}\text{ ($\exists$-right)} \qquad \dfrac{G' \mid \Sigma, B(a) \vdash C}{G' \mid \Sigma, \exists x B(x) \vdash C}\text{ ($\exists$-left)}}{G \mid G' \mid \Gamma, \Sigma \vdash C}\text{ (cut)}$$

Let $\mathcal{P}(a)$ and $\mathcal{Q}(a)$ be the proofs ending with $G \mid \Gamma \vdash B(a)$ and $G' \mid \Sigma, B(t) \vdash C$, respectively. By repeatedly applying Lemma 1 one can replace the previous proof by the following one:

$$\dfrac{\Gamma \vdash B(t) \qquad \Sigma, B(t) \vdash C}{\Gamma, \Sigma \vdash C}\text{ (cut)}$$

that contains a cut whose complexity of the cut formula is smaller than that one of the cut in the previous proof. Then it can be eliminated by induction hypothesis.

Cases involving the quantifier $\forall$ can be treated similarly.

As an easy consequence of cut-elimination we obtain the decidabilty of the fragments of $\mathbf{C}$, $\mathbf{C}^*$, and $\mathbf{G}_\infty$, respectively, where quantifiers occur only strongly:

Corollary 4. *Let H consist of sequents $\Gamma_i \vdash C_i$, where the formulas of Γ_i are purely existential and C_i is purely universal. The derivability of H in $\mathbf{HC}$, $\mathbf{HC}^*$, as well as $\mathbf{GLC}$ is decidable.*

References

1. A. Avron. Hypersequents, logical consequence and intermediate logics for concurrency. *Annals for Mathematics and Artificial Intelligence*, 4(199):225–248, 1991.

2. A. Avron. The method of hypersequents in the proof theory of propositional nonclassical logics. In *Logic: from Foundations to Applications, European Logic Colloquium*, pages 1–32. Oxford Science Publications. Clarendon Press. Oxford, 1996.

3. M. Baaz, A. Ciabattoni, C. Fermüller, and H. Veith. Proof theory of fuzzy logics: Urquhart's C and related logics. In *Mathematical Foundations of Computer Science 1998, 23rd International Symposium, MFCS'98, Brno, Czech Republic, August 24-28, 1998, Proceedings*, volume 1450 of *LNCS*, pages 203–212. Springer-Verlag, 1998.

4. A. Ciabattoni. Bounded contraction in systems with linearity. In N. Murray, editor, *Automated Reasoning with Analytic Tableaux and Related Methods, International Conference, TABLEAUX'99, Saratoga Springs*, volume 1617 of *LNAI*, pages 113–128. Springer-Verlag, 1999. To appear.

5. M. Dummett. A propositional calculus with a denumerable matrix. *Journal of Symbolic Logic*, 24:96–107, 1959.

6. J.Y. Girard. Linear logic. *Theoretical Computer Science*, 50:1–101, 1987.

7. K. Gödel. Zum intuitionistischen Aussagenkalkül. *Anzeiger der Akademie der Wissenschaften in Wien*, 69:65–66, 1932.

8. P. Hájek. *Metamathematics of Fuzzy Logic*. Kluwer, 1998.

9. P. Hájek, L. Godo, and F. Esteva. A complete many-valued logic with product-conjunction. *Archive for Math. Logic*, 35:191–208, 1996.

10. E. Kiriyama and H. Ono. The contraction rule in decision problems for logics without structural rules. *Studia Logica*, 50/2:299–319, 1991.

11. Y. Komori. Predicate logics without structural rules. *Studia Logica*, 45:393–404, 1985.

12. J. Łukasiewicz. Philosophische Bemerkungen zu mehrwertigen Systemen des Aussagenkalküls. *Comptes rendus des séances de la Société des Sciences et des Lettres de Varsovie Cl. III*, 23:51–77, 1930.

13. J.M. Mendez and F. Salto. Urquhart's C with intuitionistic negation: Dummett's LC without the contraction axiom. *Notre Dame Journal of Formal Logic*, 36/3:407–413, 1995.

14. H. Ono and Y. Komori. Logics without the contraction rule. *Journal of Symbolic Logic*, 50/1:169–201, 1985.

15. G. Takeuti. *Proof Theory*. North-Holland, Amsterdam, 2nd edition, 1987.

16. A. Urquhart. Many-valued logic. In D. Gabbay and F. Guenthner, editors, *Handbook of Philosophical Logic*, volume III. Reidel, Dordrecht, 1984.

17. A. Urquhart. Basic Many-valued logic. Manuscript, private communication, 1999.

How to Compute with DNA*

Lila Kari[1], Mark Daley[1], Greg Gloor[2], Rani Siromoney[3], and
Laura F. Landweber[4]

[1] Dept. of Computer Sci., Univ. of Western Ontario, London, ON N6A 5B7 Canada
lila@csd.uwo.ca, www.csd.uwo.ca/~lila
daley@csd.uwo.ca, www.csd.uwo.ca/~daley
[2] Dept. of Biochemistry, Univ. of Western Ontario, London ON N6A 5C1 Canada
ggloor@julian.uwo.ca, www.biochem.uwo.ca/fac/~gloor
[3] Dept. of Computer Sci., Madras Christian College, Madras 600 059 India
ranisiro@satyam.net.in
[4] Dept. of Ecology & Evolutionary Biology, Princeton Univ., NJ 08544-1003 USA
lfl@princeton.edu, www.princeton.edu/~lfl

Abstract. This paper addresses two main aspects of DNA computing research: DNA computing *in vitro* and *in vivo*. We first present a model of DNA computation developed in [5]: the circular insertion/deletion system. We review the result obtained in [5] stating that this system has the computational power of a Turing machine, and present the outcome of a molecular biology laboratory experiment from [5] that implements a small instance of such a system. This shows that rewriting systems of the circular insertion/deletion type are viable alternatives in DNA computation *in vitro*. In the second half of the paper we address DNA computing *in vivo* by presenting a model proposed in [17] and developed in [18] for the homologous recombinations that take place during gene rearrangement in ciliates. Such a model has universal computational power which indicates that, in principle, some unicellular organisms may have the capacity to perform any computation carried out by an electronic computer.

1 Introduction

Electronic computers are only the latest in a long chain of man's attempts to use the best technology available for doing computations. While it is true that their appearance, some 50 years ago, has revolutionized computing, computing does not start with electronic computers, and there is no reason why it should end with them. Indeed, even electronic computers have their limitations: there is only so much data they can store and their speed thresholds determined by physical laws will soon be reached. The latest attempt to break down these barriers is to replace, once more, the tools for doing computations: instead of electrical use biological ones.[13]

Research in this area was started by Leonard Adleman in 1994, [1], when he surprised the scientific community by using the tools of molecular biology to solve a hard computational problem. Adleman's experiment, solving an instance

---

C. Pandu Rangan, V. Raman, R. Ramanujam (Eds.): FSTTCS'99, LNCS 1738, pp. 269–282, 1999.

of the Directed Hamiltonian Path Problem solely by manipulating DNA strands, marked the first instance of a mathematical problem being solved by biological means. The experiment provoked an avalanche of computer science/molecular biology/biochemistry/physics research, while generating at the same time a multitude of open problems.[13]

The excitement DNA computing incited was mainly caused by its capability of massively parallel searches. This, in turn, showed its potential to yield tremendous advantages from the point of view of *speed, energy consumption* and *density of stored information*. For example, in Adleman's model, [2], the number of operations per second was up to 1.2×10^{18}. This is approximately 1,200,000 times faster than the fastest supercomputer. While existing supercomputers execute 10^9 operations per Joule, the energy efficiency of a DNA computer could be 2×10^{19} operations per Joule, that is, a DNA computer could be about 10^{10} times more energy efficient (see [1]). Finally, according to [1], storing information in molecules of DNA could allow for an information density of approximately 1 bit per cubic nanometer, while existing storage media store information at a density of approximately 1 bit per 10^{12} nm^3. As estimated in [3], a single DNA memory could hold more words than all the computer memories ever made.[12]

A few more words, as to why we should prefer biomolecules to electricity for doing computation: the short answer is that it seems more natural to do so. We could look at the electronic technology as just a technology that was in the right place at the right time; indeed, electricity hardly seems a suitable and intuitive means for storing information and for computations. For these purposes, nature prefers instead a medium consisting of biomolecules: DNA has been used for millions of years as storage for genetic information, while the functioning of living organisms requires computations. Such considerations seem to indicate that using biomolecules for computations is a promising new avenue, and that DNA computers might soon coexist with electronic computers.[13]

The research in the field has, from the beginning, had both experimental and theoretical aspects; for an overview of the research on DNA computing see [12]. This paper addresses both aspects. After introducing the basic notions about DNA in Section 2, in Section 3 we present a model of DNA computation developed in [5]: the circular insertion/deletion system. We show that this system has the computational power of a Turing machine and also present the results of a molecular biology laboratory experiment that implements a small instance of such a system. This shows that rewriting systems of the circular insertion/deletion type are viable alternatives in DNA computation *in vitro*. Section 4 introduces DNA computing *in vivo* by presenting a model proposed in [17], [18] for the homologous recombinations that take place during gene rearrangement in ciliates. We prove that such a model has universal computational power which indicates that, in principle, some unicellular organisms may have the capacity to perform any computation carried out by an electronic computer.

2   What is DNA?

DNA (deoxyribonucleic acid) is found in every cellular organism as the storage medium for genetic information. It is composed of units called nucleotides, distinguished by the chemical group, or base, attached to them. The four bases are *adenine, guanine, cytosine* and *thymine*, abbreviated as A, G, C, and T. (The names of the bases are also commonly used to refer to the nucleotides that contain them.) Single nucleotides are linked together end–to–end to form DNA strands. A short single-stranded polynucleotide chain, usually less than 30 nucleotides long, is called an *oligonucleotide* (or, shortly, oligo). The DNA sequence has a *polarity*: a sequence of DNA is distinct from its reverse. The two distinct ends of a DNA sequence are known under the name of the 5′ end and the 3′ end, respectively. Taken as pairs, the nucleotides A and T and the nucleotides C and G are said to be *complementary*. Two complementary single-stranded DNA sequences with opposite polarity will join together to form a double helix in a process called *base-pairing* or *annealing*. The reverse process – a double helix coming apart to yield its two constituent single strands – is called *melting*.[12]

A single strand of DNA can be likened to a string consisting of a combination of four different symbols, A, G, C, T. Mathematically, this means we have at our disposal a 4-letter alphabet $X = \{A, G, C, T\}$ to encode information. As concerning the operations that can be performed on DNA strands, the existing models of DNA computation are based on various combinations of the following primitive *bio-operations*, [12]:

– *Synthesizing* a desired polynomial-length strand.

– *Mixing:* pour the contents of two test-tubes into a third.

– *Annealing (hybridization):* bond together two single-stranded complementary DNA sequences by cooling the solution.

– *Melting (denaturation):* break apart a double-stranded DNA into its single-stranded components by heating the solution.

– *Amplifying (copying):* make copies of DNA strands by using the Polymerase Chain Reaction (PCR).

– *Separating* the strands by size using a technique called gel electrophoresis.

– *Extracting* those strands that contain a given pattern as a substring by using affinity purification.

– *Cutting* DNA double-strands at specific sites by using commercially available restriction enzymes.

– *Ligating:* paste DNA strands with compatible sticky ends by using DNA ligases.

– *Substituting:* substitute, insert or delete DNA sequences by using PCR site-specific oligonucleotide mutagenesis.

– *Detecting and Reading* a DNA sequence from a solution.

The bio-operations listed above and possibly others, will then be used to write "programs" which receive a tube containing DNA strands as input and return as output a set of tubes. A computation consists of a sequence of tubes containing DNA strands.

For further details of molecular biology terminology, the reader is referred to [12], [16].

3   How to Compute with DNA: Circular Insertions and Deletions

One of the aspects of the theoretical side of the DNA computing research comprises attempts to find a suitable model and to give it a mathematical foundation. This aspect is exemplified below by the *circular contextual insertion/deletion system*, [5] a formal language model of DNA computing. We mention the result obtained in [5] that the circular insertions/deletion systems are capable of universal computations. We also give the results of an experimental laboratory implementation of our model. This shows that rewriting systems of the circular insertion/deletion type are viable alternatives in DNA computation.

Insertions and deletions of small circular strands of DNA into/from long linear strands happen frequently in all types of cells and constitute also one of the methods used by some viri to infect a host. We describe here a generalization of insertions and deletions of words, [11], that aims to model these processes. (Note that circular DNA strings have been studied in the literature in the context of the splicing system model in [8], [9], [21], [24], [25].)[5]

In order to introduce our model, we first need some formal language definitions and notations. Throughout this paper, X represents an alphabet (a finite nonempty set), λ represents the empty word (the word containing 0 letters), $\bullet v$ represents a circular string v (a set containing every circular permutation of the linear string v). The length of a word v, denoted by $|v|$, is the number of occurrences of letters in v, counting repetitions. For a language L, by $\bullet L$ we denote the set of all words $\bullet v$ where $v \in L$. For further formal language definitions and notations the reader is referred to [22], [23].

In the style of [15], we define a *circular insertion/deletion system*, [5], as a tuple

$$ID^\bullet = (X, T, I^\bullet, D, \mathcal{A})$$

where X is an alphabet, $\mathrm{card}(X) \geq 2$, $T \subseteq X$ is the terminal alphabet, $I^\bullet \subseteq (X^*)^5$ is the finite set of circular insertion rules, $D \subseteq (X^*)^3$ is the finite set of deletion rules, and $\mathcal{A} \subseteq X^+$ is a linear strand called the *axiom*.

A circular insertion rule in $I^\bullet$ is written as $(c_1, g_1, \bullet x, g_2, c_2)_I$ where (c_1, c_2) represents the *context of the insertion*, $\bullet x$ is the string to be inserted and (g_1, g_2) are the *guides*, i.e. the location where $\bullet x$ is cut.

Given the rule above, the guided contextual circular insertion of the circular string $\bullet x$ into a linear string u is performed as follows. The circular word $\bullet x$ is linearized by cutting it between g_1 and g_2 (provided $g_1 g_2$ occurs as a subword in x) and reading it clockwise starting from g_1 and ending at g_2. The resulting linear strand is then inserted into the linear word u, between c_1 and c_2. If $c_1 c_2$ does not occur as a subword in u no insertion can take place. An example of circular insertion is illustrated in Figure 1.

A deletion rule in D is written as $(c_1, x, c_2)_D$ where (c_1, c_2) represents the context of deletion and x is the string to be deleted.

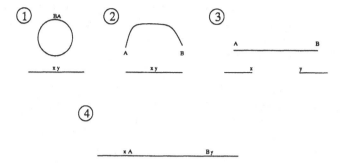

Fig. 1. Graphical representation of a circular insertion in the context (x, y), where the circular string is cut at the site (A, B). (From [5].)

Given the rule above, the linear contextual deletion of x from a linear word u accomplishes the excision of the linear strand x from u, provided x occurs in u flanked by c_1 on its left side and by c_2 on its right side.

If $u, v \in X^*$, we say that u derives v according to ID$^\bullet$ and we write $u \Rightarrow v$, [5], if v is obtained from u by either a guided contextual circular insertion or by a linear contextual deletion, i.e.,

– either $u = \alpha c_1 c_2 \beta$, $v = \alpha c_1 g_1 x' g_2 c_2 \beta$ and $I^\bullet$ contains the circular insertion rule $(c_1, g_1, \bullet x, g_2, c_2)_I$ where $g_1 x' g_2 \in \bullet x$, or

– $u = \alpha c_1 x c_2 \beta$, $v = \alpha c_1 c_2 \beta$ and D contains the linear deletion rule $(c_1, x, c_2)_D$.

A sequence of direct derivations

$$u_1 \Rightarrow u_2 \Rightarrow \ldots \Rightarrow u_k, k \geq 0$$

is denoted by $u_1 \Rightarrow^* u_k$ and u_k is said to be derived from u_1.

The language $L(ID^\bullet)$, [5], accepted by the circular insertion/deletion system ID$^\bullet$ is defined as

$$L(ID^\bullet) = \{v \in T^* \mid v \Rightarrow^* \mathcal{A}, \mathcal{A} \text{ is the axiom }\}$$

Recall that, [23], a rewriting system $(S, X \cup \{\#\}, F)$ is called a *Turing machine* iff the following conditions are satisfied.

(i) S and $X \cup \{\#\}$ (with $\# \notin X$ and $X \neq \emptyset$) are two disjoint alphabets referred to as the *state* and *tape* alphabet.

(ii) Elements $s_0 \in S$, $b \in X$, and a subset $S_f \subseteq S$ are specified, namely, the *initial state*, the *blank symbol*, and the *final state set*. A subset $V_f \subseteq X$ is specified as the *final alphabet*.

(iii) The productions in F are of the forms

(1) $s_i a \rightarrow s_j b$ overprint
(2) $s_i ac \rightarrow as_j c$ move right
(3) $s_i a\# \rightarrow as_j b\#$ move right and extend workspace
(4) $cs_i a \rightarrow s_j ca$ move left
(5) $\#s_i a \rightarrow \#s_j ba$ move left and extend workspace

where $s_i, s_j \in S$ and $a, b, c \in X$. Furthermore, for each $s_i, s_j \in S$ and $a \in X$, F either contains no productions (2) and (3) (resp. (4) and (5)) or else contains both (2) and (3) (respectively (4), (5)) for every $c \in X$. For no $s_i \in S$ and $a \in X$, the word $s_i a$ is a subword of the left side in two productions of the forms (1), (3) and (5).

We say that a word sw, where $s \in S$ and $w \in (X \cup \{\#\})^*$ is *final* iff w does not begin with a letter a such that sa is a subword of the left side of some production in F. The language *accepted* by a Turing machine TM is defined by

$$L(TM) = \{w \in V_f^* \mid \#s_0 w \# \Rightarrow^* \#w_1 s_f w_2 \# \text{ for some}$$

$$s_f \in S_f, w_1, w_2 \in X^* \text{ such that } s_f w_2 \# \text{ is final}\}$$

where $\Rightarrow$ denotes derivation according to the rewriting rules (1) – (5) of the Turing machine. A language is *acceptable* by a Turing machine iff $L = L(TM)$ for some TM. It is to be noted that TM is *deterministic*: at each step of the rewriting process, at most one production is applicable.

The following result proved in [5] shows that the circular insertion/deletion systems defined above have the computational power of a Turing machine.

Theorem 1. *If a language is acceptable by a Turing machine TM, then there exists a circular insertion/deletion system ID* accepting the same language.*

To test the empirical validity of our theoretical model, we implemented, [5], a small circular insertion/deletion system in the laboratory. The purpose of this implementation was to show that in vitro circular insertion is possible and not overwhelmingly difficult.

The following circular insertion/deletion system was chosen:

$$ID^\bullet = (X, T, I^\bullet, D, u)$$

where the alphabets are $T, X = \{A, C, G, T\}$, there are no deletion rules, i.e. $D = \emptyset$, the axiom u is a small DNA segment from the *Drosophila Melanogaster* genome and $I^\bullet = (G, G, \bullet v, TCGAC, TCGAC)$ where $\bullet v$ is a commercially available plasmid (circular strand). Note that A, C, G, T correspond to the four bases that occur in natural DNA, and that the sequence $G|TCGAC$ is the restriction (cut) site for the *Sal I* enzyme.[5]

To begin the experiment, we synthesized the linear axiom u in which we would then insert. This was accomplished by taking DNA from *Drosophila* (fruit fly) and performing PCR with the primers BC^+ and cd^-. The result was the amplification of a particular 682bp (basepair) linear sequence of DNA which became the axiom u of the circular insertion/deletion system. The 682bp linear strand was chosen to contain exactly one restriction site for the enzyme *Sal I*, corresponding to the context of insertion $(G, TCGAC)$. For the circular string $\bullet v$ to be inserted we chose pK18h, a commercially available plasmid having one restriction site for *Sal I*, corresponding to the guides $(G, TCGAC)$ in the insertion rule. [5]

After verifying that the PCR had worked correctly and we had indeed obtained the desired 682bp linear axiom u, we cut u with *Sal I*, cleaving it into two new linear strands denoted by L and R, i.e. $u = L|R$. The product was checked by gel electrophoresis to ensure the presence of bands corresponding to the sizes of L (188bp) and R (493bp), as seen from the first band in the gel of Figure 2. The plasmid $\bullet v$ was also cut and linearized in the same fashion resulting in the linear strand v. [5]

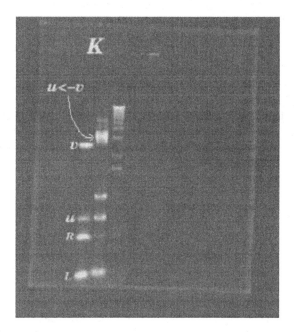

Fig. 2. The first vertical lane of this gel consists of bands corresponding to the unreacted linearized plasmid v, the linear strand u, and the two fragments of the cut linear strand (R, respectively L). The second vertical lane shows a band corresponding to the product obtained after reaction: the insertion of v into u, i.e., $u \leftarrow v$. The third lane contains a standard 1kb (kilobase) ladder used to measure the others. (From [5].)

At this point the linear strands L and R were combined with the linearized pK18h, i.e. v, and ligase was added to reconnect the strands of DNA. After allowing time for ligation, a gel was run to determine the products. The second band from the gel shown in Figure 2 indicates that in addition to the desired $L|plasmid|R$, we also obtain $R|R$, $L|R$, plasmid|plasmid and even plasmid | plasmid | plasmid. [5]

Note that the band corresponding to the approximate size of $L|\text{plasmid}|R$ can be seen as a smear. This could suggest the presence of $R|\text{plasmid}|R$ or of any other combination of two linear fragments and a plasmid which failed to separate clearly from one another due to the large size. Thus further analysis was required to ensure the presence of the desired product $L|\text{plasmid}|R$. [5]

In order to amplify the amount of DNA available at this point, the DNA was recircularized and introduced into *E. Coli* bacteria. (The complex details of this process are omitted here.)

Prior to sequencing, a restriction digest was performed on small amounts of product isolated from each of the several bacterial colonies. If the starting sample were a heterogeneous mixture of DNA molecules, each colony would yield a different product. Consequently, the restriction digest of DNA samples (each isolated from a particular colony) with enzymes *Sal I, Stu I* and *Xba* , resulted in bands indicating different size distributions. Of these, one sample corresponded to the size of $L|\text{plasmid}|R$ and the identity of the product was confirmed by sequencing. [5]

This experiment demonstrates that it is possible to insert a plasmid into a linear strand in vitro, implementing thus a circular insertion/deletion system. Future experimental work would ideally include a much larger system to test the scalability of this approach. [5]

4   How do Cells Compute?

The previous section presented one of the existing models of DNA computation and presented a toy experimental laboratory implementation. Despite the progress achieved in this direction of research, the main obstacles to creating an *in vitro DNA computer* still remain to be overcome. These obstacles are mainly practical, arising from difficulties in coping with the error rates of bio-operations, and with scaling up the existing systems. However, note that similar issues of actively adjusting the concentrations of reactions and fault detection and tolerance are all addressed by biological systems in nature: cells. This leads to another direction of research, *DNA computing in vivo*, which addresses the computational capabilities of cellular organisms.

Here we describe a model proposed in [17] and developed in [18] for the homologous recombinations that take place during gene rearrangement in ciliates and prove that such a model has the computational power of a Turing machine. This indicates that, in principle, these unicellular organisms may have the capacity to, perform at least any computation carried out by an electronic computer.

Ciliates are a diverse group of 8000 or more unicellular eukaryotes (nucleated cells) named for their wisp-like covering of cilia. They possess two types of nuclei: an active *macronucleus* (soma) and a functionally inert *micronucleus* (germline) which contributes only to sexual reproduction. The somatically active macronucleus forms from the germline micronucleus after sexual reproduction, during the course of development. The genomic copies of some protein-coding genes in the

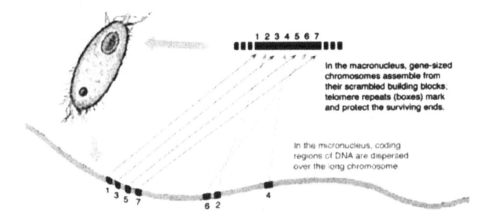

Fig. 3. Overview of gene unscrambling. Dispersed coding MDSs 1-7 reassemble during macronuclear development to form the functional gene copy (top), complete with telomere addition to mark and protect both ends of the gene. (From [17].)

micronucleus of hypotrichous ciliates are obscured by the presence of intervening non-protein-coding DNA sequence elements (internally eliminated sequences, or *IES*s). These must be removed before the assembly of a functional copy of the gene in the somatic macronucleus. Furthermore, the protein-coding DNA segments (macronuclear destined sequences, or *MDS*s) in species of *Oxytricha* and *Stylonychia* are sometimes present in a permuted order relative to their final position in the macronuclear copy. For example, in *O. nova*, the micronuclear copy of three genes (Actin I, α-telomere binding protein, and DNA polymerase α) must be reordered and intervening DNA sequences removed in order to construct functional macronuclear genes. Most impressively, the gene encoding DNA polymerase α (DNA pol α) in *O. trifallax* is apparently scrambled in 50 or more pieces in its germline nucleus [10]. Destined to unscramble its micronuclear genes by putting the pieces together again, *O. trifallax* routinely solves a potentially complicated computational problem when rewriting its genomic sequences to form the macronuclear copies. [18]

This process of unscrambling bears a remarkable resemblance to the DNA algorithm Adleman [1] used to solve a seven-city instance of the Directed Hamiltonian Path Problem. The developing ciliate macronuclear "computer" (Figure 3) apparently relies on the information contained in short direct repeat sequences to act as minimal guides in a series of homologous recombination events. These guide-sequences act as splints, and the process of recombination results in linking the protein-encoding segments (MDSs, or "cities") that belong next to each other in the final protein coding sequence. As such, the unscrambling of sequences that encode DNA polymerase α accomplishes an astounding feat of cellular computation. Other structural components of the ciliate chromatin presumably play a significant role, but the exact details of the mechanism are still unknown.[18]

In this section we define the notion of a *guided recombination system*, [18]), that models the process taking place during gene rearrangement, and prove that such systems have the computational power of a Turing machine, the most widely used theoretical model of electronic computers.

The following strand operations generalize the intra- and intermolecular recombinations defined in [17] and illustrated in Figure 4 by assuming that homologous recombination is influenced by the presence of certain contexts, i.e., either the presence of an IES or an MDS flanking a junction sequence. The observed dependence on the old macronuclear sequence for correct IES removal in *Paramecium* suggests that this is the case ([19]). This restriction captures the fact that the guide sequences do not contain all the information for accurate splicing during gene unscrambling. [18]

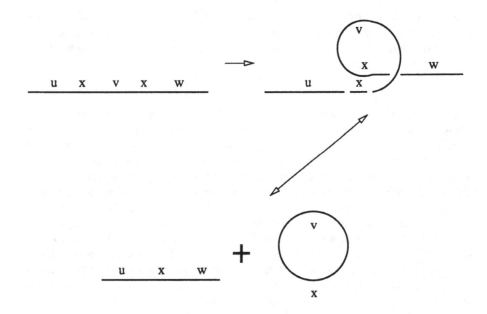

Fig. 4. Intra- and intermolecular recombinations using repeats x. During intramolecular recombination, after x finds its second occurrence in $uxvxw$, the molecule undergoes a strand exchange in x that leads to the formation of two new molecules: a linear DNA molecule uxw and a circular one $\bullet vx$. The reverse operation is intermolecular recombination. (From [14].)

Using an approach developed in [15] we use contexts to restrict the use of recombinations. A *splicing scheme*, [7], [8] is a pair $(X, \sim)$ where X is the alphabet and $\sim$, the pairing relation of the scheme, is a binary relation between triplets of nonempty words satisfying the following condition: If $(p, x, q) \sim (p', y, q')$ then $x = y$. In the splicing scheme $(X, \sim)$ pairs $(p, x, q) \sim (p', x, q')$ now define the contexts necessary for a recombination between the repeats x. Then we define *contextual intramolecular recombination*, [18] as

$$\{uxwxv\} \Rightarrow \{uxv, \bullet wx\}, \text{ where } u = u'p, w = qw' = w''p', v = q'v'.$$

This constrains intramolecular recombination within $uxwxv$ to occur only if the restrictions of the splicing scheme concerning x are fulfilled, i.e., the first occurrence of x is preceded by p and followed by q and its second occurrence is preceded by p' and followed by q'. (See Figure 4.)

Similarly, if $(p, x, q) \sim (p', x, q')$, then we define *contextual intermolecular recombination*, [18]), as

$$\{uxv, \bullet wx\} \Rightarrow \{uxwxv\} \text{ where } u = u'p, v = qv', w = w'p' = q'w''.$$

Informally, intermolecular recombination between the linear strand uxv and the circular strand $\bullet wx$ may take place only if the occurrence of x in the linear strand is flanked by p and q and its occurrence in the circular strand is flanked by p' and q'. Note that sequences p, x, q, p', q' are nonempty, and that both contextual intra- and intermolecular recombinations are reversible by introducing pairs $(p, x, q') \sim (p', x, q)$ in $\sim$. (See Figure 4.)

The above operations resemble the "splicing operation" introduced by Head in [7] and "circular splicing" ([8], [24], [21]). [20], [4] and subsequently [25] showed that these models have the computational power of a universal Turing machine. (See [22] for a review.)

The operations defined in [17] are particular cases of guided recombination, where all the contexts are empty, i.e., $(\lambda, x, \lambda) \sim (\lambda, x, \lambda)$ for all $x \in X^+$. This corresponds to the case where recombination may occur between every repeat sequence, regardless of the contexts. These unguided (context-free) recombinations are computationally not very powerful: we can prove that they can only generate regular languages. [18]

If we use the classical notion of a set, we can assume that the strings entering a recombination are available for multiple operations. Similarly, there would be no restriction on the number of copies of each strand produced by recombination. However, we can also assume some strings are only available in a limited number of copies. Mathematically this translates into using *multisets*, where one keeps track of the number of copies of a string at each moment. In the style of [6], if $\mathbf{N}$ is the set of natural numbers, a multiset of X^* is a mapping $M : X^* \longrightarrow \mathbf{N} \cup \{\infty\}$, where, for a word $w \in X^*$, $M(w)$ represents the number of occurrences of w. Here, $M(w) = \infty$ means that there are unboundedly many copies of the string w. The set $\text{supp}(M) = \{w \in X^* | M(w) \neq 0\}$, the *support of* M, consists of the strings that are present at least once in the multiset M. [12]

We now define a *guided recombination system* that captures the series of dispersed homologous recombination events that take place during scrambled gene rearrangements in ciliates.

Definition. ([18]) *A guided recombination system is a triple* $R = (X, \sim, \mathcal{A})$ *where* $(X, \sim)$ *is a splicing scheme, and* $\mathcal{A} \in X^+$ *is a linear string called the axiom.*

A guided recombination system R defines a *derivation relation*, [18], that produces a new multiset from a given multiset of linear and circular strands, as follows. Starting from a "collection" (multiset) of strings with a certain number of available copies of each string, the next multiset is *derived from* the first one by an intra- or inter-molecular recombination between existing strings. The strands participating in the recombination are "consumed" (their multiplicity decreases by 1) whereas the products of the recombination are added to the multiset (their multiplicity increases by 1).

For two multisets S and S' in $X^* \cup X^\bullet$, we say that S derives S' and we write $S \Rightarrow_R S'$, iff one of the following two cases hold, [18]:

(1) there exist $\alpha \in \mathrm{supp}(S)$, $\beta, \bullet\gamma \in \mathrm{supp}(S')$ such that
  – $\{\alpha\} \Rightarrow \{\beta, \bullet\gamma\}$ according to an intramolecular recombination step in R,
  – $S'(\alpha) = S(\alpha) - 1$, $S'(\beta) = S(\beta) + 1$, $S'(\bullet\gamma) = S(\bullet\gamma) + 1$;

(2) there exist $\alpha', \bullet\beta' \in \mathrm{supp}(S)$, $\gamma' \in \mathrm{supp}(S')$ such that
  – $\{\alpha', \bullet\beta'\} \Rightarrow \{\gamma'\}$ according to an intermolecular recombination step in R,
  – $S'(\alpha') = S(\alpha') - 1$, $S'(\bullet\beta') = S(\bullet\beta') - 1$, $S'(\gamma') = S(\gamma') + 1$.

Those strands which, by repeated recombinations with initial and intermediate strands eventually produce the axiom, form the language *accepted* by the guided recombination system. Formally, [18],

$$L_a^k(R) = \{w \in X^* |\ \{w\} \Rightarrow_R^* S \text{ and } \mathcal{A} \in \mathrm{supp}(S)\},$$

where the the multiplicity of w equals k. Note that $L_a^k(R) \subseteq L_a^{k+1}(R)$ for any $k \geq 1$.

Theorem.([18])*Let L be a language over T^* accepted by a Turing machine $TM = (S, X \cup \{\#\}, P)$ as above. Then there exist an alphabet X', a sequence $\pi \in X'^*$, depending on L, and a recombination system R such that a word w over T^* is in L if and only if $\#^6 s_0 w \#^6 \pi$ belongs to $L_a^k(R)$ for some $k \geq 1$.*

The preceding theorem implies that if a word $w \in T^*$ is in $L(TM)$, then $\#^6 s_0 w \#^6 \pi$ belongs to $L_a^k(R)$ for some k and therefore it belongs to $L_a^i(R)$ for any $i \geq k$. This means that, in order to simulate a computation of the Turing machine on w, any sufficiently large number of copies of the initial strand will do. The assumption that sufficiently many copies of the input strand are present at the beginning of the computation is in accordance with the fact that there are multiple copies of each strand available during the (polytene chromosome) stage where unscrambling occurs. Note that the preceding result is valid even if we allow interactions between circular strands or within a circular strand, particular cases of which have been formally defined in [17].[18]

The proof that a guided recombination system can simulate the computation of a Turing machine suggests that the micronuclear gene, present in multiple copies, consists of a sequence encoding the input data, combined with a sequence encoding a program, i.e., a list of encoded computation instructions. The "computation instructions" can be excised from the micronuclear gene and become circular "rules" that can recombine with the data. The process continues then by multiple intermolecular recombination steps involving the linear strand and circular "rules", as well as intramolecular recombinations within the linear strand itself. The resulting linear strand, which is the functional macronuclear copy of the gene, can then be viewed as the output of the computation performed on the input data following the computation instructions excised as circular strands. [18]

The last step, telomere addition and the excision of the strands between the telomere addition sites, can easily be added to our model as a final step consisting of the deletion of all the markers, rule delimiters and remaining rules from the output of the computation. This would result in a strand that contains only the output of the computation (macronuclear copy of the gene) flanked by end markers (telomere repeats). This also provides a new interpretation for some of the vast quantity of non-encoding DNA found in micronuclear genes.[14]

In conclusion, in this section we presented a model proposed in [17] for the process of gene unscrambling in hypotrichous ciliates. While the model is consistent with our limited knowledge of this biological process, it needs to be rigorously tested using molecular genetics. We have shown, however, that the model is capable of universal computation. This both hints at future avenues for exploring biological computation and opens our eyes to the range of complex behaviors that may be possible in ciliates, and potentially available to other evolving genetic systems. [18]

References

1. L.Adleman. Molecular computation of solutions to combinatorial problems. *Science* v.266, Nov.1994, 1021–1024.
2. L.Adleman. On constructing a molecular computer. *1st DIMACS workshop on DNA based computers*, Princeton, 1995. In *DIMACS series*, vol.27 (1996), 1–21.
3. E.Baum. Building an associative memory vastly larger than the brain. *Science*, vol.268, April 1995, 583–585.
4. E. Csuhaj-Varju, R.Freund, L.Kari, and G. Păun. DNA computing based on splicing: universality results. In Hunter, L. and T. Klein (editors). *Proceedings of 1st Pacific Symposium on Biocomputing*. World Scientific Publ., Singapore, 1996, 179-190.
5. M.Daley, L.Kari, G.Gloor, R.Siromoney. Circular contextual insertions/deletions with applications to biomolecular computation. Proceedings of *String Processing and Information REtrieval '99*, Mexico, IEEE CS Press, 1999, in press.
6. S.Eilenberg. *Automata, Languages and Machines*. Academic Press, New York, 1984.
7. T.Head. Formal language theory and DNA: an analysis of the generative capacity of recombinant behaviors. *Bulletin of Mathematical Biology*, 49(1987), 737–759.
8. T. Head. Splicing schemes and DNA. *Lindenmayer systems*, G.Rozenberg and A. Salomaa eds., Springer Verlag, Berlin, 1991, 371–383.

9. T.Head, G.Păun, D.Pixton. Language theory and genetics. Generative mechanisms suggested by DNA recombination. In *Handbook of Formal Languages* (G.Rozenberg, A.Salomaa eds.), Springer Verlag, 1996.

10. D.C.Hoffman, and D.M. Prescott. Evolution of internal eliminated segments and scrambling in the micronuclear gene encoding DNA polymerase α in two *Oxytricha* species. *Nucl. Acids Res.* 25(1997), 1883-1889.

11. L.Kari. On insertions and deletions in formal languages. *Ph.D. thesis*, University of Turku, Finland, 1991.

12. L.Kari. DNA computing: arrival of biological mathematics. *The Mathematical Intelligencer*, vol.19, nr.2, Spring 1997, 9–22.

13. L.Kari. From Micro-Soft to Bio-Soft: Computing with DNA. Proceedings of BCEC'97 (Bio-Computing and Emergent Computation) Skovde, Sweden, World Scientific Publishing Co., 146–164.

14. L.Kari, L.F.Landweber. Computational power of gene rearrangement. Proceedings of *DNA Based Computers V*, E.Winfree, D.Gifford eds., MIT, Boston, June 1999, 203-213.

15. L.Kari, G.Thierrin. Contextual insertions/deletions and computability. *Information and Computation*, 131, no.1 (1996), 47–61.

16. J.Kendrew et al., eds. *The Encyclopedia of Molecular Biology*, Blackwell Science, Oxford, 1994.

17. L.F.Landweber, L.Kari. The evolution of cellular computing: nature's solution to a computational problem. Proceedings of 4th DIMACS meeting on DNA based computers, Philadephia, 1998, 3-15.

18. L.F.Landweber, L.Kari. Universal molecular computation in ciliates. In *Evolution as Computation*, L.F.Landweber, E,Winfree, Eds., Springer Verlag, 1999.

19. E.Meyer,and S.Duharcourt. Epigenetic Programming of Developmental Genome Rearrangements in Ciliates. *Cell* (1996) 87, 9-12.

20. G.Păun. On the power of the splicing operation. *Int. J. Comp. Math* 59(1995), 27-35.

21. D.Pixton. Linear and circular splicing systems. Proceedings of the *First International Symposium on Intelligence in Neural and Biological Systems* , IEEE Computer Society Press, Los Alamos, 1995, 181–188.

22. G.Rozenberg, and A.Salomaa eds. *Handbook of Formal Languages*, Springer Verlag, Berlin, 1997.

23. A.Salomaa. *Formal Languages*. Academic Press, New York, 1973.

24. R. Siromoney, K.G. Subramanian and Dare Rajkumar, Circular DNA and splicing systems. In *Parallel Image Analysis. Lecture Notes in Computer Science* 654, Springer Verlag, Berlin, 1992, 260–273.

25. T. Yokomori, S. Kobayashi and C. Ferretti. Circular splicing systems and DNA computability Proc. of *IEEE International Conference on Evolutionary Computation'97*, 1997, 219–224.

A High Girth Graph Construction and a Lower Bound for Hitting Set Size for Combinatorial Rectangles*

L. Sunil Chandran

Indian Institute of Science, Bangalore
sunil@csa.iisc.ernet.in

Abstract. We give the following two results.
First, we give a deterministic algorithm which constructs a graph of girth $\log_k(n) + O(1)$ and minimum degree $k - 1$, taking number of nodes n and the number of edges $e = \lfloor nk/2 \rfloor$ as input. The graphs constructed by our algorithm are expanders of sub-linear sized subsets, that is subsets of size at most n^δ, where $\delta < \frac{1}{4}$. Though methods which construct high girth graphs are known, the proof of our construction uses only very simple counting arguments in comparison. Also our algorithm works for all values of n or k.
We also give a lower bound of $m/8\epsilon$ for the size of hitting sets for combinatorial rectangles of volume ϵ. This result is an improvement of the previously known lower bound, namely $\Omega(m + 1/\epsilon + \log(d))$. The known upper bound for the size of the hitting set is $m \, poly(\log(d)/\epsilon)$. [LLSZ].

1 Introduction

Expander graphs are those in which the neighbourhood of any not too large subset of vertices has large cardinality. These graphs have been used in several applications, e.g., parallel sorting networks[AKS], coding[SS], hitting combinatorial rectangles [LLSZ]. Explicit construction of regular expander graphs was a hard problem. Some of the explicit constructions are given in [Marg73,Marg88,GabGal] [LPS] etc.

Our original aim in this paper was to see if simple rules like successively adding edges to an initially empty graph with each edge connecting vertices at a large distance in the current graph, could give provable expansion properties. The main result we obtain is that simple rules of the above form do indeed yield graphs which have large expansion provably, but for *sublinear-sized* subsets. In fact, the graphs we obtain will actually be high *girth* graphs, where girth is defined to be the length of the shortest cycle. Expansion on sublinear-sized subsets will follow from a result of Kahale's[Kah].

---

* Dept. of Computer Science and Automation, Indian Institute of Science, Bangalore, 560012. This work is based on the author's masters thesis.

C. Pandu Rangan, V. Raman, R. Ramanujam (Eds.): FSTTCS'99, LNCS 1738, pp. 283–290, 1999.

The Main Result.

We give an algorithm which takes a positive integer n and an "expected degree" k as input and creates a graph $G = (V, E)$ with n nodes, $\lfloor nk/2 \rfloor$ edges and whose girth g satisfies the relation

$$\frac{n}{2} < 1 + \frac{(k+1)(k^{g-1} - 1)}{(k-1)}$$

It follows that $g > \log_k(n) + O(1)$. We prove that the degree of any node in the graph constructed by our algorithm will be $k-1, k$, or $k+1$. Thus given the values of girth (say g) and minimum degree (say t) our algorithm can be used to construct graphs with at most $2 + \frac{2(t+2)\left((t+1)^{g-1} - 1\right)}{t}$ nodes. Note that this bound is comparable to the existential results of Erdos and Sach[ES] and Sauer[S], who showed that the minimum number of vertices $n(g, t)$ required for the girth to be $\geq g$ and minimum degree $\geq t$ satisfies:

$$n(g, t) \leq 2 \frac{(t-1)^{g-1} - 1}{t-2}, if \ g \ is \ odd$$

$$n(g, t) \leq 4 \frac{(t-1)^{g-2} - 1}{t-2}, if \ g \ is \ even$$

(See also [Boll] page 107).

That our high-girth graphs have good expansion on sublinear-sized subsets is shown as follows. Let X be a subset of V. $N(X) = \{v \in V : \exists u \in X, \{u, v\} \in E\}$, is defined to be the set of neighbours of X. The expansion of X is defined to be $\frac{|N(X)|}{|X|}$. In [Kah], Kahale proves that if G is a k regular graph of girth $(\frac{4}{3} + o(1)) \log_{k-1} n$ then for any subset X of V of size at most n^δ (where $0 < \delta < 1$),

$$\frac{|N(X)|}{|X|} \geq \left[k - (k-1)^{(3+o(1))\delta} \right]$$

It is implicit in the proof of this theorem that if we use a graph of girth $\log_k(n) + O(1)$ and minimum degree $= k - 1$, then every subset X of size at most n^δ will have expansion $\beta = \frac{|N(X)|}{|X|}$ such that the following inequality is satisfied.

$$n^\delta > (k - 1 - \lceil \beta \rceil)^{\frac{1}{4}(\log_k(n) + O(1))}$$

Solving this inequality we get $\beta > (k-2) - k^{4\delta}$. Thus the graph constructed by our algorithm is such that each subset X, $|X| \leq n^\delta$, has expansion $> (k-2) - k^{4\delta}$. When $\delta < \frac{1}{4}$, the effect of the second term becomes negligible for sufficiently large k.

Other Known High-Girth Graph Constructions.

Two other constructions which give values of girth greater than what we achieve are Ramanujan graphs by Lubotzky, Philip and Sarnak [LPS] and cubic sextet graphs [Weiss,BH]. For example, the girth of the graphs constructed in [LPS] is $\geq (\frac{4}{3} + o(1)) \log_{k-1} n$. Our achievement is that the proof of our algorithm uses only elementary counting arguments. Also our construction works for all n, and all $k < n$. In comparison, Ramanujan graph construction takes two unequal primes p and q, both congruent to 1 mod 4, and constructs $p + 1$-regular graphs with either t or $t/2$ vertices where $t = q(q^2 - 1)$. But these graphs are regular while ours is only approximately regular. The construction of cubic-sextet graphs given by [Weiss,BH] gives high girth graphs which are 3-regular, but doesn't generalize to arbitrary specified minimum degree. We have presented our algorithm assuming that n is even. Later we describe how to generalize it to the case of odd n also.

An Auxiliary Result: Lower Bounding Hitting Set Size.

As mentioned above, one of the applications of Expander Graphs is in the construction of small hitting sets for combinatorial rectangles, defined as follows.

Let d and m be positive integers. Let U be $\{1, 2, \cdots, m\}^d$. That is the universe U consists of all d dimensional lattice points with all coordinates in $[m] = \{1, 2, \cdots, m\}$. A combinatorial rectangle is defined as a set of the form $R = R_1 \times R_2 \times R_3 \times \cdots \times R_d$, where $R_i \subseteq [m]$ for all $i \in \{1, 2, \cdots, d\}$. The volume of R is defined as $\frac{|R|}{|U|} = \frac{\prod_{i \in [d]} |R_i|}{m^d}$.

An (m, d, ϵ) hitting set is defined as a subset $S \subseteq U$, such that for any combinatorial rectangle R, with volume $vol(R) \geq \epsilon$, $S \cap R \neq \emptyset$.

It is proved in [LLSZ] that, if S_d is an (m, d, ϵ) hitting set, $|S_d| = \Omega(m + 1/\epsilon + \log d)$ when $m^{-d} \leq \epsilon \leq 2/9$. The value of ϵ should be in this range in order to prove the $\log d$ term. In this paper, we strengthen this lower bound by observing that a suitable graph associated with any hitting set must have expansion properties. Using this observation, we prove a lower bound of $m/8\epsilon$, when $1/m^{d-1} \leq 4\epsilon \leq 1$ and $d \geq 2$.

2   High-Girth Graph Construction

The following algorithm takes the number of nodes n and the average degree k as input and constructs a graph of girth at least $\log_k(n) + O(1)$. All the nodes in the graph will have degree $k - 1, k$ or $k + 1$.

The Algorithm.

Let n be an even integer. (This is just for convenience. We will describe the case when n is odd shortly). Assume that in the beginning, we have a perfect matching on the n nodes. That is we start with a graph having $\frac{n}{2}$ edges, the degree of each node being 1. Do for $i = \frac{n}{2} + 1$ to $\frac{kn}{2}$:

1. Let $S = \{u \in V : degree(u) \leq degree(v), \forall v \in V\}$.
2. Let $T = \{(u, v) \in S \times V : distance(u, v) \geq distance(x, y), \forall (x, y) \in S \times V\}$
3. If there is a pair $(u, v) \in T$, such that $degree(v) \leq j$, where $j = \lceil \frac{2i}{n} \rceil$, and the edge $\{u, v\}$ is already not there in the graph, introduce a new edge $\{u, v\}$ and go to start of the loop. If there are several such pairs pick one arbitrarily. Else go to 4.

4. Let $\rho = distance(u, v), (u, v) \in T$. Put $\rho = \rho - 1$. Now assign $T = \{(u, v) \in S \times V : distance(u, v) = \rho\}$. Go to 3.

(The above algorithm does the following. In step 1, it collects in S all the vertices having the least degree in the graph. Next it collects in T, all the pairs of nodes (u, v) such that $distance(u, v)$ is maximum from the set of all pairs (u, v) with u in S. Put an edge (if it is not already there) between one such pair from T, making sure that both u and v have degree less than or equal to $j = \lceil \frac{2i}{n} \rceil$. If no (u, v) satisfies this degree requirement let T be redefined as the set of pairs (u, v), such that u is in S and $distance(u, v) = \rho - 1$, where $\rho = $ distance between any pair (u, v) which was in T earlier).

Lemma 1. *The graph created by the above algorithm will be such that $\forall v \in V$, degree(v) will be $k - 1, k, $ or $k + 1$. If V_d denotes the set of vertices with degree d in the graph, $|V_{k-1}| = |V_{k+1}| \leq \frac{n}{2}$.*

Proof. Let us use induction, as each batch of $\frac{n}{2}$ edges are introduced. That is as the average degree of the graph goes up by 1. (Note that the parameter $j = \lceil \frac{2i}{n} \rceil$ remains constant as a batch of $\frac{n}{2}$ edges are introduced and when the next batch starts, it gets incremented by 1).

Consider the following induction hypothesis.

When $i = \frac{dn}{2}$ iterations of the loop are over, the degree of any node in the graph will be $d - 1, d$ or $d + 1$. Let X be the set of vertices with degree $d - 1$, Y that of vertices with degree d and Z that of vertices with degree $d + 1$. Then $|X| = |Z|$.

In the beginning of the algorithm, average degree $= 1$ and the statement is true because there are only vertices of degree 1 in the graph. We have $|X| = |Z| = 0$.

Now assuming the induction hypothesis after d batches of $\frac{n}{2}$ edges are introduced, let us prove that it is true even after the next batch is introduced. We note that since $|X| = |Z|$, $|X| \leq \frac{n}{2}$. Thus when $\frac{n}{2}$ edges are introduced, each vertex in the set X gets a chance to increase its degree (because minimum degree vertices have preference). So after $\frac{n}{2}$ edges are introduced there will be no vertices of degree $d - 1$ left. Further no vertex will have degree $> (d + 2)$, because while these edges are introduced the parameter j will remain equal to $d + 1$. Observe that since the sum of degrees during this stage can reach a maximum of $(d + 1)n$ only, there is always at least 2 nodes of degree less than $d + 2$ and therefore the algorithm will add an edge on each iteration. Thus after the new $\frac{n}{2}$ edges are introduced we retain the induction hypothesis statement that the degrees are $d, d + 1$ or $d + 2$. Now let X_1, Y_1, Z_1 be the new sets of $d, d + 1$ and $d + 2$ degree vertices respectively. It suffices to show that $X_1 = Z_1$. We know that

$$|X_1|d + |Y_1|(d + 1) + |Z_1|(d + 2) = (d + 1)n \qquad (1)$$

We also have $|X_1| + |Y_1| + |Z_1| = n$
Therefore, $|Z_1| = |X_1|$. Thus we get back the induction hypotheses. Hence the result.

Construction for Odd n.

If n is odd, one may start with a n-length cycle C^n, instead of a matching. Then at the beginning of the loop $i = n+1$. Again, when we start the algorithm, every node has a degree of 2, which is equal to the average degree. Now think of introducing batches of $\lfloor \frac{n}{2} \rfloor$ and $\lceil \frac{n}{2} \rceil$ edges alternately. That is each odd numbered batch will contain $\lfloor \frac{n}{2} \rfloor$ edges and every even numbered batch will contain $\lceil \frac{n}{2} \rceil$ edges. We can consider a slightly different induction hypothesis. If k is even, that is after we have introduced $\frac{nk}{2}$ edges every node in the graph will be of degree $k-1, k$ or $k+1$. Moreover $V_{k+1} = V_{k-1}$ implying that $|V_{k+1}| = |V_{k-1}| \leq \frac{n}{2}$. If k is odd, that is after introducing $\lfloor \frac{kn}{2} \rfloor$ edges, every node in the graph will be of degree $k-1, k$ or $k+1$ as before. But $|V_{k+1}| = |V_{k-1}| - 1$. We still have $|V_{k+1}| \leq \frac{n}{2}$. The induction hypothesis also implies that just after each odd numbered batch of edges is introduced $|V_{k-1}| \leq \lceil \frac{n}{2} \rceil$ and just after every even numbered batch is introduced $|V_{k-1}| \leq \lfloor \frac{n}{2} \rfloor$. Thus when a new batch of edges is introduced, each node in the set V_{k-1} will increase its degree as before. Also no node can attain a degree $> (k+2)$. Thus when $k+1$ is even, we get back the same equation as equation 1. When $k+1$ is odd, since we have only introduced $\lfloor \frac{n(k+1)}{2} \rfloor$ edges, the equation becomes

$$|V_k|k + |V_{k+1}|(k+1) + |V_{k+2}|(k+2) = (k+1)n - 1 \tag{2}$$

Solving this we get $|V_{k+2}| = |V_k| - 1$. Thus again $|V_{k+2}| \leq \frac{n}{2}$.

Theorem 1. *The above algorithm creates a graph which has a girth*

$$g > \log_k(n) + O(1)$$

Proof. Look at the final graph. Let the girth be g. This cycle (girdle) closed sometime back in the process. Go back to the stage, just before closing the smallest cycle. Let $d = \lceil \frac{2i}{n} \rceil$, where i is the loop iteration number at that time. For that iteration, we had selected a current minimum degree vertex, u. That is we had selected a pair (u,v), with $u \in S$. Let $B = \{x \in V : distance(u,x) \geq g\}$. Why did not we select a vertex from B to be connected with u? Because those vertices, if any, had already achieved a degree of $(d+1)$. The algorithm prohibits to connect to them. But we know by lemma 1 that the number vertices of degree $d+1$ can at most be $\frac{n}{2}$. Thus $V - B$ contains at least $\frac{n}{2}$ nodes. Thus using the fact that $k+1 \geq d+1$, ($d+1$ being the maximum degree of the graph at that stage), the maximum number of nodes possible in $V - B$ is $1 + (k+1) + (k+1)k + \cdots + (k+1)k^{(g-2)}$. Combining the lower and upper bounds for $|V - B|$, we have,

$$\frac{n}{2} \leq 1 + (k+1) + (k+1)k + \cdots + (k+1)k^{(g-2)}$$

$$\frac{n}{2} < 1 + \frac{(k+1)(k^{g-1} - 1)}{(k-1)}$$

which gives

$$\log_k(n) + O(1) < g$$

3   A Lower Bound for Hitting Set Size

Let d and m be positive integers. U be $\{1, 2, \cdots, m\}^d$. That is the universe U consists of all d dimensional lattice points with all coordinates in $[m] = \{1, 2, \cdots, m\}$. A combinatorial rectangle is defined as a set of the form $R = R_1 \times R_2 \times R_3 \times \cdots \times R_d$, where $R_i \subseteq [m]$ for all $i \in \{1, 2, \cdots, d\}$. The volume of R is defined as $\frac{|R|}{|U|} = \frac{\prod_{i \in [d]} |R_i|}{m^d}$. We will call a combinatorial rectangle with volume ϵ in U a (m, d, ϵ) combinatorial rectangle.

An (m, d, ϵ) hitting set is defined as a subset $S_d \subseteq U$, such that for any combinatorial rectangle R, with volume $vol(R) \geq \epsilon$, $S_d \bigcap R \neq \emptyset$. In the following theorem we give a lower bound for the size of a (m, d, ϵ) hitting set.

Theorem 2.

$$|S_d| \geq m/8\epsilon$$

when $d \geq 2$ and $(1/m^{d-1}) \leq 4\epsilon \leq 1$.

Proof. Any (m, d, ϵ) hitting set S_d can be viewed as a bipartite graph between the sets $A = [m]^{d-1}$ and $B = [m]$. Introduce an edge from $(x_1, x_2, \cdots, x_{d-1}) \in A$ to $x_d \in B$ if and only if $(x_1, x_2, \cdots, x_d) \in S_d$. Now observe that the number of edges starting from any $(m, d-1, 2\epsilon)$ combinatorial rectangle R in A should be at least $\frac{m}{2}$. Otherwise there will be a subset C of B, $|C| \geq \frac{m}{2}$ such that $N(R) \bigcap C = \emptyset$, where $N(R)$ is the set of neighbours of R. Then $R \times C$ will be a combinatorial rectangle in $[m]^d$ with volume $\geq \epsilon$, which is not hit by S_d, which will be a contradiction. Thus if there are $1/4\epsilon$ non-intersecting combinatorial rectangles of volume 2ϵ in $[m]^{d-1}$ it is obvious from the above argument that $|S_d|$ should be at least $m/8\epsilon$. Instead of proving this directly, we choose to prove the result as follows.

It is easy to see that one can choose positive integers $T_1, T_2, \cdots, T_{d-1}$, all at most m, such that $4\epsilon m^{d-1} \geq T_1 T_2 \cdots T_{d-1} \geq 2\epsilon m^{d-1}$. Select subsets $R_1, R_2, \cdots, R_{d-1}$ of $[m]$ where R_i is chosen uniformly at random from among all subsets of size T_i. Then for any given edge in the bipartite graph, the probability of it starting from a point in $R = R_1 \times R_2 \times \cdots \times R_{d-1} \subset A$ will be $\frac{T_1 T_2 \cdots T_{d-1}}{m^{d-1}} \leq 4\epsilon$. Thus the expected number of edges starting from points in R will be at most $4\epsilon |S_d|$. As we have seen above this expectation should be at least $\frac{m}{2}$. So $|S_d| \geq m/8\epsilon$. This completes the proof.

4   Conclusion

The high girth graph construction was invented while we were trying to explicitly construct expander graphs. The problem of explicitly constructing an asymptotic family of expander graphs which guarantees expansion for linear sized subsets is considered to be a tough problem. When Margulis gave the first explicit construction it was considered as a breakthrough [Marg73]. Other known explicit constructions are given in [GabGal,LPS,Marg88] etc. It is known that graphs whose second smallest eigen value of their Laplacian matrix is well separated from 0 are good expanders [Alon]. Some limited experiments we carried out suggest that the construction we have given in fact construct expanders of linear sized sets, though we couldn't prove it. But we point out that high girth and high minimum degree alone cannot guarantee expansion for linear sized subsets. For example, let G be a high girth graph with n nodes. Let the girth of this graph be $\Omega(\log n)$. Suppose we are interested in the expansion of subsets of size $\frac{1}{q}$ or less of the total nodes. Then construct a graph $\widehat{G}$ of order $qn+1$ by connecting q copies of G to a central node as shown in figure 1. Note that girth of $\widehat{G}$ will be again $\Omega(\log(qn+1))$, but the required expansion is not available.

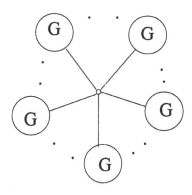

Fig. 1.

References

AKS.   M.Ajtai, J.Komloš and E.Szemeredi. Sorting in $c \log n$ parallel steps. *Combinatorica 3 (1983), 1-19.*

Alon.   N.Alon. Eigen values and expanders. *Combinatorica 6 (1986), no. 2, 83-96.*

BH.   N.L.Biggs and M.J.Hoare. The sextet construction for cubic graphs. *Combinatorica 3 (1983), 153-165.*

Boll.   B.Bollobas. Extremal Graph theory. *Academic press, London,1978.*

ES.   P.Erdos and H.Sachs. Regulare graphe gegebener taillenweite mit minimaler knotenzahl. *Math. Nat. 12 (1963), no. 3, 251-258.*

GabGal. O.Gabber and Z.Galil, Explicit constructions of linear sized super concentrators. *J. Comput. System Scie. 22 (1981), no.3,407-420.*

Kah.   N.Kahale. PhD Thesis. *Massachusetts Institute of Technology 1993.*

LLSZ.   N.Linial, M.Luby, M.Saks, D.Zuckerman. Efficient construction of a small hitting set for combinatorial rectangles in high dimension. *Combinatorica. 17(2) 1997, 215-234.*

LPS.   A.Lubotzky, R.Philip, and P.Sarnak. Ramanujan graphs. *Combinatorica 8 (1988) no. 3, 261-271.*

Marg73. G.A.Margulis. Explicit constructions of concentrators. *Problemy peredaci informacii 9 (1973) no.4,71-80.*

Marg88. G.A.Margulis. Explicit group theoretical constructions of combinatorial schemes and their application to the design of expanders and concentrators. *Problemy peredaci informacii 24 (1988) no.1,51-60.*

NZ.   N.Nisan and D.Zuckerman. More deterministic simulation in logspace *Proc. of the 25th ACM Symposium on the Theory of Computation 1993, pp 235-244.*

S.   N.Sauer. Extremaleigenschaften regularer Graphen gegebener Taillenweite I and II *Sitzungberichte Osterreich. Akad. Wiss. Math. Vatur. Kl. S-B II. 176 (1967) 9-25, ibid 176 (1967) 27-43.*

SS.   M. Sipser and D. Spielman. Expander Codes. *IEEE Transactions on Information Theory, 1996, Vol 42, No 6, 1710-1722.*

Weiss.   A.I.Weiss. Girth of bipartite sextet graphs *Combinatorica 4 (1984), 241-245.*

Protecting Facets in Layered Manufacturing*

Jörg Schwerdt[1], Michiel Smid[1], Ravi Janardan[2], Eric Johnson[2], and
Jayanth Majhi[3]

[1] Department of Computer Science, University of Magdeburg,
Magdeburg, Germany
{schwerdt,michiel}@isg.cs.uni-magdeburg.de
[2] Department of Computer Science and Engineering, University of Minnesota,
Minneapolis, MN 55455, USA
{janardan,johnson}@cs.umn.edu
[3] Mentor Graphics Corporation,
8005 S.W. Boeckman Road, Wilsonville, OR 97070, USA
jayanth_majhi@mentorg.com

1 Introduction

Layered Manufacturing (LM) is an emerging technology that is gaining importance in the manufacturing industry. (See e.g. the book by Jacobs [7].) This
technology makes it possible to rapidly build three-dimensional objects directly
from their computer representations on a desktop-sized machine connected to
a workstation. A specific process of LM, that is widely in use, is StereoLithography. The input to this process is the triangulated boundary of a polyhedral
CAD model. This model is first sliced by horizontal planes into layers. Then,
the object is built layer by layer in the following way. The StereoLithography
apparatus consists of a vat of photocurable liquid resin, a platform, and a laser.
Initially, the platform is below the surface of the resin at a depth equal to the
layer thickness. The laser traces out the contour of the first slice on the surface and then hatches the interior, which hardens to a depth equal to the layer
thickness. In this way, the first layer is created; it rests on the platform. Then,
the platform is lowered by the layer thickness and the just-vacated region is
re-coated with resin. The subsequent layers are then built in the same way.

It may happen that the current layer overhangs the previous one. Since this
leads to instabilities during the process, so-called *support structures* are generated to prop up the portions of the current layer that overhang the previous
layer. (See Figure 1 for an illustration in two dimensions.) These support structures are computed before the process starts. They are also sliced into layers,
and built simultaneously with the object. After the object has been built, the
supports are removed. Finally, the object is postprocessed in order to remove
residual traces of the supports.

---

* This work was funded in part by a joint research grant by DAAD and by NSF. RJ,
EJ and JM were also supported in part by NSF grant CCR-9712226. Part of this
work was done while JS and MS visited the University of Minnesota in Minneapolis.

C. Pandu Rangan, V. Raman, R. Ramanujam (Eds.): FSTTCS'99, LNCS 1738, pp. 291–303, 1999.

An important issue in this process is choosing an orientation of the model so that it can be built in the vertical direction. Equivalently, we can keep the model fixed, and choose a direction in which the model is built layer by layer. This direction is called the *build direction*. It affects the number of layers, the surface finish, the quantity of support structures used, and their location on the object being built.

1.1   Overview of Our Results

Let $\mathcal{P}$ be the three-dimensional polyhedron that we want to build using LM. $\mathcal{P}$ may have holes, but we assume that it is non-self-intersecting. In this paper, we consider the problem of computing all build directions for $\mathcal{P}$ for which a pre-scribed facet is *not* in contact with supports (i.e., protected from supports). This is an important problem because support removal from a facet can affect surface quality and accuracy adversely, thereby impacting the functional properties of critical facets, such as, for instance, facets on gear teeth. This problem, which we describe below, arose from discussions with engineers at Stratasys, Inc.—a Minnesota-based world leader in LM. To our knowledge, the work presented here constitutes the first provably correct, complete, and efficient solution to this important problem; current practice in industry is based on trial and error. Throughout, we assume that the facets of $\mathcal{P}$ are triangles. (This is the standard STL format used in industry.) The number of facets of $\mathcal{P}$ is denoted by n. We solve the following problems:

Problem 1. Given a facet F of $\mathcal{P}$, compute a description of all build directions for which F is not in contact with supports.

Problem 2. Compute a description of all build directions for which the total area of all facets of $\mathcal{P}$ that are not in contact with supports is maximum.

In Section 2, we give a formal definition of the notion of a facet being in contact with supports. In Section 3, we give an algorithm that solves Problem 1 in $O(n^2)$ time. This result also implies an $O(k^2n^2)$-time algorithm to protect any set of $k \geq 1$ facets of $\mathcal{P}$ from supports. We have implemented a simplified version of this algorithm. Test results on models obtained from industry are given in Section 3.3. In Section 3.4, we show that Problem 2 can be solved in $O(n^4)$ time. Section 4 concludes with some open problems.

The algorithms solving Problems 1 and 2 use fundamental concepts from computational geometry, such as convex hulls, arrangements of line segments, and the overlay of planar graphs. These concepts, however, are applied to points and segments on the unit sphere. A complete version of this paper appears as [14].

1.2   Prior Related Work

The problem of computing a "good" build direction has been considered in the literature. Asberg et al.[2] give efficient algorithms that decide if a model can be made by StereoLithography without using support structures. Allen and

Dutta [1] consider the problem of minimizing the total area of all parts of the model that are in contact with support structures. They give a heuristic for this problem, but without any analysis of the running time or the quality of the approximation. Bablani and Bagchi [3] present heuristics for improving the accuracy and finish of the part.

In our previous work [8,9,10,15], we have used techniques from computational geometry to compute build directions that optimize various design criteria. In [9], algorithms are given that minimize, for convex polyhedra, the volume of support structures used, and, independently, the total area of those parts of the model that are in contact with support structures. Both algorithms have a running time that is bounded by $O(n^2)$, where n is the number of facets. For general polyhedra, it is shown that a build direction that minimizes the so-called stair-step error can be computed in $O(n \log n)$ time. In [10], algorithms are given that minimize a combination of these measures. (For all measures that involve support structures, the algorithms only work for convex polyhedra.) In [8], algorithms are given that minimize support structures for two-dimensional simple polygons. Finally, in [15], the implementation of an algorithm that minimizes the number of layers, is discussed. This algorithm works for general polyhedra.

While writing the final version of the current paper, we became aware of related work by Nurmi and Sack [12]. (Private communication from J.-R. Sack.) They consider the following problem: Given a convex polyhedron A and a set of convex polyhedral obstacles, compute all directions of translations that move A arbitrarily far away such that no collision occurs between A and any of the obstacles. If we take for A a facet F of a polyhedron $\mathcal{P}$, and for the obstacles the other facets of $\mathcal{P}$, then we basically get Problem 1. Our algorithm for solving Problem 1 is similar to that of Nurmi and Sack. However, our algorithm is tailored to our particular application, and takes advantage of the fact that the objects of interest are the facets of a polyhedron. Moreover, we give rigorous proofs that handle all boundary cases—something that is crucial when deploying our algorithm in an actual LM application, since such boundary cases are very common in real-world STL files.

2   Geometric Preliminaries

The number of facets (i.e., triangles) of $\mathcal{P}$ is denoted by n. We consider each facet and edge of $\mathcal{P}$ as being closed. We allow that a vertex of one facet is in the interior of an edge e of another facet.

The *unit sphere*, i.e., the boundary of the three-dimensional ball centered at the origin and having radius one, is denoted by $\mathbb{S}^2$. We consider *directions* as points—or unit vectors—on $\mathbb{S}^2$. For any point $x \in \mathbb{R}^3$, and any direction $\mathbf{d} \in \mathbb{S}^2$, we denote by $r_{x\mathbf{d}}$, the ray emanating from x having direction $\mathbf{d}$.

We now formally define the notion of a point or facet *being in contact with supports* for a given build direction. (See also Figure 1 for the two-dimensional variants.) It turns out to be convenient to distinguish three cases.

Let F be a facet of $\mathcal{P}$, and $\mathbf{d} \in \mathbb{S}^2$ a direction. Let α be the angle between $\mathbf{d}$ and the outer normal of F. Note that $0 < \alpha < \pi$. If $\alpha < \pi/2$, then we say that F

is a *front facet w.r.t.* **d**. Similarly, if $\alpha > \pi/2$, then we say that F is a *back facet w.r.t.* **d**. Finally, if $\alpha = \pi/2$, then we say that F is a *parallel facet w.r.t.* **d**, or that **d** is *parallel to* F.

Definition 1. Let F be a facet of $\mathcal{P}$, x a point on F, and **d** a direction on $\mathbb{S}^2$. Point x is *in contact with supports for build direction* **d**, if one of the following three conditions holds.

1. F is a back facet w.r.t. **d**.
2. F is a front facet w.r.t. **d**, and the ray $r_{x\mathbf{d}}$ intersects the boundary of $\mathcal{P}$ in a point that is not on facet F.
3. F is a parallel facet w.r.t. **d**, and there is a facet G such that (a) the ray $r_{x\mathbf{d}}$ intersects G, and (b) at least one of the vertices of G is strictly on the same side of the plane through F as the outer normal of F.

Definition 2. Let F be a facet of $\mathcal{P}$, and **d** a direction on $\mathbb{S}^2$. We say that F is *in contact with supports for build direction* **d**, if there is a point in the interior of F that is in contact with supports for build direction **d**.

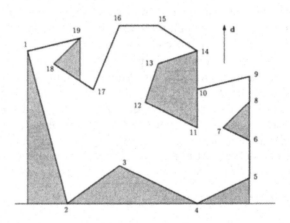

Fig. 1. *Illustrating the two-dimensional variant of Definitions 1 and 2 for a planar simple polygon with 19 vertices. The shaded regions are the supports for the vertical build direction* **d**. *No interior point of the vertical edges* (5, 6) *and* (8, 9) *is in contact with supports. On the other hand, all points of the vertical edge* (10, 11) *are in contact with supports. For build direction* **d**, *the following edges are in contact with supports:* (1, 2), (2, 3), (3, 4), (4, 5), (6, 7), (7, 8), (10, 11), (11, 12), (12, 13), (13, 14), (17, 18), *and* (18, 19).

In this paper, we will need two notions of convexity. The (standard) convex hull of the points $p_1, p_2, \ldots, p_k$ will be denoted by by $CH(p_1, p_2, \ldots, p_k)$. This notion can be generalized to *spherical convexity* for directions on $\mathbb{S}^2$ in a natural way. (See also Chen and Woo [4].) Note that if a set of directions contains two antipodal points, their spherical convex hull is the entire unit sphere. We say that a set of directions on $\mathbb{S}^2$ is *hemispherical*, if there is a three-dimensional plane H through the origin, such that all elements of D are strictly on one side of H.

3   Protecting One Facet from Supports

Throughout this section, F denotes a fixed facet of $\mathcal{P}$. We consider the problem of computing a description of all build directions $\mathbf{d}$, such that facet F is not in contact with supports, when $\mathcal{P}$ is built in direction $\mathbf{d}$.

The idea is as follows. For each facet G, we define a set $C_{FG} \subseteq \mathbb{S}^2$ of directions, such that for each $\mathbf{d} \in C_{FG}$, facet F is in contact with supports for build direction $\mathbf{d}$ "because of" facet G. That is, there is a point x in the interior of F, such that the ray $r_{x\mathbf{d}}$ emanating from x and having direction $\mathbf{d}$ intersects facet G. Hence, for each direction in the complement of the union of all sets C_{FG}, facet F is not in contact with supports. It will turn out, however, that we have to be careful with directions that are on the boundary of a set C_{FG}.

For any facet G of $\mathcal{P}$, and any point $x \in \mathbb{R}^3$, we define

$$R_{xG} := \{\mathbf{d} \in \mathbb{S}^2 : (r_{x\mathbf{d}} \cap G) \setminus \{x\} \neq \emptyset\}.$$

Hence, if $\mathbf{d} \in R_{xG}$, then the ray from x in direction $\mathbf{d}$ intersects facet G in a point that is not equal to x. For any facet G of $\mathcal{P}$, we define

$$C_{FG} := \bigcup_{x \in F} R_{xG}.$$

For any facet G, we denote by P_G the great circle consisting of all directions parallel to G.

Lemma 1. *Let G be a facet that is not coplanar with F. Assume that (i) F and G have more than one point in common, or (ii) F and G intersect in a single point, which is a vertex of one facet and in the interior of an edge of the other facet. Also, assume that for each vertex of G, one of the following is true: It is in the plane through F or on the same side of the plane through F as the outer normal of F. Then C_{FG} is the set of all directions that are (1) on or on the same side of P_F as the outer normal of facet F, and (2) on or on the same side of P_G as the inner normal of facet G.*   □

Let G be a facet of $\mathcal{P}$. Assume that either F and G are disjoint, or intersect in a single point which is a vertex of both facets. It can be shown that C_{FG} is hemispherical. We will show now how to compute C_{FG}.

First, we introduce some notation. Let $s = (s_x, s_y, s_z)$ and $t = (t_x, t_y, t_z)$ be two distinct points in $\mathbb{R}^3$, and let ℓ be the Euclidean length of the point $t - s := (t_x - s_x, t_y - s_y, t_z - s_z)$. That is, ℓ is the Euclidean distance between $t - s$

and the origin. We will denote by $\mathbf{d}_{st}$ the point on $\mathbb{S}^2$ having the same direction as the directed line segment from s to t. That is, $\mathbf{d}_{st}$ is the point on $\mathbb{S}^2$ having coordinates $\mathbf{d}_{st} = ((t_x - s_x)/\ell, (t_y - s_y)/\ell, (t_z - s_z)/\ell)$.

Lemma 2. *Let G be a facet of $\mathcal{P}$. Assume that F and G are disjoint, or intersect in a single point which is a vertex of both facets. Let*

$$D_{FG} := \{\mathbf{d}_{st} \in \mathbb{S}^2 : s \text{ is a vertex of } F, t \text{ is a vertex of } G, s \neq t\}.$$

Then C_{FG} is the spherical convex hull of the at most nine directions in D_{FG}.

Proof. We denote the spherical convex hull of the elements of D_{FG} by SCH_{FG}. We will prove that (i) $C_{FG} \subseteq SCH_{FG}$, and (ii) C_{FG} is spherically convex. Since SCH_{FG} is the smallest spherically convex set that contains the elements of D_{FG}, the fact that $D_{FG} \subseteq C_{FG}$, together with (i) and (ii), imply that $C_{FG} = SCH_{FG}$.

To prove (i), let $\mathbf{d} \in C_{FG}$. We will show that $\mathbf{d} \in SCH_{FG}$. Since $\mathbf{d} \in C_{FG}$, there is a point x on F such that $\mathbf{d} \in R_{xG}$. Let y be any point such that $y \neq x$ and $y \in r_{xd} \cap G$.

Let a, b, and c denote the three vertices of F, and u, v, and w the three vertices of G. Since $x \in CH(a, b, c)$, we have $y - x \in CH(y - a, y - b, y - c)$. Similarly, since $y \in CH(u, v, w)$, we have $y - a \in CH(u - a, v - a, w - a) =: CH_1$, $y - b \in CH(u - b, v - b, w - b) =: CH_2$, and $y - c \in CH(u - c, v - c, w - c) =: CH_3$. Therefore, $CH(y - a, y - b, y - c) \subseteq CH(CH_1, CH_2, CH_3)$. Hence, if we denote the (standard) convex hull of the point set $\{t - s : s \text{ is a vertex of } F, t \text{ is a vertex of } G\}$ by CH_{FG}, then we have shown that $y - x \in CH(CH_1, CH_2, CH_3) = CH_{FG}$. Since $y - x \neq 0$, and the ray from the origin through point $y - x$ has direction $\mathbf{d}$, it follows that $\mathbf{d} \in SCH_{FG}$.

To prove (ii), let $\mathbf{d}_1$ and $\mathbf{d}_2$ be two distinct directions in C_{FG}, and let $\mathbf{d}$ be any direction on the shortest great arc that connects $\mathbf{d}_1$ and $\mathbf{d}_2$. We have to show that $\mathbf{d} \in C_{FG}$. Let H be the plane through $\mathbf{d}_1$, $\mathbf{d}_2$, and the origin. Then $\mathbf{d}$ is also contained in H. Let $d' \in \mathbb{R}^3$ be the intersection between the line through $\mathbf{d}_1$ and $\mathbf{d}_2$, and the line through the origin and $\mathbf{d}$. (Note that d' is not the origin.) Let $0 \leq \lambda \leq 1$ be such that $d' = \lambda \mathbf{d}_1 + (1 - \lambda)\mathbf{d}_2$. Since $\mathbf{d}_1 \in C_{FG}$, there is a point x_1 on F such that the ray from x_1 in direction $\mathbf{d}_1$ intersects G in a point that is not equal to x_1. Let y_1 be any such intersection point. Similarly, there are points x_2 on F, and y_2 on G, such that $y_2 \neq x_2$ and the ray from x_2 in direction $\mathbf{d}_2$ intersects G in y_2. Let α, α_1, and α_2 be the positive real numbers such that $\mathbf{d} = \alpha d'$, $y_1 - x_1 = \alpha_1 \mathbf{d}_1$, and $y_2 - x_2 = \alpha_2 \mathbf{d}_2$. Then

$$\mathbf{d} = \left(\frac{\alpha\lambda}{\alpha_1}y_1 + \frac{\alpha(1 - \lambda)}{\alpha_2}y_2\right) - \left(\frac{\alpha\lambda}{\alpha_1}x_1 + \frac{\alpha(1 - \lambda)}{\alpha_2}x_2\right).$$

Let μ such that $\mu(\alpha\lambda/\alpha_1 + \alpha(1 - \lambda)/\alpha_2) = 1$. Define $x := (\mu\alpha\lambda/\alpha_1)x_1 + (\mu\alpha(1 - \lambda)/\alpha_2)x_2$, and $y := (\mu\alpha\lambda/\alpha_1)y_1 + (\mu\alpha(1-\lambda)/\alpha_2)y_2$. Then $\mu > 0$, and $\mu\mathbf{d} = y - x$. Hence x is a convex combination of x_1 and x_2, and y is a convex combination of y_1 and y_2. It follows that x and y are points on the facets F and G, respectively. Moreover, $y \neq x$. Hence, $\mathbf{d} \in C_{FG}$.   □

3.1   More Properties of the Sets C_{FG}

We say that a facet G is *below* facet F, if each vertex of G is in the plane through F or on the same side of this plane as the inner normal of F.

Lemma 3. *Let* $\mathbf{d}$ *be a direction on* $\mathbb{S}^2$ *such that* F *is a front facet w.r.t.* $\mathbf{d}$. *Assume that* F *is in contact with supports for build direction* $\mathbf{d}$. *Then there is a facet* G, *such that (i)* G *is not below* F, *and (ii)* $\mathbf{d}$ *is in the interior of* C_{FG}.

Proof. Assume w.l.o.g. that $\mathbf{d}$ is the vertically upwards direction. Hence, F is not vertical. There is a disk D_0 in the interior of F, having positive radius, such that each point of D_0 is in contact with supports for build direction $\mathbf{d}$. That is, the vertical ray that emanates from any point of D_0 intersects the boundary of $\mathcal{P}$ in a point that is not on F. Thus, there is a disk D of positive radius in the interior of D_0, and a facet G, $G \neq F$, such that for each $x \in D$, the ray $r_{x\mathbf{d}}$ intersects G. Put differently, let $VC := \{r_{x\mathbf{d}} : x \in D\}$, i.e., VC is the vertical "cylinder" which is unbounded in direction $\mathbf{d}$ and is bounded from below by D. Since F is not vertical, the cylinder VC is not contained in a plane. If we move the disk D vertically upwards, then it stays in the cylinder VC, and each point of the moving disk passes through G. Clearly, facet G is not below F.

Let c and ϵ be the center and radius of D, respectively. Since D is not contained in a vertical plane, and $\epsilon > 0$, there is a spherical disk SD on $\mathbb{S}^2$, centered at $\mathbf{d}$ and having positive radius, such that for each $\mathbf{d}' \in SD$, the ray $r_{c\mathbf{d}'}$ intersects G in a point that is not equal to c. This implies that SD is completely contained in C_{FG}. Since $\mathbf{d}$ is in the interior of SD, it follows that $\mathbf{d}$ is in the interior of C_{FG}.   □

The following lemma is the converse of Lemma 3.

Lemma 4. *Let* G *be a facet of* $\mathcal{P}$ *that is not below* F. *Let* $\mathbf{d}$ *be a direction that is not parallel to* F *and that is in the interior of* C_{FG}. *Then* F *is in contact with supports for build direction* $\mathbf{d}$.   □

We denote by $\mathcal{U}_F$ the union of the sets C_{FG}, where G ranges over all facets that are not below F. Lemma 3 immediately implies the following two lemmas, which state when a front facet F is not in contact with supports.

Lemma 5. *Let* $\mathbf{d}$ *be a direction on* $\mathbb{S}^2$ *such that* F *is a front facet w.r.t.* $\mathbf{d}$. *If* $\mathbf{d}$ *is not in the interior of* $\mathcal{U}_F$, *then* F *is not in contact with supports for build direction* $\mathbf{d}$.   □

Lemma 6. *Let* $\mathbf{d}$ *be a direction on* $\mathbb{S}^2$ *such that* F *is a front facet w.r.t.* $\mathbf{d}$. *Assume that (i)* $\mathbf{d}$ *is in the interior of* $\mathcal{U}_F$, *and (ii)* $\mathbf{d}$ *is not in the interior of* C_{FG}, *for all facets* G *that are not below* F. *Then* F *is not in contact with supports for build direction* $\mathbf{d}$.   □

The following three lemmas treat certain "boundary" cases, which involve directions that are parallel to F.

Lemma 7. *Let* **d** *be a direction on the great circle* P_F *that is not contained in any of the sets* C_{FG}, *where* G *ranges over all facets that are not below* F. *Then facet* F *is not in contact with supports for build direction* **d**.   □

Lemma 8. *Let* G *be a facet that is not below* F. *Let* **d** *be a direction on the great circle* P_F, *such that (i) either* **d** *is in the interior of the set* C_{FG}, *or (ii)* **d** *is in the interior of an edge of* C_{FG}, *and this edge is contained in* P_F. *Then facet* F *is in contact with supports for build direction* **d**.   □

Let $\mathcal{A}$ be the arrangement on $\mathbb{S}^2$ defined by the great circle P_F and the boundaries of the sets C_{FG}, where G ranges over all facets that are not below F.

Lemmas 7 and 8 do not consider directions on P_F that are vertices of $\mathcal{A}$. For these directions, we will use the following lemma to decide if F is in contact with supports.

Lemma 9. *Let* **d** *be a direction on the great circle* P_F, $W := \{r_{x\mathbf{d}} : x \in F\}$, *and* $\mathcal{I}$ *the set of all facets* G *such that (i)* G *is not below* F, *and (ii) the intersection of* G *and the interior of* W *is non-empty. Then* F *is in contact with supports for build direction* **d** *if and only if* $\mathcal{I} \neq \emptyset$.   □

3.2   The Facet Protection Algorithm

The algorithm that computes a description of all build directions for which facet F is not in contact with supports is based on the previous results.

Step 1: Following Lemma 1, we do the following. For each facet G of $\mathcal{P}$ that is not below F, and such that (i) either F and G have more than one point in common, or (ii) F and G intersect in a single point, which is a vertex of one facet and in the interior of an edge of the other facet, do the following: Compute the boundary of the set C_{FG} as the set of all directions that are on or on the same side of P_F as the outer normal of facet F, and on or on the same side of P_G as the inner normal of facet G.

Step 2: Following Lemma 2, we do the following. For each facet G of $\mathcal{P}$ that is not below F, and such that (i) either F and G are disjoint, or (ii) F and G intersect in a single point which is a vertex of both facets, do the following. Compute the boundary of the set C_{FG} as the spherical convex hull of the (at most nine) directions $\mathbf{d}_{st} \neq 0$, where s and t are vertices of F and G, respectively.

Step 3: Compute the arrangement $\mathcal{A}$ on $\mathbb{S}^2$ that is defined by the great circle P_F and the bounding edges of all sets C_{FG} that were computed in Steps 1 and 2. Let $\mathcal{B}$ be the arrangement on $\mathbb{S}^2$ consisting of all vertices, edges, and faces of $\mathcal{A}$ that are on P_F or on the same side of P_F as the outer normal of F. Give each edge e of $\mathcal{B}$ an orientation, implying the notions of being to the "left" and "right" of e. For edges that are not contained in the great circle P_F, these orientations are chosen arbitrarily. For each edge e of $\mathcal{B}$ that is contained in P_F, we choose the orientation such that the outer normal of F is to the "left" of e. For each edge e of $\mathcal{B}$, compute the following three values:

1. l_e (resp. r_e), which is one if the interior of the face of $\mathcal{B}$ to the left (resp. right) of e is contained in some set C_{FG} that was computed in Step 1 or 2, and zero otherwise. If e is contained in P_F, then the value of r_e is not defined.
2. i_e, which is one if the interior of e is in the interior of some set C_{FG} that was computed in Step 1 or 2, and zero otherwise.

Moreover, for each vertex v of $\mathcal{B}$ that is not on P_F, compute the value i_v, which is one if v is in the interior of some set C_{FG} that was computed in Step 1 or 2, and zero otherwise. We will show later how Step 3 can be implemented.

Step 4: Select all edges e of $\mathcal{B}$ that are not contained in P_F, and for which $l_e = 1$ and $r_e = 0$, or $l_e = 0$ and $r_e = 1$. Also, select all edges e of $\mathcal{B}$ that are contained in P_F, and for which $l_e = 0$ and $i_e = 0$.

By Lemmas 5 and 7, these edges define polygonal regions on $\mathbb{S}^2$ that represent build directions for which facet F is not in contact with supports.

Step 5: Select all edges e of $\mathcal{B}$ that are not contained in P_F, and for which $l_e = r_e = 1$ and $i_e = 0$. Similarly, select all vertices v of $\mathcal{B}$ that are not on P_F, for which $i_v = 0$, and having the property that $l_e = r_e = 1$ for all edges e of $\mathcal{B}$ that have v as one of their vertices.

By Lemma 6, these vertices and the interiors of these edges represent build directions for which facet F is not in contact with supports.

Step 6: Let D be the set of all vertices of $\mathcal{B}$ that are on P_F. For each direction $\mathbf{d} \in D$, decide if facet F is in contact with supports for build direction $\mathbf{d}$. This can be done by using an algorithm that is immediately implied by Lemma 9.

This algorithm reports a collection of spherical polygons, great arcs (the edges computed in Step 5), and single directions (the vertices computed in Steps 5 and 6). It follows from the previous results that this collection represents all build directions for which facet F is not in contact with supports. We now consider Step 3 in more detail.

After Steps 1 and 2, we have a collection of at most $n-1$ spherical polygons, each having $O(1)$ edges. For each such edge e, let K_e be the great circle that contains e. Using an incremental algorithm, we compute the arrangement $\mathcal{A}'$ on $\mathbb{S}^2$ of the $O(n)$ great circles K_e, and the great circle P_F. By removing from $\mathcal{A}'$ all vertices and edges that are strictly on the same side of P_F as the inner normal of facet F, we obtain an arrangement which we denote by $\mathcal{B}'$. We give each edge of $\mathcal{B}'$ a direction. We will show how the values l_e, r_e, i_e, and i_v for all edges e and vertices v of $\mathcal{B}'$ can be computed. Since the arrangement $\mathcal{B}$ is obtained from $\mathcal{B}'$ by removing all vertices and edges that are not contained in edges of our original polygons C_{FG}, this solves our problem.

We introduce the following notation. For each vertex v of $\mathcal{B}'$, let I_v be the set of all facets G that are not below F and for which v is in the interior of C_{FG}. For each edge e of $\mathcal{B}'$, let L_e be the set of all facets G that are not below F and for which the interior of the face of $\mathcal{B}'$ to the left of e is contained in C_{FG}. Similarly, let R_e be the set of all facets G that are not below F and for which the interior of the face of $\mathcal{B}'$ to the right of e is contained in C_{FG}. Clearly,

1. $i_v = 1$ if and only if $I_v \neq \emptyset$,
2. $l_e = 1$ if and only if $L_e \neq \emptyset$,

3. $r_e = 1$ if and only if e is not contained in P_F and $R_e \neq \emptyset$, and
4. $i_e = 1$ if and only if $L_e \cap R_e \neq \emptyset$.

The idea is to traverse each great circle that defines the arrangement $\mathcal{B}'$, and maintain the sizes of the sets I_v, L_e, R_e, and $L_e \cap R_e$. We number the facets of $\mathcal{P}$ arbitrarily from 1 to n. Let K be any of the great circles that define $\mathcal{B}'$, and let v be a vertex of $\mathcal{B}'$ which is on K. By considering all facets G that are not below F, we compute the set I_v, and store it as a bit-vector I of length n. By traversing this array I, we compute the number of ones it contains, and deduce from this number the value i_v.

Let e be an edge of $\mathcal{B}'$ which is contained in K and has v as a vertex. By considering all edges of $\mathcal{B}'$ that have v as endpoint, we know which sets C_{FG} are entered or left, when our traversal along K leaves v and enters the interior of e. We make two copies of the bit-vector I, and call them L and R. Then, by flipping the appropriate bits in the three arrays L, R, and I, we obtain the bit-vectors for the sets L_e, R_e, and $L_e \cap R_e$, respectively, and the number of ones they contain. This gives us the values l_e, r_e, and i_e.

We now continue our traversal of the great circle K. Each time we reach or leave a vertex of $\mathcal{B}'$, we flip the appropriate bits in the arrays L, R, and I, and deduce the l, r, and i values. By the *zone theorem* [5], the running time of the entire algorithm is bounded by $O(n^2)$.

Theorem 1. *Let $\mathcal{P}$ be a polyhedron with n triangular facets, possibly with holes, and let F be a facet of $\mathcal{P}$. In $O(n^2)$ time, we can compute a description of all build directions for which F is not in contact with supports.* □

In the full paper [14], we show that the set of all build directions for which facet F is not in contact with supports can have $\Omega(n^2)$ connected components. Hence, the algorithm of Theorem 1 is worst-case optimal.

The algorithm can be extended to the case when k facets, $F_1, F_2, \ldots, F_k$ have to be protected from supports. For each F_i, we run Steps 1 and 2 of the algorithm in Section 3.2. This gives us a set of spherically convex regions of total size $O(kn)$. We then compute the arrangement of these regions and the k great circles P_{F_i}, which has size $O(k^2 n^2)$, and then traverse it in essentially the way described in the algorithm. The running time is bounded by $O(k^2 n^2)$.

3.3   Experimental Results

We have implemented a simplified version of the algorithm of Theorem 1. In this implementation, the boundary of the union $\mathcal{U}_F$ is computed incrementally, i.e., the sets C_{FG}, where G ranges over all facets that are not below F, are added one after another in a brute force manner. The program outputs a collection of spherical polygons on $\mathbb{S}^2$ such that for each direction in such a polygon, facet F is not in contact with supports. For details, see [13].

The program is written in C++ using LEDA [11], and run on a SUN Ultra (300 MHz, 512 MByte RAM). We have tested our implementation on real-world polyhedral models obtained from Stratasys, Inc.. Although the running time of

model	n	#F	#C_{FG}	$|\mathcal{U}_F|$	min	max	average
rd_yelo.stl	396	396	99	3.5	0.01	73	16
cover-5.stl	906	906	482	8.2	0.03	558	103
tod21.stl	1,128	1,128	229	3.8	0.05	281	25
stlbin2.stl	2,762	1,330	1178	20.9	0.25	2,019	363
mj.stl	2,832	1,000	641	10.1	0.26	2,270	146

Table 1. n denotes the number of facets of the model; #F denotes the number of facets F for which we ran the program independently and averaged our bounds over; #C_{FG} denotes the average number of facets G that are not below F; $|\mathcal{U}_F|$ denotes the average number of vertices on the boundary of the union $\mathcal{U}_F$ (note that this union may have no vertices at all); min, max, and average denote the minimum, maximum, and average time in seconds.

our implementation is $\Theta(n^3)$ in the worst case, the actual running time is reasonable in practice.

Table 1 gives test results for five polyhedral models. rd_yelo.stl is a long rod, with grooves cut along its length; cover-5.stl resembles a drawer for a filing cabinet; tod21.stl is a bracket, consisting of a hollow quarter-cylinder, with two flanges at the ends, and a through-hole drilled in one of the flanges; stlbin2.stl is an open rectangular box, with a hole on each side and interior flanges at the corners; mj.stl is a curved part with a base and a protrusion, shaped like a pistol.

3.4 Solving Problem 2

For each facet F of $\mathcal{P}$, we compute the sets C_{FG} for all facets G that are not below F. Let $\mathcal{C}$ be the arrangement on $\mathbb{S}^2$ defined by the great circles P_F, and by the great circles that contain an edge of some set C_{FG}. Note that $\mathcal{C}$ is defined by $O(n^2)$ great circles and, hence, consists of $O(n^4)$ vertices, edges, and faces.

This arrangement $\mathcal{C}$ has the following property. Let f (resp. e) be any face (resp. edge) of $\mathcal{C}$, and let $\mathbf{d}$ and $\mathbf{d}'$ be any two directions in the interior of f (resp. in the interior of e). Let T (resp. T') be the total area of all facets of $\mathcal{P}$ that are not in contact with supports, if $\mathcal{P}$ is built in direction $\mathbf{d}$ (resp. $\mathbf{d}'$). Then $T = T'$. Problem 2 is solved by traversing each of the $O(n^2)$ great circles that define $\mathcal{C}$.

Theorem 2. *Let $\mathcal{P}$ be a polyhedron with n triangular facets, possibly with holes. In $O(n^4)$ time, we can compute a description of all build directions for which the total area of all facets that are not in contact with supports is maximum.* □

4 Concluding Remarks

We have shown that for a fixed facet F of the polyhedron $\mathcal{P}$, a description of all build directions for which F is not in contact with supports can be computed in $O(n^2)$ time, which is worst-case optimal. A natural question is to ask for the time complexity of computing *one* build direction for which F is not in contact with supports, or decide that such a direction does not exist. This problem appears to be closely related to the following one: Given $n+1$ triangles $T_0, T_1, T_2, \ldots, T_n$ in the plane, decide if T_0 is contained in the union of $T_1, \ldots, T_n$. This problem is 3SUM-hard, see Gajentaan and Overmars [6]. Therefore, we conjecture that computing a single direction for which facet F is not in contact with supports is 3SUM-hard as well.

The algorithms of Sections 3.2 and 3.4 have running time $O(n^2)$ and $O(n^4)$, respectively. It would be interesting to design output-sensitive algorithms for solving these problems.

We are not aware of any efficient algorithm that minimizes support structures for general three-dimensional polyhedra. Is it possible to compute, in polynomial time, such a build direction?

References

1. S. Allen and D. Dutta. Determination and evaluation of support structures in layered manufacturing. *Journal of Design and Manufacturing*, 5:153–162, 1995.
2. B. Asberg, G. Blanco, P. Bose, J. Garcia-Lopez, M. Overmars, G. Toussaint, G. Wilfong, and B. Zhu. Feasibility of design in stereolithography. *Algorithmica*, 19:61–83, 1997.
3. M. Bablani and A. Bagchi. Quantification of errors in rapid prototyping processes and determination of preferred orientation of parts. In *Transactions of the 23rd North American Manufacturing Research Conference*, 1995.
4. L.-L. Chen and T. C. Woo. Computational geometry on the sphere with application to automated machining. *Journal of Mechanical Design*, 114:288–295, 1992.
5. M. de Berg, M. van Kreveld, M. Overmars, and O. Schwarzkopf. *Computational Geometry: Algorithms and Applications*. Springer-Verlag, Berlin, 1997.
6. A. Gajentaan and M. H. Overmars. On a class of $O(n^2)$ problems in computational geometry. *Comput. Geom. Theory Appl.*, 5:165–185, 1995.
7. P. F. Jacobs. *Rapid Prototyping & Manufacturing: Fundamentals of StereoLithography*. McGraw-Hill, New York, 1992.
8. J. Majhi, R. Janardan, J. Schwerdt, M. Smid, and P. Gupta. Minimizing support structures and trapped area in two-dimensional layered manufacturing. *Comput. Geom. Theory Appl.*, 12:241–267, 1999.
9. J. Majhi, R. Janardan, M. Smid, and P. Gupta. On some geometric optimization problems in layered manufacturing. *Comput. Geom. Theory Appl.*, 12:219–239, 1999.
10. J. Majhi, R. Janardan, M. Smid, and J. Schwerdt. Multi-criteria geometric optimization problems in layered manufacturing. In *Proc. 14th Annu. ACM Sympos. Comput. Geom.*, pages 19–28, 1998.
11. K. Mehlhorn and S. Näher. LEDA: a platform for combinatorial and geometric computing. *Commun. ACM*, 38:96–102, 1995.
12. O. Nurmi and J.-R. Sack. Separating a polyhedron by one translation from a set of obstacles. In *Proc. 14th Internat. Workshop Graph-Theoret. Concepts Comput. Sci. (WG '88)*, volume 344 of *Lecture Notes Comput. Sci.*, pages 202–212. Springer-Verlag, 1989.
13. J. Schwerdt, M. Smid, R. Janardan, and E. Johnson. Protecting facets in layered manufacturing: implementation and experimental results. In preparation, 1999.

14. J. Schwerdt, M. Smid, R. Janardan, E. Johnson, and J. Majhi. Protecting facets in layered manufacturing. Report 10, Department of Computer Science, University of Magdeburg, Magdeburg, Germany, 1999.

15. J. Schwerdt, M. Smid, J. Majhi, and R. Janardan. Computing the width of a three-dimensional point set: an experimental study. In *Proc. 2nd Workshop on Algorithm Engineering*, pages 62–73, Saarbrücken, 1998.

The Receptive Distributed π-Calculus*
(Extended Abstract)

Roberto M. Amadio[1], Gérard Boudol[2], and Cédric Lhoussaine[3]

[1] Université de Provence, Marseille
[2] INRIA, Sophia-Antipolis
[3] Université de Provence, Marseille

Abstract. In this paper we study an asynchronous distributed π-calculus, with constructs for localities and migration. We show that a simple static analysis ensures the receptiveness of channel names, which, together with a simple type system, guarantees that any migrating message will find an appropriate receiver at its destination locality. We argue that this receptive calculus is still expressive enough, by showing that it contains the π_1-calculus, up to weak asynchronous bisimulation.

1 Introduction

In this paper we study a simplified version of Hennessy and Riely's *distributed π-calculus* Dπ [8]. This is a calculus based on the asynchronous, polyadic π-calculus, involving explicit notions of *locality* and *migration*, that is code movement from a location to another. In this model communication is purely local, so that messages to remote resources (i.e. receivers, in π-calculus terminology) must be explicitly routed. In such a model – as opposed to other ones like the JOIN calculus [6,7] or the π_{1l}-calculus [2] where messages transparently go through the network to reach their (unique) destination – the problem arises of how to avoid the situation where a message migrates to, or stays in a locality where no receiver will ever be available. In other words, we would like to ensure that any message will find an appropriate receiver in its destination locality, a property that we call *message deliverability*.

To solve this problem, we must ensure that a resource will be available when requested. An obvious way to achieve this is to enforce *receptiveness* – following Sangiorgi's terminology [10] – of any (private) channel name, that is the property that at any time a process is able of offering an input on that channel name. However, the kind of receptiveness we are looking for is neither the "uniform", nor the "linear" one described in [10]. Indeed, requiring each (private) name to be either uniform or linear receptive would result in a dramatic loss of expressive power. Then we are seeking for a formalization of something like the (unique) naming of a persistent "object", which may encapsulate a state – we also wish to ensure the desirable property of unicity of receivers, which turns out to be needed for our expressiveness result.

---

* Work partially supported by the RNRT project MARVEL.

C. Pandu Rangan, V. Raman, R. Ramanujam (Eds.): FSTTCS'99, LNCS 1738, pp. 304–315, 1999.

It is actually not very difficult to design a simple inference system for statically checking processes, in such a way that private names are (uniquely) receptive inside their scope: the system is a refinement of the one of π_{1l}, where we impose that recursive processes have a unique input parameter. Then for instance an input cannot be nested inside another, unless they are inputs on the same name. This is not enough to entail the message deliverability property however, since we not only have channel names, but also location names, and other names, called keys, that may be compared for equality but cannot be used as communication channels – the latter are crucially needed to retain the expressive power, as we shall see. In particular, the restriction operation $(\nu u)P$ of the π-calculus must be refined, because we do not demand a location name or a key to be receptive – there is no notion of a receiver for such a name – and still we want such a name to be sometimes restricted in scope. Types seem to be the obvious solution for this particular problem.

The type system we consider in this paper appears to be a simplified version of the simple type system of [8]: we use location types, that record the names and types of channels that may be used to communicate inside a locality, and we use "located types" for channels that are sent together with their location, instead of the existential types of [8]. We show the usual "subject reduction property", which holds for the system for checking receptiveness too. Then we are able to prove that typed receptive processes do not run into the kind of deadlocks we wished to avoid, where a message never finds a corresponding receiver.

The issue of avoiding deadlocks by statically inferring some properties has already been addressed, most notably by Kobayashi who studied in [9] means to avoid deadlocks in the π-calculus (see also [4]). His approach is quite different from ours, however: he uses a sophisticated type system where types involve time tags and where typing contexts record an ordering on the usage of names, whereas we only use a very simple information – the set of receiver's names. Since we regard receivers as passive entities, providing resources for names, we are not seeking to avoid the situation where no message is sent to a receiver – in this situation the resource provider should be garbage collected.

The question arises whether the nice properties of our receptive distributed π-calculus are obtained at the expense of its expressivity. Indeed, receptiveness is both a strong safety property and a quite heavy constraint. We therefore establish a result that proves that expressive power is not lost: we show that the π_1-calculus of Amadio [2], in which the JOIN calculus can be encoded [3], may be translated into our receptive π_1-calculus, in a way which is fully abstract with respect to weak asynchronous bisimulation.

2 Distributed Processes

In this section we introduce our calculus of distributed processes, which may be seen as a simplification of the distributed π-calculus of Hennessy and Riely [8], and as an extension of the π_1-calculus of Amadio [2,3]. This is basically the asynchronous, polyadic π-calculus, with some primitives for spatial distribution of processes and migration.

As usual, we assume given a denumerable set $\mathcal{N}$ of (simple) *names*, ranged over by $a, b, c, \ldots$ We will informally distinguish a subset of names, that we denote $\ell, \ell', \ldots$, which are supposed to name *localities* – formally, there will be different kinds of types for names. Then the processes may also send and receive *compound names*, which are pairs of simple names that we write $a@\ell$, meaning "the (channel) name a used at location ℓ". We use $u, v, w, \ldots$ to denote simple or compound names, and a (possibly empty) vector of such names is written $\vec{u}$. We shall often use the operation $\_@\ell$ on (compound) names, defined by

$$u@\ell = \begin{cases} a@\ell & \text{if } u = a \\ u & \text{if } u = a@\ell' \end{cases}$$

This operation is extended to sets of names in the obvious way. We use $A, B, \ldots$ to denote process identifiers, belonging to a denumerable set $\mathcal{P}$, disjoint from $\mathcal{N}$. These are parametric on names, and we shall write a recursive call as $A(a; \vec{u})$. The intended meaning is that the name a is the only one that may – and must – be used as an input channel in the body of a recursive definition of A. The syntax of distributed processes involves *types*, but we defer any other consideration about types to the next section. Then the grammar for terms is as follows, where τ is a type:

$$u, v \ldots ::= a \mid a@\ell$$
$$w ::= u \mid a : \tau$$
$$P, Q, R \ldots ::= \overline{a}\vec{u} \mid a(\vec{u}).P \mid (P \mid Q) \mid [a = b]P, Q \mid (\nu w)P$$
$$\mid A(a; \vec{u}) \mid (\text{rec } A(a; \vec{u}).P)(b; \vec{v}) \mid [\ell :: P]$$

For any w, we define its subject $\text{subj}(w)$ as follows: $\text{subj}(a) = a = \text{subj}(a@\ell) = \text{subj}(a : \tau)$. In $(\nu w)P$, the subject of w is the only name occurring in w which is bound. We make the standard assumption that recursion is *guarded*, that is in $(\text{rec } A(a; \vec{u}).P)(b; \vec{v})$ all recursive calls to A in P occur under an input guard. We shall say that P is *closed* if it does not contain any free process identifier.

The need for compound names is explained by Hennessy and Riely in [8], although they use more sophisticated "dependent names". In our syntax we do not distinguish "the process located at ℓ" from "the process P moving towards ℓ", written respectively $\ell[\![P]\!]$ and $\ell :: P$ in [8], and $\{P\}\ell$ and $\text{spawn}(\ell, P)$ in [2]: both are denoted by $[\ell :: P]$. Another difference with [8] is that our underlying π-calculus is asynchronous, and we use explicit recursion instead of replication.

In order to define a notion of bisimulation, we describe the *behaviour* of processes by means of a labelled transition system for the calculus. That is, processes perform *actions* and evolve to new processes in doing so, which is denoted $P \xrightarrow{\alpha} P'$. The set of actions is given by:

$$\alpha ::= \tau \mid u\vec{v} \mid (\nu\vec{w})\overline{u}\vec{v}$$

meaning respectively internal communication (there should be no confusion between τ as a type and τ as an action), input of names $\vec{v}$ on the name u – if $u = a@\ell$, this means input on channel a at location ℓ – and output of names,

$$(out) \; \frac{}{\overrightarrow{au} \xrightarrow{\overrightarrow{au}} 0}$$

$$(in) \; \frac{}{a(\overrightarrow{u}).P \xrightarrow{a\overrightarrow{v}} [\overrightarrow{v}/\overrightarrow{u}]P}$$

$$(ext) \; \frac{P \xrightarrow{\alpha} P'}{(\nu w)P \xrightarrow{(\nu w)\alpha} P'} \; (*)$$

$$(\nu) \; \frac{P \xrightarrow{\alpha} P'}{(\nu w)P \xrightarrow{\alpha} (\nu w)P'} \; (**)$$

$$(cm) \; \frac{P \xrightarrow{(\nu \overrightarrow{w})\overrightarrow{u}\,\overrightarrow{v}} P' \,, \, Q \xrightarrow{u\overrightarrow{v}} Q'}{P \mid Q \xrightarrow{\tau} (\nu\overrightarrow{w})(P' \mid Q')} \; \mathrm{subj}(\overrightarrow{w}) \cap \mathrm{fn}(Q) = \emptyset$$

$$(cp) \; \frac{P \xrightarrow{\alpha} P'}{P \mid Q \xrightarrow{\alpha} P' \mid Q} \; \mathrm{bn}(\alpha) \cap \mathrm{fn}(Q) = \emptyset$$

$$(@) \; \frac{P \xrightarrow{\alpha} P'}{[\ell :: P] \xrightarrow{\alpha@\ell} [\ell :: P']}$$

$$(m_t) \; \frac{P \xrightarrow{\alpha} P'}{[a = a]P, Q \xrightarrow{\alpha} P'}$$

$$(m_f) \; \frac{Q \xrightarrow{\alpha} Q'}{[a = b]P, Q \xrightarrow{\alpha} Q'} \; a \neq b$$

$$(rec) \; \frac{[\mathrm{rec}\,A(a; \overrightarrow{u}).P/A, a'; \overrightarrow{u}'/a; \overrightarrow{u}]P \xrightarrow{\alpha} P'}{(\mathrm{rec}\,A(a; \overrightarrow{u}).P)(b; \overrightarrow{v}) \xrightarrow{\alpha} P'}$$

$$(cg) \; \frac{P =_\alpha P' \quad P' \xrightarrow{\alpha} Q' \quad Q' =_\alpha Q}{P \xrightarrow{\alpha} Q}$$

$(*) \; \mathrm{subj}(w) \in \mathrm{fn}(\alpha) - \mathrm{nm}(\mathrm{subj}(\alpha))$

$(**) \; \mathrm{subj}(w) \notin \mathrm{nm}(\alpha)$

Fig. 1. Labelled transition system

---

some of them possibly being private. We also call subject of α, denoted $\mathrm{subj}(\alpha)$, the name u whenever $\alpha = u\overrightarrow{v}$ or $\alpha = (\nu\overrightarrow{w})\overrightarrow{u}\,\overrightarrow{v}$. We define the sets $\mathrm{fn}(\alpha)$ and $\mathrm{bn}(\alpha)$ of names occurring respectively free and bound in the action α in the obvious way (recall that in $(\nu a@\ell)$, a is bound while ℓ is free), and $\mathrm{nm}(\alpha)$ denotes the union of these two sets.

The rules of the transition system, given in Figure 1, extend the usual ones for the π-calculus. In the rule (out), the term 0 is $(\nu a)((\mathrm{rec}\,A(a;).a().A(a;))(a))$, that is a process which cannot perform any action. In the rules (in) and (rec), it is understood that the substitution operation involves some pattern matching, e.g. $[a@b/x@y]P$ is $[a/x, b/y]P$. In the rule (ext), the action α must be an output action (otherwise $(\nu w)\alpha$ would not be an action). There are also rules symmetric to (cm) and (cp), which are omitted. In the rule (cg), the equivalence $=_\alpha$ stands for syntactic identity up to renaming of bound names. In the rule $(@)$ for located processes, we extend the operation $\_@\ell$ to actions, as follows:

$$\alpha@\ell = \begin{cases} u@\ell \, \overrightarrow{v} & \text{if } \alpha = u\overrightarrow{v} \\ \overline{u@\ell} \, \overrightarrow{v} & \text{if } \alpha = \overline{u}\overrightarrow{v} \\ (\nu u@\ell)(\alpha'@\ell) & \text{if } \alpha = (\nu u)\alpha' \\ (\nu a : \tau)(\alpha'@\ell) & \text{if } \alpha = (\nu a : \tau)\alpha' \end{cases}$$

where in the last two cases we use α-conversion to ensure that $\ell \neq \mathrm{subj}(u)$ or $\ell \neq a$. As one can see, any action of the located process $[\ell :: P]$ is located – that is, its subject is a compound name –, but not necessarily at ℓ: if for instance this was an action of a sub-process located at ℓ', then its location is not modified, since $\alpha@\ell'@\ell = \alpha@\ell'$, as one can easily check. As a matter of fact, it is easy to see that the following equation holds, where equality is the usual strong bisimilarity:

$$[\ell' :: [\ell :: P]] \sim [\ell :: P]$$

This means that our operational semantics expresses the fact that distributed systems have a flat structure, like in [2,8], but unlike in the distributed JOIN calculus [7] or the Ambient calculus [5] where "domains", or "sites", may be nested. Moreover, there is only one "domain" with a given name, that is

$$[\ell :: P] \mid [\ell :: Q] \sim [\ell :: P \mid Q]$$

This makes a difference with the model of Ambients. These equalities allow us to interpret $[\ell :: P]$ as "the process P migrating to ℓ". For instance the process $[\ell' :: [\ell :: P] \mid Q] \mid [\ell :: R]$ behaves in the same way as $[\ell' :: Q] \mid [\ell :: P \mid R]$. Finally one can see that in the communication rule (cm) the two complementary actions must share the same subject u. This means that communication may only occur on the same channel (a if $u = a$ or $u = a@\ell$), at the same location (ℓ if $u = a@\ell$). In other words, unlike in [2,7] where messages can transparently go through domains to reach the corresponding receiver, we have the "go and communicate" semantics of the Dπ-calculus, which is also the one of [5]. Typically, no transition can arise from $[\ell :: \overline{a}\vec{u}] \mid [\ell' :: a(\vec{v}).P]$ if $\ell \neq \ell'$, and one has to explicitly move and meet in order to communicate, like in $[\ell :: [\ell' :: \overline{a}\vec{u}]] \mid [\ell' :: a(\vec{v}).P]$.

3   A Simple Type System

In the polyadic π-calculus, the names are normally used according to some typing discipline, and the processes are checked to obey this discipline. We will do the same here, requiring the processes to conform to some typing assumption about names, since this will be needed to ensure the message deliverability property we are seeking for. The types we use generalize the usual sorts of the π-calculus, and in particular channel sorts will be types. We also have to assign types to location names, and to compound names. The idea here is that a location type should record the names and types of channels on which communication is possible inside a locality. Therefore, a location type is just a typing context of a certain kind, as in [8], while the type of a located channel is a "located channel type" $\gamma^{@}$ – which is simpler than an existential type, as used in [8]. We also introduce a type val, for names that may be tested for equality, but are not used as communication channels or location names – we allow the latter to be compared too. Then the types (for names) are as follows:

$$
\begin{array}{llll}
\tau, \sigma \ldots & ::= & \zeta \mid \psi \mid \gamma^{@} & \textit{types} \\
& \zeta & ::= val \mid \gamma & \textit{values and channel types} \\
\gamma, \delta \ldots & ::= & Ch(\tau_1, \ldots, \tau_n) & \textit{channel types} \\
\psi, \phi \ldots & ::= & \{a_1 : \gamma_1, \ldots, a_n : \gamma_n\} & \textit{location types}
\end{array}
$$

$$a : val \vdash_\ell a : val \qquad a@\ell : \gamma \vdash_\ell a : \gamma \qquad \psi@\ell \vdash_{\ell'} \ell : \psi$$

$$\frac{}{a@\ell : \gamma \vdash_{\ell'} a@\ell : \gamma^@} \qquad \frac{\Psi \vdash_\ell \vec{u} : \vec{\tau} \quad , \quad \Phi \vdash_\ell \vec{v} : \vec{\sigma}}{\Psi , \Phi \vdash_\ell \vec{u} : \vec{\tau}, \vec{v} : \vec{\sigma}}$$

Fig. 2. Type system for names

where in location types the a_i's are distinct names, and the order of items $a_i : \gamma_i$ is irrelevant. The typing judgements have the form $\Psi \vdash_\ell P$, meaning that P, when located at ℓ, uses the names as prescribed by the typing context Ψ. A typing context Ψ is a pair (Ψ_{typ}, Ψ_{loc}) of mappings from a finite subset $\text{dom}(\Psi)$ of $\mathcal{N} \cup \mathcal{P}$, satisfying the following constraints – where we write $\ell \notin \Psi$ to mean that ℓ is neither in the domain of Ψ nor in the types assigned by Ψ:

1. Ψ_{typ} assigns to each $x \in \text{dom}(\Psi)$ a value or channel type (i.e. a ζ type).
2. Ψ_{loc} assigns to each $x \in \text{dom}(\Psi)$ a finite set of names, such that $\ell \in \Psi_{loc}(x) \Rightarrow \ell \notin \Psi$.
3. if $\Psi_{typ}(x) = val$ or $x \in \mathcal{P}$ then $\Psi_{loc}(x) = \emptyset$.

A context Ψ such that $\text{dom}(\Psi) = \{x_1, \ldots, x_n\}$ with $\Psi_{typ}(x_i) = \zeta_i$ and $\Psi_{loc}(x_i) = L_i$ will be written:

$$x_1@L_1 : \zeta_1, \ldots, x_n@L_n : \zeta_n$$

This means that the name x_i is used at the locations contained in L_i, uniformly with type ζ_i. We abbreviate $x@\{\ell\} : \gamma$ as $x@\ell : \gamma$, and we simply write $x_i : \zeta_i$ if $L_i = \emptyset$. In the typing rules we use the notation Ψ, Φ for the context defined as follows:

$$\text{dom}(\Psi, \Phi) = \text{dom}(\Psi) \cup \text{dom}(\Phi)$$

$$(\Psi, \Phi)_{typ}(x) = \begin{cases} \Psi_{typ}(x) \text{ if } x \in \text{dom}(\Psi) \\ \Phi_{typ}(x) \text{ if } x \in \text{dom}(\Phi) \end{cases}$$

$$(\Psi, \Phi)_{loc}(x) = \Psi_{loc}(x) \cup \Phi_{loc}(x)$$

Notice that this context is undefined if $\Psi_{typ}(x) \neq \Phi_{typ}(x)$ for some $x \in \text{dom}(\Psi, \Phi)$, or if there exist $x \in \text{dom}(\Psi, \Phi)$ and $\ell \in (\Psi_{loc}(x) \cup \Phi_{loc}(x))$ such that $\ell \in \text{dom}(\Psi, \Phi)$ or ℓ occurs in some type assigned by Ψ or Φ. To state the typing rules for $\Psi \vdash_\ell P$, we first need to introduce a system for establishing sequents of the form

$$\Psi \vdash_\ell u_1 : \tau_1, \ldots, u_n : \tau_n$$

that is for computing the types assigned to names by the context Ψ at the current location ℓ. The rules for inferring these judgements are given in Figure 2, where we write $\psi@\ell$ for $a_1@\ell : \gamma_1, \ldots, a_n@\ell : \gamma_n$ if $\psi = \{a_1 : \gamma_1, \ldots, a_n : \gamma_n\}$. The typing rules for processes are collected in Figure 3, where we use the following:

$$\frac{\Psi \vdash_\ell \vec{u} : \vec{\tau}}{a@\ell : Ch(\vec{\tau}), \Psi, \Phi \vdash_\ell \overline{a}\vec{u}} \qquad \frac{a@\ell : Ch(\vec{\tau}), \Psi, \Phi \vdash_\ell P \quad , \quad \Psi \vdash_\ell \vec{u} : \vec{\tau}}{a@\ell : Ch(\vec{\tau}), \Phi \vdash_\ell a(\vec{u}).P}$$

$$\frac{\Psi \vdash_\ell P \;,\; \Psi \vdash_\ell Q}{\Psi \vdash_\ell P \mid Q} \qquad \frac{\Psi \vdash_\ell P \;,\; \Psi \vdash_\ell Q}{\Psi \vdash_\ell [a = b]P, Q} \; (*) \qquad \frac{a@\ell : \gamma, \Psi \vdash_\ell P}{\Psi \vdash_\ell (\nu a)P}$$

$$\frac{a@\ell : \gamma, \Psi \vdash_{\ell'} P}{\Psi \vdash_{\ell'} (\nu a@\ell)P} \qquad \frac{a : val, \Psi \vdash_\ell P}{\Psi \vdash_\ell (\nu a : val)P} \qquad \frac{\psi@\ell, \Psi \vdash_{\ell'} P}{\psi, \Psi \vdash_{\ell'} (\nu \ell : \psi)P}$$

$$\frac{\Psi \vdash_\ell P}{\Psi \vdash_{\ell'} [\ell :: P]} \; \ell' \notin \Psi \qquad \frac{\Psi \vdash_\ell \vec{u} : \vec{\tau}}{a@\ell : \gamma, A : Ch(\gamma, \vec{\tau}), \Psi, \Phi \vdash_\ell A(a; \vec{u})}$$

$$\frac{a@\ell : \gamma, A : Ch(\gamma, \vec{\tau}), \Psi, \Phi \vdash_\ell P \quad , \quad \Psi \vdash_\ell \vec{u} : \vec{\tau} \quad , \quad \Psi' \vdash_\ell \vec{v} : \vec{\tau}}{b@\ell : \gamma, \Psi', \Phi \vdash_\ell (\text{rec } A(a; \vec{u}).P)(b; \vec{v})}$$

(*) $\Psi_{typ}(a) = val = \Psi_{typ}(b)$ or $a, b \notin \Psi$.

Fig. 3. Type system for terms

CONVENTION. *In the rules for the binding constructs, that is input, restriction and recursion, we implicitly have the usual condition that the bound names do not occur in the resulting context.*

Let us comment some of the rules of the system. We see that a simple name may only be used if it is of type *val*, or if it is a channel located at the current location, or if it is a location name. In this latter case, its type is the collection of channel names, together with their type, located at that location. A compound name may be used at any locality, but with a located type. In the rule for name comparison, we see that we can compare both names of value type and location names since $\ell \notin \Psi$ is true if ℓ is used as a locality in Ψ. To type $[\ell :: P]$ one must be able to type P at locality ℓ, while the resulting current locality of $[\ell :: P]$ is arbitrary. Finally in the rules for recursive processes, we note that, to ensure that the parameters are used in a consistent way, we assign to the process identifier a (channel) type.

Our type system is very close to the simple type system for Dπ presented in [8], except that we use located types $\gamma^@$ instead of existential types. Our main result about typing is the standard "subject reduction property". This result is needed to establish the message deliverability property (Theorem 4.4).

$$\frac{}{\Vdash \overline{a}\vec{u}} \qquad \frac{a \Vdash P}{a \Vdash a(\vec{u}).P} \, a \notin nm(\vec{u}) \qquad \frac{I \Vdash P \,,\, I' \Vdash Q}{I, I' \Vdash (P \mid Q)} \, I \cap I' = \emptyset$$

$$\frac{u, I \Vdash P}{I \Vdash (\nu u)P} \qquad \frac{I \Vdash P}{I \Vdash (\nu a : val)P} \qquad \frac{dom(\psi)@\ell,\, I \Vdash P}{I \Vdash (\nu \ell : \psi)P} \qquad \frac{I \Vdash P \,,\, I \Vdash Q}{I \Vdash [a = b]P, Q}$$

$$\frac{}{a \Vdash A(a; \vec{u})} \qquad \frac{a \Vdash P}{b \Vdash (rec\, A(a; \vec{u}).P)(b; \vec{v})} \qquad \frac{I \Vdash P}{I@\ell \Vdash [\ell :: P]} \, (*)$$

$(*) \quad \{ a \mid a, a@\ell \in I \} = \emptyset$

Fig. 4. Well-formed terms

Theorem 1 (Preservation Of Typing). *If* $\Psi \vdash_\ell P$ *and* $P \xrightarrow{\alpha} P'$ *then*

(i) *if* $\alpha = u\vec{v}$ *then* $\Psi = a@\ell' : Ch(\vec{\tau}), \Phi$ *with* $a@\ell' = u$ *or* $a = u$ *and* $\ell' = \ell$, *and* $\Psi', \Psi \vdash_\ell P'$ *where* $\Psi' \vdash_{\ell'} \vec{v} : \vec{\tau}$,

(ii) *if* $\alpha = (\nu\vec{w})\overline{u}\vec{v}$ *then* $\Psi = a@\ell' : Ch(\vec{\tau}), \Psi', \Phi$ *with* $a@\ell' = u$ *or* $a = u$ *and* $\ell' = \ell$, *and there exists* Ψ'' *such that* $\Psi', \Psi'' \vdash_{\ell'} \vec{v} : \vec{\tau}$ *and* $\Psi', \Psi'', \Phi \vdash_\ell P'$,

(iii) *if* $\alpha = \tau$ *then* $\Psi \vdash_\ell P'$.

4   Interfaces and Receptive Processes

It is easily seen that the type system does not guarantee the kind of safety property we are looking for: we can obviously type non-receptive processes, and we can also type processes sending a message that will never find a corresponding receiver, like for instance $(\nu a)\overline{a}$. Then we introduce an inference system for checking "well-formedness" of processes. Basically, to be well-formed a process must not contain nested inputs on different names. Moreover, any input must be involved in a recursive process that makes the receiver reincarnate itself, possibly in a different state, after being consumed. In addition, we shall impose, as in π_1 [2], that there is a unique receiver for each name, that is, two parallel components of a process are not allowed to receive on a same name. Last but not least, we demand that to restrict the scope of a name of channel type, we know for a fact that a resource is provided – that is, a receiver exists – for that name.

The well-formed processes are the ones for which a statement $I \Vdash P$, that is "P is well-formed with interface I", can be proved. In this statement the interface I is a finite set of names on which a process may perform an input. We present the rules for well-formedness in Figure 4, where we use the same convention as for the typing regarding the binding constructs, and where, as usual, a set $I = \{u_1, \ldots, u_n\}$ is represented as the list $u_1, \ldots, u_n$ of its elements, and union is written I, I'. Our first result about well-formedness is, again, a "subject reduction property":

Proposition 1 (Preservation of Well-Formedness). *If $I \Vdash P$ and $P \xrightarrow{\alpha} P'$ then*

(i) $I \Vdash P'$ *if $\alpha = \tau$ or $\alpha = u\vec{v}$,*

(ii) *if $\alpha = (\nu\vec{w})\overline{u}\vec{v}$ and $J = \{u \mid \exists i.\ w_i = u\} \cup \{a@\ell \mid \exists i.\ w_i = \ell : \psi\ \&\ a \in \mathrm{dom}(\psi)\}$ then $I, J \Vdash P$.*

We can now establish the receptiveness property for our distributed calculus. To state this property, let us define the predicate $P \downarrow u$, meaning "P may perform an input with subject u", that is:

$$P \downarrow u \quad \Leftrightarrow_{\mathrm{def}} \quad \exists\vec{v}\ \exists P'.\ P \xrightarrow{u\vec{v}} P'$$

Proposition 2 ((Receptiveness). *Let P be a closed well-formed term. Then:*

(i) *if $I \Vdash P$ then $P \downarrow u$ iff $u \in I$,*

(ii) *if $P \downarrow u$ and $P \xrightarrow{\alpha} P'$ then $P' \downarrow u$.*

This result suggests the denomination "*distributed (asynchronous) receptive π-calculus, with unique receivers*", in short the Dπ_1^r-calculus, for the set of well-formed closed processes, which is closed by labelled transitions. Similarly we call π_1^r, that is "*the receptive π_1-calculus*", the sub-calculus where we do not use any locality based feature.

We now turn to the issue of message delivery. We aim at showing that, if a message is sent (at some locality) on a channel of a known scope in a well-formed and typed process, then the process contains a receiver for this message. Let us denote by $P \downarrow \overline{u}$ the fact that P performs an output action with subject u, and let us define

$$P \simeq Q \quad \Leftrightarrow_{\mathrm{def}} \quad \begin{cases} P \sim Q & \text{and} \\ \forall \Psi, \ell.\ \Psi \vdash_\ell P \Leftrightarrow \Psi \vdash_\ell Q & \text{and} \\ \forall I.\ I \Vdash P \Leftrightarrow I \Vdash Q \end{cases}$$

As a preliminary result, we show that any process that is able to send a message is equivalent to another having a special form:

Lemma 1. $P \downarrow \overline{a}$ *iff $P \simeq (\nu\vec{w})(\overline{a}\vec{v} \mid R)$ for some $\vec{w}$, $\vec{v}$ and R with $a \notin \mathrm{subj}(\vec{w})$. $P \downarrow \overline{a@\ell}$ iff $P \simeq (\nu\vec{w})([\ell :: \overline{a}\vec{v}] \mid R)$ for some $\vec{w}$, $\vec{v}$ and R with $a \notin \mathrm{subj}(\vec{w})$.*

Now we can prove our main result, where $\to$ stands for $\xrightarrow{\tau}$:

Theorem 2 (Message Deliverability). *Let P be a closed well-formed and typed process with $I \Vdash P$ and $\Psi \vdash_\ell P$. If $P \xrightarrow{*} P'$ then*

(i) *if $P' \simeq (\nu\vec{w})(\overline{a}\vec{v} \mid R)$ with $a \in I \cup \mathrm{subj}(\vec{w})$ then $R \downarrow a$ or $R \downarrow a@\ell$.*

(ii) *if $P' \simeq (\nu\vec{w})([\ell' :: \overline{a}\vec{v}] \mid R)$ with $a@\ell' \in I$ or $a \in \mathrm{subj}(\vec{w})$ then $R \downarrow a@\ell'$ or $\ell' = \ell$ and $R \downarrow a$.*

Note that this result does not hold for untyped terms. For instance we have $\Vdash (\nu a : val)\overline{a}$ or $\Vdash (\nu a : \emptyset)\overline{a}$, and these terms contain a message that cannot be delivered.

5   Encoding the π_1-Calculus

In this section, we show that there is a translation from the π-calculus to $D\pi_1^r$. It is shown in [3] that the joined input of the join-calculus [6] can be defined in the π_1-calculus up to weak asynchronous bisimulation. On the other hand, it has been shown in [6] that there is a fully abstract translation of the asynchronous π-calculus in the join-calculus. Therefore, if we can translate the π_1-calculus in the $D\pi_1^r$-calculus – or rather in the π_1^r-calculus –, we can reasonably claim that the nice properties of the $D\pi_1^r$-calculus are not obtained at the expense of its expressive power. We now give such a translation, that we show to be fully abstract with respect to a refined form of asynchronous bisimulation [1,2], defined as follows:

Definition 1 (Asynchronous Bisimulation). *A symmetric relation S is an asynchronous bisimulation if $P\,S\,Q$ implies*

(i) *there exists I such that $I \Vdash P$ and $I \Vdash Q$, and*

(ii) *if $P \xrightarrow{\tau} P'$ then $Q \xrightarrow{\tau} Q'$ for some Q' such that $P'\,S\,Q'$,*

(iii) *if $P \xrightarrow{(\nu\vec{w})\overline{u}\,\vec{v}} P'$ with $u \notin I$ and $\mathsf{subj}(\vec{w}) \cap \mathsf{fn}(Q) = \emptyset$ then $Q \xrightarrow{(\nu\vec{w})\overline{u}\,\vec{v}} Q'$ for some Q' such that $P'\,S\,Q'$,*

(iv) *if $P \xrightarrow{u\,\vec{v}} P'$ then either $Q \xrightarrow{u\,\vec{v}} Q'$ with $P'\,S\,Q'$, or $Q \xrightarrow{\tau} Q'$ with $P'\,S\,(Q' \mid R)$ where $R = \overline{a}\vec{v}$ if $u = a$ and $R = [\ell :: \overline{a}\vec{v}]$ if $u = a@\ell$.*

We denote with $\sim_a$ the greatest asynchronous bisimulation. The notion of weak asynchronous bisimulation is obtained by replacing everywhere transitions with weak transitions. We denote with $\approx_a$ the greatest weak asynchronous bisimulation.

As a source calculus, we will consider the π_1-calculus with guarded recursion. The idea of the encoding is quite standard and simple (see [2,3,6] for similar encodings). We turn any message on a channel a into a request to a *channel manager* $CM(a)$ for a, sending the arguments of the message together with a key *out*. Symmetrically, we turn any input on a into a request to $CM(a)$, sending a key *in* and a private return channel to actually receive something. The channel manager will filter the messages according to the keys, and act as appropriate. However, there is an attack on this encoding which compromises abstraction: the environment can send a request for input to the channel manager. We then authenticate the requests for input by introducing a restricted key in_a for every channel manager (of a) which is known only by the process that can actually input on the channel a. To formulate our encoding, we will use several notational conventions and abbreviations. Let us first define the *identity* agent, and recall the *input once* construct of [2], given by

$$Id_a =_{\mathrm{def}} (\mathsf{rec}\, A(a;).a(\vec{u}).(\overline{a}\vec{u} \mid A(a;)))(a;)$$
$$a(\vec{u}){:}P =_{\mathrm{def}} a(\vec{u}).(P \mid Id_a)$$

Then

$$
\begin{array}{rcl}
\operatorname{rec} A(a; \vec{u}).P & \text{stands for} & (\operatorname{rec} A(a; \vec{u}).P)(a; \vec{u}) \\
(\nu a)P & \text{for} & (\nu a)(P \mid Id_a) \qquad (a \text{ not free in } P) \\
\bar{a}(\vec{u}, \_, \vec{v}) & \text{for} & (\nu c)(\bar{a}(\vec{u}, c, \vec{v}) \mid Id_c) \\
a(\vec{u}, \_, \vec{v}).P & \text{for} & a(\vec{u}, c, \vec{v}).P \qquad (c \text{ not in } P) \\
(\operatorname{rec} A(; \vec{b}).P)(; \vec{c}) & \text{for} & (\nu a'')(\operatorname{rec} A(a; \vec{b}).(Id_a \mid P'))(a''; \vec{c})
\end{array}
$$

where in the last clause P' is P where every free occurrence of $A(; \vec{c})$ is replaced with $(\nu a')A(a'; \vec{c})$. We shall also need a kind of internal choice (similar to the one given in [2] by means of "booleans"), $P \oplus_a Q$ where a is a channel name of type $Ch(\vec{\tau}, val, val)$, and P and Q are such that $a \Vdash P, Q$. This is defined as follows – where $c \neq c'$, $[x \neq y]P, Q$ is $[x = y]Q, P$ and, as usual, all the introduced names are fresh:

$$
\begin{aligned}
P \oplus_a Q =_{\text{def}} (\nu c \colon val)\big(\bar{a}(\_, c, c) \mid (\nu c' \colon val)\bar{a}(\_, c, c') \mid \operatorname{rec} A(a;).a(\vec{u}, x, y). \\
[x \neq c](\bar{a}(\vec{u}, x, y) \mid A(a;)), \\
\operatorname{rec} A'(a;).a(\vec{u}, x', y').[x' \neq c](\bar{a}(\vec{u}, x', y') \mid A'(a;)), \\
[x = y]P, Q\big)
\end{aligned}
$$

It is easy to see that this is a well-formed term, with interface $\{a\}$. In the following translation, the new names are assumed to be fresh. In particular, for every name a we assume a fresh name in_a (that is, not in the set of names of π_1). The name in_a is used as the key of the channel a. We translate a well-formed term $I \Vdash P$ (where P respects some sorting) of the π_1-calculus, where $I = \{a_1, \dots, a_n\}$, in the following process of the $D\pi_1^r$-calculus:

$$
\llbracket I \Vdash P \rrbracket = (\nu\, in_{a_1} \colon val) \cdots (\nu\, in_{a_n} \colon val)(CM(a_1;) \mid \cdots \mid CM(a_n;) \mid \llbracket P \rrbracket)
$$

which turns out to be also well-formed in the context I. The type of a channel name, say $Ch(\vec{\tau})$, is transformed into $Ch(val, \vec{\tau}, Ch(\vec{\tau}), val, val)$: the first argument is the input/output key of the channel, then we have the arguments of the message to be delivered, followed by the type of the return channel to which they are actually sent, and then we have two keys for internal choice. The channel manager $CM(a;)$ is given by:

$$
CM(a;) = \operatorname{rec} A(a;).a(j, \vec{b}, s, c_1, c_1').a(i, \vec{d}, r, c_2, c_2').
$$
$$
L \oplus_a \text{ if } j = in_a \text{ then } L \text{ else if } i \neq in_a \text{ then } L \text{ else } (\vec{r}\,\vec{b} \mid A(a;)) \qquad \text{where}
$$

$$
L = A(a;) \mid \bar{a}(j, \vec{b}, s, c_1, c_1') \mid \bar{a}(i, \vec{d}, r, c_2, c_2')
$$

The process $\llbracket P \rrbracket$ is defined as follows:

$$
\llbracket \bar{a}\,b \rrbracket = \bar{a}(\_, \vec{b}, \_, \_, \_)
$$
$$
\llbracket a(\vec{b}).P \rrbracket = (\nu r)(\bar{a}(in_a, \_, r, \_, \_) \mid r(\vec{b}) \colon \llbracket P \rrbracket)
$$
$$
\llbracket P \mid Q \rrbracket = (\llbracket P \rrbracket \mid \llbracket Q \rrbracket)
$$
$$
\llbracket [a = b]P, Q \rrbracket = [a = b]\llbracket P \rrbracket, \llbracket Q \rrbracket
$$
$$
\llbracket (\nu a)P \rrbracket = (\nu a)(\nu\, in_a \colon val)(CM(a;) \mid \llbracket P \rrbracket)
$$
$$
\llbracket (\operatorname{rec} A(\vec{a}; \vec{b}).P)(\vec{c}; \vec{d}) \rrbracket = (\operatorname{rec} A(; \vec{a}, \overrightarrow{in_a}, \vec{b}).\llbracket P \rrbracket)(; \vec{c}, \overrightarrow{in_c}, \vec{d})
$$
$$
\llbracket A(\vec{a}; \vec{b}) \rrbracket = A(; \vec{a}, \overrightarrow{in_a}, \vec{b})
$$

Our main result about this translation of the π_1-calculus into the receptive π_1-calculus is that it is fully abstract with respect to weak asynchronous bisimulation:

Theorem 3. *Assume that $I \Vdash P$ and $I \Vdash Q$ in the π_1-calculus. Then*

$$P \approx_a Q \;\Leftrightarrow\; [\![I \Vdash P]\!] \approx_a [\![I \Vdash Q]\!]$$

The proof, which is to be found in the full version of the paper (together with examples showing the expressivity of our calculus with respect to distributed systems), goes through an analysis of the various actions that may be performed by $[\![I \Vdash P]\!]$. We then show that an exact simulation of the behaviour of π_1 processes is provided by the translation.

References

1. R. AMADIO, I. CASTELLANI and D. SANGIORGI, *On bisimulations for the asynchronous calculus*, CONCUR'96, Springer Lect. Notes in Comp. Sci. 1119 (1996) 147-162.

2. R. AMADIO, *An asynchronous model of locality, failure, and process mobility*, In Proc. COORDINATION'97, Springer Lect. Notes in Comp. Sci. 1282 (1997). Extended version appeared as Res. Report INRIA 3109.

3. R. AMADIO, *On modeling mobility*, Journal of Theoretical Computer Science, to appear (1999).

4. G. BOUDOL, *Typing the use of resources in a concurrent calculus*, Proc. ASIAN 97, Springer Lect. Notes in Comp. Sci. 1345 (1997) 239-253.

5. L. CARDELLI and A. GORDON, *Mobile ambients*, In Proc. FoSSaCS, ETAPS'98, Springer Lect. Notes in Comp. Sci. 1378 (1998) 140-155.

6. C. FOURNET and G. GONTHIER, *The reflexive CHAM and the join-calculus*, In Proc. ACM Principles of Prog. Lang. (1996) 372-385.

7. C. FOURNET, G. GONTHIER, J.-J. LÉVY, L. MARANGET, and D. RÉMY, *A calculus of mobile agents*, In Proc. CONCUR'96, Springer Lect. Notes in Comp. Sci. 1119 (1996) 406-421.

8. M. HENNESSY and J. RIELY, *Resource access control in systems of mobile agents*, Techn. Report 2/98, School of Cognitive and Computer Sciences, University of Sussex (1998).

9. N. KOBAYASHI, *A partially deadlock-free typed process calculus*, ACM TOPLAS Vol. 20 No 2 (1998) 436-482.

10. D. SANGIORGI, *The name discipline of uniform receptiveness*, In Proc. ICALP'97, Springer Lect. Notes in Comp. Sci. 1256 (1997) 303-313.

Series and Parallel Operations on Pomsets

Zoltán Ésik[1]* and Satoshi Okawa[2]

[1] Dept. of Computer Science, A. József University
Aradi v. tere 1., 6720 Szeged Hungary
esik@inf.u-szeged.hu
[2] Department of Computer Software, The University of Aizu
Aizu-Wakamatsu-City, Fukushima 965-8580 Japan
okawa@u-aizu.ac.jp

Abstract. We consider two-sorted algebras of pomsets (isomorphism classes of labeled partial orders) equipped with the operations of series and parallel product and series and parallel omega power. The main results show that these algebras possess a non-finitely based polynomial time decidable equational theory, which can be axiomatized by an infinite set of simple equations. Along the way of proving these results, we show that the free algebras in the corresponding variety can be described by generalized series-parallel pomsets. We also provide a graph theoretic characterization of the generalized series-parallel pomsets.

1 Introduction

Partially ordered structures, and in particular isomorphism classes of labeled partial orders, called *pomsets* for Partially Ordered MultiSETs, have been used extensively to give semantics to concurrent languages [14,5,13,7,1,2], both in the operational and denotational (ordered and metric) framework, and to Petri nets [9,20,12,21,19], to mention a few references. (Pomsets are called partial words in [9].) The paper [6] deals with the relation between pomsets and Mazurkiewicz traces. For automata accepting pomset languages, i.e., sets of pomsets, we refer to [11]. The partial order is usually interpreted as a causal dependence between the events. The event structures of Winskel [22] are pomsets enriched with a conflict relation subject to certain conditions. In [5], the computations determined by event structures are modeled by pomsets. Some authors only allow for finite pomsets, while others also use pomsets of infinite size, but very often place some restrictions on the partial order or labeling. There is an extensive literature dealing with the pomsets themselves, and pomsets in relation to languages, see, e.g., [17,8,16,15,3,4].

A wide variety of operations has been defined on pomsets. The definitions are motivated by the intended applications. However, there are two operations that play a central role, the series and the parallel product. The series product $P \cdot Q$ of two pomsets P and Q is obtained by taking the disjoint union of P

* Partially supported by the grants OTKA T30511 and FKFP 247/1999.

C. Pandu Rangan, V. Raman, R. Ramanujam (Eds.): FSTTCS'99, LNCS 1738, pp. 316–328, 1999.
© Springer-Verlag Berlin Heidelberg 1999

and Q and making each vertex of Q larger than any vertex of P. The parallel product $P \otimes Q$ is just disjoint union. There are also other names used in the literature, e.g., concatenation and concurrence or concurrent composition [14,8], sequential and parallel composition [5,7,1], sequencing and concurrency [6], concatenation and disjoint union [15], sequential or serial product and parallel or shuffle product [9,3]. The motivation for using the term shuffle is due to that in the simplest language model of concurrency parallel composition is modeled by shuffle. A mathematical justification of the term is given by the result, proved independently in [16] and [3], that languages equipped with concatenation and shuffle satisfy the same set of equations as pomsets equipped with series and parallel product. In fact, these equations can be captured by the bisemigroup[1] axioms expressing that both operations are associative and parallel product is commutative. The series-parallel pomsets, i.e., those pomsets that can be generated from the singletons by series and parallel product, have a well-known graph theoretic characterization [9,17].

In this paper, we will consider non-empty countable pomsets. Since pomsets model the behavior of processes that can be executed in at most ω steps, we restrict the operation of series product to instances $P \cdot Q$ where P is finite. (See also the last paragraph of Section 8. We could also restrict ourselves to pomsets which have a linearization to an ω-chain, this would not alter our results.) Given a pomset P, we also define the series and parallel omega powers $P^\omega = P \cdot P \cdot \ldots$ and $P^{(\omega)} = P \otimes P \otimes \ldots$ Thus, P^ω and $P^{(\omega)}$ solve the fixed-point equations $X = P \cdot X$ and $Y = P \otimes Y$, respectively. The series omega power P^ω is used here only when P is finite. The pomsets P^ω and $P^{(\omega)}$ represent (sequential and parallel) infinite looping behaviors.

Since some of the operations require that an argument is a finite pomset, we will work with two-sorted algebras of pomsets. The domains corresponding to the two sorts, the finite and the infinite sort, consist of the finite non-empty and the countably infinite pomsets whose action labels are in a given set A. We present a simple infinite equational basis E of these algebras (Corollary 2) and show that the equational theory is not finitely based (Theorem 8). We also show that the equational theory is decidable in polynomial time (Theorem 9). Along the way of establishing these results, in Theorems 3 and 7 we give a concrete description by generalized series-parallel pomsets of the free algebras in the variety $\mathcal{V}$ axiomatized by the equations E. We also give a graph theoretic characterization of the generalized series-parallel pomsets (Theorem 6).

The series omega power operation $^\omega$, in conjunction with series and parallel product, has already been studied in the recent paper [4]. Some of the results of the present paper extend corresponding results in [4].

---

[1] The empty pomset is neutral element for both series and parallel product. Bisemigroups with a neutral element are called double monoids in [9], dioids in [5], and bimonoids in [3].

2   Pomsets

We will consider finite non-empty and countably infinite posets $P = (P, \leq_P, \ell_P)$ whose elements, called vertices, are labeled in a set A of actions, or labels, so that ℓ_P is a function $P \to A$. An isomorphism of A-labeled posets is an order isomorphism which preserves the labeling. An *A-labeled pomset*, or just pomset [14], for short, is an isomorphism class of A-labeled posets. Below we will identify isomorphic labeled posets with the pomset they represent.

Some notation For each non-negative integer n we denote the set $\{1, \ldots, n\}$ by $[n]$.

Suppose that $P = (P, \leq_P, \ell_P)$ and $Q = (Q, \leq_Q, \ell_Q)$ are pomsets. We define several operations, some of which will require that P is finite.

SERIES PRODUCT. If P is finite, then the series product of P and Q is constructed by taking the disjoint union of P and Q and making each vertex of Q larger than any vertex of P. Thus, assuming without loss of generality that P and Q are disjoint, $P \cdot Q = (P \cup Q, \leq_{P \cdot Q}, \ell_{P \cdot Q})$, where, for any $u, v \in P \cup Q$,

$$u \leq_{P \cdot Q} v \Leftrightarrow (u \in P \text{ and } v \in Q) \text{ or } u \leq_P v \text{ or } u \leq_Q v$$

$$\ell_{P \cdot Q}(u) = \begin{cases} \ell_P(u) \text{ if } u \in P \\ \ell_Q(u) \text{ if } u \in Q. \end{cases}$$

PARALLEL PRODUCT. The parallel product of P and Q is constructed as the disjoint union of P and Q. Thus, $P \otimes Q = (P \cup Q, \leq_{P \otimes Q}, \ell_{P \otimes Q})$, where we again assume that P and Q are disjoint. Moreover, for any $u, v \in P \cup Q$,

$$u \leq_{P \otimes Q} v \Leftrightarrow u \leq_P v \text{ or } u \leq_Q v.$$

The function $\ell_{P \otimes Q}$ is defined as $\ell_{P \cdot Q}$ above.

SERIES OMEGA POWER. Assume that P is finite. The series product of ω copies of P is called the series omega power of P. Thus, denoting $N = \{1, 2, \ldots\}$, $P^\omega = (P \times N, \leq_{P^\omega}, \ell_{P^\omega})$, where

$$(u, i) \leq_{P^\omega} (v, j) \Leftrightarrow i < j \text{ or } (i = j \text{ and } u \leq_P v)$$

$$\ell_{P^\omega}((u, i)) = \ell_P(u), \quad \text{for all } (u, i), (v, j) \in P \times N.$$

PARALLEL OMEGA POWER. The parallel omega power of P is the disjoint sum of P with itself ω-times. Thus, $P^{(\omega)} = (P \times N, \leq_{P^{(\omega)}}, \ell_{P^{(\omega)}})$, where

$$(u, i) \leq_{P^{(\omega)}} (v, j) \Leftrightarrow i = j \text{ and } u \leq_P v$$

$$\ell_{P^{(\omega)}}((u, i)) = \ell_P(u), \quad \text{for all } (u, i), (v, j) \in P \times N.$$

Below we will sometimes write ωP for $P^{(\omega)}$. Similarly, for each integer $n \geq 1$, we define nP to be the n-fold parallel product of P with itself.

Equipped with these operations, the (countable non-empty) A-labeled pomsets form a two-sorted algebra $\omega \mathbf{Pom}(A) = (\mathbf{Pom}_F(A), \mathbf{Pom}_I(A), \cdot, \otimes, ^\omega, ^{(\omega)})$, where $\mathbf{Pom}_F(A)$ and $\mathbf{Pom}_I(A)$ denote the collections of all finite non-empty and countably infinite A-labeled pomsets, respectively. We would like to know the equations satisfied by the algebras $\omega \mathbf{Pom}(A)$.

Proposition 1. *The following equations hold in any algebra $\omega\mathbf{Pom}(A)$.*

$$x \cdot (y \cdot u) = (x \cdot y) \cdot u \tag{1}$$

$$u \otimes (v \otimes w) = (u \otimes v) \otimes w \tag{2}$$

$$u \otimes v = v \otimes u \tag{3}$$

$$(x \cdot y)^\omega = x \cdot (y \cdot x)^\omega \tag{4}$$

$$(x^n)^\omega = x^\omega, \quad n \geq 2 \tag{5}$$

$$u \otimes \omega u = \omega u \tag{6}$$

$$\omega(u \otimes v) = \omega u \otimes \omega v \tag{7}$$

$$\omega u \otimes \omega u = \omega u \tag{8}$$

$$\omega(\omega u) = \omega u, \tag{9}$$

where x, y range over finite pomsets and u, v, w range over finite and infinite pomsets.

Recall from the Introduction that one-sorted structures equipped with operations $\cdot$ and $\otimes$ satisfying the equations (1), (2) and (3) are called *bisemigroups*.

Some subalgebras of $\omega\mathbf{Pom}(A)$ are also of interest. Let $\mathbf{SP}_A$ denote the set of all (finite) pomsets generated from the singleton pomsets corresponding to the elements of A by the operations of series and parallel product. Pomsets in $\mathbf{SP}_A$ are called *series-parallel*. Moreover, let $\mathbf{SP}_A^{\omega,(\omega)}$ denote the set of all infinite pomsets generated from the singletons by the operations of series and parallel product, and series and parallel omega power, and let $\mathbf{SP}_A^\omega$ denote the set of all infinite pomsets that can be constructed from the singletons by the two product operations and the series omega power operation. It follows by a straightforward induction on the number of applications of the operations that any pomset in $\mathbf{SP}_A^{\omega,(\omega)}$ has a linearization to an ω-chain. The elements of $\mathbf{SP}_A^{\omega,(\omega)}$ are called *generalized series-parallel pomsets*. We usually identify any letter $a \in A$ with the corresponding singleton pomset labeled a. For further reference, we recall

Theorem 1. Grabowski [9] $\mathbf{SP}_A$ *is the free bisemigroup generated by A.*

Theorem 2. Bloom and Ésik [4] *The algebra $(\mathbf{SP}_A, \mathbf{SP}_A^\omega, \cdot, \otimes, {}^\omega)$ is freely generated by A in the variety of all algebras equipped with the two product operations and the series omega power operation satisfying the equations (1) – (5).*

More precisely, in [4] the parallel product of two pomsets was defined only if both pomsets are finite or both are infinite, causing only a little change in the above result and its proof.

3   Freeness

Suppose that $C = (C_F, C_I, \cdot, \otimes, {}^\omega, {}^{(\omega)})$ is a two-sorted algebra, where the binary product operation $\cdot$ is defined only if its first argument is in C_F and the unary $^\omega$ operation only if its argument is in C_F. The arguments of the other binary product operation $\otimes$ as well as the argument of the unary $^{(\omega)}$ operation may come from both C_F and C_I. The result of applying the $^\omega$ or $^{(\omega)}$ operation is always in C_I. The result of applying a binary operation is in C_I iff one of the arguments of the operation is in C_I. We say that C satisfies the equations (1) – (9), or that these equations hold in C, if these equations hold in C when arbitrary elements of C_F and C_I are substituted for the variables with the proviso that elements of C_F are substituted for the variables x, y of finite sort. Let $\mathcal{V}$ denote the variety of all two-sorted algebras C equipped with the above operations satisfying the equations (1) – (9). Note that the equation $x \cdot x^\omega = x^\omega$ also holds in $\mathcal{V}$, where x is a variable of finite sort. Also, note that it is sufficient to require the power identities (5) only for prime numbers n.

Theorem 3. *The algebra* $\omega\mathbf{SP}_A = (\mathbf{SP}_A, \mathbf{SP}_A^{\omega,(\omega)}, \cdot, \otimes, ^\omega, ^{(\omega)})$ *is freely generated by A in $\mathcal{V}$.*

Proof. In our argument, we make use of the *rank* of a pomset $P \in \mathbf{SP}_A \cup \mathbf{SP}_A^{\omega,(\omega)}$, denoted rank$(P)$, defined to be the smallest number of applications of the operations by which P can be generated from the singletons. We will also make use of normal representations of A-labeled pomsets.

Suppose that P is an A-labeled pomset. A *series normal representation* of P is a representation $P = P_1 \cdot \ldots \cdot P_k$ or $P = R_1 \cdot \ldots \cdot R_m \cdot (S_1 \cdot \ldots \cdot S_n)^\omega$, where $k, n \geq 1$ and $m \geq 0$, and where the pomsets P_i, R_j and S_t, called the *components* of the representation, are serially indecomposable, i.e., none of them can be written as the series product of two pomsets.[2] Moreover, we require that $R_m \neq S_n$ if $m \geq 1$, and that $S_1 \cdot \ldots \cdot S_n$ cannot be written as a non-trivial series power of any pomset, or equivalently, as $(S_1 \cdot \ldots \cdot S_i)^{n/i}$, where i is a proper divisor of n. (Indeed, otherwise the representation could be simplified by the identity (4) or (5).)

A *parallel normal representation* of P is $P = n_1 P_1 \otimes \ldots \otimes n_k P_k$, where $k \geq 1$, $n_i \in \{1, 2, \ldots, \omega\}$, for all $i \in [k]$, and where the P_i are pairwise distinct and connected, i.e., parallelly indecomposable. Again, the pomsets P_i are called the components of the representation. If a pomset P is connected, a *normal representation* of P is a series normal representation of P. If P is disconnected, a normal representation of P is a parallel normal representation.

Note that if P is a singleton, or more generally, if P is serially (parallelly, respectively) indecomposable, then its unique series (parallel) normal representation is P. It is clear that each A-labeled pomset has at most one series and, up to a rearrangement of the components, at most one parallel normal representation. Thus, when it exists, we can refer to *the* series or parallel normal representation of a pomset. Suppose that $P \in \mathbf{SP}_A \cup \mathbf{SP}_A^{\omega,(\omega)}$. Then one can argue by a induction on rank(P) to prove that P has both a series and a parallel normal representation. Moreover, all components of the series and parallel normal representation of P are in $\mathbf{SP}_A \cup \mathbf{SP}_A^{\omega,(\omega)}$, and the rank of any component of the normal representation of P is strictly less than the rank of P.

---

[2] Recall that all pomsets considered in this paper are non-empty.

To complete the proof of Theorem 3, suppose that we are given a two-sorted algebra $C = (C_F, C_I, \cdot, \otimes, {}^\omega, {}^{(\omega)})$ in $\mathcal{V}$ together with a function $h : A \to C_F$. We need to show that h extends to a homomorphism $h^\sharp = (h_F^\sharp, h_I^\sharp) : \omega\mathbf{SP}_A \to C$, where $h_F^\sharp : \mathbf{SP}_A \to C_F$ and $h_I^\sharp : \mathbf{SP}_A^{\omega,(\omega)} \to C_I$. Of course, $h_F^\sharp$ is just the unique bisemigroup homomorphism $\mathbf{SP}_A \to C_F$ extending h_F which exists by Theorem 1. But we need to define $h_I^\sharp(P)$ for each $P \in \mathbf{SP}_A^{\omega,(\omega)}$. This is done by induction on $r = \mathrm{rank}\,(P)$. (The reader will probably be relieved that from now on we will omit the indices I and F.) We start the induction by $r = 0$ in which case there is nothing to prove, since there is no pomset in $\mathbf{SP}_A^{\omega,(\omega)}$ whose rank is 0. Suppose that $r > 0$. If P is connected, then let $P = P_1 \cdot \ldots \cdot P_k$ or $P = R_1 \cdot \ldots \cdot R_m \cdot (S_1 \cdot \ldots \cdot S_n)^\omega$ be its series normal representation. Since the rank of each component is less than r, it makes sense to define $h^\sharp(P) = h^\sharp(P_1) \cdot \ldots \cdot h^\sharp(P_k)$, in the first case, and $h^\sharp(P) = h^\sharp(R_1) \cdot \ldots \cdot h^\sharp(R_m) \cdot (h^\sharp(S_1) \cdot \ldots \cdot h^\sharp(S_n))^\omega$, in the second. If P is disconnected, then take its parallel normal representation $P = k_1 P_1 \otimes \ldots \otimes k_n P_n$. We define $h^\sharp(P) = k_1 h^\sharp(P_1) \otimes \ldots \otimes k_n h^\sharp(P_n)$. By the preceding observations and the associativity of the product operations and the commutativity of $\otimes$, function $h^\sharp$ is well-defined. The equations (1) – (9) ensure that $h^\sharp$ preserves the operations. The details are routine.   □

4   A Characterization

Ideals and *filters* of a pomset (or poset) P are defined as usual. Each non-empty subset of P is included in a smallest ideal and in a smallest filter, respectively called the ideal and the filter generated by the set. An ideal (filter, resp.) generated by a singleton set is called a *principal ideal* (*principal filter*, resp.). A filter F is connected if F is connected as a partial ordered set, equipped with the induced partial order. Below each filter and ideal of a pomset P will be considered to be a pomset determined by the partial order and labeling inherited from P.

Theorem 4. Grabowski [9], Valdes, Lawler and Tarjan [17] *An A-labeled pomset P belongs to $\mathbf{SP}_A$ iff P is finite and satisfies the N-condition, i.e., P does not have a four-element subposet $\{u_1, u_2, u_3, u_4\}$ whose non-trivial order relations are given by $u_1 < u_3$, $u_2 < u_3$ and $u_2 < u_4$.*

The following facts are clear.

Lemma 1. *Any subpomset determined by a non-empty subset of a pomset satisfying the N-condition also satisfies this condition.*

Lemma 2. *Suppose that a connected pomset P satisfies the N-condition. Then any two vertices of P have an upper or a lower bound.*

We also recall

Theorem 5. Bloom and Ésik [4] *An A-labeled pomset P belongs to $\mathbf{SP}_A^\omega$ iff P is (countably) infinite and the following hold: 1. P satisfies the N-condition. 2. Each principal ideal of P is finite. 3. Up to isomorphism P has a finite number of filters.*

The *width* of a pomset P is the maximum number of pairwise parallel vertices of P. If P has infinitely many pairwise incomparable vertices, then its width is ω. (Recall that we only consider non-empty countable pomsets.) Note that any pomset satisfying the last condition of Theorem 5 has finite width. The second condition is present in much of the literature on event structures. Also, if a pomset P satisfies this condition, then every finitely generated ideal of P is finite, and each non-empty subset of P contains a minimal element. Moreover, we have $u < v$ for two distinct vertices u and v iff there exists a sequence $u = u_1 < u_2 < \ldots < u_k = v$ such that u_{i+1} is an immediate successor of u_i, for each $i \in [k-1]$. (Of course, vertex w' is an immediate successor of w if $w < w'$ and there exists no z with $w < z < w'$.) Moreover, each vertex $u \in P$ has a finite height n, for some integer $n \geq 0$, i.e., there is a longest path from a minimal vertex to u, and the length of this path is n. (Minimal vertices have height 0.)

An *ω-branch* is a poset which is isomorphic to the poset whose vertices are the ordered pairs (i, j), where i is a non-negative integer and j is 0 or 1. Moreover, the immediate successors of a vertex $(i, 0)$ are $(i+1, 0)$ and $(i, 1)$, and vertices of the form $(i, 1)$ are maximal. Our result is:

Theorem 6. *An A-labeled pomset P is in $\mathbf{SP}_A^{\omega,(\omega)}$ if and only if it is (countably) infinite and the following conditions hold: 1. P satisfies the N-condition. 2. Each principal ideal of P is finite. 3. Up to isomorphism P has a finite number of connected filters. 4. P has no ω-branch.*

In our proof of Theorem 6, we will make use of several observations. The following fact depends on our assumption that pomsets are countable.

Proposition 2. *A pomset $P \in \mathbf{Pom}_I(A)$ has a linearization to an ω-chain iff each principal ideal of P is finite.*

Lemma 3. *Any filter of an A-labeled pomset satisfying the four conditions of Theorem 6 also satisfies these conditions.*

Lemma 4. *If P is an A-labeled pomset which has finite width and, up to isomorphism, a finite number of connected filters, and if each principal ideal of P is finite, then P has, up to isomorphism, a finite number of filters.*

Proof of Theorem 6. It can be argued by induction on the rank of the pomset $P \in \mathbf{SP}_A \cup \mathbf{SP}_A^{\omega,(\omega)}$ to show that P satisfies all conditions of Theorem 6. To prove the other direction, suppose that a countably infinite A-labeled pomset P satisfies all four conditions. If P has finite width, then, by Lemma 4, P has, up to isomorphism, a finite number of filters. Thus, by Theorem 5, P is generated from the singletons by the operations of series and parallel product and series omega power. Suppose now that the width of P is ω. We argue by induction on the number of non-isomorphic connected filters to show that P is in $\mathbf{SP}_A^{\omega,(\omega)}$. If P has, up to isomorphism, only one connected filter, then this filter F is a principal filter, and any connected component of P is isomorphic to F. Moreover, no two parallel vertices of P have an upper bound, since otherwise P would have a one-generated and a two-generated connected filter. Also, no two parallel vertices have a lower bound. Indeed, if $u_0 < u_1$ and $u_0 < u_2$, for some vertices u_0, u_1, u_2 such that u_1 and u_2 are parallel, then since the filter generated by u_1 is isomorphic to the filter generated by u_0, there are parallel vertices u_3 and u_4 with $u_1 < u_3$ and $u_1 < u_4$. Since no two parallel vertices have an upper bound, it holds that u_2 is parallel to both u_3 and u_4. Continuing in this way, there results an ω-branch, contrary to our assumptions on P. Thus, no two parallel vertices of P have an upper or a lower bound. Since P has infinite width, it follows now that P is the parallel omega power of a linearly ordered labeled pomset with itself. Since the principal ideal generated by any vertex is finite, and since P has, up to isomorphism, a single principal filter, it follows that $P = \omega a$ or $P = \omega a^\omega$, for some $a \in A$.

In the induction step, we assume that P has, up to isomorphism, $n > 1$ connected filters and that our claim is true for pomsets having, up to isomorphism, at most $n - 1$ connected filters. Note that P cannot be directed.[3] Indeed, by assumption P has an infinite number of pairwise parallel vertices, say $u_1, u_2, \ldots$. Thus, if P were directed, then for each $m \geq 1$, the vertices $u_1, u_2, \ldots, u_m$ would generate a connected filter. But any two of these filters are non-isomorphic. It follows now that P is either disconnected or eventually disconnected, i.e., there exists a least integer $k \geq 0$ such that the vertices of height k or more form a disconnected pomset. Indeed, if z_1 and z_2 do not have an upper bound and the height of z_1 is less than or equal to the height of z_2, then by Lemma 2 the vertices whose height is at least the height of z_1 form a disconnected pomset. We prove that when $k > 0$, any vertex of height k is over each vertex of height $k - 1$. To establish this fact, first we show that if v_1 and v_2 are distinct height k vertices which do not have an upper bound, then any height $k - 1$ vertex u_1 below v_1 is also below v_2. Indeed, if $u_1 < v_2$ does not hold, then take a height $k - 1$ vertex u_2 with $u_2 < v_2$. By the N-condition, v_1 and u_2 are parallel. Since the vertices of height $k - 1$ or more form a connected filter, by Lemma 2 u_1 and u_2 have an upper bound w. Since $v_1 < w$ or $v_2 < w$ does not hold, either the vertices u_1, u_2, v_1, w, or the vertices u_1, u_2, w, v_2 form an N, a contradiction. Suppose now that v_1 has height k and u has height $k - 1$. To show $u < v_1$, let v_2 be a second height k vertex such that v_1 and v_2 have no upper bound. Such a vertex exists since the vertices of height k or more are disconnected. Let u_1, u_2 have height $k - 1$, $u_i < v_i$, $i = 1, 2$. By the preceding argument, $u_2 < v_1$ and $u_1 < v_2$. To obtain a contradiction, suppose that $u < v_1$ does not hold. Then u is distinct from u_1, u_2. Using Lemma 2 and the fact that the vertices of height $k - 1$ or more are connected, it follows by the N-condition that there exists an upper bound w for u, u_1, u_2. Indeed, if $u_1 = u_2$, then this is immediate from Lemma 2. If $u_1 \neq u_2$, then let z_i be an upper bound for u and u_i, $i = 1, 2$. If $z_1 \leq z_2$ or $z_2 \leq z_1$ then we are done, let $w = z_2$ or $w = z_1$, respectively. If z_1 and z_2 are parallel, then by the N-condition, $u_1 \leq z_2$ and $u_2 \leq z_1$, so that we may again let $w = z_1$ or $w = z_2$. But since u is not below v_1 and v_1 or v_2 is not below w, either u_1, u, v_1, w or u, u_2, w, v_2 or u, v_1, w, v_2 form an N. Since this is impossible,

---

[3] A pomset P is directed if any two elements of P have an upper bound.

we have established the fact that any vertex of height k is over all of the height $k-1$ vertices.

Thus, if $k > 0$, then each vertex in S is over any vertex in R, where R is the collection of all vertices of height $k-1$ or less, and $S = P - R$. The pomset R is of finite width. Indeed, if $u_1, u_2, \ldots$ were pairwise parallel vertices of R, then an infinite family of pairwise non-isomorphic connected filters of P would result by taking, for each $m \geq 1$, the filter generated by the m vertices $u_1, \ldots, u_m$. Since both the height and the width of R are finite, it follows now by the assumptions that R is itself finite, so that $R \in \mathbf{SP}_A$, since R satisfies the N-condition. (See Lemma 1.) By Lemma 3, the pomset S also satisfies the conditions involved in the theorem. Clearly, any connected filter of S is a connected filter of P. Thus, if S has, up to isomorphism, n connected filters, then it has a filter isomorphic to P. Using this fact it follows now easily that P has an ω-branch, contradicting our assumptions on P. Thus, S has, up to isomorphism, at most $n-1$ connected filters, so that $S \in \mathbf{SP}_{A,B}^{\omega,(\omega)}$, by the induction assumption. Since $P = R \cdot S$, we have $P \in \mathbf{SP}_A^{\omega,(\omega)}$.

Suppose finally that $k = 0$, i.e., that P is disconnected. Since any connected component of P is a connected filter, it follows that P has, up to isomorphism, a finite number of connected components. Thus, since P is countable, we can write $P = P_1 \otimes \ldots \otimes P_m \otimes Q_1^{(\omega)} \otimes \ldots \otimes Q_s^{(\omega)}$, for some connected pomsets P_i and Q_j. By Lemma 3, each of these pomsets satisfies the conditions of the theorem. Also, each has at most n connected filters. Thus, by the above argument, it follows that each is in $\mathbf{SP}_A^{\omega,(\omega)}$, so that P is also in $\mathbf{SP}_A^{\omega,(\omega)}$.   □

5   Free Algebras, Revisited

Since the algebras we are dealing with are two-sorted, to get a complete description of the free algebras in the variety $\mathcal{V}$, we also need to describe the structure of the free algebras generated by pairs of sets (A, B), where A is the set of generators of finite sort, and B is a set of generators of infinite sort.

We will describe the free algebra generated by (A, B) as a subalgebra of $\omega \mathbf{SP}_{A \cup B}$, where without loss of generality we may assume that the sets A and B are disjoint. Let $\omega \mathbf{SP}_{A,B} = (\mathbf{SP}_A, \mathbf{SP}_{A,B}^{\omega,(\omega)}, \cdot, \otimes, ^\omega, ^{(\omega)})$ denote the subalgebra of $\omega \mathbf{SP}_{A \cup B}$ generated by the singleton pomsets a, for $a \in A$, and the pomsets b^ω, for $b \in B$.

Proposition 3. *Suppose that $P \in \mathbf{SP}_{A \cup B}^{\omega,(\omega)}$. Then P belongs to $\mathbf{SP}_{A,B}^{\omega,(\omega)}$ iff the following conditions hold: 1. The principal ideal generated by any vertex labeled b, for some $b \in B$, is isomorphic to b^ω. 2. If two parallel vertices have an upper bound, then both are labeled in A.*

Another representation of the algebra $\omega \mathbf{SP}_{A,B}$ can be obtained by allowing maximal vertices of a pomset to be labeled by elements of B. Formally, let $\mathbf{Pom}_I(A, B)$ denote the collection of all countable non-empty $(A \cup B)$-labeled pomsets P with the property that every vertex labeled in B is maximal, and such that P is either infinite or contains a vertex labeled in B. The set of pomsets $\mathbf{Pom}_F(A)$ was defined at the beginning of Section 2. The operations of series and parallel product and series and parallel omega power can be generalized to pomsets in $\mathbf{Pom}_F(A)$ and $\mathbf{Pom}_I(A, B)$ in a straightforward way, so that we get a two-sorted algebra $\omega\mathbf{Pom}(A, B) = (\mathbf{Pom}_F(A), \mathbf{Pom}_I(A, B), \cdot, \otimes, {}^\omega, {}^{(\omega)})$. This algebra also satisfies the identities (1) – (9), i.e., $\omega\mathbf{Pom}(A, B)$ is in $\mathcal{V}$.

Proposition 4. *The algebra $\omega\mathbf{Pom}(A, B)$ can be embedded into $\omega\mathbf{Pom}(A \cup B)$. The subalgebra of $\omega\mathbf{Pom}(A, B)$ generated by the singleton pomsets corresponding to the elements of $A \cup B$ is isomorphic to the algebra $\omega\mathbf{SP}_{A,B}$ described above.*

In fact, an embedding can be obtained by taking the identity map on the finite pomsets $\mathbf{Pom}_F(A)$ and mapping each pomset $P \in \mathbf{Pom}_I(A, B)$ to the pomset Q that results by replacing each vertex of P labeled b, for some $b \in B$, by the pomset b^ω.

Theorem 7. *The algebra $\omega\mathbf{SP}_{A,B}$ is freely generated by (A, B) in $\mathcal{V}$.*

Corollary 1. *The variety $\mathcal{V}$ is generated by either of the following classes of algebras. 1. The algebras $\omega\mathbf{Pom}(A)$ or $\omega\mathbf{Pom}(A, B)$. 2. The algebras $\omega\mathbf{SP}_A$ or $\omega\mathbf{SP}_{A,B}$.*

Corollary 2. *The following conditions are equivalent for a sorted equation $t = t'$. 1. $t = t'$ holds in $\mathcal{V}$. 2. $t = t'$ holds in all algebras $\omega\mathbf{Pom}(A)$ or $\omega\mathbf{Pom}(A, B)$. 3. $t = t'$ holds in all algebras $\omega\mathbf{SP}_A$ or $\omega\mathbf{SP}_{A,B}$.*

6   No Finite Axiomatization

By the compactness theorem, $\mathcal{V}$ has a finite axiomatization iff the equations (1) – (9) contain a finite subsystem which forms a complete axiomatization of $\mathcal{V}$.

Theorem 8. *For any finite subset E of the identities (1) – (9) there is a two-sorted algebra which is a model of E but fails to satisfy all of the power identities (5). Indeed, for any prime p there is a model $C_p = (F_p, I_p)$ which satisfies the equations (1) – (4) and (6) – (9) as well as all of the power identities (5) for $n < p$, but such that the identity $(x^p)^\omega = x^\omega$ fails in C_p. Thus $\mathcal{V}$ has no finite axiomatization.*

Proof. Given a prime p, let F_p be the set of positive integers and let $I_p = \{1, p, \top\}$. For all $n \in F_p$, let $\rho(n) = 1$ if p does not divide n, and let $\rho(n) = p$ if p divides n. Define the operations in C_p as follows: for all $a, b \in F_p$, $u, v \in I_p$, $a \cdot b = a + b$, $a \otimes b = a + b$, $a \cdot u = u$, $a \otimes u = u \otimes a = u \otimes v = \top$, $a^\omega = \rho(a)$, $a^{(\omega)} = \top$, $u^{(\omega)} = \top$. It is straightforward to check that the identities (1) – (4) and (6) – (9) hold, together with all of the power identities $(a^n)^\omega = a^\omega$, for integers n not divisible by p. However, $(1^p)^\omega = p^\omega = p$ and $1^\omega = 1$. $\qquad\square$

7 Complexity

Suppose that t and t' are two terms in the variables $X = \{x_1, x_2, \ldots\}$ of finite sort and $Y = \{y_1, y_2, \ldots\}$ of infinite sort. In order to decide whether the equation $t = t'$ holds in $\mathcal{V}$, one needs to construct the pomsets $|t|$ and $|t'|$ denoted by t and t' in the free algebra $\omega\mathbf{SP}_{X,Y}$, where each variable x_i and y_j is evaluated by the corresponding singleton pomset. By the freeness of $\omega\mathbf{SP}_{X,Y}$, $t = t'$ holds in $\mathcal{V}$ iff $|t|$ is isomorphic to $|t'|$. Since these pomsets may be infinite, it is not immediately clear that this condition is decidable.

Theorem 9. *There exists a polynomial time algorithm to decide for a given equation $t_1 = t_2$ whether $t_1 = t_2$ holds in $\mathcal{V}$.*

The simple proof is based on a polynomial time transformation of terms to normal form terms (or rather "normal form trees") corresponding to the normal representation of the pomsets in $\omega\mathbf{SP}_{A,B}$ defined in Section 3. Since each term has a unique normal form tree, the decision problem reduces to checking whether two labeled trees are isomorphic, which can be done in polynomial time, cf. [10].

8 Some Further Remarks

Adding the empty pomset 1 can be done in at least two different ways. First, from the geometric point of view, it makes sense to define both omega powers of the empty pomset to be the empty pomset. But since the omega powers should be of infinite sort, we must add the empty pomset also to the pomsets of infinite sort. Moreover, since for each finite P, the pomset $P \cdot 1^\omega$ is of infinite sort, each finite pomset has to be included in the carrier of pomsets of infinite sort. The resulting pomset algebras satisfy the following equations involving the empty pomset:

$$1 \cdot u = u \tag{10}$$
$$x \cdot 1 = x \tag{11}$$
$$u \otimes 1 = u \tag{12}$$
$$y \otimes 1^\omega = y \tag{13}$$
$$1^\omega = 1^{(\omega)} \tag{14}$$
$$x \cdot 1^\omega = x \otimes 1^\omega, \tag{15}$$

where x is of finite sort, y is of infinite sort, and the sort of u can be both finite and infinite. In fact, these equations and the axioms (1) – (9) form a basis of identities of the above pomset algebras. The free algebras can be described as the algebras $\omega\mathbf{SP}_{A,B}$ with the empty pomset and the finite series-parallel A-labeled pomsets contained in both carriers.

From the point of view of processes, both 1^ω and $1^{(\omega)}$ should be interpreted as an infinite non-terminating process. Supposing that these processes cannot be distinguished, it make sense to define the corresponding pomset algebra as follows. Let $\perp$ be a new symbol which is not in $A \cup B$. Then the pomsets in $\mathbf{Pom}_I^\perp(A, B)$ are those $A \cup B \cup \{\perp\}$-labeled pomsets satisfying the condition that every vertex labeled in $B \cup \{\perp\}$ is maximal and which are either infinite or contain a vertex labeled in $B \cup \{\perp\}$. In the algebra $\omega\mathbf{Pom}^\perp(A, B) = (\mathbf{Pom}_F(A), \mathbf{Pom}_I^\perp(A, B), \cdot, \otimes, 1, ^\omega, ^{(\omega)})$, the operations are defined as before, except that we define $1^\omega = 1^{(\omega)} = \perp$. The variety $\mathcal{V}_\perp$ generated by these algebras can be axiomatized by the equations (1) – (9) and (10), (11), (12), (14). The free algebra in $\mathcal{V}_\perp$ generated by a pair of sets (A, B) can be described as the algebra of "generalized series parallel pomsets" in $\omega\mathbf{Pom}^\perp(A, B)$ containing both the empty pomset and the pomset $\perp$.

Besides pomsets, there are other structures of interest that satisfy the equations (1) – (9). Let A denote a non-empty set, and let A^+ denote the collection of all finite non-empty words, and A^ω the collection of all ω-words over A. Moreover, let $A^* = A^+ \cup \{\epsilon\}$, where ϵ is the empty word, and consider the structure $L_A = (P(A^+), P(A^\omega), \cdot, \otimes, ^\omega, ^{(\omega)})$, where P denotes the power set operator, and where $\cdot$ is concatenation, $\otimes$ is shuffle, $^\omega$ is omega power and $^{(\omega)}$ is shuffle omega power. Thus, for all $K \subseteq A^+$ and $U, V \subseteq A^+ \cup A^\omega$,

$$K \cdot U = \{xu : x \in K,\ u \in U\}$$
$$U \otimes V = \{u_1v_1u_2v_2\ldots : u_1u_2\ldots \in U,\ v_1v_2\ldots \in V\ u_i, v_i \in A^*,\ i \geq 1\}$$
$$K^\omega = \{x_1x_2\ldots : x_i \in K\}$$
$$U^{(\omega)} = \{u_{11}u_{21}u_{22}u_{31}u_{32}u_{33}\ldots : u_{i1}u_{i2}\ldots \in U,\ u_{ij} \in A^*,\ i \geq 1\}.$$

Thus, K^ω and $U^{(\omega)}$ are in $P(A^\omega)$, while the result obtained by applying a binary operation is in $P(A^\omega)$ iff one of the two arguments is in this set. (Since both $P(A^+)$ and $P(A^\omega)$ contain the empty set, distinction should be made whether an empty set argument is considered to be a member of $P(A^+)$ or a member of $P(A^\omega)$.) It is straightforward to show that L_A satisfies the equations (1) – (9), so that L_A is in fact in $\mathcal{V}$. (The equations not involving $\otimes$ and $^{(\omega)}$ define the binoids of [18] and give a complete axiomatization of the equational theory of the corresponding language structures involving only concatenation and omega power.) However, the language structures L_A also satisfy equations that do not hold in all algebras belonging to $\mathcal{V}$. A simple example is the equation $x^2 \otimes x^{(\omega)} = x^{(\omega)}$. It is an open problem to find a characterization of the equations that hold in all language structures L_A.

Due to our motivation in concurrency, we have not allowed series products $P \cdot Q$ and series omega powers P^ω for infinite P. This restriction was achieved by considering two-sorted pomset algebras, making it possible to avoid the heavier machinery of partial algebras. Another possibility is to define $P \cdot Q = P$ and $P^\omega = P$, for infinite P. However, in this case, the free algebras and the valid equations do not seem to have a nice description. From the mathematical point of view, it is also of interest to place no restriction on the applicability of the series product and omega power operations. We will address the one-sorted pomset models that arise in this way in a forthcoming paper.

References

1. L. Aceto. Full abstraction for series-parallel pomsets. In: *TAPSOFT 91*, LNCS 493, 1–25, Springer-Verlag, 1991.
2. Ch. Baier and M. E. Majster-Cederbaum. Denotational semantics in the cpo and metric approach. *Theoret. Comput. Sci.*, 135:171–220, 1994.
3. S. L. Bloom and Z. Ésik. Free shuffle algebras in language varieties. *Theoret. Comput. Sci.*, 163:55–98, 1996.
4. S. L. Bloom and Z. Ésik. Shuffle binoids. *Theoret. Inform. Appl.*, 32:175–198, 1998.
5. G. Boudol and I. Castellani. Concurrency and atomicity. *Theoret. Comput. Sci.*, 59:1988, 25–84.
6. B. Bloom and M. Kwiatkowska. Trade-offs in true concurrency: Pomsets and Mazurkiewicz traces. In: *MFPS 91*, LNCS 598, 350–375, Springer-Verlag, 1992.
7. J. W. de Bakker and J. H. A. Warmerdam. Metric pomset semantics for a concurrent language with recursion. In: *Semantics of Systems of Concurrent Processes*, LNCS 469, 21–49, Springer-Verlag, 1990.
8. J. L. Gischer. The equational theory of pomsets. *Theoret. Comput. Sci.*, 61:199–224, 1988.
9. J. Grabowski. On partial languages. *Fund. Inform.*, 4:427–498, 1981.
10. L. Kucera. *Combinatorial Algorithms*. Adam Hilger, Bristol and Philadelphia, 1990.
11. K. Lodaya and P. Weil. Series-parallel posets: algebra, automata and languages. In: *STACS 98*, LNCS 1373, 555–565, Springer-Verlag, 1998.
12. A. Mazurkiewicz. Concurrency, modularity and synchronization. In: *MFCS 89*, LNCS 379, 577–598, Springer-Verlag, 1989.
13. J.-J. Ch. Meyer and E. P. de Vink. Pomset semantics for true concurrency with synchronization and recursion. In: *MFCS 89*, LNCS 379, 360–369, 1989.
14. V. Pratt. Modeling concurrency with partial orders. *Internat. J. Parallel Processing*, 15:33–71, 1986.
15. A. Rensink. Algebra and theory of order-deterministic pomsets. *Notre Dam J. Formal Logic*, 37:283–320, 1996.
16. S. T. Tschantz. Languages under concatenation and shuffling. *Math. Structures Comput. Sci.*, 4:505–511, 1994.
17. J. Valdes, R. E. Tarjan, and E. L. Lawler. The recognition of series-parallel digraphs. *SIAM Journal of Computing*, 11(2):298–313, 1982.
18. Th. Wilke. An Eilenberg theorem for ∞-languages. In: *ICALP 91*, LNCS 510, 588–599, 1991.
19. H. Wimmel and L. Priese. Algebraic characterization of Petri net pomset semantics. In: *CONCUR 97*, LNCS 1243, 406–420, Springer-Verlag, 1997.
20. J. Winkowski. Behaviours of concurrent systems. *Theoret. Comput. Sci.*, 12:39–60, 1980.
21. I. Winkowski. Concatenable weighted pomsets and their applications to modelling processes of Petri nets. *Fund. Inform.*, 28:403–421, 1996.
22. G. Winskel. Event structures. In: *Petri Nets: Applications and Relationships to Other Models of Concurrency, Advances in Petri Nets 1986, Part II, Proceedings of an Advanced Course*, LNCS 255, 325–392, Springer-Verlag, 1987.

Unreliable Failure Detectors with Limited Scope Accuracy and an Application to Consensus

Achour Mostéfaoui and Michel Raynal

IRISA - Campus de Beaulieu, 35042 Rennes Cedex, France
{mostefaoui,raynal}@irisa.fr

Abstract. Let the *scope* of the accuracy property of an unreliable failure detector be the minimum number (k) of processes that may not erroneously suspect a correct process to have crashed. Classical failure detectors implicitly consider a scope equal to n (the total number of processes). This paper investigates accuracy properties with limited scope, thereby giving rise to the S_k and $\diamond S_k$ classes of failure detectors.

A reduction protocol transforming any failure detector belonging to S_k (resp. $\diamond S_k$) into a failure detector (without limited scope) of the class S (resp. $\diamond S$) is given. This reduction protocol requires $f < k$, where f is the maximum number of process crashes. (This leaves open the problem to prove/disprove that this condition is necessary.)

Then, the paper studies the consensus problem in asynchronous distributed message-passing systems equipped with a failure detector of the class $\diamond S_k$. It presents a simple consensus protocol that is explicitly based on $\diamond S_k$. This protocol requires $f < min(k, n/2)$.

1 Introduction

Several crucial practical problems (such as *atomic broadcast* and *atomic commit*) encountered in the design of reliable applications built on top of unreliable asynchronous distributed systems, actually belong to a same family of problems, namely, the family of *agreement problems*. This family of problems can be characterized by a single problem, namely the *Consensus* problem, that is their "*greatest common denominator*". That is why the consensus problem is considered as a fundamental problem. This is practically and theoretically very important. From a practical point of view, this means that any solution to consensus can be used as a building block on top of which solutions to particular agreement problems can be designed. From a theoretical point of view, this means that an agreement problem cannot be solved in systems where consensus cannot be solved.

Informally, the consensus problem can be defined in the following way. Each process proposes a value and all correct processes have to decide on the same value, which has to be one of the proposed values. Solving the consensus problem in asynchronous distributed systems where processes may crash is far from being a trivial task. It has been shown by Fischer, Lynch and Paterson [4] that there is no deterministic solution to the consensus problem in those systems as soon

C. Pandu Rangan, V. Raman, R. Ramanujam (Eds.): FSTTCS'99, LNCS 1738, pp. 329–340, 1999.

as processes (even only one) may crash. This impossibility result comes from the fact that, due to the uncertainty created by asynchrony and failures, it is impossible to distinguish a "slow" process from a crashed process or from a process with which communications are very slow. So, to be able to solve agreement problems in asynchronous distributed systems, those systems have to be "augmented" with additional assumptions that make consensus solvable in such improved systems. A major and determining advance in this direction has been done by Chandra and Toueg who have proposed the *Unreliable Failure Detector* concept [2].

A failure detector can informally be seen as a set of *oracles*, one per process. The failure detector module (oracle) associated with a process provides it with a list of processes it guesses to have crashed. A failure detector can make mistakes by not suspecting a crashed process, or by erroneously suspecting a correct process. In their seminal paper [2], Chandra and Toueg have defined two types of property to characterize classes of failure detectors. A class is defined by a *Completeness* property and an *Accuracy* property. A completeness property is on the actual detection of crashes. The completeness property we are interested in basically states that "every crashed process is eventually suspected by every correct process". An accuracy property limits the mistakes a failure detector can make. In this paper, we are mainly interested in *Weak Accuracy*. Such a property basically states that "there is a correct process that is not suspected". Weak accuracy is *perpetual* if it has to be satisfied from the beginning. It is *eventual* if it is allowed to be satisfied only after some (unknown but finite) time. The class of failure detectors satisfying completeness and perpetual (resp. eventual) weak accuracy is denoted S (resp. $\Diamond S$). Let n and f denote the total number of processes and the maximum number of processes that may crash, respectively. S-based consensus protocols have been proposed in [2,7]; they require $f \leq n - 1$. $\Diamond S$-based consensus protocols have been proposed in [2,6,7,9]. they require $f < n/2$ (which has been shown to be a necessary condition with eventual accuracy [2]). Consequently, agreement problems can be solved in asynchronous distributed systems augmented with unreliable failure detectors of the classes S and $\Diamond S$.

The (perpetual/eventual) weak accuracy property has actually a *scope* spanning the whole system: there is a correct process that (from the beginning or after some time) is *not suspected by the other processes*. Here, the important issue is that the "non-suspicion of a correct process" concerns all other processes. In this paper, we investigate failure detector classes whose accuracy property has a limited scope: the number of processes that have not to suspect a correct process is limited to k ($k \leq n$). The parameter k defines the *scope of the weak accuracy* property, thereby giving rise to the classes S_k and $\Diamond S_k$ of unreliable failure detectors (S_n and $\Diamond S_n$ corresponding to S and $\Diamond S$, respectively). This paper has two aims. The first is to investigate the relation between the scope of the weak accuracy property and the maximal number of failures the system can suffer. The second is the design of consensus protocols based on failure detectors with limited scope accuracy.

The paper is composed of five sections. Section 2 introduces the computation model and Chandra-Toueg's failure detectors. Section 3 first defines weak accuracy with k-limited scope and the corresponding classes of failure detectors. Then, a reduction protocol that transforms any failure detector belonging to S_k (resp. $\Diamond S_k$) into a failure detector of the class S (resp. $\Diamond S$), is described. This transformation requires $f < k$. So, it relates the scope of the accuracy property to the number of failures. It is important to note that this transformation does not require assumptions involving a majority of correct processes. Consequently, when $f < k$, the stacking of a S-based (or a $\Diamond S$-based) consensus protocol on top of the proposed transformation constitutes a solution to the consensus problem based on a failure detector with limited scope accuracy. Section 4 investigates a "direct" approach to solve the consensus problem on top of an asynchronous distributed system equipped with a $\Diamond S_k$ failure detector. The proposed protocol directly relies on $\Diamond S_k$ and requires $f < min(k, n/2)$. Finally, Section 5 concludes the paper.

2   Asynchronous Distributed Systems and Unreliable Failure Detectors

2.1   Asynchronous Distributed System with Process Crash Failures

We consider a system consisting of a finite set Π of $n > 1$ processes, namely, $\Pi = \{p_1, p_2, \dots, p_n\}$. A process can fail by *crashing*, *i.e.*, by prematurely halting. It behaves correctly (*i.e.*, according to its specification) until it (possibly) crashes. By definition, a *correct* process is a process that does not crash. CORRECT denotes the set of correct processes. A *faulty* process is a process that is not correct. As previously indicated, f denotes the maximum number of processes that can crash. Processes communicate and synchronize by sending and receiving messages through channels. Every pair of processes is connected by a channel. Channels are not required to be FIFO, but are assumed to be reliable: they do not create, alter or lose messages. There is no assumption about the relative speed of processes or message transfer delays.

2.2   Chandra-Toueg's Unreliable Failure Detectors

Informally, a failure detector consists of a set of modules, each one attached to a process: the module attached to p_i maintains a set (named *suspected$_i$*) of processes it currently suspects to have crashed. Any failure detector module is inherently unreliable: it can make mistakes by not suspecting a crashed process or by erroneously suspecting a correct one. Moreover, suspicions are not necessarily stable: a process p_j can be added to or removed from a set *suspected$_i$* according to whether p_i's failure detector module currently suspects p_j or not. As in other papers devoted to failure detectors, we say "process p_i suspects process p_j" at some time t, if at that time we have $p_j \in suspected_i$.

As indicated in the Introduction, a failure detector class is formally defined by two abstract properties, namely a *Completeness* property and an *Accuracy* property. In this paper, we consider the following completeness property [2]:

- Strong Completeness: Eventually, every process that crashes is permanently suspected by every correct process.

Among the accuracy properties defined by Chandra and Toueg [2] we consider here the two following ones:

- Perpetual Weak Accuracy: Some correct process is never suspected.
- Eventual Weak Accuracy: There is a time after which some correct process is never suspected by correct processes.

Combined with the completeness property, these accuracy properties define the following two classes of failure detectors [2]:

- S: The class of *Strong* failure detectors. This class contains all the failure detectors that satisfy the strong completeness property and the perpetual weak accuracy property.
- $\Diamond S$: The class of *Eventually Strong* failure detectors. This class contains all the failure detectors that satisfy the strong completeness property and the eventual weak accuracy property.

Clearly, $S \subset \Diamond S$. As indicated in the Introduction, S-based consensus protocols are described in [2,7], and $\Diamond S$-based consensus protocols are described in [2,6,7,9]. The $\Diamond S$-based protocols require $f < n/2$. It has been proved that this requirement is necessary [2]. So, all these protocols are optimal with respect to the maximum number of crashes they tolerate.

3   Failure Detectors with k-Limited Weak Accuracy

3.1   Definition

As noted in the Introduction, the weak accuracy property involves all the correct processes, and consequently spans the whole system. This "whole system spanning" makes the weak accuracy property more difficult to satisfy than if it involves only a subset of processes. This observation is the guideline of the following definition where the parameter k defines the scope of the accuracy property. The k-accuracy property is satisfied if there exists a set Q of processes such that:

1. $\mid Q \mid = k$ (Scope)
2. $Q \cap \text{CORRECT} \neq \emptyset$ (At least one correct process)
3. $\exists p \in Q \cap \text{CORRECT}$ such that: $\forall q \in Q$: q does not suspect p (No suspicion)

k-accuracy means that there is a correct process that is not suspected by a set of k processes (some of those k processes may be correct, some others faulty). Practically, this means there is a cluster of k processes, including at least one correct process, whose failure detector modules do not erroneously suspect one of them that is correct. It is easy to see that when the scope k is equal to n, we get traditional weak accuracy. Perpetual (resp. eventual) weak k-accuracy is satisfied if k-accuracy is satisfied from the beginning (resp. after some finite time). Given a scope parameter (k), we get the two following à la Chandra-Toueg classes of failure detectors:

- $\mathcal{S}_k$: This class contains all the failure detectors that satisfy strong completeness and perpetual weak k-accuracy.
- $\Diamond\mathcal{S}_k$: This class contains all the failure detectors that satisfy strong completeness and eventual weak k-accuracy.

3.2   From k-Accuracy to (Full) Accuracy

This section describes a protocol (Figure 1) that transforms any failure detector of the class $\mathcal{S}_k$ (resp. $\Diamond\mathcal{S}_k$) into a failure detector of the class $\mathcal{S}$ (resp. $\Diamond\mathcal{S}$).

Local variables Let a be a set. To ease the presentation, a is represented by an array such that $a[\ell] = true$ means $\ell \in a$. Each process p_i has a local boolean matrix, namely, $k\_suspect_i[1:n, 1:n]$, representing a vector of sets. Let us first consider the i-th line of this matrix, namely, $k\_suspect_i[i, *]$. This line represents the set actually provided by the underlying layer: by assumption, it satisfies the properties defining $\mathcal{S}_k$ (resp. $\Diamond\mathcal{S}_k$). If $k\_suspect_i[i, \ell]$ is true, we say "p_i k-suspects p_ℓ". Let us now consider a line $j \neq i$. The entry $k\_suspect_i[j, \ell]$ has the value $true$ if, and only if, to p_i's knowledge, p_j has not crashed and is k-suspecting p_ℓ. Finally, the set $suspected_i[1:n]$ is provided by the protocol to the upper layer; it satisfies the properties defined by $\mathcal{S}$ (resp. $\Diamond\mathcal{S}$). When $k \in suspected_i$ (i.e., $suspected_i[k] = true$) we say "p_i suspects p_k".

Local behavior Processes permanently exchange their $k\_suspect_i[i, *]$ sets (line 2), locally provided by the underlying failure detector. When p_i receives such a set from p_j (line 3), it first updates its view of p_j's k-suspicions (line 4). Then, p_i examines p_j's k-suspicions. More precisely, if p_j does not k-suspect p_ℓ, then p_i does not suspect p_ℓ (lines 5-6). If p_j k-suspects p_ℓ, then p_i tests if the number of processes it perceives non crashed and k-suspecting p_ℓ bypasses the threshold $n - k$ (line 7). If this condition is true, p_i considers p_ℓ has crashed, and consequently adds p_ℓ to $suspected_i$ (line 8), and updates the raw $k\_suspect_i[\ell, *]$ (line 9).

As the proof will show, this reduction protocol assumes $f < k$ (it is important to note that the protocol does not require a "majority of correct processes" assumption). This has an interesting practical consequence: when f is small (i.e., $\ll n$), k (the accuracy scope) can be small too ($\ll n$). More precisely, k is not required to be $O(n)$, but only $O(f)$. This is particularly attractive to face scaling problems[1].

---

[1] Moreover, the protocol can be simplified when one is interested only in the reduction from $\mathcal{S}_k$ to $\mathcal{S}$. In that case, the weak k-accuracy property is perpetual, hence, there

> (1) **init:** $\forall(x, \ell) : k\_suspect_i[x, \ell] \leftarrow false; \quad \forall\ell : suspected_i[\ell] \leftarrow false;$
>
> (2) **repeat forever:** $\forall p_j:$ **do** send K_SUSPICION($k\_suspect_i[i, *], i$) to p_j **enddo**
>
> (3) **when** K_SUSPICION(k_susp, j) **is received:**
> (4) $\quad \forall\ell : k\_suspect_i[j, \ell] \leftarrow k\_susp[\ell];$
> (5) $\quad \forall\ell:$ **if** $\neg (k\_suspect_i[j, \ell])$
> (6) $\qquad\qquad\qquad$ **then** $suspected_i[\ell] \leftarrow false$
> (7) $\qquad\qquad\qquad$ **else** **if** $| \{x \mid k\_suspect_i[x, \ell]\} | > (n - k)$
> (8) $\qquad\qquad\qquad\qquad$ **then** $suspected_i[\ell] \leftarrow true$
> (9) $\qquad\qquad\qquad\qquad\qquad$ $\forall y : k\_suspect_i[\ell, y] \leftarrow false$
> (10) $\quad$ **endif** $\qquad\qquad\qquad$ **endif**

Fig. 1. From $\mathcal{S}_k/\Diamond\mathcal{S}_k$ to $\mathcal{S}/\Diamond\mathcal{S}$

3.3 Proof

Theorem *Let $f < k$. The protocol described in Figure 1 transforms any failure detector $\in \mathcal{S}_k$ (resp. $\in \Diamond\mathcal{S}_k$) into a failure detector $\in \mathcal{S}$ (resp. $\in \Diamond\mathcal{S}$).*

Proof   The proof is made of three parts.

i. Let us first note that (by assumption) the underlying failure detector (whether it belongs to $\mathcal{S}_k$ or to $\Diamond\mathcal{S}_k$) satisfies strong completeness. We first show that the protocol preserves this property, *i.e.*, if a process p_ℓ crashes and if p_i is correct, then eventually p_ℓ remains permanently in $suspected_i$ (*i.e.*, $suspected_i[\ell]$ remains true forever).

Note that there is a time t after which (1) all the faulty processes have crashed, (2) all the correct processes permanently k-suspect all the faulty processes, and (3) all K_SUSPICION messages not k-suspecting faulty processes have been received. Note that any K_SUSPICION message sent after t includes the set of crashed processes (more precisely, if p_ℓ has crashed, the corresponding entry $k\_suspect_i[i, \ell]$ remains true forever). Let us consider a crashed process p_ℓ. As there are at least $n - f$ correct processes, for each correct process p_i, there is a time $\geq t$ after which p_i has received (line 3) K_SUSPICION messages including p_ℓ, from at least $n - f$ distinct processes. Due to $f < k$, we have $n - f > n - k$. Consequently, p_i includes p_ℓ in $suspected_i$ (line 8). Moreover, as after t, for any correct process p_j, $k\_suspect_j[j, \ell]$ remains true, and as all messages sent by faulty processes have been received, it follows that the test at line 5 will never be satisfied and consequently p_ℓ will never be suppressed from $suspected_i$ (*i.e.*, $suspected_i[\ell]$ will never be set to *false* at line 6).

---

is a correct process p_u that is never suspected by k processes. This means that, $\forall p_i$, we have $| \{x \mid k\_suspect_i[x, u]\} | \leq (n - k)$. It follows that the lines 6 and 9 can be suppressed.

ii. Let us now assume that the $k\_suspect_i[i, *]$ sets satisfy perpetual weak k-accuracy (so, the underlying failure detector belongs to $\mathcal{S}_k$). We show that the $suspected_i$ sets satisfy perpetual weak accuracy.

It follows from the weak k-accuracy property, that there is a process p_u that is never k-suspected by k processes. For p_u to be included in $suspected_i$ (line 8), it is necessary (line 7) that p_i receives at least $(n - k + 1)$ K_SUSPICION messages including p_u (*i.e.*, messages such that $k\_susp[u] = true$). As k processes never k-suspect p_u, this is impossible.

iii. Let us finally assume that the underlying failure detector satisfies eventual weak k-accuracy (so, it belongs to $\diamond\mathcal{S}_k$). We show that the $suspected_i$ sets satisfy eventual weak accuracy.

Let p_u be the correct process that, after some time t, is never k-suspected by a set Q of k processes ($p_u \in Q$). We first show that eventually, for any correct process p_i, we have the following relation $| \{x \mid k\_suspect_i[x, u]\} | \leq (n - k)$. Let us consider $p_z \in Q$.

- Case: p_z is correct. There is a time $t1_z$ after which p_i will receive from p_z only K_SUSPICION messages not including p_u (*i.e.*, messages from p_z such that $k\_susp[u]$ is equal to $false$). Due to line 4, it follows that the predicate $k\_suspect_i[z, u] = false$ eventually holds forever.
- Case: p_z is faulty. Let us first observe that there is a time after which p_i will no longer receive messages from p_z. From this time, the update of $k\_suspect_i[z, u]$ (line 4) is no more executed. Moreover, due the completeness property (point i), there is a time $t2_z$ after which all processes permanently suspect p_z. After $t2_z$, each time p_i adds p_z to $suspected_i$ (line 8), it also updates $k\_suspect_i[z, u]$ to $false$ (line 9). It follows that there is a time after which the predicate $k\_suspect_i[z, u] = false$ holds forever.

Thus, as $| Q | = k$, there is a time after which, for any correct process p_i, $| \{x \mid k\_suspect_i[x, u]\} | \leq n - k$ holds forever. Consequently, as far as p_u is concerned, the test of line 7 will always be false after that time, and p_u will never be again added to $suspected_i$.

Finally, we show that, if it belongs to $suspected_i$, p_u will be withdrawn from this set. Note that (by assumption) Q contains only processes that after some time stop suspecting p_u. From the assumption $f < k$, we conclude that, after some time, Q contains a correct process p_j that stops suspecting p_u. So, for each correct process p_i, there is a time after which p_i receives from p_j only K_SUSPICION messages not including p_u (*i.e.*, messages from p_j such that $k\_susp[u]$ is always false). Then, when such a message is received from p_j, due to lines 4-6, it follows that if p_u is included in $suspected_i$, it is definitely suppressed from it.

It follows from the previous discussion that there is a time after which there is a correct process p_u such that the boolean $suspected_i[u]$ remains always false for any correct process p_i. Hence, the $suspected_i$ sets satisfy eventual weak accuracy.

3.4   An Open Problem

The previous transformation leaves open the problem to prove/disprove that the condition $f < k$ is necessary. We conjecture this condition is necessary. If it is, $k - 1$ is an upper bound for the number of processes that may crash when reducing S_k (resp. $\diamond S_k$) to S (resp. $\diamond S$) [2]. Let us also note that this constraint is reminiscent of the impossibilty constraint attached to the k-set agreement problem [1,3].

4   A $\diamond S_k$-Based Consensus Protocol

This section is focused on the design of protocols solving the consensus problem in asynchronous message-passing distributed systems equipped with failure detectors with limited scope weak accuracy. As noted in the Introduction, a solution can be obtained by stacking a $S/\diamond S$-based consensus protocol on top of the previous transformation (Figure 1). Another approach consists in designing a consensus protocol that directly uses a failure detector with limited scope accuracy. The rest of this section presents such a $\diamond S_k$-based consensus protocol.

4.1   The Consensus Problem

In the Consensus problem, every correct process p_i *proposes* a value v_i and all correct processes have to *decide* on some value v, in relation to the set of proposed values. More precisely, the *Consensus problem* is defined by the following three properties [2,4]:

- Termination: Every correct process eventually decides on some value.
- Validity: If a process decides v, then v was proposed by some process.
- Agreement: No two correct processes decide differently.

The agreement property applies only to correct processes. So, it is possible that a process decides on a distinct value just before crashing. *Uniform Consensus* prevents such a possibility. It has the same Termination and Validity properties plus the following agreement property:

- Uniform Agreement: No two processes (correct or not) decide differently.

In the following we are interested in the Uniform Consensus problem.

4.2   Underlying Principles

As in other failure detector-based consensus protocols [2,6,7,9], each process p_i manages a local variable est_i which contains its current estimate of the decision value. Initially, est_i is set to v_i, the value proposed by p_i. Processes proceed in consecutive asynchronous rounds. (In the following, the line numbers implicitly refer to Figure 2).

---

[2] A protocol transforming failure detectors with limited accuracy into failure detectors with full accuracy is described in [5]. This protocol assumes a minority of crashes and a majority scope, *i.e.*, it works only when $f < n/2 < k$. (Moreover, this protocol does not work in all failure patterns when it reduces $\diamond S_k$ into $\diamond S$).

Round coordination.   Each round r (initially, for each process p_i, $r_i = 0$) is managed by a predetermined set ck (current kernel) of k processes. The ck set associated with round r is defined by $coord(r)$ (line 3). This function is defined as follows. Its domain is the set of natural integers, its codomain is the finite set made of all the subsets of k distinct processes, and it realizes the following mapping: when r ranges the increasing sequence of natural numbers, each set of exactly k processes is infinitely often output by $coord$.

During round r, the k processes of $coord(r)$ act a coordinator role for the round. Due to the definition of $coord$, any set of k processes will repeatedly coordinate rounds. Each round is made of two phases. The aim of the first phase is to provide processes with a single value v coming from ck (the current kernel) or with a default value $\perp$ (when the current kernel does not provide a single value). Then, the aim of the second phase is to decide a value v provided by ck, when a process has received only this value during this second phase.

The underlying principles of the protocol are close to the ones used in [7] (where each kernel is made of a single process).

First phase of a round.   To realize its aim, the first phase (lines 4-14) is made of two communication steps. During the first step (lines 4-8), the k processes of the current kernel ck execute a sub-protocol solving an instance of a problem we call TWA (Terminating Weak Agreement). This problem is defined in the context of asynchronous systems equipped with failure detectors. Each process proposes a value, and correct processes have to decide (we say "a process TWA-decides") a value in such a way that the following properties be satisfied:

- TWA-Termination: Every correct process eventually TWA-decides on some value.
- TWA-Validity: If a process TWA-decides v, then v was proposed by some process.
- TWA-Agreement: If there is a process that is not suspected by the other processes, then no two processes TWA-decide differently.

TWA is close to but weaker than consensus. A TWA protocol has to satisfy the consensus agreement property only when there is a correct process that is not suspected by the other processes[3].

So, first, the processes $\in ck$ solve an instance of the TWA problem (line 6). This instance is identified by the current round number. A process $p_i \in ck$ starts its participation to a TWA protocol by invoking TWA_k_propose(r_i, est_i) and

---

[3] Actually, the $\mathcal{S}$-based consensus protocol described in [2] solves the TWA problem. More precisely, if the underlying failure detector satisfies the properties defining $\mathcal{S}$, this protocol ensures that the correct processes decide the same (proposed) value. If the underlying failure detector does not satisfy the properties defining $\mathcal{S}$, the correct processes decide on possibly different (proposed) values. When this protocol is used to solve instances of the TWA problem in the context of the $\mathcal{S}_k/\Diamond\mathcal{S}_k$-based consensus protocol, each of its execution involves a particular set of k processes out of n, defined by $coord(r)$.

terminates it by invoking TWA_k_decide(r_i) that provides it with a TWA-decided value. Then each process $\in ck$ broadcasts (line 7) the value it has TWA-decided, and each process $\notin ck$ waits until it has received a TWA-decided value, *i.e.*, a value sent by a process $\in ck$ (line 4). After this exchange, each process has a value (kept in $ph1\_est$) that has been TWA-decided by a process $\in ck$.

The second step of the first round (lines 9-14) starts with each process $\notin ck$ broadcasting the value it has received. So, each process p_j has broadcast (either at line 7 or at line 9) a value (est_ph_j) TWA-decided by a process $\in ck$. Then, each process p_i waits until it has received $ph1\_est$ values from a majority of processes. If all those values are equal (say to v), p_i adopts v (line 12). Otherwise, p_i adopts the default value $\perp$ (line 13). A process p_i keeps the value ($v/\perp$) it has adopted in the local variable $ph2\_est_i$.

The aim of the first phase of a round is actually to ensure that the value of each $ph2\_est_i$ local variable is equal either to a same value v TWA-decided by a process of ck, or to $\perp$.

Second phase of a round.   The aim of the second phase is to force processes to decide, and for those that cannot decide, to ensure the consensus agreement property will not be violated.

This is done in a way similar to the second step of the first phase. Processes exchange their $ph2\_est_i$ values. When a process p_i has received such values from a majority set (line 16), it considers the set of received values, namely, $ph2\_rec_i$. According to the value of this set, p_i either progresses to the next round (case $ph2\_rec_i = \{\perp\}$, line 18), or decides (case $ph2\_rec_i = \{v\}$, line 19), or considers v as its new estimate est_i and progresses to the next round (case $ph2\_rec_i = \{v, \perp\}$, line 20). As $ph2\_rec_i$ contains values received from a majority set, if $ph2\_rec_i = \{v\}$, then it follows that $\forall p_j : v \in ph2\_rec_j$. Combined with lines 19-20, this guarantees that if a value v has been decided, then all estimates are then equal to v, and consequently, no other value can be decided in future rounds.

4.3   The Protocol

The protocol is formally described in Figure 2. A process p_i starts a Consensus execution by invoking Consensus(v_i). It terminates it when it executes the statement **return** which provides it with the decided value (lines 19 and 23).

It is possible that distinct processes do not decide during the same round. To prevent a process from blocking forever (*i.e.*, waiting for a value from a process that has already decided), a process that decides, uses a *Reliable Broadcast* to disseminate its decision value (similarly as protocols described in [2,6,9]). To this end the Consensus function is made of two tasks, namely, $T1$ and $T2$. $T1$ implements the computation of the decision value. Line 19 and $T2$ implement the reliable broadcast.

Due to space limitation, the proof of the Validity, Agreement and Termination properties are omitted. The interested reader will find them in [8] (They are two-pages long). The Termination proof requires $f < k$ and $f < n/2$. The Agreement

---

Function Consensus(v_i)
cobegin
(1) **task** $T1$: $r_i \leftarrow 0$; $est_i \leftarrow v_i$; % $v_i \neq \perp$ %
(2) **while** *true* **do** % Sequence of asynchronous rounds %
(3) $r_i \leftarrow r_i + 1$; $ck_i \leftarrow coord(r_i)$;% $|ck_i| = k$ %

% Phase 1 of round r_i (two steps): from n proposals to $\leq$ two values %

% Phase 1 of round r_i. Step 1: from n proposals to $1 \leq$ #proposals $\leq k$ %
(4)   **case** $(i \notin ck_i)$ **do wait** (PH1_EST$(r_i, ph1\_est)$ received from any $p \in ck_i$);
(5) $ph1\_est_i \leftarrow ph1\_est$
(6) $(i \in ck_i)$ **do** TWA_k_propose(r_i, est_i); $ph1\_est_i \leftarrow$ TWA_k_decide(r_i);
(7) $\forall j$: send PH1_EST$(r_i, ph1\_est_i)$ to p_j
(8)   **endcase**;
         % If $k = 1$ (*i.e.*, $f = 0$): early stopping: execute **return**$(ph1\_est_i)$ %

% Phase 1 of round r_i. Step 2: from $\leq k$ proposals to two values %
(9)   **if** $(i \notin ck_i)$ **then** $\forall j$: send PH1_EST$(r_i, ph1\_est_i)$ to p_j **endif**;
(10)   **wait** (PH1_EST$(r_i, ph1\_est)$ received from $\lceil (n+1)/2 \rceil$ processes);
(11)   **let** $ph1\_rec_i = \{ ph1\_est \mid$ PH1_EST$(r_i, ph1\_est)$ received at line 4 or 10 $\}$;
                % $1 \leq |ph1\_rec_i| \leq min(k, \lceil (n+1)/2 \rceil)$ %
(12)   **case** $(|ph1\_rec_i| = 1)$ **do** $ph2\_est_i \leftarrow v$ where $\{v\} = ph1\_rec_i$
(13) $(|ph1\_rec_i| > 1)$ **do** $ph2\_est_i \leftarrow \perp$ (default value)
(14)   **endcase**;

% Phase 2 of round r_i (a single step):
     try to converge from $\leq$ two values to a single decision %

---

(15) $\forall j$: send PH2_EST$(r_i, ph2\_est_i)$ to p_j;
(16)   **wait** (PH2_EST$(r_i, ph2\_est)$ received from $\lceil (n+1)/2 \rceil$ processes);
(17)   **let** $ph2\_rec_i = \{ ph2\_est \mid$ PH2_EST$(r_i, ph2\_est)$ received at line 16 $\}$;
                % $ph2\_rec_i = \{\perp\}$ or $\{v\}$ or $\{v, \perp\}$ where $v \neq \perp$ %
(18)   **case** $(ph2\_rec_i = \{\perp\})$ **do skip**
(19) $(ph2\_rec_i = \{v\})$ **do** $\forall j \neq i$: send DECIDE(v) to p_j; **return**(v)
(20) $(ph2\_rec_i = \{v, \perp\})$**do** $est_i \leftarrow v$
(21)   **endcase**
(22) **endwhile**

(23) **task** $T2$: **upon reception of** DECIDE(v):
(24) $\forall j \neq i$: send DECIDE(v) to p_j; **return**(v)
coend

Fig. 2. A $\diamond S_k$-Based Consensus Protocol $(f < min(k, n/2))$

proof requires $f < n/2$. So, the protocol requires $f < min(k, n/2)$. Note that $f < n/2$ is a requirement necessary to solve consensus in an asynchronous distributed ssytem equipped with a failure detector that provides only *eventual* accuracy [2]. The constraint $f < k$ is due to the use of a failure detector with *limited scope* accuracy.

5   Conclusion

This paper has investigated unreliable failure detectors with limited *scope* accuracy. Such a scope is defined as the number k of processes that have not to suspect a correct process. Classical failure detectors implicitly consider a scope equal to n (the total number of processes). A reduction protocol transforming any failure detector belonging to $\mathcal{S}_k$ (resp. $\Diamond \mathcal{S}_k$) into a failure detector (without limited scope) of the class $\mathcal{S}$ (resp. $\Diamond \mathcal{S}$) has been presented. This reduction protocol requires $f < k$ (where f is the maximum number of process crashes). Then, the paper has studied solutions to the consensus problem in asynchronous distributed message-passing systems equipped with failure detectors of the class $\Diamond \mathcal{S}_k$. A simple $\Diamond \mathcal{S}_k$-based consensus protocol has been presented. It has been shown that this protocol requires $f < min(k, n/2)$.

References

1. Borowsky E. and Gafni E., Generalized FLP Impossibility Results for t-Resilient Asynchronous Computations. *Proc. 25th ACM Symposium on Theory of Computation*, California (USA), pp. 91-100, 1993.
2. Chandra T. and Toueg S., Unreliable Failure Detectors for Reliable Distributed Systems. *Journal of the ACM*, 43(2):225-267, March 1996.
3. Chaudhuri S., Agreement is Harder than Consensus: Set Consensus Problems in Totally Asynchronous Systems. *Proc. 9th ACM Symposium on Principles of Distributed Computing*, Québec (Canada), pp. 311-324, 1990.
4. Fischer M.J., Lynch N. and Paterson M.S., Impossibility of Distributed Consensus with One Faulty Process. *Journal of the ACM*, 32(2):374–382, April 1985.
5. Guerraoui R. and Schiper A., Γ-Accurate Failure Detectors. *Proc. 10th Workshop on Distributed Algorithms (WDAG'96)*, Bologna (Italy), Springer Verlag LNCS #1151, pp. 269-285, 1996.
6. Hurfin M. and Raynal M., A Simple and Fast Asynchronous Consensus Protocol Based on a Weak Failure Detector. *Distributed Computing*, 12(4), 1999.
7. Mostefaoui A. and Raynal M., Solving Consensus Using Chandra-Toueg's Unreliable Failure Detectors: a General Quorum-Based Approach. *Proc. 13th Symposium on DIstributed Computing (DISC'99), (Formerly WDAG)*, Bratislava (Slovakia), Springer Verlag LNCS #1693 (P. Jayanti Ed.), September 1999.
8. Mostefaoui A. and Raynal M., Unreliable Failure Detectors with Limited Scope Accuracy and an Application to Consensus. *Tech. Report #1255*, IRISA, Université de Rennes, France, July 1999, 15 pages.
9. Schiper A., Early Consensus in an Asynchronous System with a Weak Failure Detector. *Distributed Computing*, 10:149-157, 1997.

Graph Isomorphism: Its Complexity and Algorithms

Seinosuke Toda

Dept. Applied Mathematics, College of Humanities and Sciences, Nihon University,
3-25-40 Sakurajyousui, Setagaya-ku, Tokyo 156, JAPAN

It seems to be widely believed that the Graph Isomorphism problem in general case would not be NP-complete . The results that the problem is low for Σ_2^P and for PP have supported this belief. Furthermore, it is also known that the problem is polynomial-time equivalent to its counting problem. This result also supported the belief since it is conversely believed that any NP-complete problem would not have this property. On the other hand, it is unknown whether the problem can be solved in deterministic polynomial-time. From these situations, it is widely believed that the problem seems to have an intermediate complexity between deterministic polynomial-time and NP-completeness. In this talk, I will first give a breif survey on a current status of the computational complexity of Graph Isomorphism problem.

When restricting the class of graphs to be dealt with, there are many polynomial-time algorithms on the problem. For instance, it has been shown that the problem can be solved in linear time for the class of trees, the class of interval graphs, and in polynomial-time for the class of graphs with bounded degree and for the class of partial k-trees. Related to these results, I will present a recent work co-operated with T.Nagoya and S.Tani that the problem of counting the number of isomorphisms between two partial k-trees can be solved in time $O(n^{k+3})$. From this, we observe that the Graph Isomorphism itself can be also solved in time $O(n^{k+3})$ for partial k-trees.

C. Pandu Rangan, V. Raman, R. Ramanujam (Eds.): FSTTCS'99, LNCS 1738, pp. 341–341, 1999.
© Springer-Verlag Berlin Heidelberg 1999

Computing with Restricted Nondeterminism: The Dependence of the OBDD Size on the Number of Nondeterministic Variables

Martin Sauerhoff[*]

FB Informatik, LS 2, Univ. Dortmund, 44221 Dortmund, Germany
sauerhoff@ls2.cs.uni-dortmund.de

Abstract. It is well-known that an arbitrary nondeterministic Turing machine can be simulated with polynomial overhead by a so-called *guess-and-verify* machine. It is an open question whether an analogous simulation exists in the context of space-bounded computation. In this paper, a negative answer to this question is given for nondeterministic OBDDs. If we require that all nondeterministic variables are tested at the top of the OBDD, i.e., at the beginning of the computation, this may blow-up the size exponentially.

This is a consequence of the following main result of the paper. There is a sequence of Boolean functions $f_n \colon \{0,1\}^n \to \{0,1\}$ such that f_n has nondeterministic OBDDs of polynomial size with $O(n^{1/3} \log n)$ nondeterministic variables, but f_n requires exponential size if only at most $O(\log n)$ nondeterministic variables may be used.

1 Introduction and Definitions

So far, there are only few models of computation for which it has been possible to analyze the power of nondeterminism and randomness. Apart from the obvious question whether or not nondeterminism or randomization helps at all to decrease the complexity of problems, we may also be interested in the following, more sophisticated questions:

- How much nondeterminism or randomness is required to exploit the full power of the respective model of computation? Is there a general upper bound on the amount of these resources which we can make use of?
- How does the complexity of concrete problems depend on the amount of available nondeterminism or randomness?

For Turing machines, we even do not know whether nondeterminism or randomness helps at all to solve more problems in polynomial time, and we seem to be far away from answers to questions of the above type. On the other hand, nondeterminism and randomness are well-understood, e.g., in the context of two-party communication complexity. It is a challenging task to fill the gap in our knowledge between the latter model and the more "complicated" ones.

---

[*] This work has been supported by DFG grant We 1066/8-2.

C. Pandu Rangan, V. Raman, R. Ramanujam (Eds.): FSTTCS'99, LNCS 1738, pp. 342–355, 1999.

In this paper, we consider the scenario of space-bounded computation of Boolean functions. Besides Boolean circuits and formulae, branching programs are one of the standard representations for Boolean functions.

Definition 1.

(1) A (deterministic) branching program (BP) G on the set of input variables $X_n = \{x_1, \ldots, x_n\}$ is a directed acyclic graph with one source and two sinks, the latter labeled by the constants 0 and 1, resp. Interior nodes are labeled by variables from X_n and have two outgoing edges labeled by 0 and 1, resp. This graph represents a function $f_n \colon \{0, 1\}^n \to \{0, 1\}$ on X_n as follows. In order to evaluate f_n for a given assignment $a \in \{0, 1\}^n$ of the input variables, one follows a path starting at the source. At an interior node labeled by x_i, the path continues with the edge labeled by a_i. The output for a is the label of the sink reached in this way.

(2) A nondeterministic branching program is syntactically a deterministic branching program on "usual" variables from X_n and some "special" variables from $Y_n = \{y_1, \ldots, y_{r(n)}\}$, called nondeterministic variables. Nodes labeled by the latter variables are called nondeterministic nodes. On each path from the source to one of the sinks, each nondeterministic variable is allowed to appear at most once. The nondeterministic branching program computes 1 on an assignment a to the variables in X_n iff there is an assignment b to the nondeterministic variables such that the 1-sink is reached for the path belonging to the complete assignment consisting of a and b. Such a path to the 1-sink is called accepting path.

(3) The size of a deterministic or nondeterministic branching program G, $|G|$, is the number of nodes in G. The deterministic (nondeterministic) branching program size of a function f_n is the minimum size of a deterministic (nondeterministic, resp.) branching program representing f_n.

For a history of results on branching programs, we have to refer to the literature, e. g., the monograph of Wegener [15] (and the forthcoming new monograph [16]).

It is a fundamental open problem to prove superpolynomial lower bounds on the size of branching programs for explicitly defined Boolean functions even in the deterministic case. The nondeterministic case seems to be still harder (see, e. g., the survey of Razborov [12]). Nevertheless, several interesting restricted types of branching programs could be analyzed quite successfully, and for some of these models even exponential lower bounds could be proven. The goal in complexity theory is to provide proof techniques for more and more general types of branching programs, and on the other hand, to extend these techniques also to the nondeterministic or randomized case.

Here we deal with OBDDs (ordered binary decision diagrams) which are one of the restricted types of branching programs whose structure is especially well-understood.

Definition 2. *Let $X_n = \{x_1, \ldots, x_n\}$, and let $\pi \colon \{1, \ldots, n\} \to X_n$ describe a permutation of X_n. A π-OBDD on X_n is a branching program with the property that for each path from the source to one of the sinks, the list of variables on this* *path is a sublist of* $\pi(1), \ldots, \pi(n)$. *We call a graph an* OBDD *if it is a* π-*OBDD for some permutation* π. *The permutation* π *is called the* variable ordering *of the OBDD.*

OBDDs have been invented as a data structure for the representation of Boolean functions and have proven to be very useful in various fields of application. For many applications it is crucial that one can work with OBDDs of small size for the functions which have to be represented. Hence, lower and upper bounds on the size of OBDDs are also of practical relevance. We consider the following two nondeterministic variants of OBDDs:

Definition 3.

(1) Let G *be a nondeterministic branching program on variables from* $X_n \cup Y_n$, *where* Y_n *contains the nondeterministic variables, and let* π *be a permutation of the variables in* X_n. *We call* G *a nondeterministic* π-OBDD *if the order of the* X_n-*variables on all paths from the source to one of the sinks in* G *is consistent with* π.*

(2) If there is even a permutation π' *on all variables* $X_n \cup Y_n$ *such that* G *is syntactically a* π'-*OBDD (according to Definition 2), then we call* G *a* synchronous nondeterministic π'-OBDD.

Synchronous nondeterministic OBDDs have the desirable property that they may be manipulated using the well-known efficient algorithms for deterministic OBDDs. On the other hand, the more general definition (1) is the natural one for complexity theory, since it allows OBDDs to use nondeterminism in the same way as done by Turing machines. (Observe that the time steps when a nondeterministic Turing machine may access its advice bits need not be fixed in advance for all computations.)

We review some known facts on the size complexity of the different variants of OBDDs. There is a well-understood technique which allows a straightforward application of results on one-way communication complexity to prove lower bounds on the OBDD size (we will describe this formally in the next section). This technique works in the deterministic as well as the nondeterministic and randomized case and has yielded several exponential lower bounds on the size of OBDDs (see, e. g., [1,2,3,7,8,13,14]). Having understood how simple (exponential) lower bounds on the size of OBDDs can be proven, we can try to analyze the dependence of the size on the resources nondeterminism and randomness in more detail.

The dependence of the OBDD size on the resource randomness has been dealt with to some extent in [13]. Analogous to a well-known result of Newman [11] for randomized communication complexity, it could be shown that $O(\log n)$ random bits (where n is the input length) are always sufficient to exploit the full power of randomness for OBDDs. On the other hand, there is also an example of a function for which this amount of randomness is necessary to have randomized OBDDs of polynomial size (see [13] for details).

What do we know about the resource nondeterminism? In the context of communication complexity theory, Hromkovič and Schnitger [6] have proven that, contrary to the randomized case, the number of nondeterministic advice bits cannot be bounded by $O(\log n)$ without restricting the computational power of the model. An asymptotically exact tradeoff between one-way communication complexity and the number of advice bits has been proven by Hromkovič and the author [5].

These results lead to the conjecture that there should be sequences of functions which have nondeterministic OBDDs of polynomial size, but which require exponential size if the amount of nondeterminism, measured in the number of nondeterministic variables, is limited to $O(\log n)$, where n is the input length. But although lower bounds on the size of nondeterministic OBDDs can immediately be obtained by using the results on one-way communication complexity, the upper bounds do not carry over. Proving a tradeoff between size and nondeterminism turns out to be a difficult task even for OBDDs.

In this paper, we present a sequence of functions $f_n\colon \{0,1\}^n \to \{0,1\}$ with the following properties:

- f_n has *synchronous* nondeterministic OBDDs of polynomial size which use $r(n) \cdot (1 + o(1))$ nondeterministic variables, where $r(n) := (1/3) \cdot (n/3)^{1/3} \log n$;
- f_n requires exponential size in the *synchronous* model if at most $(1 - \varepsilon) \cdot r(n)$ nondeterministic variables may be used; $\varepsilon > 0$ an arbitrarily small constant; and
- f_n has exponential size even for general nondeterministic OBDDs if the number of nondeterministic variables is limited to $O(\log n)$.

The rest of the paper is organized as follows. In Section 2, we present some tools from communication complexity theory. In Section 3, the main results are stated and proven. We conclude the paper with a discussion of the results and an open problem.

2 Tools from Communication Complexity Theory

In this section, we define deterministic and nondeterministic communication protocols and state two lemmas required for the proof of the main theorem of the paper. For a thorough introduction to communication complexity theory, we refer to the monographs of Hromkovič [4] and Kushilevitz and Nisan [10].

A *deterministic two-party communication protocol* is an algorithm by which two players, called Alice and Bob, cooperatively evaluate a function $f\colon X \times Y \to \{0,1\}$, where X and Y are finite sets. Alice obtains an input $x \in X$ and Bob an input $y \in Y$. The players determine $f(x,y)$ by sending messages to each other. Each player is assumed to have unlimited (but deterministic) computational power to compute her (his) messages. The *(deterministic) communication complexity of f* is the minimal number of bits exchanged by a communication protocol by which Alice and Bob compute $f(x,y)$ for each input $(x,y) \in X \times Y$.

Here we only consider protocols with one round of communication, so-called *one-way communication protocols*. We use $D^{A \to B}(f)$ to denote the *(determinis-*

tic) one-way communication complexity of f, by which we mean the minimum number of bits sent by Alice in a deterministic one-way protocol for f.

Furthermore, we also have to deal with the nondeterministic mode of computation for protocols. We directly consider the special case of nondeterministic one-way protocols.

Definition 4. *A nondeterministic one-way communication protocol for* $f: X \times Y \rightarrow \{0,1\}$ *is a collection of deterministic one-way protocols* $P_1, \ldots, P_d$, *where* $d = 2^r$, *with* $f(x,y) = 1$ *iff there is an* $i \in \{1, \ldots, d\}$ *such that* $P_i(x,y) = 1$. *The number* r *is called the number of* advice bits *of* P.

The number of witnesses *of* P *for an input* (x,y), $\mathrm{acc}_P(x,y)$, *is defined by* $\mathrm{acc}_P(x,y) := |\{i \mid P_i(x,y) = 1\}|$. *Furthermore, let the* nondeterministic complexity *of* P *be defined by* $N(P) := r + \max_{1 \le i \le d} D(P_i)$, *where* $D(P_i)$ *denotes the deterministic complexity of* P_i.

The nondeterministic one-way complexity *of* f, $N^{\mathrm{A} \rightarrow \mathrm{B}}(f)$, *is the minimum of* $N(P)$ *over all protocols* P *as described above. By the* nondeterministic one-way complexity *of* f *with restriction to* r *advice bits and* w *witnesses for 1-inputs,* $N_{r,w}^{\mathrm{A} \rightarrow \mathrm{B}}(f)$, *we mean the minimum complexity of a nondeterministic protocol* P *for* f *which uses* r *advice bits and which fulfills* $\mathrm{acc}_P(x,y) \ge w$ *for all* $(x,y) \in f^{-1}(1)$. *Finally, we define* $N_r^{\mathrm{A} \rightarrow \mathrm{B}}(f) := N_{r,1}^{\mathrm{A} \rightarrow \mathrm{B}}(f)$.

The following well-known function plays a central role in this paper.

Definition 5. *Let* $|a|$ *denote the value of an arbitrary Boolean vector* a *as the binary representation of a number. Define* $\mathrm{IND}_n: \{0,1\}^n \times \{0,1\}^{\lceil \log n \rceil} \rightarrow \{0,1\}$ *on inputs* $x = (x_1, \ldots, x_n)$ *and* $y = (y_1, \ldots, y_{\lceil \log n \rceil})$ *by* $\mathrm{IND}_n(x,y) := x_{|y|+1}$, *if* $|y| \in \{0, \ldots, n-1\}$, *and* $\mathrm{IND}_n(x,y) := 0$, *otherwise.*

This function is referred to as "index" or "pointer function" in the literature. We may interpret it as the description of direct storage access: The x-vector plays the role of the "memory contents," and the y-vector is an "address" in the memory.

Kremer, Nisan, and Ron [9] have shown that IND_n has complexity $\Omega(n)$ for randomized one-way communication protocols with bounded error. It is easy to see that essentially $\log n$ bits of communication are sufficient and also necessary to compute IND_n by nondeterministic one-way protocols.

Here we will require the following more precise lower bound on the nondeterministic one-way communication complexity of IND_n which also takes the number of advice bits and the number of witnesses for 1-inputs into account:

Theorem 1. $N_{r,w}^{\mathrm{A} \rightarrow \mathrm{B}}(\mathrm{IND}_n) \ge nw \cdot 2^{-r-1} + r$.

This is essentially a special case of a result derived by Hromkovič and the author [5] for a more complicated function. Because Theorem 1 is an important ingredient in the proof of our new result presented here, we give an outline of the proof.

Sketch of Proof. Let P be a nondeterministic one-way protocol for IND_n with r advice bits and $\text{acc}_P(x,y) \geq w$ for all $(x,y) \in \text{IND}_n^{-1}(1)$. Hence, there are $d = 2^r$ deterministic one-way protocols $P_1, \ldots, P_d$ which compute f as described in Definition 4. For $i = 1, \ldots, d$, let g_i be the function computed by P_i. Obviously, it holds that $g_i \leq \text{IND}_n$.

By a simple counting argument (used in the same way by Hromkovič and Schnitger in [6] and originally due to Yao [17]), we can conclude that there is an index $i_0 \in \{1, \ldots, d\}$ such that $|g_{i_0}^{-1}(1)| \geq w|\text{IND}_n^{-1}(1)|/d$. It is easy to see that $|\text{IND}_n^{-1}(1)| = n \cdot 2^{n-1}$, hence, $|g_{i_0}^{-1}(1)| \geq nw \cdot 2^{n-r-1}$.

It now remains to prove a lower bound on the deterministic one-way communication complexity of an arbitrary function g with $g \leq \text{IND}_n$ in terms of the number of 1-inputs of this function g. More precisely, we claim that $D^{A \rightarrow B}(g) \geq |g^{-1}(1)|/2^n$. From this, the theorem follows.

For the proof of this lower bound on the deterministic one-way communication complexity, we consider the communication matrix M_g of g, which is the $2^n \times 2^{\lceil \log n \rceil}$-matrix with 0- and 1-entries defined by $M_g(x,y) := g(x,y)$ for $x \in \{0,1\}^n$ and $y \in \{0,1\}^{\lceil \log n \rceil}$. It is a well-known fact that $D^{A \rightarrow B}(g) = \lceil \log(\text{nrows}(M_g)) \rceil$, where $\text{nrows}(M_g)$ is the number of different rows of M_g.

Hence, it is sufficient to prove that $\log(\text{nrows}(M_g)) \geq |g^{-1}(1)|/2^n$. The key observation to prove this is that a vector $a \in \{0,1\}^n$ with k ones can occur in at most 2^{n-k} rows of M_g since $g \leq \text{IND}_n$. Thus, we need many different rows in M_g if we want to have many 1-inputs for g. The proof can be completed by turning the last statement into a precise numerical estimate (this part is omitted here). □

Finally, we briefly introduce the well-known standard technique for proving lower bounds on the OBDD size in form of a technical lemma. In the following, the technique is called *reduction technique* for easier reference.

The reduction technique has appeared in various disguises in several papers. We use the formalism from [14] for our description. This makes use of the following standard reducibility concept from communication complexity theory.

Definition 6 (Rectangular reduction). *Let X_f, Y_f and X_g, Y_g be finite sets. Let $f \colon X_f \times Y_f \rightarrow \{0,1\}$ and $g \colon X_g \times Y_g \rightarrow \{0,1\}$ be arbitrary functions. Then we call a pair (φ_1, φ_2) of functions $\varphi_1 \colon X_f \rightarrow X_g$ and $\varphi_2 \colon Y_f \rightarrow Y_g$ a rectangular reduction from f to g if $g(\varphi_1(x), \varphi_2(y)) = f(x,y)$ for all $(x,y) \in X_f \times Y_f$. If such a pair of functions exists for f and g, we say that f is reducible to g.*

We describe the reduction technique only in the special case which is important for this paper. The technique works for the general definition of nondeterministic OBDDs.

Lemma 1 (Reduction Technique). *Let the function $g \colon \{0,1\}^n \rightarrow \{0,1\}$ be defined on the variable set $X = \{x_1, \ldots, x_n\}$. Let π be a variable ordering on X. W. l. o. g. (by renumbering) we may assume that π is described by $x_1, \ldots, x_n$.*

Assume that there is a function $f \colon U \times V \rightarrow \{0,1\}$, where U and V are finite sets, and a parameter p with $1 \leq p \leq n-1$ such that f is reducible to

$g: \{0,1\}^p \times \{0,1\}^{n-p} \to \{0,1\}$. *Let G be a (general) nondeterministic π-OBDD for g which uses at most r nondeterministic variables and has w accepting paths for each 1-input of g. Then it holds that*

$$\lceil \log |G| \rceil \geq N_{r,w}^{A \to B}(f) - r.$$

3   Results

Now we are ready to present the main result of the paper. We consider the function defined below.

Definition 7.

(1) *Let k and n be positive integers and $m := n + \lceil \log n \rceil$. We define the function $\text{IND}_n^*: \{0,1\}^{3km} \to \{0,1\}$ on the (disjoint) variable vectors $s = (s_1, \ldots, s_{km})$, $t = (t_1, \ldots, t_{km})$, and $v = (v_1, \ldots, v_{km})$. The vectors s and t are "bit masks" by which variables from v are selected. If the vectors s and t do not contain exactly n and $\lceil \log n \rceil$ ones, resp., then we let $\text{IND}_n^*(s,t,v) := 0$. Otherwise, let $p_1 < \cdots < p_n$ and $q_1 < \cdots < q_{\lceil \log n \rceil}$ be the positions of ones in the s- and t-vector, resp., and define*

$$\text{IND}_n^*(s,t,v) := \text{IND}_n\left((v_{p_1}, \ldots, v_{p_n}), (v_{q_1}, \ldots, v_{q_{\lceil \log n \rceil}})\right).$$

(2) *Let $N = 3k^2 m$, $m = (n + \lceil \log n \rceil)$. We define the function $\text{MIND}_{k,n}$ ("multiple index with bit masks") on k variable vectors $b_i = (s^i, t^i, v^i)$, where $i = 1, \ldots, k$ and $s^i, t^i, v^i \in \{0,1\}^{km}$, by*

$$\text{MIND}_{k,n}(b_1, \ldots, b_k) := \text{IND}_n^*(s^1, t^1, v^1) \wedge \cdots \wedge \text{IND}_n^*(s^k, t^k, v^k).$$

Theorem 2.

(1) *The function $\text{MIND}_{k,n}$ can be represented by synchronous nondeterministic OBDDs using $k\lceil \log n \rceil$ nondeterministic variables in size $O(k^2 n^3 \log n)$;*

(2) *every synchronous nondeterministic OBDD representing the function $\text{MIND}_{k,n}$ and using at most r nondeterministic variables has size $2^{\Omega(n \cdot 2^{-r/k})}$.*

Corollary 1. *Let $N = 3n^2(n + \lceil \log n \rceil)$ (the input length of $\text{MIND}_{n,n}$).*

(1) *The function $\text{MIND}_{n,n}$ can be represented by synchronous nondeterministic OBDDs using $n\lceil \log n \rceil = (1/3) \cdot (N/3)^{1/3} \log N \cdot (1 + o(1))$ nondeterministic variables in size $O(N^{5/3} \log N)$;*

(2) *every synchronous nondeterministic OBDD representing $\text{MIND}_{n,n}$ with at most $O(\log N)$ nondeterministic variables requires size $2^{\Omega(N^{1/3})}$. It still requires size $2^{\Omega(N^{\epsilon/3})}$ if only $(1-\epsilon) \cdot (1/3) \cdot (N/3)^{1/3} \log N$ nondeterministic variables may be used, where $\epsilon > 0$ is an arbitrarily small constant.*

This follows by substituting $k = n$ in Theorem 2 and some easy calculations (we have to omit the proof here).

Corollary 2. *The function* $\text{MIND}_{n,n}$ *with input length* N *requires exponential size in* N *for (general) nondeterministic OBDDs with* $O(\log N)$ *nondeterministic variables. Furthermore, it also requires exponential size for (general) nondeterministic OBDDs with the restriction that no nondeterministic variable may appear after a usual variable on a path from the source to the sinks.*

Sketch of Proof. The first claim follows from the fact that it is possible to move all nondeterministic decisions to the top of a nondeterministic OBDD with only a polynomial increase of the size if the number of nondeterministic variables is at most $O(\log n)$.

For the second claim, let G be a nondeterministic OBDD of the described type representing $\text{MIND}_{n,n}$. Then G has a part consisting of nondeterministic nodes at the top by which one of the deterministic nodes of G is chosen. Let $r(n)$ be the minimal number of nondeterministic variables needed for the nondeterministic choice of the nodes. Then $2^{r(n)} \leq |G|$. Now either $r(n) > (1-\varepsilon) \cdot n \log n$ and we immediately obtain an exponential lower bound, or $r(n) \leq (1-\varepsilon) \cdot n \log n$ and Theorem 2 can be applied. $\square$

In the remainder of the section, we prove Theorem 2. We start with the easier upper bound.

Proof of Theorem 2(1)—Upper Bound. The function $\text{MIND}_{k,n}$ is the conjunction of k copies of the "masked index function" IND_n^*. The essence of the proof is to construct sub-OBDDs for the k copies of IND_n^* in $\text{MIND}_{k,n}$ and to combine these sub-OBDDs afterwards by identifying the 1-sink of the ith copy with the source of the $(i+1)$-th one, for $i = 1, \ldots, k-1$. Hence, it is sufficient to describe the construction of a nondeterministic OBDD for a single function IND_n^*,

We use the ordering $y_1, \ldots, y_{\lceil \log n \rceil}, s_1, t_1, v_1, s_2, t_2, v_2, \ldots, s_{km}, t_{km}, v_{km}$ in the synchronous nondeterministic OBDD G for IND_n^*, where $y_1, \ldots, y_{\lceil \log n \rceil}$ are the nondeterministic variables. With a tree of nondeterministic nodes labeled by the y-variables at the top of G, a deterministic sub-OBDD G_d from $G_1, \ldots, G_n$ is chosen. The number $d \in \{1, \ldots, n\}$ is interpreted as a "guess" of the address for the index function IND_n.

In G_d, we count the number of ones in the s- and t-vector already seen in order to find the variables $v_{p_1}, \ldots, v_{p_n}$ and $v_{q_1}, \ldots, v_{q_{\lceil \log n \rceil}}$ for the evaluation of IND_n^* (see Def. 7). We compare the "real" address with the guessed address d and output the addressed bit v_{p_d} in the positive case. It is easy to see how this can be done using $O(k(n + \log n) \cdot n \log n)$ nodes in G_d. Thus, the overall size of G is $O(kn^3 \log n)$. The OBDD for $\text{MIND}_{k,n}$ contains k copies of OBDDs of this type and therefore has size $O(k^2 n^3 \log n)$. $\square$

We now turn to the proof of the lower bound. As a first idea, we could try to apply the reduction technique (Lemma 1). The following function appears to be a suitable candidate for a rectangular reduction to $\text{MIND}_{k,n}$.

Definition 8. *Define* $\text{IND}_{k,n} \colon \{0,1\}^{kn} \times \{0,1\}^{k \lceil \log n \rceil} \to \{0,1\}$ *on inputs* $x^i = (x_1^i, \ldots, x_n^i)$ *and* $y^i = (y_1^i, \ldots, y_{\lceil \log n \rceil}^i)$, *where* $i = 1, \ldots, k$, *by*

$$\text{IND}_{k,n}\left((x^1, \ldots, x^k), (y^1, \ldots, y^k)\right) := \text{IND}_n(x^1, y^1) \wedge \cdots \wedge \text{IND}_n(x^k, y^k).$$

We would like to consider nondeterministic one-way protocols for $\mathrm{IND}_{k,n}$ according to the partition of the variables where Alice has $(x^1, \ldots, x^k)$ and Bob $(y^1, \ldots, y^k)$, i.e., where the variables of all copies of IND_n are "split" between the players. It has been shown in [5] that in this case, the players essentially cannot do better than evaluate all k copies of IND_n independently which requires $k\lceil \log n \rceil$ nondeterministic advice bits.

Unfortunately, we cannot assume that the variable ordering of the given nondeterministic OBDD for $\mathrm{MIND}_{k,n}$ allows a reduction according to this partition of the variables. It turns out that the described idea does not work for variable orderings of the following type.

Definition 9. *Let $f\colon \{0,1\}^{kn} \to \{0,1\}$ be defined on variable vectors $x^i = (x_1^i, \ldots, x_n^i)$, $i = 1, \ldots, k$. An ordering π of the variables of f is called* blockwise *with respect to $x^1, \ldots, x^k$ if there are permutations $(b_1, \ldots, b_k)$ of $\{1, \ldots, k\}$ and $(j_{i,1}, \ldots, j_{i,n})$ of $\{1, \ldots, n\}$ for $i = 1, \ldots, k$ such that π is the ordering*

$$x_{j_{1,1}}^{b_1}, \ldots, x_{j_{1,n}}^{b_1}, \ x_{j_{2,1}}^{b_2}, \ldots, x_{j_{2,n}}^{b_2}, \ \ldots, \ x_{j_{k,1}}^{b_k}, \ldots, x_{j_{k,n}}^{b_k}.$$

The x^i are called blocks *in this context. For the ease of notation, we may assume that the blocks are simply ordered according to $x^1, \ldots, x^k$ in π and that the variables within each block are ordered as in the definition of x^i.*

For $\mathrm{MIND}_{k,n}$ we consider orderings which are blockwise with respect to (s^i, t^i, v^i) if we ignore the nondeterministic variables. Such an ordering is used in the proof of the upper bound of $\mathrm{MIND}_{k,n}$. One can verify that our first idea to reduce $\mathrm{IND}_{k,n}$ to $\mathrm{MIND}_{k,n}$ does no longer work in the case of blockwise orderings of this type.

Nevertheless, we claim that a synchronous nondeterministic OBDD for $\mathrm{MIND}_{k,n}$ will become large if there are too few nondeterministic variables, even if we choose a blockwise ordering. First, we show that it is sufficient to consider blockwise variable orderings, which seem to constitute the "hardest case" for the proof of the lower bound. By the definition of $\mathrm{MIND}_{k,n}$, we have ensured that we can select a blockwise subordering of an arbitrary variable ordering by fixing the bit mask vectors in an appropriate way.

Lemma 2. *Let G be a synchronous nondeterministic OBDD for $\mathrm{MIND}_{k,n}$. Let π be the subordering of the usual variables in G. Then there are assignments to the s- and t-variables of $\mathrm{MIND}_{k,n}$ such that by applying these assignments to G one obtains a synchronous nondeterministic OBDD G' for the function $\mathrm{IND}_{k,n}$ which is no larger than G, uses at most as many nondeterministic variables as G, and where the usual variables are ordered according to π_b described by*

$$x_1^1, \ldots, x_n^1, y_1^1, \ldots, y_{\lceil \log n \rceil}^1, \ \ldots, \ x_1^k, \ldots, x_n^k, y_1^k, \ldots, y_{\lceil \log n \rceil}^k$$

after renaming the selected v-variables.

Proof. Let L be the list of the v-variables of $\mathrm{MIND}_{k,n}$ ordered according to π. Only by deleting variables, we obtain a sublist of L where the variables appear in a blockwise ordering with respect to the v^i as blocks. This is done in steps $t = 1, \ldots, k$. Let L_t be the list of variables we are still working with at the beginning of step t, and let m_t be the minimum number of variables in all blocks which have not been completely removed in the list L_t. We start with $L_1 = L$ and $m_1 = km$.

In step t, we define p as the smallest index in L_t such that the sublist of elements with indices $1, \ldots, p$ contains exactly $m = n + \lceil \log n \rceil$ variables of a block v^i. Define $b_t := i$ and choose indices $j_{t,1}, \ldots, j_{t,m}$ such that the variables $v_{j_{t,1}}^{b_t}, \ldots, v_{j_{t,m}}^{b_t}$ are under the first p variables of L_t. Afterwards, delete the first p variables and all variables of the block v^{b_t} from L_t to obtain L_{t+1}.

It is easy to verify that $m_t \geq (k - t + 1)m$ for $t = 1, \ldots, k$, and hence, the above algorithm can really be carried out sufficiently often. Let

$$v_{j_{1,1}}^{b_1}, \ldots, v_{j_{1,m}}^{b_1}, \ v_{j_{2,1}}^{b_2}, \ldots, v_{j_{2,m}}^{b_2}, \ \ldots, \ v_{j_{k,1}}^{b_k}, \ldots, v_{j_{k,m}}^{b_k}$$

be the obtained sublist of variables. For $i = 1, \ldots, k$, fix s^i such that it contains ones exactly at the positions $j_{i,1}, \ldots, j_{i,n}$, and fix t^i such that it contains ones exactly at the positions $j_{i,n+1}, \ldots, j_{i,n+\lceil \log n \rceil}$. It is a simple observation that, in general, assigning constants to variables may only reduce the OBDD size and the number of variables, and the OBDD obtained by the assignment (nondeterministically) represents the restricted function.   □

Hence, we are left with the task to prove a lower bound on the size of synchronous nondeterministic OBDDs with "few" nondeterministic variables for the function $\text{IND}_{k,n}$ and the blockwise variable ordering π_b on the usual variables. Essentially, our plan is again to decompose the whole OBDD into sub-OBDDs which are responsible for the evaluation of the single copies of IND_n. The following central lemma will be used for the decomposition. It is crucial for the proof of this lemma that we consider *synchronous* nondeterministic OBDDs.

Lemma 3 (The Nondeterministic Partition Lemma). *Let $f_n \colon \{0,1\}^n \to \{0,1\}$ be an arbitrary function, and let $f_{k,n} \colon \{0,1\}^{kn} \to \{0,1\}$ be defined on variable vectors $x^1, \ldots, x^k$ by $f_{k,n}(x^1, \ldots, x^k) := f_n(x^1) \wedge \cdots \wedge f_n(x^k)$. Let π_b be a blockwise variable ordering with respect to $x^1, \ldots, x^k$. Let G be a synchronous nondeterministic OBDD for $f_{k,n}$ where the usual variables are ordered according to π_b and which uses r nondeterministic variables.*

Then there are synchronous nondeterministic OBDDs G_1 and G_2 with the ordering π_b on the usual variables and numbers $r_1 \in \{0, \ldots, r\}$ and $w \in \{1, \ldots, 2^{r_1}\}$ such that:

(1) $|G_1| \leq |G|$ and $|G_2| \leq |G|$;

(2) G_1 represents f_n, uses at most r_1 nondeterministic variables, and there are at least w accepting paths in G_1 for each 1-input of f_n;

(3) G_2 represents $f_{k-1,n}$ and uses at most $r - r_1 + \lceil \log w \rceil$ nondeterministic variables.

Proof. Let G be as described in the lemma. We assume that the ordering π_b of the usual variables is given by $x_1^1, \ldots, x_n^1, \ldots, x_1^k, \ldots, x_n^k$. Let r_1 be the number of nondeterministic variables tested before x_n^1 (thus, $r - r_1$ nondeterministic variables are tested after x_n^1). Let y^1 and y^2 be vectors with the nondeterministic variables tested before and after x_n^1, resp.

For the construction of G_1, we consider the set of nodes in G reachable by assignments to x^1 and y^1. We replace such a node by the 0-sink, if the 1-sink of G is not reachable from it by assignments to $x^2, \ldots, x^k$ and y^2, and by the 1-sink, otherwise. The resulting graph is called G_1. It can easily be verified that it represents f_n. We define w as the minimum of the number of accepting paths in G_1 for a 1-input of f_n. Thus G_1 fulfills the requirements of the lemma.

The synchronous nondeterministic OBDD G_2 is constructed as follows. Choose an assignment $a \in f^{-1}(1)$ to x^1 such that G_1 has exactly w accepting paths for a. Let G_a be the nondeterministic OBDD on $y^1, x^2, \ldots, x^k$ and y^2 obtained from G by fixing the x^1-variables to a. The top of this graph consists of nondeterministic nodes labeled by y^1-variables. Call the nodes reached by assignments to y^1 "cut nodes." W.l.o.g., we may assume that none of the cut nodes represents the 0-function. (Otherwise, we remove the node, as well as all nodes used to reach it and the nodes only reachable from it. This does not change the represented function.)

By the choice of a and the above assumption, there are at most w paths belonging to assignments to y^1 by which cut nodes are reached, hence, the number of cut nodes is also bounded by w. Now we rearrange the top of the graph G_a consisting of the nodes labeled by y^1-variables such that only the minimal number of nondeterministic variables is used. Obviously, $\lceil \log w \rceil$ nondeterministic variables are sufficient for this. Call the resulting graph G_2. This is a synchronous nondeterministic OBDD which obviously represents $f_{k-1,n}$ and uses at most $r - r_1 + \lceil \log w \rceil$ nondeterministic variables.   □

We use this lemma to complete the proof of the main theorem.

Proof of Theorem 2(2)—Lower Bound.   Let G be a synchronous nondeterministic OBDD for $\mathrm{MIND}_{k,n}$ with r nondeterministic variables. We apply Lemma 2 to obtain a synchronous nondeterministic OBDD G' for $\mathrm{IND}_{k,n}$ with $|G'| \leq |G|$, at most r nondeterministic variables and the blockwise ordering π_b on the usual variables. The variable blocks are (x^i, y^i), where $x^i = (x_1^i, \ldots, x_n^i)$ and $y^i = (y_1^i, \ldots, y_{\lceil \log n \rceil}^i)$.

Define $s_{k,r}(n)$ as the minimal size of a synchronous nondeterministic OBDD for $\mathrm{IND}_{k,n}$ with at most r nondeterministic variables and the ordering π_b for the usual variables. We claim that

$$\lceil \log s_{k,r}(n) \rceil \geq 2^{1/k-2} \cdot n \cdot 2^{-r/k}.$$

From this, we obtain the lower bound claimed in the theorem. We prove the above inequality by induction on k, using the Partition Lemma for the induction step. The required lower bounds on the size of sub-OBDDs will be derived by the reduction technique.

Case $k = 1$: By Theorem 1, $N_r^{A \to B}(\mathrm{IND}_n) \geq n \cdot 2^{-r-1} + r$. By a standard application of the reduction technique, we immediately get $\lceil \log s_{1,r}(n) \rceil \geq 2^{-1} \cdot n \cdot 2^{-r}$.

Case $k > 1$: We assume that the claim has been shown for $s_{k-1,r'}$, for arbitrary r'. Let G' be a synchronous nondeterministic OBDD for $\mathrm{IND}_{k,n}$ with r nondeterministic variables and ordering π_b on the usual variables.

We first apply the Partition Lemma to obtain synchronous nondeterministic OBDDs G_1' and G_2' with their usual variables ordered according to π_b and numbers r_1 and w with the following properties:

- G_1' represents IND_n, uses at most r_1 nondeterministic variables, and there are at least w accepting paths for each 1-input of IND_n;
- G_2' represents $\mathrm{IND}_{k-1,n}$ and uses at most $r - r_1 + \lceil \log w \rceil$ nondeterministic variables.

Furthermore, $|G_1'| \leq |G'|$ and $|G_2'| \leq |G'|$. By Theorem 1, $N_{r_1,w}^{A \to B}(\mathrm{IND}_n) \geq nw \cdot 2^{-r_1-1} + r_1$. Applying the reduction technique, we get a lower bound on $|G_1'|$. Together with the induction hypothesis we have

$$\lceil \log |G_1| \rceil \geq nw \cdot 2^{-r_1-1} \quad \text{and}$$
$$\lceil \log |G_2| \rceil \geq 2^{1/(k-1)-2} \cdot n \cdot 2^{-(r-r_1+\lceil \log w \rceil)/(k-1)}.$$

It follows that

$$\lceil \log s_{k,r}(n) \rceil \geq \max\{nw \cdot 2^{-r_1-1}, 2^{1/(k-1)-2} \cdot n \cdot 2^{-(r-r_1+\log w+1)/(k-1)}\},$$

where we have removed the ceiling using $\lceil \log w \rceil \geq \log w + 1$. The two functions within the maximum expression are monotonous increasing and decreasing in w, resp. Thus, the minimum with respect to w is attained if

$$nw \cdot 2^{-r_1-1} = 2^{1/(k-1)-2} \cdot n \cdot 2^{-(r-r_1+\log w+1)/(k-1)}.$$

Solving for w, we obtain $w = 2^{1/k-2} \cdot 2^{-r/k} \cdot 2^{r_1+1}$. By substituting this into the above estimate for $\lceil \log s_{k,r}(n) \rceil$, we obtain the desired result,

$$\lceil \log s_{k,r}(n) \rceil \geq 2^{1/k-2} \cdot n \cdot 2^{-r/k}.$$

$\square$

Discussion of the Results and an Open Problem

We have shown that the number of nondeterministic variables in nondeterministic OBDDs cannot be bounded by $O(\log n)$ without restricting the power of the model. Requiring that the nondeterministic variables are tested at the top of the graph may cause an exponential blow-up of the OBDD size. As a by-product, these results have also led to a deeper understanding of the structure of nondeterministic OBDDs.

A natural question which could not be investigated here due to the lack of space is how the synchronous and the general model of nondeterministic OBDDs are related to each other. It is easy to see that general nondeterministic OBDDs can be made synchronous (while maintaining polynomial size) by spending additional nondeterministic variables. Using the ideas in this paper, one can also show that there are functions for which this increase of the number of variables is indeed unavoidable.

The next step can be to analyze the influence of the resource nondeterminism for more general types of branching programs in greater detail. It is already a challenging task to try to prove a tradeoff between the size of *nondeterministic read-once branching programs* and the number of nondeterministic variables. In a deterministic read-once branching program, each variable may appear at most once on each path from the source to one of the sinks. A nondeterministic read-once branching program fulfills this restriction for the usual and for the nondeterministic variables.

Open Problem. Find a sequence of functions $f_n : \{0,1\}^n \to \{0,1\}$ such that f_n has polynomial size for unrestricted nondeterministic read-once BPs, but requires exponential size if only $O(\log n)$ nondeterministic variables may be used.

One may again try the conjunction of several copies of a function which is easy for nondeterministic read-once branching programs, but a new approach is required for the proof of the respective lower bound.

Acknowledgement

Thanks to Ingo Wegener for proofreading and improving the upper bound of Theorem 2, to Juraj Hromkovič, Detlef Sieling, and Ingo Wegener for helpful discussions, and finally to the referees for careful reading and useful hints.

References

1. F. Ablayev. Randomization and nondeterminism are incomparable for polynomial ordered binary decision diagrams. In *Proc. of the 24th Int. Coll. on Automata, Languages, and Programming (ICALP), LNCS 1256*, 195–202. Springer, 1997.
2. R. E. Bryant. Graph-based algorithms for Boolean function manipulation. *IEEE Trans. Computers*, C-35(8):677–691, Aug. 1986.
3. K. Hosaka, Y. Takenaga, and S. Yajima. On the size of ordered binary decision diagrams representing threshold functions. In *Proc. of the 5th Int. Symp. on Algorithms and Computation (ISAAC), LNCS 834*, 584 – 592. Springer, 1994.
4. J. Hromkovič. *Communication Complexity and Parallel Computing*. EATCS Texts in Theoretical Computer Science. Springer, Berlin, 1997.
5. J. Hromkovič and M. Sauerhoff. Communication with restricted nondeterminism and applications to branching program complexity. Manuscript, 1999.
6. J. Hromkovič and G. Schnitger. Nondeterministic communication with a limited number of advice bits. In *Proc. of the 28th Ann. ACM Symp. on Theory of Computing (STOC)*, 551 – 560, 1996.

7. S. P. Jukna. Entropy of contact circuits and lower bounds on their complexity. *Theoretical Computer Science*, 57:113 – 129, 1988.
8. M. Krause. Lower bounds for depth-restricted branching programs. *Information and Computation*, 91(1):1–14, Mar. 1991.
9. I. Kremer, N. Nisan, and D. Ron. On randomized one-round communication complexity. In *Proc. of the 27th Ann. ACM Symp. on Theory of Computing (STOC)*, 596 – 605, 1995.
10. E. Kushilevitz and N. Nisan. *Communication Complexity*. Cambridge University Press, Cambridge, 1997.
11. I. Newman. Private vs. common random bits in communication complexity. *Information Processing Letters*, 39:67 – 71, 1991.
12. A. A. Razborov. Lower bounds for deterministic and nondeterministic branching programs. In *Proc. of Fundamentals of Computation Theory (FCT)*, *LNCS 529*, 47–60. Springer, 1991.
13. M. Sauerhoff. *Complexity Theoretical Results for Randomized Branching Programs*. PhD thesis, Univ. of Dortmund. Shaker, 1999.
14. M. Sauerhoff. On the size of randomized OBDDs and read-once branching programs for k-stable functions. In *Proc. of the 16th Ann. Symp. on Theoretical Aspects of Computer Science (STACS)*, *LNCS 1563*, 488–499. Springer, 1999.
15. I. Wegener. *The Complexity of Boolean Functions*. Wiley-Teubner, 1987.
16. I. Wegener. *Branching Programs and Binary Decision Diagrams—Theory and Applications*. Monographs on Discrete and Applied Mathematics. SIAM, 1999. To appear.
17. A. C. Yao. Lower bounds by probabilistic arguments. In *Proc. of the 24th IEEE Symp. on Foundations of Computer Science (FOCS)*, 420 – 428, 1983.

Lower Bounds for Linear Transformed OBDDs and FBDDs

(Extended Abstract)

Detlef Sieling[*]

FB Informatik, LS 2, Univ. Dortmund,
44221 Dortmund, Germany
sieling@Ls2.cs.uni-dortmund.de

Abstract. Linear Transformed OBDDs (LTOBDDs) have been suggested as a generalization of OBDDs for the representation and manipulation of Boolean functions. Instead of variables as in the case of OBDDs parities of variables may be tested at the nodes of an LTOBDD. By this extension it is possible to represent functions in polynomial size that do not have polynomial size OBDDs, e.g., the characteristic functions of linear codes. In this paper lower bound methods for LTOBDDs and some generalizations of LTOBDDs are presented and applied to explicitly defined functions. By the lower bound results it is possible to compare the set of functions with polynomial size LTOBDDs and their generalizations with the set of functions with polynomial size representations for many other restrictions of BDDs.

1 Introduction

Branching Programs or Binary Decision Diagrams (BDDs) are a representation of Boolean functions with applications in complexity theory and in programs for hardware design and verification as well. In complexity theory branching programs are considered as a model of sequential computation. The goal is to prove upper and lower bounds on the branching program size for particular Boolean functions in order to obtain upper and lower bounds on the sequential space complexity of these functions. Since for unrestricted branching programs no method to obtain exponential lower bounds is known, a lot of restricted variants of branching programs has been considered; for an overview see e.g. Razborov [16].

In hardware design and verification data structures for the representation and manipulation of Boolean functions are needed. The most popular data structure are Ordered Binary Decision Diagrams (OBDDs), which were introduced by Bryant [3]. They allow the compact representation and the efficient manipulation of many important functions. However, there are a lot of other important functions for which OBDDs are much too large. For this reason many generalizations of OBDDs have been proposed as a data structure for Boolean functions.

---

[*] The author was supported in part by DFG grant We 1066/8.

C. Pandu Rangan, V. Raman, R. Ramanujam (Eds.): FSTTCS'99, LNCS 1738, pp. 356–368, 1999.
© Springer-Verlag Berlin Heidelberg 1999

Most of these generalizations are restricted branching programs that are also investigated in complexity theory. Hence, the lower and upper bound results and methods from complexity theory are also useful in order to compare the classes of functions for which the different extensions of OBDDs have polynomial size.

In this paper we consider several variants of branching programs that are obtained by introducing linear transformations in OBDDs or generalizations of OBDDs. In order to explain the differences between ordinary OBDDs and OBDDs with linear transformations we first repeat the definition of BDDs or branching programs, respectively, and of OBDDs. A Binary Decision Diagram (BDD) or Branching Program for a function $f(x_1, \ldots, x_n)$ is a directed acyclic graph with one source node and two sinks. The sinks are labeled by the Boolean constants 0 and 1. Each internal node is labeled by a variable x_i and has an outgoing 0-edge and an outgoing 1-edge. For each input $a = (a_1, \ldots, a_n)$ there is a computation path from the source to a sink. The computation path starts at the source and at each internal node labeled by x_i the next edge of the computation path is the outgoing a_i-edge. The label of the sink reached by the computation path for a is equal to $f(a)$. The size of a branching program or BDD is the number of internal nodes. In OBDDs the variables have to be tested on each computation path at most once and according to a fixed ordering.

In the following we call an expression $x_{i(1)} \oplus \cdots \oplus x_{i(k)}$ a *linear test*. A *generalized variable ordering* over $x_1, \ldots, x_n$ is a sequence of n linear independent linear tests. In Linear Transformed OBDDs (LTOBDDs) the internal nodes may be labeled by linear tests instead of single variables as in the case of OBDDs. However, on each computation path the tests have to be arranged according to a fixed generalized variable ordering. The function f represented by an LTOBDD is evaluated in the obvious way: The computation path for some input $a = (a_1, \ldots, a_n)$ starts at the source. At an internal node labeled by $x_{i(1)} \oplus \cdots \oplus x_{i(k)}$ the outgoing edge labeled by $a_{i(1)} \oplus \cdots \oplus a_{i(k)}$ has to be chosen. The label of the sink at the end of the computation path is equal to $f(a)$.

In Section 2 we present an alternate definition of LTOBDDs. There we also define several extensions of LTOBDDs. An example of an LTOBDD is shown in the left of Figure 1. We remark that the linear independence of the linear tests of a generalized variable ordering is necessary, since otherwise not all inputs can be distinguished by the LTOBDD so that not all functions can be represented.

The evaluation of linear tests instead of single variables at the nodes of BDDs was already suggested by Aborhey [1] who, however, only considers decision trees. LTOBDDs have been suggested as a generalization of OBDDs (Meinel, Somenzi and Theobald [14]), since they are a more compact representation of Boolean functions than OBDDs. The results of Bern, Meinel and Slobodová [2] on Transformed BDDs imply that the algorithms for the manipulation of OBDDs can also be applied to LTOBDDs so that existing OBDD packages can easily be extended to LTOBDDs. Günther and Drechsler [7] present an algorithm for computing optimal generalized variable orderings.

An example that shows the power of LTOBDDs are the characteristic functions of linear codes. It is easy to see that all characteristic functions of linear

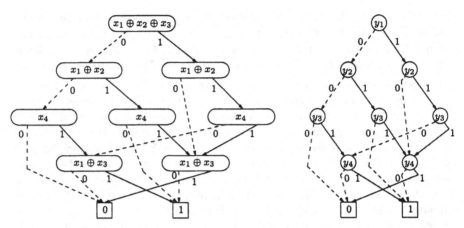

Fig. 1. An example of an LTOBDD with the generalized variable ordering $x_1 \oplus x_2 \oplus x_3$, $x_1 \oplus x_2$, x_4, $x_1 \oplus x_3$ for some function f and of an OBDD for some function g so that $f(x) = g(A \cdot x)$.

codes can be represented by LTOBDDs of linear size: In order to check whether a word x belongs to a linear code it suffices to test whether the inner product of x and each row of the parity check matrix of the code is equal to 0. For each row we can choose a linear test that is equal to the inner product of the row and the input. Since the rows of the parity check matrix are linearly independent, we can choose these linear tests as a generalized variable ordering of an LTOBDD, and an LTOBDD computing the NOR of these linear tests also computes the characteristic function of the code. On the other hand, exponential lower bounds on the size of many restrictions of branching programs are known for the characteristic functions of certain linear codes: Exponential lower bounds for syntactic read-k-times branching programs are proved by Okol'nishnikova [15], for nondeterministic syntactic read-k-times branching programs by Jukna [9], for semantic $(1, +k)$-branching programs by Jukna and Razborov [11], and for $\oplus$OBDDs by Jukna [10].

The aim of this paper is to present methods to prove exponential lower bounds on the size of LTOBDDs and some generalizations of LTOBDDs. Many lower bounds for restricted BDDs have been proved by arguments based on communication complexity theory. Roughly, the BDD is cut into two parts so that some part of the input is known only in the first part of the BDD and the other part of the input only in the second part of the BDD. If the computation of the considered function requires the exchange of a large amount of information between those two parts of the input, the cut through the BDD and, therefore, also the BDD has to be large. For LTOBDDs this approach is more difficult to apply. If in one part of the LTOBDD $x_1 \oplus x_2$ is tested and in the other part $x_1 \oplus x_3$, one can hardly say that nothing about x_1 is known in one of the parts of the LTOBDD. The main result of this paper is to show how to overcome this problem.

The paper is organized as follows. In Section 2 we repeat the definitions of several variants of BDDs and define the corresponding variants of LTOBDDs. In Section 3 we present lower bound methods for LTFBDDs and in Section 4 for ⊕LTOBDDs. Finally, we summarize our results and compare the classes of functions with polynomial size LTOBDDs with the corresponding classes for other variants of BDDs. Due to lack of space all upper bound proofs and some details of the lower bound proofs are omitted and can be found in the full version [17].

2   Further Definitions

Before we generalize the definition of LTOBDDs we discuss an alternate definition of LTOBDDs, which is equivalent to that given in the Introduction and will be useful in our lower bound proofs. In order to simplify the notation we always assume that vectors are column vectors and we also use vectors as arguments of functions. Furthermore we only consider vector spaces over $\mathbb{F}_2$. Then an LTOBDD for some function f consists of an OBDD for some function g and a nonsingular matrix A, so that $f(x) = g(A \cdot x)$. In order to illustrate this definition Figure 1 shows an LTOBDD for some function f and an isomorphic OBDD for some function g, so that $f(x) = g(A \cdot x)$ for

$$A = \begin{pmatrix} 1\,1\,1\,0 \\ 1\,1\,0\,0 \\ 0\,0\,0\,1 \\ 1\,0\,1\,0 \end{pmatrix}.$$

We now define some generalizations of OBDDs and the linear transformed variants of these generalizations. In FBDDs (Free BDDs, also called read-once branching programs) on each computation path each variable is tested at most once. This property is also called the read-once property. An LTFBDD for some function f consists of an FBDD for some function g and a nonsingular matrix A so that $f(x) = g(A \cdot x)$. If we draw LTFBDDs as FBDDs with linear tests at the internal nodes, we see that at most n different linear tests may occur in an LTFBDD. Another possibility to define LTFBDDs is to allow an arbitrary number of different linear tests. We call the resulting variant of LTFBDDs *strong LTFBDDs:* In a strong LTFBDD the linear tests of each computation path have to be linearly independent. This definition is quite natural, since the term "free" in the name FBDD means that a path leading from the source to some node v can be extended to a computation path (a path corresponding to some input) via the 0-edge leaving v and the 1-edge as well. In a BDD this is obviously equivalent to the read-once property. In linear transformed BDDs this is possible iff on each path the linear tests performed on this path are linearly independent.

Obviously, an LTFBDD is also a strong LTFBDD, while the opposite is not true. We shall even see in the following section that polynomial size strong

LTFBDDs are more powerful than polynomial size LTFBDDs so that the name strong LTFBDD is justified.

A nondeterministic OBDD is an OBDD where each internal node may have an arbitrary number of outgoing 0-edges and 1-edges. Hence, for each input there may be more than one computation path. A function represented by a nondeterministic OBDD takes the value 1 on the input a, if there is at least one computation path for a that leads to the 1-sink. $\oplus$OBDDs are syntactically defined as nondeterministic OBDDs. However, a $\oplus$OBDD computes the value 1 on the input a, if the number of computation paths for a from the source to the 1-sink is odd. We may define nondeterministic LTOBDDs and $\oplus$LTOBDDs by introducing a nonsingular transformation matrix A as described above or by allowing linear tests at the internal nodes, where a generalized variable ordering has to be respected.

The investigation of $\oplus$OBDDs is motivated by polynomial time algorithms for several important operations on Boolean functions, which are presented by Gergov and Meinel [5] and Waack [18]. It is straightforward to extend most of these algorithms to $\oplus$LTOBDDs. We shall see that polynomial size $\oplus$LTOBDDs can represent a larger class of functions than $\oplus$OBDDs. On the other hand, we also obtain exponential lower bounds for $\oplus$LTOBDDs.

3   Lower Bounds for LTFBDDs and a Comparison of LTFBDDs and Strong LTFBDDs

Lower bounds for FBDDs can be proved by cut-and-paste arguments as shown by Wegener [19] and Žák [20]. The following lemma, which we present without proof, describes an extension of the cut-and-paste method that is suitable for LTFBDDs.

Lemma 1. Let $f : \{0,1\}^n \to \{0,1\}$ and let $k \geq 1$. If for all $k \times n$ matrices D with linearly independent rows, for all $c = (c_0, \ldots, c_{k-1}) \in \{0,1\}^k$ and for all $z \in \{0,1\}^n$, $z \neq (0, \ldots, 0)$, there is an $x \in \{0,1\}^n$ so that $D \cdot x = c$ and $f(x) \neq f(x \oplus z)$, the LTFBDD size for f is bounded below by $2^{k-1} - 1$.

We show how to apply this method to a particular function. We call the function defined in the following the matrix storage access function MSA. We remark that a similar function was considered by Jukna, Razborov, Savický and Wegener [12]. Let n be a power of 2 and let $b = \lfloor \lfloor (n-1)/\log n \rfloor^{1/2} \rfloor$. Let the input $x_0, \ldots, x_{n-1}$ be partitioned into x_0, into $t = \log n$ matrices $C^0, \ldots, C^{t-1}$ of size $b \times b$ and possibly some remaining variables. Let $s_i(x) = 1$ if the matrix C^i contains a row consisting of ones only, and let $s_i(x) = 0$ otherwise. Let $s(x)$ be the value of $(s_{t-1}(x), \ldots, s_0(x))$ interpreted as a binary number. Then

$$MSA(x_0, \ldots, x_{n-1}) = \begin{cases} x_0 & \text{if } s(x) = 0, \\ x_0 \oplus x_{s(x)} & \text{if } s(x) > 0. \end{cases}$$

The following upper and lower bound results for MSA show that polynomial size strong LTFBDDs are more powerful than polynomial size LTFBDDs.

Theorem 2. *There is a strong LTFBDD for MSA with $O(n^2/\log n)$ nodes.*

Theorem 3. *LTFBDDs for MSA have size $2^{\Omega((n/\log n)^{1/2})}$.*

Sketch of Proof. It suffices to show that for $k = b-2$ the assumptions of Lemma 1 are fulfilled. Let a $k \times n$-matrix D, a vector $c = (c_0, \ldots, c_{k-1})$ and a vector $z \in \{0,1\}^n$, $z \neq (0, \ldots, 0)$, be given. We are going to construct an input x for which $D \cdot x = c$ and $MSA(x) \neq MSA(x \oplus z)$.

If $z_0 = 1$, we choose $s^* = 0$. Otherwise we choose some s^* for which $z_{s^*} = 1$. We shall construct an input x so that $s(x) = s^*$ and $s(x \oplus z) = s^*$ as well. If $s^* = 0$, it holds that $MSA(x) = x_0 \neq x_0 \oplus z_0 = MSA(x \oplus z)$ and, if $s^* \neq 0$, we have $MSA(x) = x_0 \oplus x_{s(x)} \neq (x_0 \oplus z_0) \oplus (x_{s(x \oplus z)} \oplus z_{s(x \oplus z)}) = MSA(x \oplus z)$ as required.

Let $(s^*_{t-1}, \ldots, s^*_0)$ be the representation of the chosen value for s^* as a binary number. For $i = 0, \ldots, t - 1$ we successively construct linear equations of the form $x_j = 0$ or $x_j = 1$, which make sure that for x and $x \oplus z$ the matrix C^i contains a row consisting of ones only, if $s^*_i = 1$, or that C^i contains a column consisting of zeros only, if $s^*_i = 0$. Hence, for a solution of the system of equations the number $s(x)$ takes the value s^* and $MSA(x) \neq MSA(x \oplus z)$. However, we also have to make sure that the equations $D \cdot x = c$ are fulfilled. Hence, we shall choose the equations $x_j = 0$ or $x_j = 1$ in such a way that the vectors of coefficients of all equations together with the rows of D are a linearly independent set. Then there is a solution x for the system of all considered linear equations so that $D \cdot x = c$ and $MSA(x) \neq MSA(x \oplus z)$.

For each $i = 0, \ldots, t - 1$ we inductively construct $2b$ equations where the left-hand-sides are single variables from the matrix C^i. These equations make sure that $s_i(x)$ takes the value s^*_i. W.l.o.g. let $s^*_i = 1$. By rank arguments it can be shown that we can choose two rows of C^i so that the set of $2b$ equations, which have as left-hand side the single variables of those two rows, together with all previously constructed equations are linearly independent. For one of those rows we choose equations that make sure that for the input x this row is a row only consisting of entries 1. Similarly for the other row equations are chosen that make sure that for the input $x \oplus z$ this row is a row only consisting of entries 1. Here we remember that z is fixed. If $s^*_i = 0$ by the same arguments two column can be constructed so that the first one only consists of entries 0 for x and the second one only consists of entries 0 for $x \oplus z$.

Altogether, we obtain a system of linear equations where the set of vectors of coefficients is linearly independent. Let x be a solution. Then the linear equations enforce that $D \cdot x = c$, that $s(x) = s(x \oplus z)$, and, by the case distinction above, that $MSA(x) \neq MSA(x \oplus z)$.   □

4 Lower Bounds for LTOBDDs, ⊕LTOBDDs and Nondeterministic LTOBDDs

LTOBDDs, ⊕LTOBDDs and nondeterministic LTOBDDs have in common that they respect a generalized variable ordering. Hence, we shall apply communication complexity based arguments in order to prove lower bounds. For an introduction into communication complexity theory we refer to the monographs of Hromkovič [8] and Kushilevitz and Nisan [13]. We are going to prove lower bounds on the communication complexity by constructing large fooling sets. In order to introduce the notation we repeat the definition of fooling sets.

Definition 4. *Let $f : \{0,1\}^n \to \{0,1\}$ be a Boolean function. Let (L,R) be a partition of the set of input variables. For an input x let $x^{(l)}$ denote the assignment to the variables in L according to x and let $x^{(r)}$ denote the assignment to the variables in R according to x. Let $(x^{(l)}, y^{(r)})$ be the input consisting of $x^{(l)}$ and $y^{(r)}$. A fooling set for f and the partition (L,R) of the input variables is a set $M \subseteq \{0,1\}^n$ of inputs which has for some $c \in \{0,1\}$ the following properties.*

1. *$\forall x \in M : f(x) = c$.*
2. *$\forall x, y \in M, x \neq y : [f(x^{(l)}, y^{(r)}) = \bar{c}] \vee [f(y^{(l)}, x^{(r)}) = \bar{c}]$.*

We say that M is a strong fooling set if it has the following property 2' instead of property 2 from above.

2'. *$\forall x, y \in M, x \neq y : [f(x^{(l)}, y^{(r)}) = \bar{c}] \wedge [f(y^{(l)}, x^{(r)}) = \bar{c}]$.*

We call M a 1-fooling set or strong 1-fooling set, respectively, if it has the above properties for $c = 1$.

It is well-known that the size of a fooling set for a function f and a partition (L,R) is a lower bound on the size of OBDDs for f and all variable orderings where the variables in L are arranged before the variables in R. However, in an LTOBDD for f and the transformation matrix A the function $g(y) = f(A^{-1} \cdot y)$ is represented. Hence, we have to construct large fooling sets for g in order to obtain lower bounds on the LTOBDD size for f. In order to simplify the notation let $B = A^{-1}$ throughout this section. Furthermore, let the number n of variables be an even number. We always partition the set $\{y_0, \ldots, y_{n-1}\}$, which g depends on, into $L = \{y_0, \ldots, y_{n/2-1}\}$ and $R = \{y_{n/2}, \ldots, y_{n-1}\}$. Furthermore, we use the notation $y^{(l)}$ and $y^{(r)}$ to denote $(y_0, \ldots, y_{n/2-1})$ and $(y_{n/2}, \ldots, y_{n-1})$, respectively. We shall apply the following lemmas to prove the lower bounds. Lemma 6 is inspired by the presentation of Dietzfelbinger and Savický [4]. The proofs of the lemmas are omitted.

Lemma 5. *If for all nonsingular matrices B there is a fooling set of size at least b for the function $g(y) = f(B \cdot y)$ and the partition (L,R), the LTOBDD size for f and all generalized variable orderings is at least b.*

Lemma 6. *If for all nonsingular matrices B there is a fooling set of size at least b for the function $g(y) = f(B \cdot y)$ and the partition (L,R), the $\oplus$LTOBDD size for f and all generalized variable orderings is at least $b^{1/2} - 1$. If for all nonsingular matrices B there is even a strong fooling set of size at least b' for the function $g(y) = f(B \cdot y)$ and the partition (L,R), the $\oplus$LTOBDD size for f and all generalized variable orderings is at least b'.*

Lemma 7. *If for all nonsingular matrices B there is a 1-fooling set of size at least b for the function $g(y) = f(B \cdot y)$ and the partition (L, R), the nondeterministic LTOBDD size for f and all generalized variable orderings is at least b.*

It remains the problem to apply the lemmas to obtain an exponential lower bound for an explicitly defined function. In the following we define the function *INDEX-EQ*, a combination of the functions *INDEX* and *EQ*, which are both well-known functions in communication complexity theory. We get lower bounds on the size of LTOBDDs, $\oplus$LTOBDDs and nondeterministic LTOBDDs by constructing large fooling sets which are even simultaneously strong fooling sets and 1-fooling sets.

Definition 8. *Let k be a power of 2 and let $N = 2^k$. The function INDEX-EQ is defined on $n = 3N/2$ variables $x_0, \ldots, x_{n-1}$. The variables $x_0, \ldots, x_{N-1}$ are interpreted as a memory and the $N/2$ variables $x_N, \ldots, x_{n-1}$ are interpreted as $N/(2 \log N)$ pointers each consisting of $\log N$ bits. Let $m = N/(4 \log N)$. Let $a(1), \ldots, a(m), b(1), \ldots, b(m)$ denote the values of the pointers. Then INDEX-EQ$(x_0, \ldots, x_{n-1})$ takes the value 1 iff the following conditions hold.*

1. $\forall i \in \{1, \ldots, m\} : x_{a(i)} = x_{b(i)}$.
2. $a(1) < \cdots < a(m)$ and $b(1) < \cdots < b(m)$.
3. $a(m) < b(1)$ or $b(m) < a(1)$.

Because of the first condition the computation of the function includes the test whether the words whose bits are addressed by the pointers are equal. The second and the third condition ensure that the equality test has only to be performed if the pointers are ordered and if either all a-pointers are smaller than all b-pointers or vice versa. We remark that the last two conditions are not necessary for the proof of the lower bound. These conditions allow to prove a polynomial upper bound on the FBDD size of *INDEX-EQ*, which we state without proof. Afterwards we prove the lower bound.

Theorem 9. *There are FBDDs of size $O(n^6)$ for INDEX-EQ.*

Theorem 10. *The size of LTOBDDs, $\oplus$LTOBDDs and nondeterministic LTOBDDs for the function INDEX-EQ is bounded below by $2^{\Omega(n/\log n)}$.*

Sketch of Proof. By Lemmas 5–7 it suffices to show that for all nonsingular $n \times n$ matrices B there is a strong 1-fooling set of size at least 2^m for the function $g(y) = INDEX\text{-}EQ_n(B \cdot y)$ and the partition $(\{y_0, \ldots, y_{n/2-1}\}, \{y_{n/2}, \ldots, y_{n-1}\})$. The construction of the strong 1-fooling set essentially consists of the following steps. In the first one we construct sets I and J of indices of memory variables. In the second step we construct a set which will be the fooling set. The set I has the property that for all inputs of the fooling set the values of the variables with indices in I only depend on the results of the linear tests in $\{y_0, \ldots, y_{n/2-1}\}$. Similarly, for the inputs of the fooling set the values of the memory variables with indices in J only depend on the linear tests in $\{y_{n/2}, \ldots, y_{n-1}\}$. Finally, it has to be shown that the constructed set is really a fooling set.

Let B a nonsingular $n \times n$ matrix. We always keep in mind that B is the matrix for which $(x_0, \ldots, x_{n-1}) = B \cdot (y_0, \ldots, y_{n-1})$. In particular, each row of B corresponds to one of the x-variables and each column of B to one of the y-variables.

We use the following notation. Let $B^{(l)}$ be the left half of B, i.e., the $n \times n/2$ matrix consisting of the first $n/2$ elements of each row of B. Similarly, let $B^{(r)}$ be the right half of B. Let $B[x_i]$ denote the ith row of B (the row corresponding to x_i), and let $B^{(l)}[x_i]$ and $B^{(r)}[x_i]$ be the left and right half of this row, respectively. Let each pointer $a(i)$ consist of the $k = \log N$ bits $a_{k-1}(i), \ldots, a_0(i)$ which are interpreted as a binary number. Similarly let each pointer $b(i)$ consist of the bits $b_{k-1}(i), \ldots, b_0(i)$. We shall use the notations x_j and $a_l(i)$ simultaneously even if both denote the same bit. Then $B[a_l(i)]$ denotes the row of B corresponding to the bit $a_l(i)$ of the input.

The choice of I and J can to be done in such a way that I and J have the following properties. We omit the details how to choose I and J.

(P1) $|I| = N/16$, $|J| = N/16$ and $I, J \subseteq \{0, \ldots, N-1\}$.
(P2) The set $\{B^{(l)}[x_i] \mid i \in I\}$ is linearly independent and
$\text{span}\{B^{(l)}[x_i]) \mid i \in I\} \cap \text{span}\{B^{(l)}[x_j] \mid j \in J \vee N \le j \le n-1\} = \{(0, \ldots, 0)\}$.
(P3) The set $\{B^{(r)}[x_j] \mid j \in J\}$ is linearly independent and
$\text{span}\{B^{(r)}[x_j] \mid j \in J\} \cap \text{span}\{B^{(r)}[x_i] \mid i \in I \vee N \le i \le n-1\} = \{(0, \ldots, 0)\}$.

We shall apply property (P2) (and similarly (P3)) in order to prove that a system of linear equations whose vectors of coefficients are $B^{(l)}[x_i]$, where $i \in I \cup J \cup \{j \mid N \le j \le n-1\}$, has a solution. (P2) and (P3) also imply that $I \cap J = \emptyset$.

Let i^* be the $N/32$ smallest element of I and let j^* be the $N/32$ smallest element of J. If $i^* < j^*$, we choose for I^* the m smallest elements of I and for J^* the m largest elements of J. W.l.o.g. $k \ge 32$. Then $m < N/32$ and all elements of I are smaller than all elements of J. If $i^* > j^*$, we choose for I^* the m largest elements of I and for J^* the m smallest elements of J. Then all elements of I are larger than all elements of J. Let $I^* = \{i(1), \ldots, i(m)\}$ and $J^* = \{j(1), \ldots, j(m)\}$ such that $i(1) < \cdots < i(m)$ and $j(1) < \cdots < j(m)$. We shall only construct inputs where the chosen addresses $a(1), \ldots, a(m), b(1), \ldots, b(m)$ are equal to $i(1), \ldots, i(m), j(1), \ldots, j(m)$. Hence, for all considered inputs the second and the third condition of the definition of INDEX-EQ are fulfilled so that the value of INDEX-EQ on these inputs only depends on the first condition.

Let $i_\nu(\alpha)$ and $j_\nu(\alpha)$ denote the νth bit of $i(\alpha)$ and $j(\alpha)$, respectively. Let $s = (s_0, \ldots, s_{n-1})$ be an arbitrary solution of the system of linear equations that consists for all $\alpha \in \{1, \ldots, m\}$ and all $\nu \in \{0, \ldots, k-1\}$ of the following equations.

$$B[a_\nu(\alpha)] \cdot s = i_\nu(\alpha)$$
$$B[b_\nu(\alpha)] \cdot s = j_\nu(\alpha) \tag{1}$$

This system of linear equations has a solution since all rows of B are linearly independent.

Now we construct the fooling set. For all $(c(1), \ldots, c(m)) \in \{0,1\}^m$ we construct a system of linear equations. We shall prove that this system has at least one solution. We select an arbitrary solution and include it into the fooling set M. It can be shown that for different assignments to $c(1), \ldots, c(m)$ we get different systems of linear equations which have disjoint sets of solutions so that we obtain a set M of size 2^m.

In the following system of linear equations the only variables are denoted by y; all other identifiers denote constants. The linear equations are arranged as a table which shows the connections between the different equations. Note that the left column only contains variables in $y^{(l)}$ and the right column only variables in $y^{(r)}$. Hence, we may also consider the equations as two independent systems, one determining the values of $y^{(l)}$ and the other one the values of $y^{(r)}$.

1st block: For all $\alpha \in \{1, \ldots, m\}$:
$$B^{(l)}[x_{i(\alpha)}]y^{(l)} = c(\alpha) \oplus B^{(r)}[x_{i(\alpha)}]s^{(r)} \text{ and } B^{(r)}[x_{i(\alpha)}]y^{(r)} = B^{(r)}[x_{i(\alpha)}]s^{(r)}$$

2nd block: For all $\alpha \in \{1, \ldots, m\}$:
$$B^{(l)}[x_{j(\alpha)}]y^{(l)} = B^{(l)}[x_{j(\alpha)}]s^{(l)} \qquad \text{and } B^{(r)}[x_{j(\alpha)}]y^{(r)} = c(\alpha) \oplus B^{(l)}[x_{j(\alpha)}]s^{(l)}$$

3rd block: For all $\alpha \in \{1, \ldots, m\}$ and for all $\nu \in \{0, \ldots, k-1\}$:
$$B^{(l)}[a_\nu(\alpha)]y^{(l)} = B^{(l)}[a_\nu(\alpha)]s^{(l)} \qquad \text{and } B^{(r)}[a_\nu(\alpha)]y^{(r)} = B^{(r)}[a_\nu(\alpha)]s^{(r)}$$
$$B^{(l)}[b_\nu(\alpha)]y^{(l)} = B^{(l)}[b_\nu(\alpha)]s^{(l)} \qquad \text{and } B^{(r)}[b_\nu(\alpha)]y^{(r)} = B^{(r)}[b_\nu(\alpha)]s^{(r)}$$

We only summarize the remaining steps of the proof. Using the Properties (P2) and (P3) it can be shown that the system of linear equations has a solution. The claim that M is a strong 1-fooling set can be proved by combining the system of equations of the definition of the fooling set, the definition of B and the system of equations (1). □

5   A Comparison of Complexity Classes of Linear Transformed BDDs

Let P-LTOBDD, P-LTFBDD, P-sLTFBDD, NP-LTOBDD and ⊕P-LTOBDD denote the classes of functions with polynomial size LTOBDDs, polynomial size LTFBDDs, polynomial size strong LTFBDDs, polynomial size nondeterministic LTOBDDs and polynomial size ⊕LTOBDDs, respectively. Let P-OBDD, P-FBDD, ... be defined similarly. In Figure 2 some inclusions between these classes are summarized. $A \to B$ means that A is a proper subset of B, and a dotted lines between classes A and B means that these classes are not comparable, i.e., $A \not\subseteq B$ and $B \not\subseteq A$. The numbers in the figure refer to the following list of functions proving that the corresponding inclusion is proper or proving that the classes are not comparable. In order to make Figure 2 clearer, the relations between P-LTOBDD and NP-LTOBDD and some related classes are drawn separately.

First we remark that it is easy to see that all inclusions shown in Figure 2 hold. Besides the functions mentioned in the following, in the literature a lot of functions can be found that witness that the inclusions (1), (3) and (15)

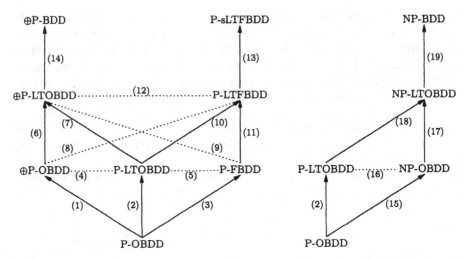

Fig. 2. Comparison of the complexity classes of polynomial size LTOBDDs and related BDD variants.

are proper. Our results on the function *MSA* prove that the inclusions (1), (7), (13), (15) and (18) are proper. The polynomial upper bounds for nondeterministic OBDDs and ⊕OBDDs are straightforward, and the upper bound for strong LTFBDDs is stated in Theorem 2. The lower bounds are stated in Theorem 3. The characteristic functions of linear codes prove that the inclusions (2), (6), (11) and (17) are proper. The upper bound and references for the lower bounds are given in the Introduction. In particular the exponential lower bound for nondeterministic OBDDs follows from the fact that nondeterministic OBDDs are simultaneously (syntactic) nondeterministic read-k-times branching programs. We remark that Günther and Drechsler [6] presented a different function to prove that (2) is a proper inclusion. Their results implicitly imply that also (6) and (17) are proper inclusions. Our results for *INDEX-EQ* prove that (3), (10), (14) and (19) are proper inclusions (Theorems 9 and 10). Here we use the fact that *INDEX-EQ* has polynomial size ⊕BDDs and polynomial size nondeterministic BDDs because it has polynomial size FBDDs. It remains to discuss the incomparability results. (4) and (16) follow from the bounds on *MSA* and on the characteristic functions of linear codes, (5) follows from the bounds on *INDEX-EQ* and the characteristic functions of linear codes, and (8), (9) and (12) follow from our results on *INDEX-EQ* and *MSA*.

We conclude that the methods presented in this paper allow to prove exponential lower bounds for several variants of linear transformed BDDs. In particular, it is possible to separate the classes of functions with polynomial size representations for many variants of linear transformed BDDs. It remains an open problem to prove exponential lower bounds for strong LTFBDDs and to prove exponential lower bounds on the size of LTOBDDs for practically important functions like multiplication.

Acknowledgment

I thank Beate Bollig, Rolf Drechsler, Wolfgang Günther, Martin Sauerhoff, Stephan Waack and Ingo Wegener for fruitful discussions and helpful comments.

References

1. Aborhey, S. (1988). Binary decision tree test functions. *IEEE Transactions on Computers* 37, 1461–1465.
2. Bern, J., Meinel, C. and Slobodová, A. (1995). Efficient OBDD-based Boolean manipulation in CAD beyond current limits. In *Proc. of 32nd Design Automation Conference*, 408–413.
3. Bryant, R.E. (1986). Graph-based algorithms for Boolean function manipulation. *IEEE Transactions on Computers* 35, 677–691.
4. Dietzfelbinger, M. and Savický, P. (1997). Parity OBDDs cannot represent the multiplication succinctly. Preprint Universität Dortmund.
5. Gergov, J. and Meinel, C. (1996). Mod-2-OBDDs—a data structure that generalizes EXOR-sum-of-products and ordered binary decision diagrams. *Formal Methods in System Design* 8, 273–282.
6. Günther, W. and Drechsler, R. (1998). BDD minimization by linear transformations. In *Proc. of Advanced Computer Systems*, Szczecin, Poland, 525–532.
7. Günther, W. and Drechsler, R. (1998). Linear transformations and exact minimization of BDDs. In *Proc. of IEEE Great Lakes Symposium on VLSI*, 325–330.
8. Hromkovič, J. (1997). *Communication Complexity and Parallel Computing.* Springer.
9. Jukna, S. (1995). A note on read-k times branching programs. *RAIRO Theoretical Informatics and Applications* 29, 75–83.
10. Jukna, S. (1999). Linear codes are hard for oblivious read-once parity branching programs. *Information Processing Letters* 69, 267–269.
11. Jukna, S. and Razborov, A. (1998). Neither reading few bits twice nor reading illegally helps much. *Discrete Applied Mathematics* 85, 223–238.
12. Jukna, S., Razborov, A., Savický, P. and Wegener, I. (1997). On P versus $NP \cap co\text{-}NP$ for decision trees and read-once branching programs. In *Proc. of Mathematical Foundations of Computer Science*, LNCS 1295, 319–326.
13. Kushilevitz, E. and Nisan, N. (1997). *Communication Complexity.* Cambridge University Press.
14. Meinel, C., Somenzi, F. and Theobald, T. (1997). Linear sifting of decision diagrams. In *Proc. of 34th Design Automation Conference*, 202–207.
15. Okol'nishnikova, E.A. (1991). On lower bounds for branching programs. *Metody Diskretnogo Analiza* 51, 61–83 (in Russian). English Translation in *Siberian Advances in Mathematics* 3, 152–166, 1993.
16. Razborov, A.A. (1991). Lower bounds for deterministic and nondeterministic branching programs. In *Proc. of Fundamentals of Computing Theory*, LNCS 529, 47–60.
17. Sieling, D. (1999). Lower bounds for linear transformed OBDDs and FBDDs. Preprint Universität Dortmund.
18. Waack, S. (1997). On the descriptive and algorithmic power of parity ordered binary decision diagrams. In *Proc. of Symposium on Theoretical Aspects of Computer Science*, LNCS 1200, 201–212.

19. Wegener, I. (1988). On the complexity of branching programs and decision trees for clique functions. *Journal of the Association for Computing Machinery* 35, 461–471.
20. Žák, S. (1984). An exponential lower bound for one-time-only branching programs. In *Proc. of Mathematical Foundations of Computer Science*, LNCS 176, 562–566.

A Unifying Framework for Model Checking Labeled Kripke Structures, Modal Transition Systems, and Interval Transition Systems

Michael Huth

Department of Computing and Information Sciences,
Kansas State University, Manhattan, KS 66506-2302,
huth@cis.ksu.edu
www.cis.ksu.edu/~huth

Abstract. We build on the established work on modal transition systems and probabilistic specifications to sketch a framework in which system description, abstraction, and finite-state model checking all have a uniform presentation across various levels of qualitative and quantitative views together with mediating abstraction and concretization maps. We prove safety results for abstractions within and across such views for the entire modal mu-calculus and show that such abstractions allow for some compositional reasoning with respect to a uniform family of process algebras à la CCS.

1 Introduction and Motivation

Process algebras such as Milner's CCS [16] and modular guarded command languages such as McMillan's SMV [15] are important description languages for a wide range of computer systems. The operational meaning of such descriptions is typically captured by a triple $\mathcal{M} = (S, R, L)$, where S is a set of states, R the state-transition relation, and L contains atomic state information; the latter is usually trivial in an event-based setting. The analysis of such descriptions can be done in a variety of ways. In model checking [3,20], the idea is to have a finite-state system $\mathcal{M}$ and a specification ϕ in some temporal logic, together with an efficient algorithm for deciding whether ϕ holds for all initial states of $\mathcal{M}$. Due to the typical exponential blow-up of the state set in the number of "parallel" components, one often requires abstraction techniques for data and even control flow paths in order to bring S down to smaller size [4]. One then needs to make certain that a positive model check of ϕ for the abstracted system $\mathcal{M}'$ means that the original system $\mathcal{M}$ satisfies the specification ϕ as well.

While this triad of system description, abstraction, and verification formalism has been well established and successful for qualitative system design and analysis, its transfer to quantitative system descriptions has, by and large, been problematic. On the conceptual side, in moving from a qualitative to a quantitative view of a system, one ordinarily has to change the description language, the

C. Pandu Rangan, V. Raman, R. Ramanujam (Eds.): FSTTCS'99, LNCS 1738, pp. 369–380, 1999.
© Springer-Verlag Berlin Heidelberg 1999

notion of abstraction, and the verification engine completely; for a notable exception see J. Hillston's work in [8]. Such changes not only necessitate the knowledge of sophisticated and computationally expensive concepts, such as measure theory [7] and probabilistic bisimulation [13,1], but also make it hard to embed the qualitative description into such a quantitative view, or to re-interpret quantitative results as qualitative judgments. Ideally, one would like to have a *uniform* family of such triads with view-mediating maps across all three dimensions: description, abstraction, and verification.

While our paper is an initial contribution toward crafting such a family, it also proposes the development of such a model checking framework for *loosely* specifying and verifying qualitative and quantitative systems. Systems often cannot be described in complete detail and usually we would like to give an implementor more flexibility in how to realize a specified system. Note that these comments apply to qualitative systems, such as concurrency protocols, as well as to quantitative ones, like loose Markov chains, where the actual state-transition probabilities may only be known to be within some interval. Our work extends and builds upon the work on modal transition systems by K. G. Larsen and B. Thomsen [14,12], and probabilistic specifications [10] by B. Jonsson and K. G. Larsen. Both approaches have in common that transitions $s \to^a s'$ are *loosely* specified, meaning that the system description does not determine the actual implementation fully. For modal transition systems, transitions $s \to^a s'$ are either *guaranteed* ($s \to^a_\Box s'$), or *possible* ($s \to^a_\Diamond s'$) [14]. For probabilistic specifications, we have transitions of the form $s \to^a_\mathcal{P} s'$, where $\mathcal{P}$ is a "set of probabilities" [10] which we will always assume to be a closed interval $[x, y]$ with $0 \le x$ and $y \le 1$. Conceptually, the latter models are interesting because they do not commit themselves to being "reactive", "generative", or even "probabilistic" right away. Such interpretations enter via the chosen notion of refinement, where "total" refinements, our implementations, exhibit such desired properties. In this paper, we study a modal and a probabilistic interpretation of such models. In the context of model checking loosely specified systems, "safety" now means that the information computed for a property ϕ is a consistent and valid approximation of the information that we could compute for any possible refinement, or implementation.

In the next section, we present three different views of a system along with their corresponding refinement notions. In Section 3, each such view determines a semantics of the modal mu-calculus which we prove to be sound with respect to refinement. Section 4 discussed abstractions and concretizations across system views and shows how model checking results transfer across such views. In Section 5, we hint at a process algebra framework which accommodates viewing systems at three levels and prove some compositionality results of process algebra operators with respect to refinement. Section 6 briefly covers a probabilistic view, giving rise to loose Markov chains. Finally, Section 7 provides an outlook on future work.

2   Different Views of a System

To illustrate, we consider the model of an unreliable medium in Figure 1. Clearly, the fully specified qualitative system, (a), is of little use as only flawed media (with an error state) are allowed refinements up to bisimulation. The qualitative, but loosely specified system, (b), is already faring better, since an implementation may now choose not to realize the transition from state full to state error. A possible refinement would therefore be the ideal and always reliable medium obtained from (a) by removing state error and all its incoming and outgoing transitions. The fully specified quantitative system, (c), prescribes even more realistic behavior by giving probabilities for correct system behavior. This Markov chain may be analyzed further, e.g. to determine its steady-state probability distribution. A *loosely specified Markov chain*, (d), however, allows more freedom in that we only specify a range for actual state-transition probabilities. This requires a generalization of existing techniques for analyzing Markov chains.

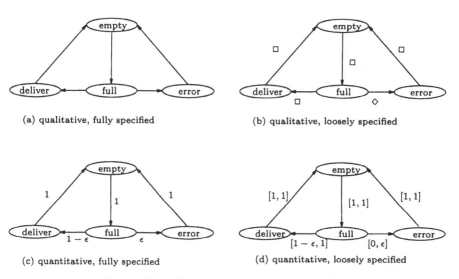

(a) qualitative, fully specified   (b) qualitative, loosely specified

(c) quantitative, fully specified   (d) quantitative, loosely specified

Fig. 1. Modeling an unreliable medium [10].

Three views of models. In general, we propose three views of models $\mathcal{M} = (S, R, L)$ with S a set of states, $R \colon S \times \text{Act} \times S \to D$ the state-transition function, and $L \colon S \times \text{AP} \to D$ the state-labeling function, where Act is a set of action labels, AP a set of atomic state predicates, and D is one of the three *domains of view*, the *type* of the model $\mathcal{M}$. We write $\mathcal{M} \triangleright D$ to indicate this relationship. If D equals K, the two-element lattice $\{\text{ff} < \text{tt}\}$, then such models are known as labeled Kripke structures (Figure 1(a); we omitted the action labels since Act is a singleton set); such structures are also known as *Doubly Labeled Transition Systems* (L²TS) in the literature [6]. If D equals the three-element poset $\{\mathtt{dk}, \mathtt{ff}, \mathtt{tt}\}$, which has $\mathtt{dk}$ ($\mathtt{dk}$ for "don't know") as a least element, and all other elements are maximal; then models are essentially the modal transition systems of [14] (Figure 1(b); $\square$ is interpreted as $\{\mathtt{tt}\}$ and $\lozenge$ as $\{\mathtt{dk}, \mathtt{tt}\}$). Finally, if D equals the interval domain I [17,21], the collection of all closed intervals $[x, y]$ with $0 \le x \le y \le 1$, ordered under reverse containment: $[u, v] \le [x, y]$ iff $u \le x \le y \le v$; then models are *interval transition systems*, a special case of the probabilistic specifications in [10] (Figure 1(d)). Note that the Markov chain in Figure 1(c) can be seen as a *maximal* interval transition system as all its behavior is fully specified with respect to the information ordering on I, by identifying any $r \in [0, 1]$ with $[r, r] \in$ I. It is helpful and insightful to also interpret $\square$ and $\lozenge$ on the domains K and I. On K, both modalities are identified with the set $\{\mathtt{tt}\}$; all possible behavior is also guaranteed. On I, we write $\square[x, y]$ iff $x > 0$, and $\lozenge[x, y]$ iff $y > 0$: x stands for the guarantee and y for the possibility of a transition $R(s, a, s') = [x, y]$. Later on, we interpret negation on D and $\square$ will not be the dual of $\lozenge$, unless D equals K.

Abstractions within system views. It is straightforward to define the sum $M + M'$ of two systems of type D. Therefore, we may reduce the concept of "system M abstracts system M'" to "state t abstracts state s in system $M + M'$" which, in turn, may be reduced to "state s refines state t in system $M + M'$". The intuitive meaning of "s refines t" is that *possible* transitions out of s are matched with possible transitions out of t, and *guaranteed* transitions out of t are matched with guaranteed transitions out of s [14]; no conditions are imposed on guaranteed transitions out of s, or possible transitions out of t. Further, a notion of refinement should be co-inductive, monotone with respect to the information ordering on D, uniform in the choice of D, and should allow for some compositional reasoning.

Definition 1. *For $M = (S, R, L) \triangleright D$, we define a functional $F_D : \mathcal{P}(S \times S) \to \mathcal{P}(S \times S)$: given $Q \subseteq S \times S$, we set $(s, t) \in F_D(Q)$ iff*

1. *For all $a \in$ Act, and all $s' \in S$, if $\lozenge R(s, a, s')$, then there is some $t' \in S$ such that $(s', t') \in Q$, $\lozenge R(t, a, t')$, and $R(t, a, t') \le R(s, a, s')$ in D.*
2. *For all $a \in$ Act, and all $t' \in S$, if $\square R(t, a, t')$, then there is some $s' \in S$ such that $(s', t') \in Q$, $\square R(s, a, s')$, and $R(t, a, t') \le R(s, a, s')$ in D.*
3. *For all $p \in$ AP, we have $L(t, p) \le L(s, p)$ in D.*

Subsets $Q \subseteq S$ satisfying $Q \subseteq F_D(Q)$ are called D-refinements.

It is easily seen that these functions F_D are monotone, so they have a greatest fixed point $\sqsubseteq_D$ which is also the greatest D-refinement. One may readily show that D-refinements are closed under all unions and relational composition. For D being K, K-refinements are simply Milner's bisimulations [16] for all event-based models (= trivial labeling function). To illustrate, consider the model in Figure 1(d). If we annotate all transitions of Figure 1(a) with $[1, 1]$ and remove the state **error**, then the resulting system is an I-refinement of the system in (d). Similarly, if we write $[p, p]$ for each state-transition probability p of the model in Figure 1(c), then this renders another I-refinement of the model in (d). Our M-refinements differ from Larsen's and Thomsen's refinement notion for modal transition systems in [14] in that they match an $R(s, a, s') = $ dk with some $R(t, a, t')$ such that $R(t, a, t') \leq$ tt, whereas we insist on a monotone match $R(t, a, t') = $ dk, since $R(t, a, t') \leq R(s, a, s')$ is enforced. This is a sharper constraint: a possible, *but not guaranteed*, transition out of the refining state s has to be matched with a possible, but not guaranteed, transition out of the refined state t. While our notion is suited for unifying it with a refinement for interval transition systems, Larsen's and Thomsen's refinement is not only sound, but also complete [12], for equivalences based on the fragment of the modal mu-calculus without fixed-points, covered in the next section.

3   Three Semantics of Temporal Logic

We use the modal mu-calculus [11] as our logic for specifying system properties; its syntax is given by $\phi ::= \mathtt{false} \mid p \mid Z \mid \neg\phi \mid \phi_1 \wedge \phi_2 \mid \langle a \rangle \phi \mid [a] \phi \mid \mu Z.\phi$ where p ranges over AP, Z over a set of variables, $a \in$ Act, and the bodies ϕ in $\mu Z.\phi$ are formally monotone. We write true and $\vee$ for the corresponding derived operators. Without fixed points and variables, its semantics for modal transition systems is implicitly given in [12] for a modal interpretation of Hennessey-Milner logic to obtain a characterization of their notion of refinement. There are several "natural" semantics for models of type I; a probabilistic interpretation is sketched in Section 6. A quantitative modal semantics is developed below. We define all these semantics uniformly and only point out the salient differences. Given a model $\mathcal{M} = (S, R, L) \triangleright D$, the meaning $[\![\phi]\!]^D$ is generally a function of type $\mathsf{Env}_D \to S \to D$, where Env_D is the space of all functions (environments ρ) which map variables Z to elements in D. For the remainder of this paper, we assume that all models $\mathcal{M} = (S, R, L) \triangleright D$ are *image-finite*: $\{s' \in S \mid \Diamond R(s, a, s')\}$ is finite for all $s \in S$ and $a \in$ Act.

Semantics for K and M. The interpretation of propositional logic over K is the usual one. For M, we extend this interpretation by $\neg$dk = dk, dk $\wedge$ ff = ff, dk $\wedge x = $ dk if $x \neq$ ff, and $\vee$ is the deMorgan dual of $\wedge$ via $\neg$. Then $[\![[a] \phi]\!]^D \rho s = \bigwedge_D \{ [\![\phi]\!]^\rho s' \mid \Diamond R(s, a, s') \}$ and $[\![\langle a \rangle \phi]\!]^D \rho s = \bigvee_D \{ [\![\phi]\!]^\rho s' \mid \Box R(s, a, s') \}$, where $\bigwedge$ is the interpretation of the *nary* $\wedge$ and $\bigvee$ that of *nary* $\vee$ on $D = $ K or M. Note that this semantics is conservative with respect to refinement as the universal modality $[a]$ quantifies over all *possible* transitions, whereas the existential modality $\langle a \rangle$ only ranges over guaranteed system moves. Except for $\neg$ on K, all operations have continuous meaning with respect to the Scott-topology on $S \to D$ (pointwise ordering). Thus, we may define the meaning of $\mu Z.\phi$ as a least fixed point (only for K do we require that ϕ be formally monotone). The semantics for K is the usual one for labeled Kripke structures, since $\Box$ and $\Diamond$ agree on K. The semantics for M, without fixed points, is essentially the one in [12]; note that $\neg \langle a \rangle \neg$ has a different interpretation than $[a]$ on M. We get safety of model checks with respect to D-refinement.

Theorem 1. *Let $\mathcal{M} \triangleright D$ be a model of type K, or M, with state set S and $s \sqsubseteq_D t$ in S. Then $[\![\phi]\!]^D \rho t \leq [\![\phi]\!]^D \rho s$ holds for all ϕ and ρ.*

Thus, model checking an abstraction will give us sound results for the model check of any of its refinements, including an actual implementation.

A modal view of I. Interval transition systems exhibit two, almost orthogonal, dimensions of non-determinism: first, which element in $\{s' \in S \mid \Diamond R(s,a,s')\}$ will be chosen for execution; second, which r in the interval $R(s,a,s')$ may an implementation realize for the transition $s \to^a s'$? These notions overlap precisely when $R(s,a,s')$ equals $[0,y]$ with $y > 0$, for then s' may, or may not, be in the set of actual a-successors of s in the implemented system. This subtlety has to be reflected in any semantics $[\![\phi]\!]^I$. We define a modal semantics $[\![\phi]\!]^I \rho s$, which is an interval $[x,y]$, such that x is the greatest lower bound *guarantee* that $s \models_\rho \phi$ holds, whereas y is the least upper bound *possibility* thereof; note that $s \models_\rho \phi$ is a shorthand for $s \in [\![\phi]\!]^K \rho$, where we turn a system of type I into one of type K, as explained in Section 4. This intuition determines the semantics uniquely up to an interpretation of set-theoretic conjunction over I, needed in the set qualifications for $[a]$ and $\langle a \rangle$. Our semantics therefore depends implicitly on a *t-norm* $T \colon [0,1] \times [0,1] \to [0,1]$, a Scott-continuous map (= preserving all directed suprema) which interprets conjunction and makes $([0,1], T, 1)$ into a commutative monoid. The interpretation of $\neg$ is $\neg[x,y] = [1-y, 1-x]$. The meaning of $\neg$ may be justified with the prescriptive intuition given above. We illustrate such reasoning for the interpretation of $\wedge$: if $[\![\phi]\!]^I \rho s = [x,y]$ and $[\![\psi]\!]^I \rho s = [u,v]$, then $[x,y] \wedge [u,v]$ ought to be $[\min(x,u), \min(y,v)]$. We only justify the choice of $\min(x,u)$ as the other case is argued similarly: x is a guarantee for $s \models_\rho \phi$ and u is a guarantee for $s \models_\rho \psi$; thus, we have at least the guarantee $\min(x,u)$ in either case. Using the proof rule $\wedge$-introduction, we obtain that $\min(x,u)$ is a guarantee for $s \models_\rho \phi \wedge \psi$. If a is another such guarantee, we may use the proof rules $\wedge$-elimination twice to conclude that a is also a guarantee for $s \models_\rho \phi$ and $s \models_\rho \psi$. But then $a \leq x$ and $a \leq u$ follows as x and u are least upper bounds on the guarantees of these respective properties.

Now we define the *modal* quantitative semantics of $\langle a \rangle$ and $[a]$. In the sequel, we write $\mathrm{pr}_1 \colon I \to [0,1]$ for the function $[x,y] \mapsto x$ and $\mathrm{pr}_2 \colon I \to [0,1]$ for $[x,y] \mapsto y$. According to our primary semantic guideline, we set $[\![\langle a \rangle \phi]\!]^I \rho s = [x,y]$, where $x = \bigvee_{[0,1]} \{T(\mathrm{pr}_1 R(s,a,s'), \mathrm{pr}_1 [\![\phi]\!]^I \rho s') \mid \Box R(s,a,s')\}$ and $y = \bigvee_{[0,1]} \{T(\mathrm{pr}_2 R(s,a,s'), \mathrm{pr}_2 [\![\phi]\!]^I \rho s') \mid \Diamond R(s,a,s')\}$. Since the $\mathrm{pr}_2 R(s,a,s')$ are all least upper bound possibilities for $s \to^a s'$, since $\mathrm{pr}_2 [\![\phi]\!]^I \rho s'$ is assumed to be the least upper bound possibility for $s' \models_\rho \phi$ to hold, the least upper bound for the possibility of $s \models_\rho \langle a \rangle \phi$ is a maximal $T(\mathrm{pr}_2 R(s,a,s'), \mathrm{pr}_2 [\![\phi]\!]^I \rho s')$, where the transition to s' is a possible one. A similar justification for x being the greatest lower bound guarantee for $s \models_\rho \langle a \rangle \phi$ can be given. Although the qualifications $\Box$ and $\Diamond$ are not really needed for computing the meaning of $\langle a \rangle$, they reveal a duality between $\langle a \rangle$ and $[a]$. We obtain the meaning of $[a] \phi$ from the one of $\langle a \rangle \phi$ by swapping all occurrences of $\Box$ and $\Diamond$, note that $\Box$ and $\Diamond$ are then no longer redundant, and by replacing all occurrences of $\bigvee_{[0,1]}$ with $\bigwedge_{[0,1]}$. This reflects the fact that we now have to reason about bounds for *all* possible next states. If we write $[u,v]$ for $[\![[a] \phi]\!]^I \rho s$, then $u = \bigwedge_{[0,1]} \{T(\mathrm{pr}_1 R(s,a,s'), \mathrm{pr}_1 [\![\phi]\!]^I \rho s') \mid \Diamond R(s,a,s')\}$

and $v = \bigvee_{[0,1]} \{T(\mathrm{pr}_2 R(s, a, s'), \mathrm{pr}_2 \llbracket \phi \rrbracket^{\mathrm{I}} \rho \, s') \mid \Box R(s, a, s')\}$. The justification of this semantics is dual to the one of $\langle a \rangle$ in the sense of "duality" explained above.

The justification for the least fixed-point semantics of $\llbracket \mu Z. \phi \rrbracket^{\mathrm{I}} \rho \, s$ is that we begin to unfold the recursive meaning with initial value $[0, 1]$, the bottom of I, at each state: initially, no guarantees, but all possibilities are given. The process of unfolding increases evidence for the guarantee of $s \models_\rho \mu Z. \phi$, whereas it decreases its possibility. If, and when this process stabilizes, we have established the best evidence we could find for $s \models_\rho \mu Z. \phi$ without knowing the particular implementation. Since M and I are not complete lattices, but are domains, we have to, and can, define the meaning of *greatest* fixed points $\llbracket \nu Z. \phi \rrbracket^D$ as $\llbracket \neg \mu Z. \neg \phi[\neg Z/Z] \rrbracket^D$, where $\phi[\neg Z/Z]$ is the result of replacing all free occurrences of Z in ϕ with $\neg Z$.

Theorem 2. *For the denotational semantics $\llbracket \phi \rrbracket^{\mathrm{I}}$, all its operations are Scott-continuous. In particular, the approximation of fixed points reaches its meaning at level ω. Moreover, if $\mathcal{M} \rhd \mathrm{I}$ has state set S with $s \sqsubseteq_{\mathrm{I}} t$ in S, then $\llbracket \phi \rrbracket^{\mathrm{I}} \rho \, t \leq \llbracket \phi \rrbracket^{\mathrm{I}} \rho \, s$ holds for all ϕ and ρ.*

This semantics is also continuous in the sense that the meaning $\llbracket \phi \rrbracket^{\mathrm{I}}$ will depend continuously on small changes made to R and L in an underlying model $\mathcal{M} = (S, R, L) \rhd \mathrm{I}$.

4 Abstractions Across Views

With a semantics of temporal logic formulas for each system view at hand, we need to understand whether and how such meanings transfer if we change the view of a system under consideration. We give such an account for moving between K and M, and M and I, respectively.

Abstraction and safety between K and M. The change of view between models of type K and M is different in quality from a move between models of type M and I, for $\neg$ is *not* monotone on models of type K. Moreover, the embedding $i \colon \mathrm{K} \to \mathrm{M}$, with $i(x) = x$ for all $x \in \mathrm{K}$, is *not* monotone as well. It induces embeddings of models $\mathcal{M} = (S, R, L) \rhd \mathrm{K}$ by setting $i\mathcal{M} = (S, i \circ R, i \circ L) \rhd \mathrm{M}$. Conversely, $\alpha_*, \alpha_*^{\mathrm{tr}} \colon \mathrm{M} \to \mathrm{K}$ are both monotone maps γ with $\gamma \circ i = \mathrm{id}_{\mathrm{K}}$ and $i \circ \gamma \geq \mathrm{id}_{\mathrm{M}}$; among those maps α_* and α_*^{tr} are uniquely defined by $\alpha_*(\mathrm{dk}) = \mathrm{ff}$ and $\alpha_*^{\mathrm{tr}}(\mathrm{dk}) = \mathrm{tt}$, respectively. We set $\alpha_* \mathcal{M}' = (S, \alpha_*^{\mathrm{tr}} \circ R', \alpha_* \circ L') \rhd \mathrm{K}$ for any $\mathcal{M}' = (S, R', L') \rhd \mathrm{M}$. The map α_* translates "truth values" pertaining to "propositional" information (model checks and labeling functions) and behaves well with respect to propositional logic: $\alpha_* \circ \neg \leq \neg \circ \alpha_*$, $\alpha_* \circ \wedge = \wedge \circ \alpha_* \times \alpha_*$, and $\alpha_* \circ \vee = \vee \circ \alpha_* \times \alpha_*$. Since α_*^{tr} gives us a conservative account of transitions, and since α_* preserves all suprema as a lower adjoint of α, which differs from i in that it sends ff to dk, we may relate model checks on $\mathcal{M}' \rhd \mathrm{M}$ to those on $\alpha_* \mathcal{M}' \rhd \mathrm{K}$, and, similarly, we may compare such results computed for $\mathcal{M} \rhd \mathrm{K}$ and $i\mathcal{M} \rhd \mathrm{M}$.

Theorem 3. *For all formulas ϕ of the modal mu-calculus, we have the inequality $\alpha_* \llbracket \phi \rrbracket^{\mathcal{M}' \rhd \mathrm{M}} \rho' \leq \llbracket \phi \rrbracket^{\alpha_* \mathcal{M}' \rhd \mathrm{K}} \alpha_* \circ \rho'$ and the equality $i \llbracket \phi \rrbracket^{\mathcal{M} \rhd \mathrm{K}} \rho = \llbracket \phi \rrbracket^{i\mathcal{M} \rhd \mathrm{M}} i \circ \rho$ for all models and environments of the required types.*

For $D = $ K or M, we write $s \models_\rho^{\mathcal{M}} \phi$, if $[\![\phi]\!]^{\mathcal{M} \rhd D} \rho s = $ tt; and $s \not\models_\rho^{\mathcal{M}} \phi$, if $[\![\phi]\!]^{\mathcal{M} \rhd D} \rho s = $ ff, where $\mathcal{M}$ is a model of type D and $\rho \in \text{Env}_D$. Given any $\mathcal{M}' = (S, R', L') \rhd$ M with $s \models_\rho^{\mathcal{M}'} \phi$, we may use the previous theorem to infer that tt $= \alpha_*[\![\phi]\!]^{\mathcal{M}' \rhd \text{M}} \rho s \leq [\![\phi]\!]^{\alpha_* \cdot \mathcal{M}' \rhd \text{K}} \alpha_* \circ \rho s$; but since tt is a maximal element in K, this implies $[\![\phi]\!]^{\alpha_* \cdot \mathcal{M}' \rhd \text{K}} \alpha_* \circ \rho s = $ tt. Thus, $s \models_\rho^{\mathcal{M}'} \phi$ implies $s \models_{\alpha, \rho}^{\alpha_* \cdot \mathcal{M}'} \phi$ *for all formulas of the modal mu-calculus*, where the latter is the standard notion of satisfaction for labeled Kripke structures. However, such an inference cannot be made for the negative version (at the meta-level): if $s \not\models_\rho^{\mathcal{M}'} \phi$, then both parts of our theorem provide no additional information in general. In this case, the inequality in this theorem is redundant as the left hand side, $\alpha_*[\![\phi]\!]^{\mathcal{M}' \rhd \text{M}} \rho' s$, denotes the least element of K; the theorem's equality is of no use as well, since $i \circ \alpha_*$ is not equal to, but above, the identity id_M. These results are quite similar in structure to the ones obtained in [5] and it would be of interest to establish connections to the work in loc. cit.

Abstraction and safety between M and I. We embed M into I via β such that $\beta(\text{dk}) = [0, 1]$, $\beta(\text{ff}) = [0, 0]$ and $\beta(\text{tt}) = [1, 1]$. Note that this map is monotone and matches our semantic intuition of $[x, y]$ giving guarantees and possibilities of truth. We re-translate such truth-value intervals with the upper adjoint β^* which is uniquely determined by being monotone and satisfying $\beta^* \circ \beta = \text{id}_\text{M}$ and $\beta \circ \beta^* \leq \text{id}_\text{I}$. As for the values of transitions, we define a map $\beta_{\text{tr}}^* : \text{I} \to \text{M}$ which is uniquely determined by preserving the three predicates $\Box$, $\Diamond \wedge \neg\Box$, and $\neg\Diamond$, which single out all elements of M. Note that this map also reflects $\Box$ and $\Diamond$ from M back to I. One can readily see that β^* is a homomorphism for $\neg$, $\wedge$, and $\vee$; e.g. $\beta^* \circ \neg = \neg \circ \beta^*$. As for the modalities and fixed points, we make crucial use of the fact that β^* is the upper adjoint of β. For $\mathcal{M} \rhd$ I and $\mathcal{M}' \rhd$ M, we define $\beta^* \mathcal{M} = (S, \beta_{\text{tr}}^* \circ R, \beta^* \circ L) \rhd$ M and $\beta \mathcal{M}' = (S, \beta \circ R', \beta \circ L') \rhd$ I.

Theorem 4. *Let ϕ be any formula of the modal mu-calculus and consider models $\mathcal{M} \rhd$ I and $\mathcal{M}' \rhd$ M. Then $\beta^* [\![\phi]\!]^{\mathcal{M} \rhd \text{I}} \rho \leq [\![\phi]\!]^{\beta^* \mathcal{M} \rhd \text{M}} \beta^* \circ \rho$ and $[\![\phi]\!]^{\beta \mathcal{M}' \rhd \text{I}} \rho \leq \beta[\![\phi]\!]^{\mathcal{M}' \rhd \text{M}} \beta \circ \rho$ hold for all t-norms T such that $T(a, b) = 0$ implies $a = 0$ or $b = 0$.*

The proof of this theorem reveals that the condition on the t-norm is necessary and $\text{LAND}(a, b) = \max(a + b - 1, 0)$ is an example of a Scott-continuous t-norm that does not satisfy it; take a and b to be 0.5. We also require that $T(a, b) = 1$ imply $a = b = 1$, but this holds for all t-norms, since min is known to be the *greatest* t-norm in the pointwise ordering: $T(a, b) \leq \min(a, b)$ holds for all $a, b \in [0, 1]$ and all t-norms T. The first inequality in the theorem above states that if a model check of type I results in a "truth value" $[0, 0]$ or $[1, 1]$, then that value is also the result of the same model check on the more concrete system of type M. One may now combine Theorems 3 and 4 to link model-checking results between types K and I.

5   Three Views of Description Languages

We choose process algebras as system description languages which are parametric in the domain of view D, and whose structural operational semantics can be seen as an abstract interpretation, based on D, of the concrete operational semantics for K. For sake of brevity, we only consider a fragment of Milner's CCS [16], given by the syntax $p ::= \text{nil} \mid a_d.p \mid p + p \mid p\|p \mid p\lceil B \mid x \mid \text{fix}\,x.p$, where $d \in D$, $a \in \text{Act}$, x ranges over a set of process variables, and $B \subseteq \text{Act}$. Note that the only non-standard feature is the annotation of the standard prefix, $a.p$, with a domain element. We also assume the usual involution $a \mapsto \bar{a}\colon \text{Act} \rightarrow \text{Act}$ for communication with the self-involutive symbol $\tau \notin \text{Act}$ for internal, non-observable actions. In Figure 2, the abstract interpretations $+^D$ and Par^D may well depend on the semantic interpretation one has in mind; e.g. whether one considers a modal or probabilistic semantics for $D = \text{I}$. To wit, we define $\text{Par}^D\{(d_1^a, d_2^{\bar{a}}) \mid a \in \text{Act}\} = \bigvee^D\{d_1^a \wedge^D d_2^{\bar{a}} \mid a \in \text{Act}\}$ for $D = \text{K}$ or M. For $D = \text{I}$, we choose a modal interpretation, setting $\text{pr}_2\text{Par}^D\{(d_1^a, d_2^{\bar{a}}) \mid a \in \text{Act}\} = 0$ if there is no $a \in \text{Act}$ with $\Diamond d_1^a$ and $\Diamond d_2^{\bar{a}}$; otherwise, we define $\text{pr}_2\text{Par}^D\{(d_1^a, d_2^{\bar{a}}) \mid a \in \text{Act}\} = \max\{\min(\text{pr}_2 d_1^a, \text{pr}_2 d_2^{\bar{a}}) \mid \Diamond d_1^a \text{ and } \Diamond d_2^{\bar{a}}\}$ and $\text{pr}_1\text{Par}^D\{(d_1^a, d_2^{\bar{a}}) \mid a \in \text{Act}\} = \min\{\min(\text{pr}_1 d_1^a, \text{pr}_1 d_2^{\bar{a}}) \mid \Box d_1^a \text{ and } \Box d_2^{\bar{a}}\}$. Thus, this semantics computes the worst-case, respectively, best-case evidence for observing an internal τ-move. The modalities are placed as in the semantics of $[a]$ and we could have used any Scott-continuous t-norm instead of the binary min operator, as long as $\Box$ and $\Diamond$ distribute over it. The rule for recursion indicates that these interpretations have to be continuous. Note that each process term, p, determines a model of type D: if there is a judgment $\vdash R(p, a, p') = d$, then d is the value of $R(p, a, p')$; otherwise, we set it to be $[\![\text{false}]\!]^D \rho s$. Note that $a_{\text{ff}}.p$ is bisimilar to nil for D being K.

Theorem 5. *Let Par^D be defined as above. For $D = \text{K}$, the "abstract interpretation" in Figure 2 matches the structural operational semantics of the corresponding fragment of CCS in [16]. For $D = \text{M}$, the abstract interpretation in the same figure matches the semantics of the modal process logic for the corresponding fragment in [14], where we identify $a_\Diamond.p$ and $a_\Box.p$ from [14] with $a_{\text{dk}}.p$ and $a_{\text{tt}}.p$, respectively.*

We are not aware of process algebras based on intervals in the literature, so we cannot compare our abstract interpretation for $D = \text{I}$. Since each process term p of type D determines a model of type D, we can write $p \sqsubseteq_D q$ if p D-refines q in the system formed by the sum of p and q. We can prove that refinements are compositional for some of the process algebra operators.

Theorem 6. *For all $a \in \text{Act}$, $d \in D$, and closed process algebra terms $p \sqsubseteq_D q$, $p_i \sqsubseteq_D q_i$ $(i = 1, 2)$ we have $a_d.p \sqsubseteq_D a_d.q$, $p_1\|p_2 \sqsubseteq_D q_1\|q_2$, and $p\lceil B \sqsubseteq_D q\lceil B$.*

This result can be extended to recursion with a machinery very similar to the one employed for $D = \text{K}$ in [16].

$$\frac{}{\vdash R(a_d.p, a, p) = d} \text{ Act}$$

$$\frac{\vdash R(p_1, a, p') = d_1 \text{ and } \vdash R(p_2, a, p') = d_2}{\vdash R(p_1 + p_2, a, p') = d_1 +^D d_2} \text{ Sum}$$

$$\frac{\vdash R(p_1, a, p_1') = d_1}{\vdash R(p_1 \| p_2, a, p_1' \| p_2) = d_1} \text{ Com\_1}$$

$$\frac{\vdash R(p_2, a, p_2') = d_2}{\vdash R(p_1 \| p_2, a, p_1 \| p_2') = d_2} \text{ Com\_2}$$

$$\frac{\vdash R(p, a, p') = d_a \text{ and } a \in B}{\vdash R(p \lceil B, a, p' \lceil B) = d_a} \text{ Res}$$

$$\frac{\vdash R(p_1, a, p_1') = d_1^a \text{ and } \vdash R(p_2, \bar{a}, p_2') = d_2^{\bar{a}}}{\vdash R(p_1 \| p_2, \tau, p_1' \| p_2') = \text{Par}^D \{(d_1^a, d_2^{\bar{a}}) \mid a \in \text{Act}\}} \text{ Com\_3}$$

$$\frac{\vdash R(p[\text{fix } x.p/x], a, p') = d}{\vdash R(\text{fix } x.p, a, p') = d} \text{ Rec}$$

Fig. 2. Abstract interpretation of a structural operational semantics for our three process algebras

6   Loose Markov Chains

Knowing the class of implementations may well allow the customization of our framework to such a class. For example, if interval transition systems are to specify labeled Markov chains, then we can restrict our attention to certain models of type I. A labeled Markov chain (S, P, L) satisfies $\sum_{s'} P(s, a, s') = 1$ for all $a \in \text{Act}$ and $s \in S$. We may approximate such a model with the same set of states by $\mathcal{M} = (S, R, L) \rhd \text{I}$ such that, for all $a \in \text{Act}$ and $s \in S$, we have $\sum_{s'} \text{pr}_1 R(s, a, s') \leq 1$ and $\text{pr}_2 R(s, a, s') \leq 1 - \sum_{s'' \neq s'} \text{pr}_1 R(s, a, s'')$. The first inequality says that the lower bounds for the actual state-transition probabilities form a subprobability distribution; the sum of probabilistic guarantees must not exceed 1. The second inequality is a consistency condition, saying that the upper bound on the possible probability of $s \to^a s'$ cannot be greater than 1 minus the sum of all lower bound guarantees on probabilities of moves to any other successor state of s. We call such models *loose Markov chains*. The models in Figure 1(c) and (d) are such examples and (c) is an I-refinement of (d). It would be of interest to define a probabilistic refinement which coincides with probabilistic bisimulation [13] for maximal models (Markov chains) and to compare such a notion with the work of [10]. As for a semantics of formulas ϕ, we change the modal semantics by re-interpreting $\wedge$, $\langle a \rangle$, and $[a]$. For $\wedge$, we may either use a safe t-norm, as done in [18,9,2], or develop a measure theory of measures of type $\mu \colon \Sigma(X) \to \text{I}$, where the conventional measures of type $\mu \colon \Sigma(X) \to [0,1]$ form the maximal elements of that space. As for the modalities, we identify the meaning of $\langle a \rangle$ and $[a]$ and set $\text{pr}_1 [\![\langle a \rangle \phi]\!]^P \rho s = \sum_{s'} \text{pr}_1 R(s, a, s') \cdot \text{pr}_1 [\![\phi]\!]^P \rho s'$ and $\text{pr}_2 [\![\langle a \rangle \phi]\!]^P \rho s = \min(1, \sum_{s'} \text{pr}_2 R(s, a, s') \cdot \text{pr}_2 [\![\phi]\!]^P \rho s')$; note that $\cdot$ is a Scott-continuous t-norm.

Theorem 7. *Let* $\mathcal{M} \rhd I$ *be a loose Markov chain with state set* S *and* $s \sqsubseteq_I t$ *in* S. *Then* $[\![\phi]\!]^P \rho\, t \leq [\![\phi]\!]^P \rho\, s$ *holds for all* ϕ *and* ρ *and all monotone interpretations of* $\wedge$; *in particular, this holds when* $\wedge$ *is interpreted as a probabilistically conservative t-norm.*

7 Outlook

The design and analysis of algorithms for deciding D-refinements needs to be done particularly for the case $D = I$. It would be of interest to obtain an independent *logical characterization* of these refinements. Interval transition systems should be evaluated toward their suitability of describing systems with uncertainty, or vagueness. A guarded-command language for the description of such models may provide a foundation for the formal analysis of fuzzy interval inference systems. Connections to Bayesian networks and Dempster-Shafer theories of evidence need to be explored. The models of loose Markov chains require a customized description language; their ergodic analysis should reduce to an optimization problem. The computation of conditional probabilities, however, may require a "domain theory" for probability measures, where the latter are maximal elements in a space of "measures" of range I. This needs to be a conservative extension in the sense that the "probability axioms" for the I-valued measures will reduce to the familiar axioms in case that the measure is maximal. Such work may well transfer to the *generalized probabilistic logic* (GPL) designed by N. Narashima, R. Cleaveland and P. Iyer in [19]. Loose Markov chains will also benefit from a probabilistic version of I-refinement which should coincide with a familiar probabilistic bisimulation for "maximal" models. Martin Escardo pointed out that our framework is extendible to cover infinite-state systems as well. One obtains a continuous, and computable, semantics of model checks, provided that the sets of possible ($\Diamond$) and guaranteed ($\Box$) a-successors of state s are compact for all $a \in \mathtt{Act}$ and $s \in S$, where S is a compact Hausdorff space.

Acknowledgments

A number of people have made valuable suggestions during visits, or talks, given at their institutes. Among them were Martin Escardo, Stephen Gilmore, Jane Hillston, Marta Kwiatkowska, Annabelle McIver, Carroll Morgan, and Jeff Sanders.

References

1. C. Baier. Polynomial Time Algorithms for Testing Probabilistic Bisimulation and Simulation. In *Proceedings of CAV'96*, number 1102 in Lecture Notes in Computer Science, pages 38–49. Springer Verlag, 1996.
2. C. Baier, M. Kwiatkowska, and G. Norman. Computing probability bounds for linear time formulas over concurrent probabilistic systems. *Electronic Notes in Theoretical Computer Science*, 21:19 pages, 1999.

3. E. M. Clarke and E. M. Emerson. Synthesis of synchronization skeletons for branching time temporal logic. In D. Kozen, editor, *Proc. Logic of Programs*, volume 131 of *LNCS*. Springer Verlag, 1981.
4. E. M. Clarke, O. Grumberg, and D. E. Long. Model Checking and Abstraction. In *19th Annual ACM SIGPLAN-SIGACT Symposium on Principles of Programming Languages*, pages 343–354. ACM Press, 1992.
5. Dennis Dams, Rob Gerth, and Orna Grumberg. Abstract interpretation of reactive systems. *ACM Transactions on Programming Languages and Systems*, 19(2), 1997.
6. R. de Nicola and F. Vaandrager. Three Logics for Branching Bisimulation. *Journal of the Association of Computing Machinery*, 42(2):458–487, March 1995.
7. P. R. Halmos. *Measure Theory*. D. van Norstrand Company, 1950.
8. J. Hillston. *A Compositional Approach to Performance Modelling*. Cambridge University Press. Distinguished Dissertation Series, 1996.
9. M. Huth. The Interval Domain: A Matchmaker for aCTL and aPCTL. In M. Mislove, editor, *2nd US-Brazil joint workshop on the Formal Foundations of Software Systems held at Tulane University, New Orleans, Louisiana, November 13-16, 1997*, volume 14 of *Electronic Notes in Theoretical Computer Science*. Elsevier, 1999.
10. B. Jonsson and K. G. Larsen. Specification and Refinement of Probabilistic Processes. In *Proceedings of the International Symposium on Logic in Computer Science*, pages 266–277. IEEE Computer Society, IEEE Computer Society Press, July 1991.
11. D. Kozen. Results on the propositional mu-calculus. *Theoretical Computer Science*, 27:333–354, 1983.
12. K. G. Larsen. Modal Specifications. In J. Sifakis, editor, *Automatic Verification Methods for Finite State Systems*, number 407 in Lecture Notes in Computer Science, pages 232–246. Springer Verlag, June 12–14, 1989 1989. International Workshop, Grenoble, France.
13. K. G. Larsen and A. Skou. Bisimulation through Probabilistic Testing. *Information and Computation*, 94(1):1–28, September 1991.
14. K. G. Larsen and B. Thomsen. A Modal Process Logic. In *Third Annual Symposium on Logic in Computer Science*, pages 203–210. IEEE Computer Society Press, 1988.
15. K. L. McMillan. *Symbolic Model Checking*. Kluwer Academic Publishers, 1993.
16. R. Milner. *Communication and Concurrency*. Series in Computer Science. Prentice-Hall International, 1989.
17. R. E. Moore. *Interval Analysis*. Prentice-Hall, Englewood Cliffs, 1966.
18. C. Morgan, A. McIver, and K. Seidel. Probabilistic predicate transformers. *ACM Transactions on Programming Languages and Systems*, 18(3):325–353, May 1996.
19. M. Narashima, R. Cleaveland, and P. Iyer. Probabilistic Temporal Logics via the Modal Mu-Calculus. In W. Thomas, editor, *Foundations of Software Science and Computation Structures*, volume 1578 of *Lecture Notes in Computer Science*, pages 288–305. Springer Verlag, March 1999.
20. J. P. Quielle and J. Sifakis. Specification and verification of concurrent systems in cesar. In *Proceedings of the fifth International Symposium on Programming*, 1981.
21. D. S. Scott. Lattice Theory, Data Types and Semantics. In *Formal Semantics of Programming Languages*, pages 66–106. Prentice-Hall, 1972.

Graded Modalities and Resource Bisimulation

Flavio Corradini[1], Rocco De Nicola[2], and Anna Labella[3]

[1] Dipartimento di Matematica Pura ed Applicata, Università dell'Aquila
flavio@univaq.it
[2] Dipartimento di Sistemi e Informatica, Università di Firenze
denicola@dsi.unifi.it
[3] Dipartimento di Scienze dell'Informazione, Università di Roma "La Sapienza"
labella@dsi.uniroma1.it

Abstract. The logical characterization of the strong and the weak (ignoring silent actions) versions of resource bisimulation are studied. The temporal logics we introduce are variants of Hennessy-Milner Logics that use graded modalities instead of the classical box and diamond operators. The considered strong bisimulation induces an equivalence that, when applied to labelled transition systems, permits identifying all and only those systems that give rise to isomorphic unfoldings. Strong resource bisimulation has been used to provide nondeterministic interpretation of finite regular expressions and new axiomatizations for them. Here we generalize this result to its weak variant.

1 Introduction

Modal and temporal logics have been proved useful formalisms for specifying and verifying properties of concurrent systems (see e.g. [19]), and different tools have been developed to support such activities [8,7]. However, to date, there is no general agreement on the type of logic to be used. Since a logic naturally gives rise to equivalences (two systems are equivalent if they satisfy the same formulae) often, for a better understanding and evaluation, the proposed logics have been contrasted with behavioural equivalences. The interested reader is referred to [10,17] for comparative presentations of many such equivalences.

Establishing a direct correspondence between a logic and a behavioural equivalence provides additional confidence in both approaches. A well-known result relating behavioural and logical semantics is that reported in [15]; there, a modal logic, now known as Hennessy-Milner Logic *HML*, is defined which, when interpreted over (arc-) labelled transition systems with and without silent actions, is proved to be in full agreement with two equivalences called *strong* and *weak* observational equivalence. Other correspondences have been established in [2] where two equivalences over Kripke structures (node-labelled transition systems) are related to two variants of *CTL** [13], and in [12] where three different logical characterizations are provided for another variant of bisimulation called *branching bisimulation*.

In this paper, we study the logical characterization of yet another variant of bisimulation that we call *resource bisimulation* [6]. This bisimulation takes

C. Pandu Rangan, V. Raman, R. Ramanujam (Eds.): FSTTCS'99, LNCS 1738, pp. 381–393, 1999.
© Springer-Verlag Berlin Heidelberg 1999

into account the number of choices a system has, even after it has decided the specific action to be performed. The new equivalence counts the instances of specific actions a system may perform and thus considers as different the two terms P and $P + P$; the latter representing the non-deterministic composition of a system with itself. Intuitively, this can be motivated by saying that $P + P$ duplicates the resources available in P. This permits differentiating systems also relatively to a form of fault tolerance known as "cold redundancy": $P + P$ is more tolerant to faults than P, because it can take advantage of the different instances of the available resources.

Resource bisimulation enjoys several nice properties (we refer to [6] for a comprehensive account and for additional motivations). It has been shown that resource bisimulation coincides with the kernel of resource simulation (this result is new for simulation-like semantics). Moreover, it permits identifying all and only those labelled transition systems that give rise to isomorphic unfoldings. Also, resource bisimulation, when used to provide nondeterministic interpretation of finite regular expressions, leads to a behavioural semantics that is in full agreement with a tree-based denotational semantics and is characterized via a small set of axioms obtained from Salomaa's axiomatization of regular expressions [21] by removing the axioms stating idempotence of + and distributivity of • over +, see Table 1.

Table 1. Axioms for resource bisimulation over finite regular expressions.

$$
\begin{array}{ll}
X + Y = Y + X & \text{(C1)} \\
(X + Y) + Z = X + (Y + Z) & \text{(C2)} \\
X + 0 = X & \text{(C3)} \\[1em]
(X{\cdot}Y){\cdot}Z = X{\cdot}(Y{\cdot}Z) & \text{(S1)} \\
X{\cdot}1 = X & \text{(S2)} \\
1{\cdot}X = X & \text{(S3)} \\
X{\cdot}0 = 0 & \text{(S4)} \\
0{\cdot}X = 0 & \text{(S5)} \\[1em]
(X + Y){\cdot}Z = (X{\cdot}Z) + (Y{\cdot}Z) & \text{(RD)}
\end{array}
$$

In this paper we continue our investigation on resurce bisimulation in two directions. First, we study a logical characterization of resource bisimulation, then, we provide a sound and complete axiomatization for its weak variant. The logic which characterizes resource bisimulation is obtained by replacing both the box and the diamond modalities of HML with the family of graded modalities [14], defined below, where # denotes *multisets cardinality*.

$$ p \models \langle \mu \rangle_n \psi \text{ if and only if } \#\{\!| p' \mid p \xrightarrow{\mu} p' \text{ and } p' \models \psi |\!\} = n. $$

If we define Graded HML $(GHML)$ to be the set of formulae generated by the grammar:

$$\psi \ ::= \ True \ \Big| \ \psi_1 \wedge \psi_2 \ \Big| \ \langle \mu \rangle_n \psi \qquad \text{where } \mu \in A \text{ and } 0 \leq n < \infty$$

it can be established that

$$(\forall \psi \in GHML, \ P \models \psi \Longleftrightarrow Q \models \psi) \text{ if and only if } P \sim_r Q$$

We shall also study the weak variant of resource bisimulation over regular expressions enriched with a distinct *invisible* τ-action, and we shall provide also for this new equivalence both an axiomatic and a logical characterization. The complete axiomatization will be obtained by adding the axiom

$$\alpha \bullet \tau \bullet X \ = \ \alpha \bullet X$$

to those for (strong) resource bisimulation of Table 1. The logical characterization is obtained by providing a different (*weak*) interpretation of the modal operators described above.

Due to space limitation, all proofs are omitted, they are reported in the full version of the paper.

2   Nondeterministic Expressions and Resource Bisimulation

In this section we provide an observational account of finite nondeterministic regular expressions, by interpreting them as equivalence classes of labelled transition systems. This part has been extensively treated in [6]. The proposed equivalence relies on the same recursive pattern of bisimulation but takes into account also the number of equivalent states that are reachable from a given one.

Let $A \cup \{1\}$ be a set of actions. The set of nondeterministic finite regular expressions over A is the set **PL** of terms generated by the following grammar:

$$P \ ::= \ 0 \ \Big| \ 1 \ \Big| \ a \ \Big| \ P + P \ \Big| \ P \bullet P \quad \text{where } a \text{ is in } A.$$

We give the following interpretation to nondeterministic regular expressions. Like in [1], the term 0, denotes the empty process. The term 1 denotes the process that does nothing and successfully terminates. The term a denotes a process that executes a visible action a and then successfully terminates. The operator + can be seen as describing the nondeterministic composition of agents. The operator $\bullet$ models sequential composition.

Definition 1. A labelled transition system is a triple $< Z, L, T >$ where Z is a set of states, L is a set of labels and $T = \{ \xrightarrow{n} \subseteq Z \times Z \mid n \in L \}$ is a transition relation.

Table 2. Active predicate.

$$active(1)$$

$$active(a)$$

$$active(P) \lor active(Q) \implies active(P + Q)$$

$$active(P) \land active(Q) \implies active(P \bullet Q)$$

Table 3. Operational Semantics for **PL**.

Tic)
$$\frac{}{1 \xrightarrow{<1,\epsilon>} 1}$$

Atom)
$$\frac{}{a \xrightarrow{<a,\epsilon>} 1}$$

Sum$_1$)
$$\frac{P \xrightarrow{<\mu,u>} P', \; active(Q)}{P+Q \xrightarrow{<\mu,lu>} P'}$$

Sum$_1'$)
$$\frac{P \xrightarrow{<\mu,u>} P', \; \neg active(Q)}{P+Q \xrightarrow{<\mu,u>} P'}$$

Sum$_2$)
$$\frac{Q \xrightarrow{<\mu,u>} Q', \; active(P)}{P+Q \xrightarrow{<\mu,ru>} Q'}$$

Sum$_2'$)
$$\frac{Q \xrightarrow{<\mu,u>} Q', \; \neg active(P)}{P+Q \xrightarrow{<\mu,u>} Q'}$$

Seq$_1$)
$$\frac{P \xrightarrow{<\alpha,u>} P', \; active(Q)}{P \bullet Q \xrightarrow{<\alpha,u>} P' \bullet Q}$$

Seq$_2$)
$$\frac{P \xrightarrow{<1,u>} 1, \; Q \xrightarrow{<\mu,u'>} Q'}{P \bullet Q \xrightarrow{<\mu,uu'>} Q'}$$

In our case, states are terms of **PL** and labels are pairs $< \mu, u >$ with $\mu \in A \cup \{1\}$ and u a word, called *choice sequence*, in the free monoid generated by $\{l, r\}$. The transition relation relies on the "active" predicate defined in Table 2 and is defined in Table 3. There, and in the rest of the paper, we write $z \xrightarrow{n} z'$ instead of $< z, z' > \in \xrightarrow{n}$.

We have two kinds of transitions:

- $P \xrightarrow{<a,u>} P'$: P performs an action a, possibly preceded by 1-actions with choice sequence u.
- $P \xrightarrow{<1,u>} 1$: P performs 1-actions to reach process 1 with choice sequence u.

These transitions are atomic, which means that they cannot be interrupted and keep no track of intermediate states. In both cases, u is used to keep information about the possible nondeterministic structure of P, and will permit distinguishing those transitions of P whose action label and target state have the same name but are the result of different choices. Thus for $a+a$, it is possible to record that it can perform two different a actions: $a+a \xrightarrow{<a,l>} 1$ and $a+a \xrightarrow{<a,r>} 1$; without the l and r labels, we would have only the $a+a \xrightarrow{a} 1$ transition.

The predicate *active* over **PL** processes that is used in Seq_1 allows us to detect empty processes and to avoid performing actions leading to deadlocked states.

The rules of Table 3 should be self-explanatory. We only comment on those for $+$ and $\bullet$.

The rule for $P+Q$ says that if P can perform $< \mu, u >$ to become P', and Q is not deadlocked, then $P+Q$ can perform $< \mu, lu >$ to become P' where l records that action μ has been performed by the left alternative. If Q is deadlocked, then, no track of the choice is kept in the label. The right alternative is dealt with symmetrically.

$\text{Seq}_1)$ mimics sequential composition of P and Q; it states that if P can perform $< \mu, u >$ then $P \bullet Q$ can evolve with the same label to $P' \bullet Q$. The premise $active(Q)$ of the inference rule ensures that Q can successfully terminate.

In order to abstract from choice sequences while keeping information about the alternatives a process has for performing a specific action, we introduce a new transition relation that associates to every pair $< P \in \textbf{PL}, \mu \in Act \cup \{1\} >$, a multiset M, representing all processes that are target of different $< \mu, u >$-transitions from P. The new transition relation is defined as the relation that satisfies:

$$P \xrightarrow{\mu} \{ P' \mid \exists u.\ P \xrightarrow{<\mu,u>} P' \}$$

Thus, for example, we have:

- $a + a \xrightarrow{a} \{1, 1\}$ because $-a + a \xrightarrow{<a,l>} 1$ and $-a + a \xrightarrow{<a,r>} 1$,
- $(1+1) \bullet (a + a) \xrightarrow{a} \{1, 1, 1, 1\}$ because
  - $(1+1) \bullet (a + a) \xrightarrow{<a,ll>} 1$, $-(1+1) \bullet (a + a) \xrightarrow{<a,lr>} 1$,
  - $(1+1) \bullet (a + a) \xrightarrow{<a,rl>} 1$ $-(1+1) \bullet (a + a) \xrightarrow{<a,rr>} 1$.

We shall now introduce the bimulation-based relations that identifies only those systems that have exactly the same behaviour and differ only for their syntactic structure. This equivalence relation, called *resource bisimulation* and introduced in [6], relates only those terms whose unfolding, via the operational semantics, gives rise to isomorphic labelled trees. The transition relation $\xrightarrow{\mu}$, introduced above, is the basis for defining *resource bisimulation*.

Definition 2. (*Resource Bisimulation*)

a. A relation $\Re \subseteq \textbf{PL} \times \textbf{PL}$ is a *r-bisimulation* if for each $< P, Q > \in \Re$, for each $\mu \in A \cup \{1\}$:

   i. $P \xrightarrow{\mu} M$ implies $Q \xrightarrow{\mu} M'$ and $\exists f$ injective: $M \to M'$, such that $\forall P' \in M, < P', f(P') > \in \Re$;

   ii. $Q \xrightarrow{\mu} M'$ implies $P \xrightarrow{\mu} M$ and $\exists g$ injective: $M' \to M$, such that $\forall Q' \in M', < Q', g(Q') > \in \Re$;

b. P and Q are *r-bisimilar* ($P \sim_r Q$), if there exists a *r-bisimulation* $\Re$ containing $< P, Q >$.

The above definitions should be self explanatory. We just remark that the injection $f : M \to M'$ is used to ensure that different (indexed) processes in M are simulated by different (indexed) processes in M' [1]. Thus r-bisimilarity requires the cardinality of M be less or equal to the cardinality of M'.

Since the multisets we are dealing with are finite, conditions i) and ii) of Definition 2 can be summarized as follows: $P \xrightarrow{\mu} M$ implies $Q \xrightarrow{\mu} M'$ and there exists a bijective $f : M \to M'$, s.t. for all $P' \in M$, $< P', f(P') > \in \Re$.

With standard techniques it is possible to show that $\sim_r$ is an equivalence relation and it is preserved by nondeterministic composition and sequential composition. It is not difficult to check that $a \not\sim_r a + a$, $a + b \sim_r b + a$ and $(1 + 1) \cdot a \sim_r a + a$.

3   A Logical Characterization of Resource Bisimulation

In this section, we provide a positive logic for resource bisimulation. In [15], a modal logic, now known as Hennessy-Milner Logic (HML), is defined which, when interpreted over labelled transition systems with (or without) silent actions, is proved to be in full agreement with weak (or strong) observational equivalence.

Our logic is obtained from HML by eliminating the *false* predicate and by replacing both the box and the diamond modalities (or, alternatively, both the box and the $\neg$ modality) with a family of so called *graded modalities* [14] of the form $\langle \mu \rangle_n \varphi$ where $0 \le n < \infty$. Intuitively, a process P satisfies the formula $\langle \mu \rangle_n \varphi$ if P has exactly n μ-derivatives satisfying formula φ.

Let Graded HML (GHML) be the set of formulae generated by the following grammar:

$$\varphi ::= tt \;\Big|\; \varphi \wedge \varphi \;\Big|\; \langle \mu \rangle_n \varphi \qquad \text{where } \mu \in A \cup \{1\} \text{ and } 0 \le n < \infty.$$

The satisfaction relation $\models$ for the logic over GHML formulae is given by:

$$P \models tt \qquad \text{for any } P$$
$$P \models \varphi_1 \wedge \varphi_2 \;\text{ iff } P \models \varphi_1 \text{ and } P \models \varphi_2$$
$$P \models \langle \mu \rangle_n \varphi \quad \text{iff } \#(\{\!\!\{ P' \mid \exists u.\ P \xrightarrow{<\mu, u>} P' \}\!\!\} \cap \{\!\!\{ P' \mid P' \models \varphi \}\!\!\}) = n$$

We shall let $\Re_L$ denote the binary relation over PL processes that satisfy the same set of GHML formulae:

$$\Re_L = \{(P, Q) \mid \forall \varphi \in \text{GHML} ,\ P \models \varphi \iff Q \models \varphi\}$$

and will show that $\Re_L$ is a resource bisimulation.

---

[1] Since a multiset can be seen as a set of indexed elements, an injection between multisets can be seen as an ordinary injection between sets.

Indeed, we can prove that the equivalence induced by GHML formulae coincides with resource equivalence. The proof that if $P \sim_r Q$ then, for all $\varphi \in$ GHML, it holds that $(P \models \varphi$ iff $Q \models \varphi)$ is standard and follows by induction on the syntactic structure of formulae.

Proposition 1. Let P, Q be **PL** processes. If $P \sim_r Q$ then $(\forall \varphi \in$ GHML, $P \models \varphi \Longleftrightarrow Q \models \varphi)$.

The proof of the reverse implication, namely, the proof that any two processes satisfying the same set of GHML formulae are weak resource equivalent, requires a more sophisticated proof technique. It needs to be shown that, if $\Re_L$ was not a weak resource bisimulation, then, there would exist $(P, Q) \in \Re_L$ such that for some $\mu \in Act$, $P \xrightarrow{\mu} M$ implies $Q \xrightarrow{\mu} M'$ and for all bijective $f_i \colon M \to M'$, there would exist $P_i \in M$ such that $(P_i, f_i(P_i)) \notin \Re_L$. This implies that there exists a formula $\bar{\varphi} \in$ GHML such that $P_i \models \bar{\varphi}$ but $f(P_i) \not\models \bar{\varphi}$.

We can prove that, given a multiset of processes, we can find a formula characterizing each of its bisimulation classes, in the sense that every element in a class satisfies the characteristic formula of the class and does not satisfies any formula characterizing any other class. In this way, from the hypothesis that $P \sim_r Q$ does not hold, we can obtain a formula satisfied by one of the original processes, but not by the other one.

Given a multiset M and GHML formulae $\varphi_1, \varphi_2, \dots, \varphi_n$, let $M_i^{\varphi_i}$ be the subset of M that satisfies φ_i:

$$M_i^{\varphi_i} = \{\!\!\{ P' \in M \mid P' \models \varphi_i \}\!\!\} \qquad i \in [1..n]$$

If $\{M_1^{\varphi_1}, M_2^{\varphi_2}, ..., M_n^{\varphi_n}\}$ is a *partition* of M, then we shall write $M = M_1^{\varphi_1} \uplus M_2^{\varphi_2} \uplus ... \uplus M_n^{\varphi_n}$.

Lemma 1. Given a finite multiset M and a partition $M_1 \uplus M_2 \uplus ... \uplus M_n$ of M satisfying the property that two elements of the same class satisfy the same formulae and two elements in two different classes behave differently for at least one formula, then, for every class, there is a formula (*the characteristic formula*) satisfied by all the elements of that class and by none of any other class. Therefore we can write $M = M_1^{\xi_1} \uplus M_2^{\xi_2} \uplus ... \uplus M_n^{\xi_n}$, where $\xi_1,...,\xi_n$ are such that each $P \in M_i^{\xi_i}$ satisfies ξ_i, $(P \models \xi_i)$, while each $Q \in M_j^{\xi_j}$, $j \neq i$, does not satisfy ξ_i, $(Q \not\models \xi_i)$.

The coincidence between resource bisimulation and the equivalence induced by the GHML formulae immediately follows from the lemma above.

Proposition 2. Let P, Q be **PL** processes.
If $(\forall \varphi \in$ GHML, $P \models \varphi \Longleftrightarrow Q \models \varphi)$ then $P \sim_r Q$.

4   Weak Resource Bisimulation

This section is devoted to giving expressions in presence of invisible actions.

Let $A \cup \{1\}$ be a set of visible actions and $\tau \notin A \cup \{1\}$ be the invisible action. We use $\mu, \gamma, ..., \mu', \gamma', ...$ to range over by $A \cup \{1\} \cup \{\tau\}$, $\alpha, \beta, ..., \alpha', \beta', ...$ to range over by $A \cup \{\tau\}$ and $a, b, ..., a', b', ...$ to range over by A.

The set of nondeterministic regular expressions over $A \cup \{\tau\}$ is the set of terms generated by the following grammar:

$$P ::= 0 \mid 1 \mid a \mid \tau \mid P + P \mid P \bullet P \quad \text{where } a \text{ is in } A.$$

We will refer to the set of terms above as **PL** as well. We extend the interpretation given for the τ–less case as follows: τ denotes a process which can internally evolve and then successfully terminates. For those familiar with the operational semantic of process algebras, we would like to remark that 1-actions do not play the same role of invisible τ- actions. They simply stand for successful terminated processes.

To deal with the new actions, we extend the transition relation of Table 3 by adding the rule:

$$\text{Tau)} \quad \frac{}{\tau \xrightarrow{<\tau, \epsilon>} 1}$$

It relies on the predicate *active* defined in Table 2 extended with the condition below, that is used to detect empty processes

$$active(\tau).$$

We have now three kinds of transitions:

- $P \xrightarrow{<a,u>} P'$: P performs an action a, possibly preceded by 1-actions, with choice sequence u.
- $P \xrightarrow{<1,u>} 1$: P performs 1-actions to reach process 1 with choice sequence u.
- $P \xrightarrow{<\tau,u>} P'$: P performs an action τ, possibly preceded by 1-actions, with choice sequence u.

These transitions are atomic; they cannot be interrupted and, moreover, leave no track of intermediate states. In both cases, u is used to keep information about the possible nondeterministic structure of P, and will permit distinguishing those transitions of P with identical action label and target state.

Starting from elementary transitions, weak transitions can be defined. They can be invisible or visible. Weak invisible transitions denote sequences of τ-transitions (possibly interleaved by 1's) that lead to branching nodes, while weak visible transitions denote the execution of visible actions (possibly) followed or preceded by invisible moves. As usual, we have also terminating moves, i.e., sequences of 1-actions leading to successful termination of a process; see Table 4 for their formal definitions. In order to be able to give full account of the differ-

Table 4. Weak Transitions for **PL**.

$$
\text{WT}_1)\ \ \frac{P_1 \xrightarrow{<\mu,u>} P_2}{P_1 \xRightarrow{<\mu,u>} P_2}
$$

$$
\text{WT}_2)\ \ \frac{P_1 \xRightarrow{<\tau,u_1>} P_2 \ \text{and}\ P_2 \xRightarrow{<\alpha,u_2>} P_3}{P_1 \xRightarrow{<\alpha,u_1u_2>} P_3}
$$

$$
\text{WT}_3)\ \ \frac{P_1 \xRightarrow{<\alpha,u_1>} P_2 \ \text{and}\ P_2 \xRightarrow{<\tau,u_2>} P_3}{P_1 \xRightarrow{<\alpha,u_1u_2>} P_3}
$$

ent alternatives a process has when determining the specific action to perform, we introduce a transition relation that associates a multiset M, to every pair $P \in \textbf{PL}$, $\mu \in Act \cup \{1\} \cup \{\tau\}$. M represents all processes that reacheable via (initial) weak $< \mu, u >$-transitions by P. Since we are interested in the branching structure of processes and in detecting their actual choice points, we remove from M all those processes which can perform a τ actions in a purely deterministic fashion. That is, we remove those target processes which can perform an initial τ-transition "without choice". This new transition relation is defined as the least relation such that:

$$
P \xRightarrow{\mu} \{\!\{\, P' \mid \exists u.\, P \xRightarrow{<\mu,u>} P'\,\}\!\} - \{\!\{\, P' \mid P' \xRightarrow{<\tau,v>} \text{ with } v = \epsilon\,\}\!\}.
$$

The transition relation $\xRightarrow{\mu}$ is the basis for defining *resource equivalence*.

Definition 3. (*Weak Resource Bisimulation*)

1. A relation $\Re \subseteq \textbf{PL} \times \textbf{PL}$ is a *weak resource bisimulation* if for each $< P, Q > \in \Re$, for each $\mu \in A \cup \{1\} \cup \{\tau\}$:
   (i) $P \xRightarrow{\mu} M$ implies $Q \xRightarrow{\mu} M'$ and there exists an injective $f : M \to M'$, such that for all $P' \in M$, $< P', f(P') > \in \Re$;

   (ii) $Q \xRightarrow{\mu} M'$ implies $P \xRightarrow{\mu} M$ and there exists an injective $g : M' \to M$, such that for all $Q' \in M'$, $< g(Q'), Q' > \in \Re$.
2. P and Q are *weak resource equivalent* ($P \approx_r Q$), if there exists a *weak r-bisimulation* $\Re$ containing $< P, Q >$.

Remark 1. An immediate difference between weak resource equivalence and the standard observational equivalence, see e.g. [18], is that we do not consider "empty" τ- moves, i.e., moves of the form $P \xRightarrow{<\epsilon,\epsilon>} P$; we require at least one τ to be performed. In our framework, a process cannot idle to match a transition of another process which performs invisible actions.

Below, we provide a number of examples that should give an idea of the discriminating power of weak resource equivalence:

- Processes $\tau \bullet \tau$ and τ are related, while $\tau + \tau$ and τ are taken apart. The reason for the latter differentiation is similar to that behind $1 + 1 \not\approx_r 1$. Indeed $(\tau + \tau) \bullet a$ is equal to $\tau \bullet a + \tau \bullet a$ which has to be different (in this counting setting) from $\tau \bullet a$.
- Processes $a \bullet \tau$ and a are related because the τ following action a is not relevant from the branching point of view.
- $\tau \bullet a$ and a are instead distinguished because the τ action preceding the a in the former process can influence choices when embedded in a non-deterministic context.
- Processes τ and 1 are not equivalent, again because a τ action can be ignored only after a visible move.
- Processes $(\tau + 1) \bullet a$ and $a + \tau \bullet a$ are weak resource bisimilar.
- Processes $\tau(\tau + 0)$ and $\tau \bullet \tau$ are weak resource bisimilar

The following proposition states a congruence result for weak resource bisimulation. We can prove that our equivalence is actually preserved by all **PL** operators; noticeably it is preserved by $+$. This is another interesting property of our equivalence notion; weak equivalences are usually not preserved by $+$ and additional work is needed to isolate the coarsest congruence contained in them.

Proposition 3. Weak resource equivalence is preserved by all **PL** operators.

Let us consider now the simulation relation, denoted by $\preceq_r$ and called weak resource simulation, obtained by removing item 1.(ii) from Definition 3. It can be can shown that (like for strong resource bisimulation) the kernel of $\preceq_r$, coincide with weak resource equivalence.

Proposition 4. Let $\preceq_r$ be the preorder obtained by considering one of the two items in the definition of $\approx_r$, and let P and Q, be two processes. Then, $P \approx_r Q$ iff $P \preceq_r Q$ and $Q \preceq_r P$.

The logical characterization of resource equivalence can be easily extended to the weak case. It is sufficient to extend the alphabet and to introduce a τ modality. Then, within the actual definition of the satisfaction relation

- $P \xrightarrow{<\mu,u>} P'$ has to be replaced by $P \xRightarrow{<\mu,u>} P'$;
- $\#((\{ P' \mid \exists u.\, P \xrightarrow{<\mu,u>} P' \}) \cap \{ P' \mid P' \models \varphi_1 \}) = n$ has to be replaced by $\#((\{ P' \mid \exists u.P \xRightarrow{<\mu,u>} P' \} - \{ P' \mid P' \xRightarrow{<\tau,v>} \text{ with } v = \epsilon \}) \cap \{ P' \mid P' \models \varphi_1 \}) = n$.
- $P \sim_r Q$ has to be replaced by $P \approx_r Q$.

Proposition 5. The equivalence induced by extended weak GHML formulae coincides with weak resource equivalence.

Table 5. The τ−law for **EPL**.

$$\alpha{\bullet}\tau{\bullet}X = \alpha{\bullet}X \qquad (\text{T1})$$

A sound and complete axiomatization of weak resource equivalence over **PL** processes can also be provided. We can prove that the new weak equivalence is fully characterized by the axiom of Table 5 (please remember that now $\alpha \in A\cup\{\tau\}$) together with the set of axioms of Table 1 (which soundly and completely axiomatize strong resource equivalence [6]).

Proposition 6. The axiom of Table 5 together with the set of axioms of Table 1 soundly and completely axiomatize weak resource bisimulation over **PL**.

Remark 2. Consider the axiom of Table 5 and replace action α with μ. The resulting axiom, $\mu{\bullet}\tau{\bullet}X = \mu{\bullet}X$, is not sound. Indeed, by letting $\mu = 1$ and $X = 1$ we would have that $1{\bullet}\tau{\bullet}1$ and $1{\bullet}1$ are related by the equational theory, while they are not weak resource equivalent as remarked above. Therefore the axiom $X{\bullet}\tau{\bullet}Y = X{\bullet}Y$ is not sound.

5 Conclusions

We have introduced graded modalities and used them to provide a logical characterization of the strong and weak versions of resource bisimulation, an equivalence which discriminates processes according to the number of different computation they can perform to reach specific states. As a result, resource bisimulation identifies all and only those labelled transition systems that give rise to isomorphic unfoldings. In the case of the weak variant this isomorphism is guaranteed up to ignoring the invisible τ−actions. We have also extended the complete axiomatization of strong resource bisimulation of [6] to the weak variant of the equivalence.

The results that we have obtained for regular espressions can easily be extended to full-fledged process algebras like CCS, CSP, ACP or variants thereof that are equipped with a structural operational semantics, if care is taken to properly model the choice operators.

The logic we have introduced to characterize both resource and weak resource bisimulation, can easily be related with other modal logics introduced for dealing with bisimulation. In particular, referring to [9] as a comprehensive treatment, we can describe our logic as a polymodal $\aleph_O$ graduated logic. Also in [9], bisimulation is a k-counting bisimulation, that in the case $k = \aleph_O$ coincide with our resource bisimulation. On the other hand the two points of view are completely different and, in some sense, complementary. There, one was interested in the largest logic (the more expressive one) invariant under bisimulation. Here, we are looking for the minimal logic that is sufficient for characterizing bisimilar processes. As a consequence, our logic is extremely poor in connectives (just the conjunction) as well as in atomic propositions (just tt). In this way we showed that for example negation is not necessary. We have extended also our result to the case in which a "silent" relation τ between worlds is allowed, while we have not explicitly treated infinite terms. Nonetheless it can be immediately seen, by looking to the structure of proofs, that, if we allow infinite terms corresponding to behaviours with finite branching, e.g. guarded by μ operators, all the results will still hold. This is in accordance with the fact that, in the modal logics quoted above, μ operators are introduced for positive formulae only, and our language consists of strict positive formulae.

References

1. Baeten,J.C.M., Bergstra,J.A.: Process Algebra with a Zero Object. In Proc. *Concur'90*, LNCS **458**, pp. 83-98, 1990.
2. Browne,M.C., Clarke,E., Grümberg O.: Characterizing Finite Kripke Structures in Propositional Temporal Logic. *Theoretical Computer Science* **59**(1,2), pp. 115-131, 1998.
3. Baeten,J., Weijland, P.: *Process Algebras*. Cambridge University Press, 1990.
4. Corradini,F., De Nicola,R. and Labella,A.: Fully Abstract Models for Nondeterministic Regular Expressions. In Proc. *Concur'95*, LNCS **962**, Springer Verlag, pp. 130-144, 1995.
5. Corradini,F., De Nicola,R. and Labella,A.: A Finite Axiomatization of Non deterministic Regular Expressions. *Theoretical Informatics and Applications*. To appear. Available from: ftp://rap.dsi.unifi.it/pub/papers/FinAxNDRE. Abstract in FICS, Brno, 1998.
6. Corradini,F., De Nicola,R. and Labella,A.: Models for Non deterministic Regular Expressions. *Journal of Computer and System Sciences*. To appear. Available from: ftp://rap.dsi.unifi.it/pub/papers/NDRE.
7. Clarke,E.M., Emerson,E.A., Sistla,A.P.: Automatic Verification of Finite State Concurrent Systems using Temporal Logic Specifications. *ACM Toplas* **8**(2), pp. 244-263, 1986.
8. Cleaveland,R., Parrow,J., Steffen,B.: The Concurrency Workbench. *ACM Toplas* **15**(1), pp. 36-72, 1993.
9. D'Agostino,G.: Modal Logics and non well-founded Set Theories: translation, bisimulation and interpolation. Thesis, Amsterdam, 1998.
10. De Nicola,R.: Extensional Equivalences for Transition Systems. *Acta Informatica* **24**, pp. 211-237, 1987.
11. De Nicola,R., Labella,A.: Tree Morphisms and Bisimulations, *Electronic Notes in TCS* **18**, 1998.
12. De Nicola,R., Vaandrager,F.: Three Logics for Branching Bisimulation. *Journal of ACM* **42**(2), pp. 458-487, 1995.
13. Emerson,E.H., Halpern,Y.: "Sometimes" and "not never" revisited: On branching versus linear time temporal logic. *Journal of ACM* **42**, pp. 458-487, 1995.
14. Fattorosi-Barnaba,M., De Caro,F.: Graded Modalities, I. *Studia Logica* **44**, pp. 197-221, 1985.
15. Hennessy,M., Milner,R.: Algebraic Laws for Nondeterminism and Concurrency. *Journal of ACM* **32**, pp. 137-161, 1985.

16. Hoare, C.A.R.: *Communicating Sequential Processes*, Prentice Hall, 1989.
17. van Glabbeek,R.J.: Comparative Concurrency Semantics and Refinement of Actions. Ph.D. Thesis, Free University, Amsterdam, 1990.
18. Milner,R.: *Communication and Concurrency*, Prentice Hall, 1989.
19. Manna,Z., and Pnueli,A.: *The Temporal Logic of Reactive and Concurrent Systems*. Springer Verlag, 1992.
20. Park,D.: Concurrency and Automata on Infinite sequences. In Proc. GI, LNCS **104**, pp. 167-183, 1981.
21. Salomaa,A.: Two Complete Axiom Systems for the Algebra of Regular Events. *Journal of ACM* **13**, pp. 158-169, 1966.

The Non-recursive Power of Erroneous Computation*

Christian Schindelhauer[1,3] and Andreas Jakoby[2,3]

[1] ICSI Berkeley, 1947 Center Street, Berkeley, USA
schindel@icsi.berkeley.edu
[2] Depart. of Computer Science, Univ. of Toronto, Canada
jakoby@cs.toronto.edu
[3] Med. Univ. zu Lübeck, Inst. für Theoretische Informatik,
Lübeck, Germany

Abstract. We present two new complexity classes which are based on a complexity class C and an error probability function F. The first, F-Err C, reflects the (weak) feasibility of problems that can be computed within the error bound F. As a more adequate measure to investigate lower bounds we introduce F-Err$_{io}$ C where the error is infinitely often bounded by the function F. These definitions generalize existing models of feasible erroneous computations and cryptographic intractability.

We identify meaningful bounds for the error function and derive new diagonalizing techniques. These techniques are applied to known time hierarchies to investigate the influence of error bound. It turns out that in the limit a machine with slower running time cannot predict the diagonal language within a significantly smaller error prob. than $\frac{1}{2}$.

Further, we investigate two classical non-recursive problems: the halting problem and the Kolmogorov complexity problem. We present strict lower bounds proving that any heuristic algorithm claiming to solve one of these problems makes unrecoverable errors with constant probability. Up to now it was only known that infinitely many errors will occur.

1 Introduction

The answer of the question whether $\mathcal{NP}$ equals $\mathcal{P}$ is a main goal of computational complexity theory. If they differ, which is the widely proposed case, a polynomial time bounded deterministic algorithm cannot correctly decide an $\mathcal{NP}$-complete problem. Hence, the correctness of such an algorithm is the most reasonable requirement. Nevertheless, in the desperate situation where one wants to solve an infeasible problem one may accept errors, provided their influence can be controlled somehow. The quality of such an error can be exploited in more detail.

Probabilistic error: It is common sense that $\mathcal{BPP}$ can be seen as a class of efficiently solvable problems. Unlike an $\mathcal{P}$-algorithm a $\mathcal{BPP}$-algorithm can make errors, but on every input this error probability has to be bounded by $\frac{1}{3}$. So,

---

* parts of this work are supported by a stipend of the "Gemeinsames Hochschulsonderprogramm III von Bund und Länder" through the DAAD.

C. Pandu Rangan, V. Raman, R. Ramanujam (Eds.): FSTTCS'99, LNCS 1738, pp. 394–406, 1999.

$\mathcal{BPP}$-algorithms can be modified to decrease this error probability to an arbitrarily small non-zero polynomial. Furthermore, $\mathcal{BPP}$-algorithms solve problems for which no $\mathcal{P}$-algorithm is known yet, i.e. test of primalty.

Expected time: A valid enhancement of the notion of efficiency is to measure the expectation over the running times of different inputs according to a probability distribution over the input space. E.g. it turns out that the Davis-Putnam-algorithm [DaPu 60] solves SAT for randomly chosen Boolean formulas in polynomial expected time, if the probability distribution fulfills certain requirements [PuBr 85,SLM 92]. Note that, since this algorithm is deterministic, the probability refers only to the way of choosing the input. Probabilistic algorithms (using random bits) have been investigate in this relationship, too [Hoos 98]. An algorithm which is efficient with respect to its expected time behaviors has to compute a function always correctly, but its computation time can exceed the expected time bound for some inputs, enormously.

Average complexity classes: It turns out that complexity classes defined by expected time bounds are not closed under polynomial time bounded simulations. For this reason Levin defined *polynomial on the average* [Levi 86], a superset of expected polynomial time that initiated the average-case branch of computational complexity. Using Levin's measure for the average complexity there exists a reasonable notion of $\mathcal{NP}$-*average-case-completeness*.

Traditionally, one investigates the worst case of resources over an input length needed to solve a problem. In average case complexity theory these resources are weighted by a probability distribution over the input space. As a consequence, in worst case theory it is only necessary to consider functions $f : \mathbb{N} \to \mathbb{N}$ for an entire description of the considered complexity bound. In average case complexity theory there are a variety of ways to average over the resource function. Here it is necessary to examine pairs of resource functions $f : \Sigma^* \to \mathbb{N}$ (e.g. time) and probability distributions over the set of inputs. A variety of different concepts are investigated to define average complexity measures and corresponding classes [Levi 86,Gure 91,BCG 92,ScWa 94,CaSe 99,ReSc 96]. All these concepts have in common that the running times of all possible input values with positive weights account for the average behavior. An important result in average complexity is that if *average-$\mathcal{P}$* ($\text{Av}\mathcal{P}$) covers $\mathcal{NP}$, then $\mathcal{NE} = \mathcal{E}$ [BCG 92]. Furthermore, the fraction of non-polynomial computations of an algorithms solving an $\mathcal{NP}$-problem can be bounded under this premise.

Benign faults: Here the algorithm outputs a special symbol "?" on inputs where it fails, yet produces correct outputs in polynomial time for all other inputs. An algorithm for an $\mathcal{NP}$-problem producing only a small quantity of so-called *benign faults* can be transformed into an $\text{Av}\mathcal{P}$-algorithm [Imp2 95]. This observation is intuitively clear. Since, if an algorithm "knows" that it cannot solve a certain instance of an $\mathcal{NP}$-problem, it can use the trivial exponentially time-bounded simulation of the nondeterministic Turing-machine. If the probability for these instances is exponentially small, the resulting average-case time stays polynomial.

Similar questions were considered in [Schi 96], e.g. Schindelhauer introduces the class MedDistTime(T,F) which is strongly related to the statistical p-th quantile. Here, a machine M may violate the time bound $T(|x|)$ for at most $F(\ell)$ of the ℓ most likely inputs x. But the machine has to decide on the language correctly. This setting of correctness is equivalent to benign fault when we consider time-bounded complexity classes.

Real faults: In [ImWi 98,Schi 96,ScJa 97,Yam1 96,Yam2 96] a different approach is introduced. They investigate the number of inputs for a given machine causing an erroneous computation or breaking the time limit respectively. Note that for a fix time bound exceeding the time limit causes an error.

Yamakami proposed the notion of Nearly-$\mathcal{P}$ and Nearly-$\mathcal{BPP}$ (see [Yam1 96] and [Yam2 96]). Here the error probability has to be smaller than any polynomial in the input length. More precisely, even a polynomial number of instances (chosen according to the probability distribution) induces a super-polynomial small error bound again. Thus, Nearly-$\mathcal{P}$ and Nearly-$\mathcal{BPP}$ define reasonable efficient complexity classes if the corresponding algorithm is only used for a polynomial number of inputs for each length.

Independently, Schindelhauer et al. in [Schi 96,ScJa 97] introduced a similar approach. Based on a T-time decidable languages L and an error probability function F they investigate pairs of languages L' and probability distribution μ where for all $\ell \in \mathbb{N}$ the number of the ℓ most likely inputs $x \in \Sigma^*$ (according to μ) with $L'(x) \neq L(x)$ is bounded by $F(\ell)$.

Impagliazzo and Wigderson in [ImWi 98] investigate the relationship between $\mathcal{BPP}$ and an error complexity class called HeurTime$_{\epsilon(n)}(T(n))$, where for each input length the number of unrecoverable errors of a T-time bounded algorithm is bounded by ϵ. Their main motivation for this definition is the better understanding of the relationship of $\mathcal{BPP}$, $\mathcal{P}\backslash$poly and $\mathcal{E}$.

Erroneous Computation has a practical impact in designing more efficient algorithms, see for example [Fagi 92,GeHo 94,Reif 83]. In these papers parallel algorithms, resp. circuits, for adding two large binary numbers are presented which are efficient on the average. The basic part of these strategies is a fast but erroneous algorithm with a polynomial share of inputs causing wrong outputs. This results in a double logarithmic time bound. An additional error-detection-strategy restores correctness.

Based on this work done so far, it is reasonable to extend these definitions to arbitrary classes C and to consider the general properties of error complexity classes.

Definition 1 *For a class C and a bound $F : \mathbb{N} \rightarrow [0,1]$ define the distributional complexity classes of* **F-error bounded C** *and* **infinitely often F-error bounded C** *as sets of pairs of languages $L \subseteq \Sigma^*$ and probability distributions $\mu : \Sigma^* \rightarrow [0,1]$ as follows:*

$$F\text{-}\mathbf{Err}C := \{(L,\mu) \mid \exists S \in C \; \forall n \; : \; \text{Prob}_\mu[x \in (L \bigtriangleup S) \mid x \in \Sigma^n] \; \leq \; F(n)\},$$
$$F\text{-}\mathbf{Err}_{io}C := \{(L,\mu) \mid \exists S \in C \; \exists_{io} n \; : \; \text{Prob}_\mu[x \in (L \bigtriangleup S) \mid x \in \Sigma^n] \; \leq \; F(n)\}$$

where $A \bigtriangleup B := (A \setminus B) \cup (B \setminus A)$ as the symmetric difference of sets A and B.

Figure 1 illustrates the error behavior of languages $S_1, S_2, S_3 \in \mathcal{C}$ with respect to a given language L. S_1 provides smaller error probability than F for all inputs. Hence, S_1 proves that $L \in F\text{-}\mathrm{Err}\,\mathcal{C}$. The error probability of S_2 will infinitely often fall below the error bound F. If no language in $\mathcal{C}$ with the behavior of S_1 and S_2 exists, L cannot be approximated by any language in $\mathcal{C}$. It follows that $L \notin F\text{-}\mathrm{Err}_{\mathrm{io}}\,\mathcal{C}$. Figure 2 illustrates the error probability of S w.r.t. L: $L \in F_1\text{-}\mathrm{Err}\,\{S\}$ and $L \in F_2\text{-}\mathrm{Err}_{\mathrm{io}}\,\{S\}$, but $L \notin F_2\text{-}\mathrm{Err}\,\{S\}$ and $L \notin F_3\text{-}\mathrm{Err}_{\mathrm{io}}\,\{S\}$.

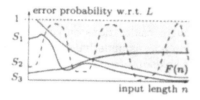

Fig. 1. The error probability of languages S_1, S_2, and S_3 with respect to L for increasing input length n.

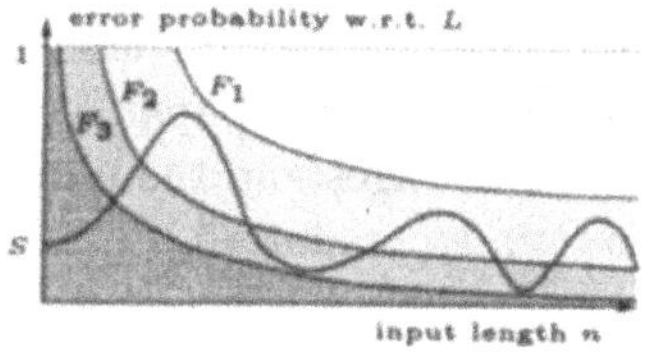

Fig. 2. F_3 gives a lower bound of the error probability of a language S with respect to the io-measure. The error bound F_1 gives an upper bound for both classes.

Using definition 1 we can generalize the classes Nearly-$\mathcal{BPP}$ (introduced by Yamakami [Yam1 96,Yam2 96]) and Nearly-$\mathcal{P}$ by **Nearly-$\mathcal{C}$** $:= n^{-\omega(1)}\text{-}\mathrm{Err}\,\mathcal{C}$ for an arbitrary class $\mathcal{C}$. Nearly-$\mathcal{C}$ represent complexity classes with a very low error probability.

Computational classes do not only describe the feasibility of problems, sometimes they are used to ensure intractability. An important application is cryptography, where one wants to prove the security of encryption algorithms, interactive protocols, digital signature schemes, etc. The security is often based on *intractability assumptions* for certain problems, e.g. factoring of large numbers or the computation of quadratic residuosity. Problems are called intractable if every polynomial time bounded algorithm outputs errors for a sufficient high number of instances.

An adequate measure of lower bounds turns out to be $F\text{-}\mathrm{Err}_{\mathrm{io}}\,\mathcal{C}$, which generalizes existing models of cryptographic intractability, e.g.

Definition 2 *[GMR 88] A function f is* **GMR-intractable** *for a probability distribution μ if for all probabilistic polynomial time bounded algorithms A it holds* $\forall c > 0 \; \forall_{ae} k \; : \; \mathrm{Prob}_\mu[A(x) = f(x) \mid x \in \Sigma^k] \leq \frac{1}{k^c}$.

Let $\mathcal{FBPP}$ be the functional extension of $\mathcal{BPP}$, or more precisely: $f \in \mathcal{FBPP}$ iff there exists a polynomial time bounded probabilistic Turing machine M such that $\mathrm{Prob}[M(x) \neq f(x)] \leq \frac{1}{3}$. Note that the error of $\frac{1}{3}$ can be decreased to any polynomial, without loosing the polynomial time behavior of M. To classify GMR-intractable functions by error complexity classes we take an appropriate generalization of $F\text{-Err}_{io} \mathcal{C}$ for functional classes $\mathcal{FC}$, i.e. $(f, \mu) \in F\text{-Err}_{io} \mathcal{FC}$ iff

$$\exists g \in \mathcal{FC} \; \exists_{io} n \; : \; \text{Prob}_\mu[f(x) \neq g(x) \mid x \in \Sigma^n] \; \leq \; F(n) \, .$$

Proposition 1 *GMR-intractable functions are not in $\left(1 - n^{-\Omega(1)}\right)$-$\text{Err}_{io}\, \mathcal{FBPP}$.*

The intractability assumption of [GoMi 82] and the *hard-core* sets of [Imp1 95] refer to non-uniform Boolean circuits. These classes can analogously be expressed by using error complexity classes.

In the rest of this paper we concentrate our considerations on complexity classes of languages and the uniform distribution μ_{uni} as underlying probability distribution, where for all $n \in \mathbb{N}$ and $x, y \in \Sigma^n$ holds $\mu_{uni}(x) = \mu_{uni}(y) > 0$. For the sake of readability we omit the distribution:

$$L \in F\text{-Err}\mathcal{C} :\Leftrightarrow (L, \mu_{uni}) \in F\text{-Err}\,\mathcal{C} \, ,$$
$$L \in F\text{-Err}_{io}\mathcal{C} :\Leftrightarrow (L, \mu_{uni}) \in F\text{-Err}_{io}\,\mathcal{C} \, .$$

In this paper we show a first classification of suitable error bounds and extend these lower bound results to time hierarchies. In section 4 we discuss in detail the error complexity of the halting problem and the Kolmogorov complexity problem. Non-recursiveness does not imply high error bounds in general. For the halting problem the Gödel-enumeration of programs has to be examined under new aspects. For some enumerations there are algorithms computing the halting problem within small error probability. We show in the following that even a standard enumeration yields a high lower error bound. In the case of Kolmogorov-complexity the lower bound is even higher and worst possible: We give a constant bound independent from the encoding.

2   Notations

For a predicate $P(n)$ let $\forall_{ae} n : P(n)$ be equivalent to $\exists n_0 \forall n \geq n_0 : P(n)$ and $\exists_{io} n : P(n)$ to $\forall n_0 \exists n \geq n_0 : P(n)$. Further, define $f(n) \leq_{ae} g(n)$ as $\forall_{ae} n : f(n) \leq g(n)$ and $f(n) \leq_{io} g(n)$ as $\exists_{io} n : f(n) \leq g(n)$.

We consider strings over an at least binary alphabet Σ, where λ denotes the empty word. Define $\Sigma^{\leq n} := \bigcup_{i \leq n} \Sigma^i$. Further, we use the lexicographical order function ord $: \Sigma^* \mapsto \mathbb{N}$ as straightforward isomorphism and its reverse function str $: \mathbb{N} \mapsto \Sigma^*$. $\mathcal{RE}$ and $\mathcal{REC}$ define the sets of all partial recursive, resp. recursive predicates. For a partial function f the domain is called $\text{dom}(f)$. We use $f(x) = \bot$ to denote $x \notin \text{dom}(f)$. Furthermore, let $L_0, L_1, L_2, \ldots$ be an enumeration of the languages in $\mathcal{RE}$ over a given alphabet Σ. For easier notation we present a language L_i resp. $L_i[a, b]$ by its **characteristic string** $L_i[a, b] := L_i(a) \cdots L_i(b)$, where $L_i(\text{ord}(w)) = 1$ if $w \in L_i$ and 0 otherwise.

For a partial recursive function φ and for all $x, y \in \Sigma^*$ define the **relative Kolmogorov complexity** as $C_\varphi(x|y) := \min\{|p| : \varphi(p, y) = x\}$. A programming system φ is called **universal** if for all partial recursive functions f holds

$\forall x, y \in \Sigma^* : \quad C_\varphi(x|y) \leq C_f(x|y) + O(1)$. Fixing such a universal programming system φ we define $C_\varphi(x) := C_\varphi(x|\lambda)$ as the (absolute) **Kolmogorov complexity** of x.

3 The Bounds of Error Complexity Classes

When the error probability tends to 1, the error complexity becomes meaningless. But for the io-error classes the corresponding probability is $\frac{1}{2}$.

Proposition 2 *Let $\mathcal{C}, \mathcal{C}'$ be complexity classes, where $\mathcal{C}$ is closed under finite variation and $\mathcal{C}'$ contains at least two complementary languages. Then for all functions $z(n) =_{ae} 0$ and $z'(n) =_{io} 0$ it holds that*

$$2^{\Sigma^*} = \left(1 - \frac{z(n)}{|\Sigma|^n}\right)\text{-Err}\,\mathcal{C} \quad and \quad 2^{\Sigma^*} = \left(\frac{1}{2} - \frac{z'(n)}{|\Sigma|^n}\right)\text{-Err}_{io}\,\mathcal{C}' .$$

This only holds for decision problems—the situation for functional classes is very different. Further note that this proposition holds for all classes $\mathcal{C}$ covering the set of regular languages. Consequently, all languages (even the non-recursive languages) can be computed by a finite automaton within these error probabilities. The first upper error bound of this proposition is sharp: Using a delayed diagonalization technique we can construct a language L that cannot be computed with an arbitrary Turing machine within an error probability $1 - \frac{e(n)}{|\Sigma|^n}$ if $e(n) \geq_{io} 1$.

Theorem 1 *There exists a language, such that for all funct. $e(n) \geq_{io} 1$ it holds*

$$L \notin \left(1 - \frac{e(n)}{|\Sigma^n|}\right)\text{-Err}\,\mathcal{RE} .$$

To prove that the upper bound of the io-error complexity measure is tight in the limit we will use the following technical lemma dealing with the **Hamming distance** $|x - y|$ of two binary strings $x, y \in \{0,1\}^n$.

Lemma 1 *Let $m(k, n) := \sum_{i=0}^{k} \binom{n}{i}$ then it holds*

1. *Let $x_1, \ldots x_k \in \{0,1\}^\ell$ and $d \in \mathbb{N}$ with $k \cdot m(d, \ell) < 2^\ell$. There exists $y \in \{0,1\}^\ell$ such that $\min_{i \in \{1,\ldots,k\}}(|x_i - y|) \geq d$.*
2. *For $\alpha < \frac{1}{2}$ and $\alpha \cdot n \in \mathbb{N}$ it holds that $m(\alpha \cdot n, n) < \binom{n}{\alpha \cdot n} \cdot \frac{1-\alpha}{1-2\alpha}$.*
3. *For $f \in \omega(\sqrt{n})$ there exists $g \in \omega(1) : \quad g(n) \cdot m(n/2 - f(n), n) <_{ae} 2^n$.*

One may wonder whether high error complexity implies high Kolmogorov complexity. At least the contrary is true.

Lemma 2

$$L \in \frac{f(n)}{|\Sigma|^n}\text{-Err}\,\mathcal{REC} \implies \exists c\, \forall n : \; C(L \cap \Sigma^n) \leq \log m(f(n), |\Sigma^n|) + c$$

$$L \in \frac{f(n)}{\Sigma^n}\text{-Err}_{io}\,\mathcal{REC} \implies \exists c\, \exists_{io} n : \; C(L \cap \Sigma^n) \leq \log m(f(n), |\Sigma^n|) + c$$

In [GGH 93] a similar result is presented. They relate average time complexity and time-bounded Kolmogorov complexity.

The Kolmogorov complexity of enumerable sets is low, since for all enumerable sets L it holds $\forall n : C(L \cap \Sigma^n) \leq n + O(1)$. For an excellent survey over this field see [LiVi 97]. We show an explicit construction of a diagonal language with low Kolmogorov complexity (comparable to an enumerable set), giving a strict lower bound for the io-error complexity. This language L will be constructed by using the Hamming distance between k sublanguages $L_0[i, i+\ell], \ldots, L_k[i, i+\ell]$ of length ℓ for increasing k and ℓ. We will show that L cannot be approximated by any Turing-machine within an io-error of c for any constant $c < 1/2$.

Theorem 2 *For any $\varepsilon \in \omega(1)$ there exists $L \notin \left(\frac{1}{2} - \frac{1}{o(|\Sigma|^{n/2})} \right)$-$\mathrm{Err}_{io}\, \mathcal{RE}$ such that $\forall n : C(L \cap \Sigma^n) \leq n + \varepsilon(n)$.*

Proof: For a function $f \in \omega(\sqrt{m})$, let $\alpha(m) := \frac{1}{2} - \frac{f(m)}{m}$, $g(m) := \frac{f(m) \cdot \sqrt{8\pi m}}{m - 2f(m)}$, and $\gamma(x) := \min\{ \ell \in \mathbb{N} \mid g(|\Sigma|^\ell) \geq x \}$.

We construct a language L which cannot be approximated by partial recursive languages $L_0, L_1, \ldots$ within an io-error probability of $\alpha(|\Sigma|^n)$ as follows: Define $b_i := |\Sigma^{<i}|$, $k_i := |\Sigma^i|$ and choose a sublanguage $S_{\ell,i}$ *lexicographically minimal* such that for all $j \leq \ell$ holds $|L_j[b_i, b_{i+1} - 1] - S_{\ell,i}| \geq \alpha(k) \cdot k_i$. From Lemma 1 we can conclude that for all $\ell \in \mathbb{N}$ and $c \geq 1$ such a language $S_{\ell, \delta(\ell)+c}$ always exists. Finally, we define the language L as the concatenation of the sublanguages $S_{\ell_i, i}$ for $i = 0, 1, 2, \ldots$ and $\ell_i := \lfloor g(|\Sigma|^i) \rfloor$. From the definition of $S_{\ell_i, i}$ it follows, that for each language L_i, there exists an index j, such that for all $n \geq j$ the Hamming distance of $L_i \cap \Sigma^n$ and $L \cap \Sigma^n$ is at least $\alpha(|\Sigma|^n) \cdot |\Sigma|^n$. That means, L_i cannot predict the language L restricted to words of length n within an error probability smaller than $\alpha(|\Sigma|^n) = \frac{1}{2} - \frac{1}{o(|\Sigma|^{n/2})}$.

On the other hand the sequence $L_0[b_n, b_{n+1} - 1] \ldots L_{g(n)}[b_n, b_{n+1} - 1]$ can be reconstructed if $g(n)$, n, and the number of elements of the corresponding sublanguages are known. Thus, it has a Kolmogorov complexity of at most $O(\log g(n)) + \log(g(n) \cdot |\Sigma^n|) + c_1$ for constant c_1. Using this sequence we can easily construct $L \cap \Sigma^n$. Hence, the Kolmogorov complexity of $L \cap \Sigma^n$ is bounded by $n + \varepsilon(n)$ for an arbitrarily small $\varepsilon \in \omega(1)$ if f is chosen appropriately. □

If we restrict ourselves to computational complexity classes we can apply the results shown so far also to classes specified by time bounded Turing machines. Let **DTime(T)** resp. **DTime$_k$(T)** be the class of all languages which can be accepted by a T-time bounded deterministic (k-tape) Turing machine. We call a function f **T-time k-tape computable**, if $f \in \mathcal{F}\mathbf{DTime}_k(T)$ and it is called **time constructible**, if $T \in \mathcal{F}\mathbf{DTime}_2(T)$. In [CaSe 99,ReSc 96] tight average time hierarchies are presented for very carefully defined average time classes. We state corresponding hierarchies for both error-classes following from Theorem 1 and 2.

Corollary 1 *Let T_1, T_2 be two time-constructible functions with $T_1 \in \omega(T_2)$ and $f \geq_{io} 1$, $f' \in o(\sqrt{\mathcal{N}})$ T_1-time computable functions. Then for $k \geq 2$ there exists* *a function $\delta \in \omega(1)$ with $\delta \cdot T_2 \in o(T_1)$ such that it holds*

$$\mathrm{DTime}_k(T_1) \quad \not\subseteq \quad \left(1 - \frac{f(n)}{|\Sigma^n|}\right)\text{-Err}\,\mathrm{DTime}_k(T_2)\ ,$$

$$\mathrm{DTime}_k(T_1) \quad \not\subseteq \quad \left(\frac{1}{2} - \frac{1}{f(|\Sigma|^{\delta(n)})}\right)\text{-Err}_{io}\,\mathrm{DTime}_k(T_2)\ .$$

Hence, there are languages computable in time T which cannot be accepted by a Turing machine with asymptotically slower running time within an error significantly smaller than $\frac{1}{2}$. Of course, these results can also be transfered to other computational resources like space or reversals. To transfer these results to $\mathrm{DTime}(T)$, an additional factor of $\log T$ for the tape reduction has to be taken into account.

4   Partial Recursive Functions with High Error Bounds

In the last section we showed that there are languages which cannot be approximated by any partial recursive language within an io-error bound significantly smaller than $1/2$. To identify some well known languages which cannot be approximated by a partial recursive language within a (nearly) constant io-error fraction we consider the Halting Problem and the Kolmogorov complexity. Note that their complements are not partial recursive.

We will show that both problems cannot be solved by a recursive function within an io-error smaller than a constant, with the constant depending on the chosen programming system. That means that any algorithm, like a universal program checker, claiming to solve one of these problems within a neglectable small error, fails.

4.1   Lower Bounds for the Halting Problem

The halting problem occurs for various models of computation. Obviously, the error complexity depends on the chosen programming system. To get a general approach, we follow the notation of [Smit 94]:

Definition 3 *A **programming system** φ is a sequence $\varphi_0, \varphi_1, \varphi_2, \ldots$ of all partial recursive functions such that there exists a universal program $u \in \mathbb{N}$ with $\varphi_u(\langle i, x\rangle) = \varphi_i(x)$ for a bijective function $\langle \cdot, \cdot \rangle : \Sigma^* \times \Sigma^* \to \Sigma^*$ called **pairing function**. The **halting problem** H_φ for a programming system φ is defined as: Given a pair $\langle i, x\rangle$, decide whether $x \in \mathrm{dom}(\varphi_i(x))$.*

The programming system highly influences the error complexity of the halting problem. Consider for example a programming system for a binary alphabet where $\psi_{2^i} \equiv \varphi_i$ and ψ_j describes the identity function for all $j \neq 2^i$. Of course this anomalous programming system allows a program to compute the halting problem within exponential small error probability. To restrict the programming systems we define:

Definition 4 *The **repetition rate** of domain equivalent partial functions $RD_{\varphi,i}$ is defined as $RD_{\varphi,i}(n) := \text{Prob}[\text{dom}(\varphi_x) = \text{dom}(\varphi_i) \mid x \in \Sigma^n]$. A programming system φ is **dense**, iff $\forall i \ \exists c > 0 : \ RD_{\varphi,i}(n) \geq_{ae} c$.*

Note that most of *real world* programming systems, like PASCAL, are dense.

The second parameter directly influencing the error complexity of the halting problem is the pairing function of the universal program. We can change the situation considerably, if the pairing function is chosen appropriately. Then we can achieve the highest possible error complexity using a diagonalization.

Theorem 3 *There exists a pairing function such that for all programming systems φ and for any function $f <_{ae} 1$ it holds $H_\varphi \notin f\text{-Err}\,\mathcal{REC}$*

This only shows that an artificial pairing function can cause high error complexity. To derive more general results we define the notion of pair-fairness.

Definition 5 *We call a pairing function **pair-fair**, if for sets $X, Y \subseteq \Sigma^*$ with $\exists c_1 > 0 \ \forall n : \frac{|X \cap \Sigma^n|}{|\Sigma^n|} \geq c_1$ and $\exists \ell_1, \ell_2 \in \mathbb{N} : Y = \{w \mid \text{ord}(w) \equiv \ell_1 \pmod{\ell_2}\}$ it holds $\exists c_2 \ \forall_{ae} n : \ \text{Prob}[x \in X \wedge y \in Y \mid \langle x, y \rangle \in \Sigma^n] \geq c_2$.*

Proposition 3 *The standard pairing $\langle x, y \rangle = x + \frac{(x+y)(x+y+1)}{2}$ is pair-fair.*

In the following we restrict our considerations to fair pairing functions $\langle \cdot, \cdot \rangle$. Before we can prove a lower bound of the error complexity of the halting problem, we have to show the following technical lemma.

Lemma 3 *For a pair-fair function $\langle \cdot, \cdot \rangle$ and a pairing function $\langle\!\langle \cdot, \cdot \rangle\!\rangle$ it holds $\forall i \ \exists j \ \forall x \ \forall y : \ \text{dom}(\varphi_x) = \text{dom}(\varphi_j) \implies \varphi_i(\langle x, \langle\!\langle x, y \rangle\!\rangle \rangle) \neq H_\varphi(\langle x, \langle\!\langle x, y \rangle\!\rangle \rangle)$.*

Choosing $\langle\!\langle x, y \rangle\!\rangle := 2^x \cdot (2y + 1) - 1$, we can conclude from Lemma 3:

Theorem 4 *For any dense programming system φ it holds $\forall M \ \exists \alpha > 0 \ \forall_{ae} n : \ \text{Prob}[M(x) \neq H_\varphi(x) \mid x \in \Sigma^n] \geq \alpha$.*

This means that every heuristic that claims to solve the halting problem makes at least a constant fraction of errors.

Corollary 2 *For any dense progr. system φ and any function $f \in \omega(1)$, it holds $H_\varphi \notin \frac{1}{f}\text{-Err}_{\text{io}}\,\mathcal{REC}$.*

The question whether or not there exists a constant lower bound for the io-error complexity of halting is still open. Perhaps the trivial constant upper bound can be improved by showing that for a sequence of Turing machines the error complexity tends to zero in the limit. The last corollary implies that $H_\varphi \notin$ Nearly-$\mathcal{REC}$. Thus, even an improved upper bound would not be helpful for practical issues.

4.2   Lower Error Bounds for Kolmogorov Complexity

Another well known non-recursive problem is Kolmogorov complexity. One of the fundamental results of Kolmogorov complexity theory is the proof of the existence of a *universal programming system*. Such a programming system is not necessarily dense, although many programming systems provide both properties. A sufficient condition for both features is the capability of the universal program to store a fixed input parameter one-to-one into the index string, that means there exists a function $s : \Sigma^* \to \Sigma^*$ such that for all x, y it holds $\varphi_u(\langle x, y \rangle) = \varphi_{s(x)}(y)$ and $|s(x)| = |x| + O(1)$. This observation implies a trivial upper bound for the Kolmogorov complexity of x.

We consider the following decision problem based on the Kolmogorov complexity for a classification of its io-error complexity.

Definition 6 *For a function* $f : \mathbb{N} \to \mathbb{N}$ *define* $C_{\leq f}$ *as the set of all inputs* x *with Kolmogorov complexity smaller than* $f(|x|)$, *i.e.* $C_{\leq f} := \{x \in \Sigma^* \mid C(x) \leq f(|x|)\}$. *For a constant* c *we define the function* $\kappa_c : \mathbb{N} \to [0, 1]$ *as* $\kappa_c(n) := \text{Prob}[C(x) \leq n - c \mid x \in \Sigma^n]$.

In general, the functions C and κ_c are not recursive. But at least $C_{\leq f}$ is recursively enumerable. In the following we will investigate the size of $C_{\leq n-c}$ and show a linear lower and upper bound.

Lemma 4 *For any constant* $c \geq 1$ *there exist constants* $k_1, k_2 > 0$ *such that* $k_1 \leq_{ae} \kappa_c \leq_{ae} 1 - k_2$.

It is well known that for small recursive functions $f, g \leq \log n$ with $f \in \Omega(g)$ and $g \in \omega(1)$ the set $C_{\leq f}$ is partially recursive. Furthermore, no infinite recursive enumerable set A is completely included in $\overline{C_{\leq f}}$, i.e. $A \cap C_{\leq f} \neq \emptyset$. The following Lemma substantiates the size of this non-empty set $A \cap C_{\leq f}$ for $f(n) = n - c$.

Lemma 5 *Let* $A \in \mathcal{RE}$ *such that* $|A \cap \Sigma^n| \geq_{ae} c_1 \cdot |\Sigma^n|$ *for constant* $c_1 > 0$. *Then it holds* $\forall c_2 > 0 \; \exists c_3 > 0 : \; \text{Prob}[C(x) \leq n - c_2 \mid x \in A \cap \Sigma^n] \geq_{ae} c_3$.

Using these lemmas the following lower bound can be shown.

Theorem 5 *For any* $c \geq 1$ *there exists a constant* $\alpha < 1$ *such that* $C_{\leq n-c} \notin \alpha\text{-Err}_{io} \mathcal{REC}$.

Proof: For a machine M define $K_n := \Sigma^n \cap C_{\leq n-c}$, $A_n := \{x \in \Sigma^n \mid M(x) = 0\}$, and $F_n := \{x \in \Sigma^n \mid M(x) \neq C_{\leq n-c}(x)\}$ for all $n \in \mathbb{N}$.

Note that $\Sigma^n \setminus (A_n \triangle K_n) \subseteq F_n$ and $A_n \cap K_n \subseteq F_n$. From Lemma 4 we can conclude that $k_1 \cdot |\Sigma^n| \leq_{ae} |K_n| \leq_{ae} (1 - k_2) \cdot |\Sigma^n|$ and therefore either $c_1 \cdot |\Sigma^n| \leq_{ae} |A_n|$ or $|A_n \triangle K_n| \leq_{ae} (1 - c_2) \cdot |\Sigma^n|$ for some constants $k_1, k_2, c_1, c_2 > 0$. Using Lemma 5 it follows that for a constant $c_3 > 0$ $|A_n \cap K_n| \geq_{ae} c_3 \cdot |\Sigma^n|$ or $|A_n \triangle K_n| \leq_{ae} (1 - c_2) \cdot |\Sigma^n|$. Finally, we can conclude that for some constant $c_4 > 0$ it holds $|F_n| \geq_{ae} c_5 \cdot |\Sigma^n|$. $\square$

It follows that there exists a fixed constant α such that no matter which algorithm tries to compute $C_{\leq n-c}$ it fails for at least a fraction of α of all inputs.

5   Discussion

One might expect that the concept of immune sets and complexity cores are suitable for showing lower bounds. Recall that a set S is called C-*immune* if it has no infinitive subset in C. S is called C-*bi-immune* if S and $\overline{S}$ are immune for C. A recursive set X is called a *complexity core* of S if for every algorithm M recognizing S and every polynomial p, the running time of M on x exceeds $p(|x|)$ on all but finitely many $x \in X$.

Orponen and Schöning [ScOr 84] observed that a set $S \notin \mathcal{P}$ is bi-immune for $\mathcal{P}$ iff Σ^* is a complexity core for S. But this does not imply reasonable lower bounds for the error complexity, since a precondition for complexity cores is the correct computation of an algorithm. Since bi-immune sets may be very sparse, even the trivial language $\emptyset$ gives a low error bound. Thus, the knowledge of a bi-immune set S for C does not result in a high error complexity. It is an open problem how density for bi-immune sets has to be defined such that reasonable results for the error complexity can be achieved.

However, we can show that the existence of a immune set of a class C corresponds to a small error bound separation:

Theorem 6 *Let C be closed under finite variation and $\emptyset \in C$. There exists an C-immune set iff $C \neq \frac{1}{|\Sigma^n|}$-$\mathrm{Err}\, C \cap \mathcal{RE}$.*

Since some elementary closure properties of C guarantee the existence of a C-immune set [BoDu 87], this restriction is not severe. On the other hand in [Yesh 83] it is shown that the structural property of *conjunctively-self-reducibility* suffices to overcome all erroneous outputs.

As shown in section 4 the Kolmogorov complexity problem, i.e. the question to decide whether a string can be compressed more than a constant, cannot be computed by any machine within a smaller error probability than a constant. It is notable that this error probability is independent from the machine. Both, the halting and the Kolmogorov problem are not in Nearly-$\mathcal{BPP}$ and remain intractable with respect to the set of recursive predicates.

Because of their strong relationship to the halting problem some other problems – like program verification or virus-program detection – are not recursive in the general case, too. So, it seems that the lower bounds proved so far influence the error bounds of these problems. The exact classification is still an open problem.

Acknowledgment

We would like to thank Karin Genther, Rüdiger Reischuk, Arfst Nickelsen, Gerhard Buntrock, Hanno Lefmann, and Stephan Weis for helpful suggestions, critics and fruitful discussions. Furthermore, we thank several unknown referees and the members of the program-comitee for their suggestions, comments and pointers to the literature.

References

BCG 92.   S. Ben-David, B. Chor, O. Goldreich, M. Luby, *On the Theory of Average Case Complexity*, Journal of Computer and System Sciences, Vol. 44, 1992, 193-219.

BoDu 87.   R. Book, D. Du, *The Existence and Density of Generalized Complexity Cores*, Journal of the ACM, Vol. 34, No. 3, 1987, 718-730.

CaSe 99.   J-Y. Cai and A. Selman, *Fine separation of average time complexity classes*, SIAM Journal on Computing, Vol. 28(4), 1999, 1310–1325.

DaPu 60.   Martin Davis, Hillary Putnam, *A Computing Procedure for Quantification Theory*, Journal of the ACM, 1960, 201-215.

Fagi 92.   B. Fagin, *Fast Addition for Large Integers*, IEEE Tr. Comput. Vol. 41, 1992, 1069-1077.

GeHo 94.   P. Gemmel, M. Horchol *Tight Bounds on Expected Time to Add Correctly and Add Mostly Correctly*, Inform. Proc. Letters, 1994, 77-83.

GGH 93.   M. Goldmann, P. Grape and J. Håstad. *On average time hierarchies*, Information Processing Letters, Vol. 49(1), 1994, 15-20.

GoMi 82.   S. Goldwasser, S. Micali, *Probabilistic Encryption & How to Play Mental Poker Keeping Secret All Partial Information*, Proc. 14th Annual ACM Symposium on Theory of Computing, 1982, 365-377.

GMR 88.   S. Goldwasser, S. Micali, R. Rivest, *A Digital Signature Scheme Secure Against Adaptive Chosen-Message Attacks*, SIAM J. Comput. Vol.17, No. 2, 1988, 281-308.

Gure 91.   Y. Gurevich, *Average Case Completeness*, Journal of Computer and System Sciences, Vol. 42, 1991, 346-398.

Hoos 98.   H. Hoos, *Stochastic Local Search - Methods, Models, Applications*, Dissertation, Technical University Darmstadt , 1998.

Imp1 95.   R. Impagliazzo, *Hard-core distributions for somewhat hard problems*, In 36th Annual Symposium on Foundations of Computer Science, 1995, 538-545.

Imp2 95.   R. Impagliazzo, *A personal view of average-case complexity*, In Proceedings of the Tenth Annual Structure in Complexity Theory Conference, 134-147, 1995.

ImWi 98.   R. Impagliazzo, A. Wigderson, *Randomness vs. Time: De-randomization under a uniform assumption*, Proc. 39th Symposium on Foundations of Computer Science, 1998, 734–743.

Levi 86.   Leonid Levin, *Average Case Complete Problems*, SIAM Journal on Computing, Vol. 15, 1986, 285-286.

LiVi 97.   M. Li, P. Vitáni, *An Introduction to Kolmogorov Complexity and its Application*, Springer, 1997.

PuBr 85.   P. W. Purdom, C. A. Brown, *The Pure Literal Rule and Polynomial Average Time*, SIAM J. Comput., 1985, 943-953.

Reif 83.   J. Reif, *Probabilistic Parallel Prefix Computation*, Comp. Math. Applic. 26, 1993, 101-110. Technical Report Havard University, 1983.

ReSc 96.   R. Reischuk, C. Schindelhauer, *An Average Complexity Measure that Yields Tight Hierarchies*, Journal on Computional Complexity, Vol. 6, 1996, 133-173.

Schi 96.   C. Schindelhauer, *Average- und Median-Komplexitätsklassen*, Dissertation, Medizinische Universität Lübeck, 1996.

ScJa 97.   C. Schindelhauer, A. Jakoby, *Computational Error Complexity Classes*, Technical Report A-97-17, Medizinische Universität Lübeck, 1997.

ScOr 84.   U. Schöning, P. Orponen, *The Structure of Polynomial Complexity Cores*, Proc. 11th Symposium MFCS, LNCS 176, 1984, 452-458.

ScWa 94.   R. Schuler, O. Watanabe, *Towards Average-Case Complexity Analysis of NP Optimization Problems*, Proc. 10th Annual IEEE Conference on Structure in Complexity Theory, 1995, 148-159.

SLM 92.   Bart Selman, Hector Levesque, David Mitchell, *Hard and Easy Distributions of SAT Problems*, Proc. 10. Nat. Conf. on Artificial Intelligence, 1992, 440-446.

Smit 94.   C. Smith, *A Recursive Introduction to the Theory of Computation*, Springer, 1994.

Yam1 96.   T. Yamakami, *Average Case Computational Complexity Theory*, Phd. Thesis, Technical Report 307/97, Department of Computer Science, University of Toronto.

Yam2 96.   T. Yamakami, *Polynomial Time Samplable Distributions*, Proc. Mathematical Foundations of Computer Science, 1996, 566-578.

Yesh 83.   Y. Yesha, *On certain polynomial-time truth-table reducibilities of complete sets to sparse sets*, SIAM Journal on Computing, Vol. 12(3), 1983, 411-425.

Analysis of Quantum Functions*
(Preliminary Version)

Tomoyuki Yamakami

Department of Computer Science, Princeton University
Princeton, New Jersey 08544

Abstract. Quantum functions are functions that are defined in terms of quantum mechanical computation. Besides quantum computable functions, we study quantum probability functions, which compute the acceptance probability of quantum computation. We also investigate quantum gap functions, which compute the gap between acceptance and rejection probabilities of quantum computation.

1 Introduction

A paradigm of quantum mechanical computers was first proposed in early 1980's [2,10] to gain more computational power over classical computers. Recent discoveries of fast quantum algorithms for a variety of problems have raised much enthusiasm among computer scientists as well as physicists. These discoveries have supplied general and useful tools in programming quantum algorithms.

We use in this paper a multi-tape *quantum Turing machine* (abbreviated a QTM) [6,4] as a mathematical model of a quantum computer. A well-formed QTM can be identified with a unitary operator, so-called a *time-evolution operator*, that performs in the infinite dimensional space of *superpositions*, which are linear combinations of configurations of the QTM.

The main theme of this paper is a study of polynomial time-bounded *quantum functions*. A *quantum computable function*, a typical quantum function, computes an output of a QTM with high probability. Let **FQP** and **FBQP** denote the collections of functions computed by polynomial-time, well-formed QTMs, respectively, with certainty and with probability at least 2/3. A *quantum probability function*, on the contrary, computes the acceptance probability of a QTM. For notational convenience, **#QP** denotes the collection of such quantum functions particularly witnessed by polynomial-time, well-formed QTMs. Another important quantum function is the one that computes the gap between the acceptance and rejection probabilities of a QTM. We call such functions *quantum gap functions* and use the notation **GapQP** to denote the collection of polynomial-time quantum gap functions.

In this paper, we explore the characteristic feature of these **FQP**-, **FBQP**-, **#QP**-, and **GapQP**-functions and study the close connection between **GapQP**-functions and classical **GapP**-functions. One of the most striking features of

* This work was partly supported by NSERC Fellowship and DIMACS Fellowship.

quantum gap functions is that if $f \in \mathbf{GapQP}$ then $f^2 \in \#\mathbf{QP}$, where $f^2(x) = (f(x))^2$.

We also study relativized quantum functions that can access oracles to help their computation. We exhibit an oracle A showing that $\mathbf{FQP}^A$ is more powerful than $\#\mathbf{P}^A$ and also show the existence of another oracle A such that $\#\mathbf{QP}^A$ is not included in non-adaptive version of $\#\mathbf{QP}^A$.

2   Basic Notions and Notation

We use the standard notions and notation that are found elsewhere. In this section, we explain only what needs special attention.

Let $\mathbb{A}$ denote the set of complex algebraic numbers and let $\tilde{\mathbb{C}}$ denote the set of complex numbers whose real and imaginary parts can be approximated to within 2^{-n} in time polynomial in n. We freely identify any natural number with its binary representation. When we discuss integers, we also identify each integer with its binary representation following a *sign bit* that indicates the (positive or negative) sign[1] of the integer. Moreover, a rational number is also identified as a pair of integers, which are further identified as binary integers.

Let $\mathbb{N}^{\Sigma^*}$ be the set of all functions that map Σ^* to $\mathbb{N}$. Similarly, we define $\{0,1\}^{\Sigma^*}$, $\mathbb{N}^{\mathbb{N}}$, etc. We identify a set S with its characteristic function, which is defined as $S(x) = 1$ if $x \in S$ and 0 otherwise. In this paper, a *polynomial with k variables* means an element in $\mathbb{N}[x_1, x_2, \ldots, x_k]$.

Assumed is the reader's familiarity with central complexity classes, such as $\mathbf{P}$, $\mathbf{NP}$, $\mathbf{BPP}$, $\mathbf{PP}$, and $\mathbf{PSPACE}$ and function classes, such as $\mathbf{FP}$, $\#\mathbf{P}$ [13], and $\mathbf{GapP}$ [8]. Note that $\mathbf{FP} \subseteq \#\mathbf{P} \subseteq \mathbf{GapP}$.

The notion of a *quantum Turing machine*—abbreviated a QTM—was originally introduced in [6] and developed in [4]. For convenience, we use in this paper a slightly more general definition of QTMs defined in [14]; a k-tape QTM M is a quintuple $(Q, \{q_0\}, Q_f, \Sigma_1 \times \Sigma_2 \times \cdots \times \Sigma_k, \delta)$, where each Σ_i is a finite alphabet with a distinguished blank symbol $\#$, Q is a finite set of states including an initial state q_0 and a set Q_f of final states, and δ is a multi-valued *quantum transition function* from $Q \times \Sigma_1 \times \Sigma_2 \times \cdots \times \Sigma_k$ to $\tilde{\mathbb{C}}^{Q \times \Sigma_1 \times \Sigma_2 \times \cdots \Sigma_k \times \{L,R,N\}^k}$. The QTM has k two-way infinite tapes of cells indexed by $\mathbb{Z}$ and its read/write heads that move along the tapes either to the left or to the right, or the heads stay still. In particular, the last tape of M is used to write an output. It is known in [4,16,14] that our model is polynomially "equivalent" to more restrictive model as in [4], which is called *conservative* in [14].

Let M be a QTM. The *running time* of M on input x is defined to be the minimal number T, if any, such that, at time T, every computation path of M on x reach a certain final configuration. We say that M on input x *halts in time T* if its running time is defined and equals T. We call M a *polynomial-time QTM* if there exists a polynomial p such that, for every x, M on input x halts in time $p(|x|)$. A QTM is *well-formed* if its time-evolution operator preserves

---

[1] For example, we set 1 for a positive integer and 0 for a negative integer.

the L_2-norm. For any x, the notation $\rho_M(x)$ denotes the *acceptance probability* of M on x: that is, the sum of every squared magnitude of the amplitude, in the final superposition of M, of any configuration whose output tape constitutes a symbol "1" (called an *accepting configuration*). For a nonempty subset K of $\mathbb{C}$, we say that M has K-*amplitudes* if the entries of its time-evolution matrix are all drawn from K. For more notions and terminology (e.g., *stationary*, *synchronous*, *normal form*), the reader refers to [4,14].

We also use an *oracle* QTM that is equipped with an extra tape, called a *query tape* and two distinguished states, a pre-query state q_p and a post-query state q_a. The oracle QTM invokes an oracle query by entering state q_p. In a single step, the content $|y \circ b\rangle$ of the query tape, where $b \in \{0,1\}$ and "$\circ$" denotes concatenation, is changed into $|y \circ (b \oplus A(y))\rangle$ and the machine enters state q_a.

For a superposition $|\phi\rangle$, the notation $M(|\phi\rangle)$ denotes the final superposition of M that starts with $|\phi\rangle$ as an initial superposition.

There are two useful unitary transforms used in this paper. The *phase shift P* maps $|0\rangle$ to $-|0\rangle$ and $|1\rangle$ to $|1\rangle$. The *Hadamard transform H* changes $|0\rangle$ into $\frac{1}{\sqrt{2}}(|0\rangle + |1\rangle)$ and $|1\rangle$ into $\frac{1}{\sqrt{2}}(|0\rangle - |1\rangle)$.

Throughout this paper, K denotes an arbitrary subset of $\mathbb{C}$ that includes $\{0, \pm 1\}$. All quantum function classes discussed in this paper depend on the choice of K-amplitudes. We find it convenient to drop script K when $K = \tilde{\mathbb{C}}$.

3   Various Quantum Functions

In this section, we formally define a variety of quantum functions and discuss their fundamental properties. Recall that $\{0, \pm 1\} \subseteq K \subseteq \mathbb{C}$

3.1 Exact Quantum Computable Functions.
We begin with quantum functions whose values are the direct outputs, with certainty, of polynomial-time, well-formed QTMs. We call them *exact quantum polynomial-time computable* in a similar fashion to polynomial-time computable functions.

Definition 1. *Let* $\mathbf{FQP}_K$ *be the set of* K-*amplitude exact quantum polynomial-time computable functions; that is, there exists a polynomial-time, well-formed QTM with* K-*amplitudes such that, on every input* x, M *outputs* $f(x)$ *with certainty.*

As noted in Section 2, we drop subscript K when $K = \tilde{\mathbb{C}}$ and write $\mathbf{FQP}$ instead of $\mathbf{FQP}_K$. Note that $\mathbf{FP} \subseteq \mathbf{FQP}_K$.

We first show a relativized separation result. For a nondeterministic TM M and a string x, the notation $\#M(x)$ denotes the number of accepting computation paths of M on input x, and let $\#\mathbf{TIME}(t)$ be the collection of all functions $\lambda x.\#M(x)$ for nondeterministic TMs M running in time at most $t(|x|)$. For a set T, set $\#\mathbf{TIME}(T) = \bigcup_{t \in T} \#\mathbf{TIME}(t)$.

Theorem 1. *There exists a set A such that* $\mathbf{FQP}^A \not\subseteq \#\mathbf{TIME}(o(2^n))^A$.

Proof. For a set A, let $f^A(x) = 2^{-2|x|} \cdot (|A \cap \Sigma^{|x|}| - |\Sigma^{|x|} \setminus A|)^2$ for every x. Consider the following oracle QTM N. On input x of length n, write $0^n \circ 0$ in a query tape and apply $H^n \oplus I$. Invoke an oracle query and then apply the phase shift P to the oracle answer qubit. Again, make a query to A and apply $H^n \oplus I$. Accept x if N observes $|0^n \circ 0\rangle$ in the query tape, and rejects x otherwise. It follows by a simple calculation that $f^A(x) = \rho_{N^A}(x)$. In particular, f^A belongs to $\mathbf{FQP}^A$ if A satisfies the condition that, for every x, either $|A \cap \Sigma^{|x|}| = |\Sigma^{|x|} \setminus A|$ or $|A \cap \Sigma^{|x|}| \cdot |\Sigma^{|x|} \setminus A| = 0$.

Let $\{M_i\}_{i \in \mathbb{N}}$ and $\{q_i\}_{i \in \mathbb{N}}$ be two enumerations of all polynomial-time nondeterministic TMs and all nondecreasing functions in $o(2^n)$, respectively, such that each M_i halts in time at most $q_i(n)$ on all inputs of length n. Initially, set $n_{-1} = 0$ and $A_{-1} = \emptyset$. At stage i of the construction of A, let n_i denote the minimal integer satisfying that $n_{i-1} < n_i$ and $q_i(n_i) < 2^{n_i - 1}$. First let $B = A_i \cup \Sigma^{n_i - 1} 1$. Clearly $f^B(0^{n_i}) = 1$. If $\#M_i^B(0^{n_i}) \neq 1$, then define A_i to be B. Assume otherwise. There exists a unique accepting computation path p of M_i on 0^{n_i}. Let Q denote the set of all words that M_i queries along path p. Since $|Q| \leq q_i(n_i) < 2^{n_i - 1}$, there is a subset C of Σ^{n_i} such that $A_{i-1} \subseteq C$ and $|C \cap \Sigma^{n_i}| = |\Sigma^{n_i} \setminus C|$. For this C, $\#M_i^C(0^{n_i}) \geq 1$ but $f^C(0^{n_i}) = 0$. □

3.2 Bounded Error Quantum Computable Functions.

By replacing exact quantum computation in Definition 1 with bounded-error quantum computation, we can define another quantum function class $\mathbf{FBQP}$.

Definition 2. *A function f is in $\mathbf{FBQP}_K$ if there exist a constant $\epsilon \in (0, \frac{1}{2}]$ and a polynomial-time, well-formed QTM with K-amplitudes that, on input x, outputs $f(x)$ with probability $\frac{1}{2} + \epsilon$. More generally, for a function t, $\mathbf{FBQTIME}(t)$ is defined similarly but by requiring M to halt in time $t(|x|)$. For a set T, define $\mathbf{FBQTIME}(T) = \bigcup_{t \in T} \mathbf{FBQTIME}(t)$.*

Clearly, $\mathbf{FQP}_K \subseteq \mathbf{FBQP}_K$ and $\mathbf{FBQP}_\mathbb{Q} \subseteq \mathbf{FP}^{\mathbf{PSPACE}}$.

Brassard et al. [5] extend Grover's database search algorithm [12] and show how to compute with ϵ-accuracy the amplitude of a given superposition in $O(\sqrt{N})$ time, where N is the size of search space, with high probability. In our terminology, $\#\mathbf{TIME}(O(n)) \subseteq \mathbf{FBQTIME}(O(2^{n/2}))$.

Bennett et al. [3], however, show that there exists an $\mathbf{NP}$-set that cannot be recognized by any well-formed QTMs running in time $o(2^{n/2})$ relative to random oracle. This immediately implies that $\#\mathbf{P}^A \not\subseteq \mathbf{FBQTIME}(o(2^{n/2}))^A$ relative to random oracle A.

3.3 Quantum Probability Functions.

It is essential in quantum complexity theory to study the behavior of the acceptance probability of a well-formed QTM. Here, we consider quantum functions that output such probabilities. We briefly call these functions *quantum probability functions*.

Definition 3. *A function f from Σ^* to $[0,1]$ is called a polynomial-time quantum probability function with K-amplitudes if there exists a polynomial-time, well-formed QTM M with K-amplitudes such that $f(x) = \rho_M(x)$ for all x.*

In short, we say that M *witnesses* in the above definition.

By abusing the existing Valiant's notation #**P**, we can coin the new notation #**QP**$_K$ for the class of polynomial-time quantum probability functions.

Definition 4. *The notation* #**QP**$_K$ *denotes the set of all polynomial-time quantum probability functions with K-amplitudes.*

The lemma below is almost trivial and its proof is left to the reader.

Lemma 1.   *1. Every $\{0,1\}$-valued* **FQP**$_K$*-function is in* #**QP**$_K$.
2. For every #**P***-function f, there exist an* **FP***-function ℓ and a* #**QP**$_Q$*-function g such that $f(x) = \ell(1^{|x|})g(x)$ for every x.*
3. Let $f \in$ #**QP** *and $s, r \in$* **FQP***. Assume that* range$(s) \subseteq (0,1) \cap \mathbb{Q}$ *and* range$(r) \subseteq (0,1) \cap \mathbb{Q}$*. There exists a* #**QP***-function g such that, for every x, $f(x) = s(x)$ iff $g(x) = r(x)$.*

Next we show several closure properties of #**QP**-functions.

Lemma 2. *Let f and g be any functions in* #**QP**$_K$ *and h an* **FQP**$_K$*-function.*
(1) If $\alpha, \beta \in K$ satisfy $|\alpha|^2 + |\beta|^2 = 1$, then $\lambda x.(|\alpha|^2 f(x) + |\beta|^2 g(x))$ is in #**QP**$_K$*.*
(2) $f \circ h \in$ #**QP**$_K$*. (3) $f \cdot g$ is in* #**QP**$_K$*. (4) If h is polynomially bounded[2], then $\lambda x. f(x)^{h(x)}$ is in* #**QP**$_K$*.*

Proof Sketch.   We prove only the last claim. Let a well-formed M_f witness f in time polynomial q. Assume that a polynomial p satisfies $h(x) \leq p(|x|)$. On input x, run M_f $h(x)$ times and idle $q(|x|)$ steps $p(|x|) - h(x)$ times to avoid the timing problem [4,14]. Accept x if all the first $h(x)$ runs of M_f reach accepting configurations; otherwise, reject x.   □

3.4 Quantum Gap Functions.   Notice that #**QP** is not closed under subtraction. To compensate the lack of this property, we can introduce the *quantum gap functions*. A quantum gap function is defined to compute the difference between the acceptance and rejection probabilities of a well-formed QTM.

Definition 5. *A function f from Σ^* to $[-1,1]$ is called a* polynomial-time quantum gap function *with K-amplitudes if there exists a polynomial-time, K-amplitude, well-formed QTM M such that, for every x, $f(x)$ is $\||\phi_x^{(1)}\rangle\|^2 - \||\phi_x^{(0)}\rangle\|^2$ when M on input x halts in a final superposition $|\phi_x^{(0)}\rangle|0\rangle + |\phi_x^{(1)}\rangle|1\rangle$, where the last qubit represents the content of the output tape of M. In other words, $f(x) = 2\rho_M(x) - 1$.*

We use the notation **GapQP**$_K$ to denote the collection of such functions.

Definition 6. **GapQP**$_K$ *is the set of all polynomial-time quantum gap functions with K-amplitudes.*

---

[2] A function from Σ^* to $\mathbb{N}$ is *polynomially bounded* if there exists a polynomial p such that $f(x) \leq p(|x|)$ for every x.

For any two sets $\mathcal{F}$ and $\mathcal{G}$ of functions, the notation $\mathcal{F} - \mathcal{G}$ denotes the set of functions of the form $f - g$, where $f \in \mathcal{F}$ and $g \in \mathcal{G}$. The following proposition shows another characterization of $\mathbf{GapQP}_K$. Recall that $\{0, \pm 1\} \subseteq K \subseteq \mathbb{C}$.

Proposition 1. $\mathbf{GapQP}_K = \#\mathbf{QP}_K - \#\mathbf{QP}_K$. *Thus,* $\#\mathbf{QP}_K \subseteq \mathbf{GapQP}_K$.

Proof.   Clearly, $\mathbf{GapQP}_K \subseteq \#\mathbf{QP}_K - \#\mathbf{QP}_K$. Conversely, assume that there exist two polynomial-time, K-amplitude, well-formed QTMs M_g and M_h satisfying $f(x) = \rho_{M_g}(x) - \rho_{M_h}(x)$ for all x. Consider the following QTM N. On input x, N first writes 0 and applies H. If $|1\rangle$ is observed, it simulates M_g. Let a_g be its output. Otherwise, N simulates M_h. Let a_h be its output. In the case where $a_g = 1 - a_h = 1$, N accepts x; otherwise, rejects x. The acceptance probability of N is exactly $\frac{1}{2}\rho_{M_g}(x) + \frac{1}{2}(1 - \rho_{M_h}(x))$. Thus, the gap $2\rho_N(x) - 1$ is exactly $\rho_{M_g}(x) - \rho_{M_h}(x)$, which equals $f(x)$.   □

4   Computational Power of Quantum Gap Functions

In the previous section, we have introduced quantum gap functions. In this section, we discuss the computational power of these functions.

4.1 Squared Function Theorem.   It is shown in [9] that, for every **GapP**-function f, there exists a polynomial-time, well-formed QTM that accepts input x with probability $2^{-p(|x|)} f(x)^2$ for a certain fixed polynomial p. This implies that if $f \in \mathbf{GapP}$ then $\lambda x.2^{-p(|x|)} f(x)^2 \in \#\mathbf{QP}$. A slightly different argument demonstrates that if $f \in \mathbf{GapQP}$ then $f^2 \in \#\mathbf{QP}$. This is a characteristic feature of quantum gap functions.

Theorem 2. *(Squared Function Theorem)   If $f \in \mathbf{GapQP}$ then $f^2 \in \#\mathbf{QP}$, where $f^2(x) = (f(x))^2$ for all x.*

It is, however, unknown whether $\mathbf{GapQP} \cap [0,1]^{\Sigma^*} \subseteq \#\mathbf{QP}$. To show Theorem 2, we utilize the following lemma, whose proof generalizes the argument used in the proof of Proposition 1. See also Lemma 5 in [14].

Lemma 3. *(Gap Squaring Lemma)   Let M be a well-formed QTM that, on input x, halts in time $T(x)$. There exists a well-formed QTM N that, on inputs x given in tape 1 and $1^{T(x)}$ in tape 2 and empty elsewhere, halts in time $O(T(x)^2)$ in a final superposition in which the amplitude of configuration $|x\rangle|1^{T(x)}\rangle|1\rangle$ is $2\rho_M(x) - 1$, where the last qubit is the content of the output tape.*

Proof.   Let M be the QTM given in the lemma. We define the desired QTM N as follows. First, N simulates M on input $(x, 1^{T(x)})$. Assume that M halts in a final superposition $|\phi\rangle = \sum_y \alpha_{x,y}|y\rangle|b_y\rangle$, where y ranges all (valid) configurations of M and the qubit $|b_y\rangle$ represents the content of the output tape of M. Note that $\rho_M(x) = \sum_{y:b_y=1} |\alpha_{x,y}|^2$. After N applies the phase shift P to $|b_y\rangle$, we have the superposition $|\phi'\rangle = \sum_{y:b_y=1} \alpha_{x,y}|y\rangle|1\rangle - \sum_{y:b_y=0} \alpha_{x,y}|y\rangle|0\rangle$.

There exists a well-formed QTM M^R that reverses the computation of M in time $O(T(x)^2)$ on input $(x, 1^{T(x)})$ [14]. Then, N simulates M^R starting with $|\phi'\rangle$. Note that if we apply M^R to $|\phi\rangle$ instead, then $M^R(|\phi\rangle) = |x\rangle|1^{T(x)}\rangle|\#\rangle$. Moreover, the inner product of $|\phi\rangle$ and $|\phi'\rangle$ is $\langle\phi|\phi'\rangle = \sum_{y:b_y=1} |\alpha_{x,y}|^2 - \sum_{y:b_y=0} |\alpha_{x,y}|^2$, which equals $2\rho_M(x) - 1$.

At the end, N outputs 1 (i.e., accept) if it observes exactly the initial configuration; otherwise, N outputs 0 (i.e., reject). Note that the acceptance probability of N is exactly $\langle N(|\phi\rangle)|N(|\phi'\rangle)\rangle$. Since N preserves the inner product, we have $\langle N(|\phi\rangle)|N(|\phi'\rangle)\rangle = \langle\phi|\phi'\rangle = 2\rho_M(x) - 1$.   □

Now we are ready to prove Theorem 2.

Proof of Theorem 2. Let f be in **GapQP** and M a polynomial-time, well-formed QTM such that $f(x) = 2\rho_M(x) - 1$ for all x. It follows from Gap Squaring Lemma that there exists a polynomial-time, well-formed QTM N that halts in a final superposition in which the squared magnitude of the amplitude of $|x\rangle|1\rangle$ is $(2\rho_M(x) - 1)^2$, which clearly equals $f^2(x)$.   □

4.2 Approximation of Quantum Gap Functions.

Quantum gap functions are closely related to their classical counterpart: gap functions [8]. The following proposition shows the close relationship between **GapQP** and **GapP**. Let $sign(a)$ be 0, 1, and -1 if $a = 0$, $a > 0$, and $a < 0$, respectively.

Theorem 3.   *1. For every $f \in$ **GapQP**$_\mathbb{C}$, there exists a function $g \in$ **GapP** such that, for every x, $f(x) = 0$ iff $g(x) = 0$.*

   *2. For every $f \in$ **GapQP** and every polynomial q, there exist two functions $k \in$ **GapP** and $\ell \in$ **FP** such that $\left|f(x) - \frac{k(x)}{\ell(1^{|x|})}\right| \le 2^{-q(|x|)}$ for all x.*

   *3. For every $f \in$ **GapQP**$_{A\cap\mathbb{R}}$, there exists a function $g \in$ **GapP** such that $sign(f(x)) = sign(g(x))$ for all x.*

   *4. For every $f \in$ **GapQP**$_\mathbb{Q}$, there exist two functions $k \in$ **GapP** and $\ell \in$ **FP** such that $f(x) = \frac{k(x)}{\ell(1^{|x|})}$ for all x.*

To show the theorem, we need the following lemma. For a QTM M, let $amp_M(x, C)$ denote the amplitude of configuration C of M on input x in a final superposition. The *(complex) conjugate* of M is the QTM M^* defined exactly as M except that its time-evolution matrix is the complex conjugate of the time-evolution matrix of M.

Lemma 4. *Let M be a well-formed, synchronous, stationary QTM in normal form with running time $T(x)$ on input x. There exists a well-formed QTM N such that, for every x, (1) N halts in time $O(T(x))$ with one symbol from $\{0, 1, ?\}$ in its output tape; (2) $\sum_{C\in D_x^1} amp_N(x, C) = \rho_M(x)$; and (3) $\sum_{C\in D_x^0} amp_N(x, C) = 1 - \rho_M(x)$, where D_x^i is the set of all final configurations, of M on x, whose output tape consists only of symbol $i \in \{0, 1\}$.*

Proof Sketch.   The desired QTM N works as follows. On input x, N simulates M on input x; when it halts in a final configuration, N starts another round of simulation of M^* in a different tape. After N reaches a final configuration, we obtain two final configurations. Then, N (reversibly) checks if both final configurations are identical. If not, N outputs symbol "?" and halts. Assume otherwise. If this unique configuration is an accepting configuration, then N outputs 1; otherwise, it outputs 0.   □

Proof of Theorem 3.   (1) First we note from [15] that, for every $g \in \#\mathbf{QP_C}$, there exists a $h \in \mathbf{GapP}$ such that, for every x, $g(x) = 0$ iff $h(x) = 0$.

Let f be any function in $\mathbf{GapQP_C}$. By Squared Function Theorem, f^2 belongs to $\#\mathbf{QP_C}$. Thus, there exists a $h \in \mathbf{GapP}$ such that, for every x, $f^2(x) = 0$ iff $h(x) = 0$. It immediately follows that $f(x) = 0$ iff $h(x) = 0$.

(2) Let $f \in \mathbf{GapQP}$. By Lemma 4, it follows that there exists a polynomial-time, well-formed QTM M such that $f(x)$ equals $\sum_{C \in D_x^1} amp_M(x, C) - \sum_{C \in D_x^0} amp_M(x, C)$, where D_x^i is defined in Lemma 4. Let r be a polynomial that bounds the running time of M and also satisfies $|D_x^0 \cup D_x^1| \leq 2^{r(|x|)}$.

Let x be any string of length n. Let $\ell(x) = 2^{2r(n)+q(n)+1}$ and $g(x, C) = amp_M(x, C)$. Assume first that there exists a $\mathbf{GapQP}$-function $\tilde{g}$ such that $|g(x, C) - \frac{\tilde{g}(x, C)}{\ell(1^n)}| \leq 2^{-r(n)-q(n)-1}$, which implies $\left|\sum_{C \in D_x^i} g(x, C) - \sum_{C \in D_x^i} \frac{\tilde{g}(x, C)}{\ell(1^n)}\right| \leq 2^{r(n)} \cdot 2^{-r(n)-q(n)-1} = 2^{-q(n)-1}$ for each $i \in \{0, 1\}$. The desired $\mathbf{GapQP}$-function k is defined as $k(x) = \sum_{C \in D_x^1} \tilde{g}(x, C) - \sum_{C \in D_x^0} \tilde{g}(x, C)$.

To complete the proof, we show the existence of such $\tilde{g}$. Note that every amplitude of the transition function of M is approximated by a polynomial-time TM. By simulating such a machine in polynomial time, we can get a $2r(n) + q(n) + 1$ bit approximation of the corresponding amplitude to within $2^{-2r(n)-q(n)-1}$ so that, for every computation path P of M on input x, we can compute the approximation $\tilde{\rho}_P$ of the amplitude ρ_P of path P to within $2^{-r(n)-q(n)-1}$. Let $h(x, P)$ be the integer satisfying $h(x, P) = \ell(1^n)|\tilde{\rho}_P|$. Set $\tilde{g}(x, C) = \sum_{P \in P_{x,C}} h(x, P)$, where $P_{x,C}$ is the collection of all paths of M on x that lead to configuration C.

(3) This follows by a modification of the proof of Lemma 6.8 in [1].

(4) In the proof of (2), since M has $\mathbb{Q}$-amplitudes, we can exactly compute the amplitude ρ_P of path P and thus, we have $h(x, P) = \ell(1^n)|\rho_P|$. Therefore, $f(x) = \frac{k(x)}{\ell(1^n)}$. See also Theorem 3.1 in [11].   □

5   Quantum Complexity Classes

Using the values of quantum functions, we can introduce a variety of quantum complexity classes. As in the previous sections, when $K = \tilde{\mathbb{C}}$, we drop script K.

5.1 GapQP-Definable Complexity Classes.   What is the common feature of known quantum complexity classes? Bernstein and Vazirani [4] introduced $\mathbf{EQP}_K$ (exact QP) as the collections of sets S such that an $\mathbf{FQP}_K$-function f satisfies $f(x) = S(x)$ for all x. Similarly, $\mathbf{BQP}_K$ (bounded-error QP) is defined by an $\mathbf{FBQP}_K$-function instead of an $\mathbf{FQP}_K$-function. Adleman et al. [1] introduced $\mathbf{NQP}_K$ (nondeterministic QP), which is defined as the collection of sets S such that there exists a $\#\mathbf{QP}_K$-function f satisfying that, for every x, $S(x) = 1$ iff $f(x) > 0$. By a simple observation, $\mathbf{EQP}_K$, $\mathbf{BQP}_K$, and $\mathbf{NQP}_K$ can be all characterized in terms of $\mathbf{GapQP}_K$-functions.

We introduce a general notion of *GapQP-definability* in similar spirit that Fenner et al. [8] defined Gap-definability.

Definition 7. *A complexity class $\mathcal{C}$ is $GapQP_K$-definable if there exist a pair of disjoint sets $A, R \subseteq \Sigma^* \times [0,1]$ such that, for any S, $S \in \mathcal{C}$ iff there exists an $f \in \mathbf{GapQP}_K$ satisfying that, for every x, (i) $x \in S$ implies $(x, f(x)) \in A$ and (ii) $x \notin S$ implies $(x, f(x)) \in R$. We write $GapQP_K(A, R)$ to denote this $\mathcal{C}$.*

Proposition 2. $\mathbf{EQP}_K$, $\mathbf{BQP}_K$, *and* $\mathbf{NQP}_K$ *are all $GapQP_K$-definable.*

An immediate challenge is to prove or disprove that the following quantum complexity class is GapQP-definable: $\mathbf{WQP}_K$ (wide QP), the collection of sets A such that there exist an $f \in \#\mathbf{QP}_K$ and a $g \in \mathbf{FQP}_K$ with range$(g) \subseteq (0,1] \cap \mathbb{Q}$ satisfying $f(x) = A(x) \cdot g(x)$ for every x. Notice that we can replace $\#\mathbf{QP}_K$ by $\mathbf{GapQP}_K$. By definition, $\mathbf{EQP}_K \subseteq \mathbf{WQP}_K \subseteq \mathbf{NQP}_K$.

5.2 Sets with Low Information.   Let $\mathcal{C}$ be a relativizable complexity class. Consider a set A satisfying $\mathcal{C}^A = \mathcal{C}$. Apparently, this set A has low information since it does not help the class $\mathcal{C}$ gain more power. Such a set is called a *low set* for $\mathcal{C}$. Let low-$\mathcal{C}$ denote the collection of all low sets for $\mathcal{C}$.

Obviously, $\mathbf{low\text{-}EQP} = \mathbf{EQP}$. Moreover, since $\mathbf{BQP}^{\mathbf{BQP}} = \mathbf{BQP}$ [3], we obtain $\mathbf{low\text{-}BQP} = \mathbf{BQP}$. However, $\mathbf{low\text{-}NQP}$ does not appear to coincide with $\mathbf{NQP}$ since $\mathbf{EQP} \subseteq \mathbf{low\text{-}NQP} \subseteq \mathbf{NQP} \cap \text{co-}\mathbf{NQP}$. Thus, if $\mathbf{NQP} \neq \text{co-}\mathbf{NQP}$, then $\mathbf{low\text{-}NQP} \neq \mathbf{NQP}$.

The following proposition is almost trivial with the fact that $\{0, \pm 1\} \subseteq K$.

Proposition 3. $\mathbf{EQP}_K = \text{low-}\#\mathbf{QP}_K = \text{low-}\mathbf{GapQP}_K$.

5.3 Quantum Complexity Classes beyond BQP.   We briefly discuss a few quantum complexity classes beyond $\mathbf{BQP}$.

We first note that any $\mathbf{BQP}$ set enjoys the *amplification property*: for every polynomial p, there exists a $\#\mathbf{QP}$-function f such that, for every x, if $x \in A$ then $f(x) \geq 1 - 2^{-p(|x|)}$, and otherwise $0 \leq f(x) \leq 2^{-p(|x|)}$ [3]. Let $\mathbf{AQP}_K$ (amplified QP) be defined similarly by replacing $\#\mathbf{QP}$ with $\mathbf{GapQP}_K$. By definition, $\mathbf{BQP} \subseteq \mathbf{AQP}$. Note that if $\mathbf{AQP} \neq \mathbf{BQP}$, then $\mathbf{GapQP} \cap [0,1]^{\Sigma^*} \neq \#\mathbf{QP}$. By Proposition 3, $\mathbf{AQP}_{\mathbb{Q}}$ remains within $\mathbf{AWPP}$, which is defined in [7].

Another natural class beyond **BQP** is **PQP**$_K$ (proper QP) defined as $GapQP_K(A, \overline{A})$, where $A = \{(x, r) \mid r > 0\}$ and $\overline{A}$ is the complement of A. Proposition 3, however, yields the coincidence between **PQP**$_Q$ and **PP**.

It is important to note that **AQP**$_C$ and **PQP**$_C$ are no longer recursive since **BQP**$_C$ is known to be non-recursive [1].

6   Computation with Oracle Queries

In this section, we study quantum functions that can access oracle sets. In what follows, r denotes an arbitrary function from $\mathbb{N}$ to $\mathbb{N}$, R a subset of $\mathbb{N}^{\mathbb{N}}$, and A a subset of Σ^*.

6.1 Adaptive and Nonadaptive Queries.   We begin with the formal definitions.

Definition 8. *A function f is in* **FQP**$^{A[r]}$ *if there exists a polynomial-time, well-formed, oracle QTM M such that, for every x, M on input x computes $f(x)$ using oracle A and makes at most $r(|x|)$ queries on each computation path. The class* **FQP**$^{A[R]}$ *is the union of such sets* **FQP**$^{A[r]}$ *over all $r \in R$.*

Note that, when $R = \mathbb{N}^{\mathbb{N}}$, **FQP**$^{A[R]}$ coincides with **FQP**^{A}.

Proposition 4. *Let C be a complexity class that is closed under union, complement, and polynomial-time conjunctive reducibility. If $f \in$ #**QP**^{C} then $\lambda x. \frac{f(x)}{2^{p(|x|)}} \in$ #**QP**$^{C[1]}$ for a certain polynomial p.*

Proof Sketch.   Assume that a given QTM M is of "canonical form"; that is, there exists a polynomial p such that M makes $p(n)$ queries of length $p(n)$ on every computation path for n the length of input [14]. Consider the following quantum algorithm. First guess future oracle answers $\{a_i\}_{1 \leq i \leq p(n)}$ and start the simulation of M using $\{a_i\}_{1 \leq i \leq p(n)}$ to answer actual queries $\{w_i\}_{1 \leq i \leq p(n)}$. When M accepts the input, make a single query $\langle w_1, \cdots, w_{p(n)}, a_1, \cdots, a_{p(n)} \rangle$ to oracle and verify the correctness of $\{a_i\}_{1 \leq i \leq p(n)}$.   □

Next, we introduce quantum functions that make *nonadaptive* (or *parallel*) queries to oracles. The functions in Definition 8, on the contrary, make *adaptive* (or *sequential*) queries.

Definition 9. *The class* **FQP**$_{\parallel}^{A[r]}$ *is the subset of* **FQP**$^{A[r]}$ *with the extra condition that, on each computation path p, just before M enters a pre-query state for the first time, it completes a query list[3] — a list of all query words (separated by a special separator in a distinguished tape) that are possibly[4] queried on path p. The notation* **FQP**$_{\parallel}^{A[R]}$ *denotes the union of* **FQP**$_{\parallel}^{A[r]}$ *over all $r \in R$.*

---

[3] When we say a "query list", we refer to the list completed just before the first query. This query list may be altered afterward to interfere with other computation paths having different query lists.

[4] All the words in the query list may not be queried but any word that is queried must be in the query list.

A similar constraint gives rise to $\mathbf{FBQP}_{\|}^{A[R]}$, $\#\mathbf{QP}_{\|}^{A[R]}$, and $\mathbf{GapQP}_{\|}^{A[R]}$.

The proof of Theorem 1 implies that $\mathbf{FQP}_{\|}^{A[2]} \not\subseteq \#\mathbf{TIME}(o(2^n))^A$ relative to a certain oracle A.

Proposition 5. $\mathbf{FQP}^{A[r]} \subseteq \mathbf{FQP}_{\|}^{A[r \cdot 2^r]}$ *for any function r in $O(\log n)$.*

Proof. Let $f \in \mathbf{FQP}^{A[r]}$ and assume that a polynomial-time, well-formed QTM M witnesses f with at most $r(|x|)$ queries to A on each computation path on any input x. Consider the following quantum algorithm.

Let x be any input of length n. We use a binary string a of length $r(n)$. Initially, we set $a = 0^{r(n)}$. In the kth round, $1 \leq k \leq 2^{r(n)}$, we simulate M on input x except that when M makes the ith query, we draw the ith bit of a as its oracle answer. In case where M makes more than $r(n)$ queries, we automatically set their oracle answers to be 0. We record all query words in an extra tape. After M halts, we increment a lexicographically by one. After $2^{r(n)}$ rounds, we have a query list of size at most $r(n)2^{r(n)}$. We then simulate M again with using oracle A. This last procedure preserves the original quantum interference. Therefore, $f \in \mathbf{FQP}_{\|}^{A[r \cdot 2^r]}$. □

In particular, we have $\mathbf{FQP}^{\mathbf{NQP}[O(\log n)]} \subseteq \mathbf{FQP}_{\|}^{\mathbf{NQP}}$, which is analogous to $\mathbf{FP}^{\mathbf{NP}[O(\log n)]} \subseteq \mathbf{FP}_{\|}^{\mathbf{NP}}$. It is open, however, whether $\mathbf{FQP}_{\|}^{A[2^r]} \subseteq \mathbf{FQP}^{A[r]}$.

6.2 Separation Result.   We show the existence of a set A that separates $\#\mathbf{QP}^A$ from $\#\mathbf{QP}_{\|}^A$. For a string y and a superposition $|\phi\rangle$ of configurations, let $q_y(|\phi\rangle)$ denote the *query magnitude* [3]; that is, the sum of squared magnitudes in $|\phi\rangle$ of configurations which has a pre-query state and query word y.

Theorem 4. *There exists a set A such that $\#\mathbf{QP}^A \cap \{0,1\}^{\Sigma^*} \not\subseteq \#\mathbf{QP}_{\|}^A$.*

Proof. Define $f^A(x) = 2^{-|x|} \cdot |\{y \in \Sigma^{|x|} \mid A(x \circ y_A) = 1\}|$ for $x \in \Sigma^*$ and $A \subseteq \Sigma^*$, where $y_A = A(y0^{|y|})A(y0^{|y|-1}1)A(y0^{|y|-2}11)\cdots A(y1^{|y|})$. We call a set A *good at n* if, for any pair $y, y' \in \Sigma^n$, $|y| = |y'|$ implies $y_A = y'_A$; A is *good* if A is good at every n. It follows that, for any good A, $f^A \in \#\mathbf{QP}^A \cap \{0,1\}^{\Sigma^*}$.

We want to construct a good set A such that $f^A \notin \#\mathbf{QP}_{\|}^A$. Let $\{M_i\}_{i \in \mathbb{N}}$ and $\{p_i\}_{i \in \mathbb{N}}$ be two enumerations of polynomial-time, well-formed QTMs and polynomials such that each M_i halts in time $p_i(n)$ on all inputs of length n. We build by stages a series of disjoint sets $\{A_i\}_{i \in \mathbb{N}}$ and then define $A = \bigcup_{i \in \mathbb{N}} A_i$.

For convenience, set $A_{-1} = \emptyset$. Consider stage i. Let n_i be the minimal integer such that $n_{i-1} < n_i$ and $8p_i(n_i)^4 < 2^{n_i}$. It suffices to show the existence of a set $A_i \subseteq \Sigma^{2n_i} \cup \Sigma^{2n_i+1}$ such that A_i is good at n_i and $A_i(0^{n_i} \circ y_{A_i}) \neq \rho_{M_i^{A_i}}(0^{n_i})$.

For readability, we omit subscript i in what follows. To draw a contradiction, we assume otherwise. Let $|\phi_j\rangle$ be the superposition of M on input 0^n at time j. For each $y \in \Sigma^n$, let $\tilde{q}_y$ be the sum of squared magnitudes in any superposition of M's configurations whose query list contains word y. Let S be the set of all $y \in \Sigma^n$ such that M on input 0^n queries $0^n y$. In general, $\sum_{y \in S} \tilde{q}_y \leq p(n)^2$ since the size of each query list is at most $p(n)$. By our assumption, however, $|S| = 2^n$.

Let y be any string in S and fix A such that $y = y_A$. Moreover, let A_y be A except that $A_y(0^n y) = 1 - A(0^n y)$. It follows by our assumption that $\rho_{M^A}(0^n) = 1 - \rho_{M^{A_y}}(0^n)$. By Theorem 3.3 in [3], since $|\rho_{M^A}(0^n) - \rho_{M^{A_y}}(0^n)| = 1$, we have $\sum_{j=1}^{p(n)-1} q_y(|\phi_j\rangle) \geq \frac{1}{8p(n)}$. Since $\sum_{j=1}^{p(n)-1} q_y(|\phi_j\rangle) \leq \tilde{q}_y \cdot p(n)$, we have $\tilde{q}_y \geq \frac{1}{8p(n)^2}$. This immediately draws the conclusion that $|S| \leq 8p(n)^4$, a contradiction. $\square$

Acknowledgments

The author would like to thank Andy Yao and Yaoyun Shi for interesting discussion on quantum complexity theory. He is also grateful to anonymous referees for their critical comments on an early draft.

References

1. L. M. Adleman, J. DeMarrais, and M. A. Huang, Quantum computability, *SIAM J. Comput.*, **26** (1997), 1524–1540.
2. P. Benioff, The computer as a Physical system: A microscopic quantum mechanical Hamiltonian model of computers as represented by Turing machines, *J. Stat. Phys.*, **22** (1980), 563–591.
3. C. H. Bennett, E. Bernstein, G. Brassard, and U. Vazirani, Strengths and weaknesses of quantum computing, *SIAM J. Comput.*, **26** (1997), 1510–1523.
4. E. Bernstein and U. Vazirani, Quantum complexity theory, *SIAM J. Comput.*, **26** (1997), 1411–1473.
5. G. Brassard, P. Høyer, and A. Tapp, Quantum counting, *Proc. 25th International Colloquium on Automata, Languages, and Programming*, Lecture Notes in Computer Science, Vol.1443, pp.820–831, 1998.
6. D. Deutsch, Quantum theory, the Church-Turing principle, and the universal quantum computer, *Proc. Roy. Soc. London*, A, **400** (1985), 97–117.
7. S. Fenner, L. Fortnow, S. Kurtz, and L. Li, An oracle builder's toolkit, *Proc. 8th IEEE Conference on Structure in Complexity Theory*, pp.120–131, 1993.
8. S. Fenner, L. Fortnow, and S. Kurtz, Gap-definable counting classes, *J. Comput. and System Sci.*, **48** (1994), 116–148.
9. S. Fenner, F. Green, S. Homer, and R. Pruim, Determining acceptance possibility for a quantum computation is hard for PH, *Proc. 6th Italian Conference on Theoretical Computer Science*, World-Scientific, Singapore, pp.241–252, 1998.
10. R. Feynman, Simulating Physics with computers, *Intern. J. Theoret. Phys.*, **21** (1982), 467–488.
11. L. Fortnow and J. Rogers, Complexity limitations on quantum computation, *Proc. 13th IEEE Conference on Computational Complexity*, pp.202–209, 1998.
12. L. Grover, A fast quantum mechanical algorithm for database search, *Proc. 28th ACM Symposium on Theory of Computing*, pp.212–219, 1996.
13. L. G. Valiant, The complexity of computing the permanent, *Theor. Comput. Sci.*, **8** (1979), 410–421.
14. T. Yamakami, A foundation of programming a multi-tape quantum Turing machine, *Proc. 24th International Symposium on Mathematical Foundations of Computer Science*, Lecture Notes in Computer Science, Vol.1672, pp.430–441, 1999. See also LANL quant-ph/9906084.

15. T. Yamakami and A. C. Yao, $NQP_C = co\text{-}C_= P$, to appear in *Inform. Process. Lett.* See also LANL quant-ph/9812032, 1998.

16. A. C. Yao, Quantum circuit complexity, *Proc. 34th IEEE Symposium on Foundation of Computer Science*, pp.352–361, 1993.

On Sets Growing Continuously

Bernhard Heinemann

Fachbereich Informatik, FernUniversität Hagen
D–58084 Hagen, Germany
phone: ++49-2331-987-2714
fax: ++49-2331-987-319
bernhard.heinemann@fernuni-hagen.de

Abstract. In the given paper we introduce a system by means of which the increasing of sets in the course of continuous (linear) time can be modelled. The system is based on a bimodal language. It originates from certain logics admitting *topological reasoning* where *shrinking* of sets is the subject of investigation, but is 'dual' to those in an obvious sense. After the motivating part we examine the system from a logical point of view. Our main results include *completeness* of a proposed axiomatisation and *decidability* of the set of validities. The intended applications concern all fields of (spatio–)temporal reasoning.

1 Introduction

Subsequently we are concerned with the change of sets in the course of time. For the moment, this is a very general approach: changing sets actually occur in many fields of computer science. Let us mention three different examples. First, considering an agent involved in a *multi–agent system*, the set of states representing its *knowledge* changes during a run of the system. Thus, in order to specify its behaviour one should have at one's disposal a tool by which one is able to treat changing sets (of this particular kind) formally. The significance of the knowledge–based approach to modelling multi–agent systems has been pointed out convincingly; see [6], e.g., or [10], where in particular *distributed systems* are emphasized. The language we introduce below in fact originates from this context, and a *knowledge oprator* is retained in it (but is mainly intended to quantify inside sets now).

The second example stems from the realm of databases, where *spatio–temporal* modelling and reasoning is an actual field of research. There one might want to specify the temporal change of geometric shapes like bad–weather regions or coastlines, for instance. Applications to weather forecast and geological prognoses, respectively, are obvious.

Finally, spatio–temporal reasoning receives much attention in AI as well. Here the picture is rather heterogeneous, and several different approaches to this field have been proposed. To get an impression of the state of the art the reader may consult, for instance, the proceedings volume of the last ECAI conference [15]

C. Pandu Rangan, V. Raman, R. Ramanujam (Eds.): FSTTCS'99, LNCS 1738, pp. 420–431, 1999.
© Springer-Verlag Berlin Heidelberg 1999

where a considerable number of papers has been contributed to the corresponding subdivision.

In what follows the reader will find a *unifying approach* to those different viewpoints. Accordingly, we extract parts of their common ground and cast them into an appropriate *logical system* towards a corresponding laying of the foundations. To this end let us briefly describe our starting position.

A system admitting *topological reasoning* – and being related to the logic of knowledge – has been developed recently [5]. It captures *shrinking* of sets and thus offers access to our topic in a certain sense. For convenience of the reader we mention its very basic features. Although originating from *modal logic* the domains of interpretation involved are not usual Kripke models, but *set spaces* $(X, \mathcal{O})$, where X is a non–empty set (of *states*, e.g.) and $\mathcal{O}$ is a set of subsets of X. In the simplest case there are two modalities included: one, designated K, which quantifies 'horizontally' over the elements of a set $U \in \mathcal{O}$, and another one, $\square$, which quantifies 'vertically' over its subsets contained in $\mathcal{O}$, expressing 'shrinking' in this way. The modalities also interact. This interaction depends on the actual model, the class of semantic structures one has in mind, and it is a challenging task in general to describe it axiomatically.

In this paper we dually deal with the *growth* of sets. Our idea is to develop a suitable 'modal' logic which modifies the $\square$–operator of the logic of set spaces appropriately. Moreover, we want to consider frames in which sets increase *continuously*, as it is mostly the case in real life. (The discrete case is treated elsewhere.) It turns out that this works in a way still preserving connections with distributed systems, the most important area of application of the logic of knowledge: our logic can be applied to *synchronous* multi–agent systems *with no learning;* see [11].

It should be remarked that there are different formalisms of computational logic dealing with dynamic aspects of sets as well: the logical treatment of *hybrid systems* [1], for instance, or the *duration calculus* [12]. Their interesting relationship with the present approach deserves a closer examination.

All in all, our aim is to provide a *modal basis* for the spatio–temporal reasoning framework described above, which focuses on continuously increasing sets presently. In this respect our exposition is of a *theoretical* nature. What we develop subsequently is a bimodal system having a strong *temporal flavour;* in fact, it may be viewed as a generalization of the temporal logic of continuous linear time (see [9], §8). Although the use of modal logics of this kind is very common in computer science, systems related to the present one are scarce; see [7], [8], [13], [14] for some examples. The treatment of *increasing sets* and *continuity* in this context is new; note that the systems considered in [4] are different from ours.

We now proceed to the technical details. In the next section we define the language underlying our system. In particular, we argue that set frames can be used for our task of modelling. Afterwards we introduce the logic and prove its soundness and completeness w.r.t. several classes of structures we have in mind.

In the final technical section we show that the set of theorems of the logic, i.e., the set of formulas which are derivable in the given calculus, is decidable.

Except for some fundamentals of modal and temporal logic the paper is self-contained. Concerning this basic material we refer the reader to the standard textbooks [3], [2], and, in particular, [9]. All new techniques and constructions are brought in, but detailed proofs are omitted due to limited space.

2 Prerequisites

The definition of the *syntax* of CG starts at a suitable finite alphabet, which in particular enables one to define a recursive set of *propositional variables, PV*. The set $\mathcal{F}$ of *CG-formulas* is the minimal set of strings satisfying

$$PV \subseteq \mathcal{F}, \text{ and } \alpha, \beta \in \mathcal{F} \Longrightarrow \neg\alpha, K\alpha, \Box\alpha, (\alpha \wedge \beta) \in \mathcal{F}.$$

The operator K is retained from the logic of knowledge [6]; presently it is intended to quantify within sets. The second operator, $\Box$, captures increasing of sets by quantifying over all supersets of the actually considered set. As it is usual we let $L\alpha :\equiv \neg K\neg\alpha$ and $\Diamond\alpha :\equiv \neg\Box\neg\alpha$.

The idea to define the *semantics* of CG is as follows. We would like to describe the continuous growing of a given set, Y. Thus certain supersets of Y have to be considered in the formal model. Consequently, we take a universe, X, in which all these sets are contained, and the system of these sets, $\mathcal{O}$, as the basic ingredients of interpreting formulas. Moreover, by means of a mapping σ we assign a truth value to the propositions depending (only) on points (and not on sets; see below). Thus we take certain triples $(X, \mathcal{O}, \sigma)$ as the semantic structures being relevant. These are specified precisely by the subsequent definition.

Definition 1. *1. Let X be a non–empty set and $\mathcal{O}$ a set of non–empty subsets of X. Then the pair $\mathcal{S} = (X, \mathcal{O})$ is called a set frame.*

  2. A set frame $\mathcal{S}$ is called densely ordered, iff $\mathcal{O}$ is linearly ordered by (reverse) proper set inclusion, and for all $U, V \in \mathcal{O}$ such that $U \supset V$ there exists $W \in \mathcal{O}$ satisfying $U \supset W \supset V$. $\mathcal{S}$ is called rational, iff $(\mathcal{O}, \supset)$ is isomorphic to $(\mathbb{Q}, >)$, and continuous, iff it is isomorphic to $(\mathbb{R}, >)$.

  3. Let $\mathcal{S} = (X, \mathcal{O})$ be a (densely ordered, rational, continuous) set frame and $\sigma : PV \times X \longrightarrow \{0, 1\}$ a mapping. Then σ is called a valuation, and the triple $\mathcal{M} = (X, \mathcal{O}, \sigma)$ is called a (densely ordered, continuous) model (based on $\mathcal{S}$).

Note that we have three 'rising' qualities of continuity given by the order type of $\mathcal{O}$; it turns out that they are *not* distinguishable modally. — We define next how to interpret formulas in models at *situations* of set frames, which are simply pairs x, U (designated without brackets mostly) such that $x \in U \in \mathcal{O}$. Only the crucial cases are mentioned.

Definition 2. *Let a model* $\mathcal{M} = (X, \mathcal{O}, \sigma)$ *and a situation* x, U *of the set frame* $(X, \mathcal{O})$ *be given. Then we define for all* $A \in PV$ *and* $\alpha \in \mathcal{F}$:

$$x, U \models_\mathcal{M} A \quad :\Longleftrightarrow \sigma(A, x) = 1$$
$$x, U \models_\mathcal{M} K\alpha :\Longleftrightarrow y, U \models_\mathcal{M} \alpha \text{ for all } y \in U$$
$$x, U \models_\mathcal{M} \Box\alpha :\Longleftrightarrow x, V \models_\mathcal{M} \alpha \text{ for all } V \supset U \text{ contained in } \mathcal{O}.$$

Notice that the definition of the validity of a propositional variable A at a situation x, U is independent of U. This is in accordance with the proceeding in topological modal logic (see [5]) and enables us to define the semantics in a quite natural way via situations; over and above that we use this definition in the completeness proof decisively.

In case $x, U \models_\mathcal{M} \alpha$ is valid we say that α *holds in* $\mathcal{M}$ *at the situation* x, U; moreover, the formula $\alpha \in \mathcal{F}$ is said to *hold in* $\mathcal{M}$ (denoted by $\models_\mathcal{M} \alpha$), iff it holds in $\mathcal{M}$ at every situation. If there is no ambiguity, we sometimes omit the index $\mathcal{M}$.

3   The Logic

In this section we present a logical system which permits of formally deriving all validities of the semantic domains we have in mind: models based on densely ordered, rational and continuous set frames, respectively. We list several axioms and a couple of rules below, constituting this system. Afterwards we show completeness of the axiomatisation w.r.t. the just mentioned classes of structures. In this way we obtain in particular a generalization of the modal system K4DLX considered in [9], p. 56 f. As axioms we have for all $A \in PV$ and $\alpha, \beta \in \mathcal{F}$:

(1) All $\mathcal{F}$–instances of propositional tautologies.
(2) $K(\alpha \to \beta) \to (K\alpha \to K\beta)$
(3) $K\alpha \to \alpha$
(4) $K\alpha \to KK\alpha$
(5) $L\alpha \to KL\alpha$
(6) $(A \to \Box A) \wedge (\neg A \to \Box\neg A)$
(7) $\Box(\alpha \to \beta) \to (\Box\alpha \to \Box\beta)$
(8) $\Box\alpha \to \Diamond\alpha$
(9) $\Box\alpha \leftrightarrow \Box\Box\alpha$
(10) $\Box(\alpha \wedge \Box\alpha \to \beta) \vee \Box(\beta \wedge \Box\beta \to \alpha)$
(11) $\Box K\alpha \to K\Box\alpha$.

Let us give some comments on these axioms. Apart from the first and the last one – the first embeds propositional logic and the last will be discussed later on – they fall into two groups apparently, each of which concerning one modal operator. The first group, given by the schemes (2) – (5), is well–known from the common logic of knowledge of a single agent; see [6]. In terms of modal logic, these S5–axioms express the properties of reflexivity, transitivity and weak symmetry, respectively, of the accessibility relation of the frame under consideration.

The 'transitivity axiom' is also present for "□": compare the left–to–right direction of the scheme (9). Axiom (10) corresponds with *weak connectedness* in this modal meaning; i.e., given arbitrary points s, t, u of a usual Kripke frame (X, R) such that $s\,R\,t$ and $s\,R\,u$, then $t\,R\,u$ or $u\,R\,t$ or $u = t$ holds iff the scheme is valid in (X, R). In set frames it is responsible for linearity of the set $\mathcal{O}$, in connection with (11). The scheme (8) tones down reflexivity to *seriality*, which means that for all $s \in X$ there is a $t \in X$ such that $s\,R\,t$. Finally, the right–to left direction of (9) encapsulates *weak density;* i.e., if $s\,R\,t$, then $s\,R\,u$ and $u\,R\,t$ holds for some u $(s, t, u \in X)$. — But the group of axioms involving only the □–operator comprises yet another peculiarity: the scheme (6). It allows us to define the semantics of propositional variables in the way we did above, namely without explicit reference to sets occurring in situations, but it implies that the system to be defined immediately is not closed under substitution. Regarding content (6) says that the atomic propositions are 'stable' or 'persistent' during increasing. In the paper the scheme also serves as an appropriate technical means in order to prove completeness.

Last but not least axiom (11) combining both modalities is associated with the growth of sets. It is very powerful, as we shall see below. — By adding the following rules we get a deductive system designated **CG**. Let $\alpha, \beta \in \mathcal{F}$.

$$(1)\ \frac{\alpha \to \beta, \alpha}{\beta} \qquad (2)\ \frac{\alpha}{K\alpha} \qquad (3)\ \frac{\alpha}{\Box\alpha}$$

So, we have *modus ponens*, K–*necessitation* and □–*necessitation*. — Soundness of the system w.r.t. the structures introduced in Definition 1(3) can easily be established.

Proposition 1. *All of the above axioms hold in every model, and the rules preserve validity.*

We are going to sketch how completeness of the system **CG** w.r.t. the class of densely ordered models is proved. Rational and continuous models are considered in a separate section afterwards. We start at the *canonical model* $\widetilde{\mathcal{M}}$ of **CG**. This is built in the usual way (see [9], §5); i.e., the domain C of $\widetilde{\mathcal{M}}$ consists of the set of all maximal **CG**–consistent sets of formulas, and the accessibility relations induced by the modal operators K and □ are defined as follows:

$$s \overset{L}{\longrightarrow} t :\iff \{\alpha \in \mathcal{F} \mid K\alpha \in s\} \subseteq t, \quad s \overset{\Diamond}{\longrightarrow} t :\iff \{\alpha \in \mathcal{F} \mid \Box\alpha \in s\} \subseteq t,$$

for all $s, t \in C$. Finally, the distinguished valuation of the canonical model is defined by $\sigma(A, s) = 1 :\iff A \in s$ $(A \in PV,\ s \in C)$. — The subsequent *truth lemma* is well–known.

Lemma 1. *Let us denote the usual satisfaction relation of multimodal logic by* $\models$, *and let* $\vdash$ *designate* **CG**–*derivability. Then it holds that*

(a) $\widetilde{\mathcal{M}} \models \alpha[s]$ *iff* $\alpha \in s$, *and (b)* $\widetilde{\mathcal{M}} \models \alpha$ *iff* $\vdash \alpha$, *for all* $\alpha \in \mathcal{F}$ *and* $s \in C$.

Parts (a) and (b) of the following proposition are likewise commonly known. Axioms (3) – (5) are responsible for (a), whereas (8) – (10) imply (b). Part (c) is a consequence of the scheme (11). Its proof is not quite immediate, but can be done in a similar fashion as that of [5], Proposition 2.2.

Proposition 2. *(a) The relation* $\xrightarrow{L}$ *is an equivalence relation on the set C.*
(b) The relation $\xrightarrow{\Diamond}$ *on C is serial, transitive, weakly dense and weakly connected.*
(c) Let $s, t, u \in C$ be given such that $s \xrightarrow{L} t \xrightarrow{\Diamond} u$. Then there exists a point $v \in C$ satisfying $s \xrightarrow{\Diamond} v \xrightarrow{L} u$.

Following a common manner of speaking in the language of subset spaces let us call the property asserted in (c) the *modified cross property*. — The next proposition reads as [8], Proposition 9(3), and is crucial to our purposes.

Proposition 3. *Let $s, t \in C$ be given such that $s \xrightarrow{L} t$ and $s \xrightarrow{\Diamond} t$ holds. Then s and t coincide.*

Later on the following consequence of this proposition and the modified cross property will be applied.

Corollary 1. *Let $s, t \in C$ be given such that $s \xrightarrow{L} t$ and $s \xrightarrow{\Diamond} s$ holds. Then also $t \xrightarrow{\Diamond} t$ is valid.*

The next result can be proved inductively with the aid of Proposition 3.

Proposition 4. *The relation* $\xrightarrow{\Diamond}$ *on C is antisymmetric.*

For every $s \in C$ let $[s]$ denote the $\xrightarrow{L}$–equivalence class of s. A relation $\prec$ on the set of all such classes is defined as follows:

$$[s] \prec [t] :\iff \text{ there are } s' \in [s],\ t' \in [t] \text{ such that } s' \xrightarrow{\Diamond} t',$$

for all $s, t \in C$. — As a consequence of the above assertions and the modified cross property we get:

Proposition 5. *The relation $\prec$ is serial, transitive, weakly dense, weakly connected and antisymmetric.*

Later on we argue that we may restrict attention to the submodel of the canonical model generated by a suitable $s \in C$ which has carrier

$$C^s = \{[s]\} \cup \bigcup_{t \in C,\, [s] \prec [t]} [t]$$

and accessibility relations the restricted ones. So, let us proceed with this model which we likewise designate $\widetilde{\mathcal{M}}$, abusing notation.

In order to arrive at a densely ordered model eventually we have to ensure that density is not caused by reflexivity at certain points. This is done by 'blowing up' the model appropriately, substituting every 'reflexive point' by a copy of $\mathbb{Q}$. To this end we let $C_r^s := \{t \in C^s \mid t \xrightarrow{\diamond} t\}$, define the set

$$C' := (C^s \setminus C_r^s) \cup \{(r,t) \mid r \in \mathbb{Q} \text{ and } t \in C_r^s\},$$

and binary relations $\xrightarrow{L}, \xrightarrow{\diamond}$ on C' by letting for all $x, y \in C'$

$$x \xrightarrow{L} y : \iff x, y \in C^s \text{ and } x \xrightarrow{L} y, \text{ or } x = (r,t), y = (r,t') \text{ and } t \xrightarrow{L} t',$$

for some $r \in \mathbb{Q}$ and $t, t' \in C^s$, and

$$x \xrightarrow{\diamond} y : \iff \begin{cases} x, y \in C^s \text{ and } x \xrightarrow{\diamond} y, \text{ or } x = (r,t), \ y \in C^s \text{ and } t \xrightarrow{\diamond} y, \text{ or} \\ x \in C^s, \ y = (r,t) \text{ and } x \xrightarrow{\diamond} t, \text{ or} \\ x = (r,t), \ y = (r',t') \text{ and } t \xrightarrow{\diamond} t', \text{ or} \\ x = (r,t), \ y = (r',t) \text{ and } r < r', \end{cases}$$

for some $r, r' \in \mathbb{Q}$ and $t, t' \in C^s$. Then all the properties stated in Proposition 2 hold for $\xrightarrow{\diamond}$ and $\xrightarrow{\diamond}$, respectively. Moreover, $\xrightarrow{\diamond}$ is irreflexive, i.e., $x \xrightarrow{L} x$ is not valid for any $x \in C'$.

Proposition 6. *(a) The relation $\xrightarrow{L}$ is an equivalence relation on the set C'.*

(b) The relation $\xrightarrow{\diamond}$ on C' is serial, irreflexive, transitive, dense and weakly connected.

(c) Let $s, t, u \in C'$ be given such that $s \xrightarrow{L} t \xrightarrow{\diamond} u$. Then there exists a point $v \in C'$ satisfying $s \xrightarrow{\diamond} v \xrightarrow{L} u$.

Designating the relation induced by $\prec$ on the set of all $\xrightarrow{L}$–equivalence classes $\prec'$ and taking advantage of Corollary 1 (among other things) we get:

Proposition 7. *The relation $\prec'$ is a dense linear order.*

The distinguished valuation on the canonical model is lifted to C' by

$$\sigma'(A,x) := \begin{cases} \sigma(A,x) & \text{if } x \in C^s \\ \sigma(A,t) & \text{if } x = (r,t) \text{ for some } r \in \mathbb{Q} \text{ and } t \in C^s, \end{cases}$$

for all $A \in PV$ and $x \in C'$. Finally, let h denote the canonical mapping from C' onto C and $\mathcal{O}'$ the preimage of $\mathcal{O}$ w.r.t. h. The subsequent lemma, in which $\mathcal{M}' := (C', \mathcal{O}', \sigma')$, is easily proved by a structural induction then.

Lemma 2. *For all $\alpha \in \mathcal{F}$ and $x \in C'$ we have $\mathcal{M}' \models \alpha[x]$ iff $\widetilde{\mathcal{M}} \models \alpha[h(x)]$.*

Now we are in a position to define a densely ordered model $\mathcal{M}$ falsifying a given non–derivable formula $\alpha \in \mathcal{F}$. For this purpose we choose a maximal CG–consistent set $s \in C$ containing $\neg \alpha$ and consider both the submodel $\widetilde{\mathcal{M}}$ of the canonical model generated by s and the model $\mathcal{M}'$ depending on $\widetilde{\mathcal{M}}$ that has been just constructed. Let $s_0 := s$, if $s \in C'$, and $s_0 := (0,s)$ otherwise. For every $x \in C'$ let $[x]$ designate the $\overset{L}{\hookrightarrow}$–equivalence class of x (this notation will not be confused with that one of the same kind on C), define for all $y \in C'$

$$[x] \preceq' [y] :\iff [x] = [y] \text{ or } [x] \prec' [y],$$

and let $C'_x := \{[y] \mid y \in C' \text{ and } [x] \preceq' [y]\}$. Furthermore, let $C'^{s_0} := \bigcup_{[y] \in C'_{s_0}} [y]$ and define a function $f_x : C'_x \longrightarrow C'^{s_0}$ for every $x \in C'$ such that $[x] \in C'_{s_0}$ by

$$f_x([x']) := \begin{cases} x & \text{if } [x'] = [x] \\ x_1 \in [x'] \text{ satisfying } x \overset{\diamond}{\hookrightarrow} x_1 & \text{otherwise,} \end{cases}$$

for all $[x'] \in C'_x$. According to our previous results every function f_x is well–defined. The set $X := \{f_x \mid [x] \in C'_{s_0}\}$ will serve as the carrier set of $\mathcal{M}$. Moreover, for every $[x] \in C'_{s_0}$ we let

$$U'_{[x]} := \{f_y \mid y \in [x]\} \text{ and } U_{[x]} := \bigcup_{[s_0] \preceq' [x'] \preceq' [x]} U'_{[x']}.$$

Subsequently we write $\mathcal{O}$ instead of $\{U_{[x]} \mid [x] \in C'_{s_0}\}$. Finally, we let a valuation σ on X be induced by σ' in an obvious sense: $\sigma(A, f_x) = 1 :\iff \sigma'(A, x) = 1$, for all propositional variables A and functions $f_x \in X$. Note that σ is correctly defined as well. — The structure $\mathcal{M} := (X, \mathcal{O}, \sigma)$ is a densely ordered model by construction, and the following *truth lemma* is valid.

Lemma 3. *For all formulas $\beta \in \mathcal{F}$ and $x, x' \in C'^{s_0}$ such that $[x] \preceq' [x']$ it holds that $f_x, U_{[x']} \models_\mathcal{M} \beta$ iff $\mathcal{M}' \models \beta [f_x([x'])]$.*

As an immediate consequence we get the first of our desired completeness results.

Theorem 1. *Every $\alpha \in \mathcal{F}$ that is not **CG**–derivable can be falsified in a densely ordered model.*

4 Continuity

Just as it is the case with the modal system K4DLX mentioned earlier we are able to prove completeness of the system **CG** w.r.t. the smaller classes of rational and continuous models simultaneously. But we have to develop new techniques for this because we cannot work with *filtrations* as in classical modal logic. This is caused by the failure of *connectedness* of any filtration of $\overset{\diamond}{\longrightarrow}$ which is due to the fact that a generated submodel of the canonical model is 'two–dimensional' in essence. — The following notions are preparatory.

Definition 3. *Let $\mathcal{I} := (I, \leq)$ be a non–empty linearly ordered set.*

1. *A subset $\emptyset \neq J \subseteq I$ is called a segment of $\mathcal{I}$, iff there is no $i \in I \setminus J$ strictly between any two elements of J.*

2. *A partition of I into segments is called a* segmentation *of $\mathcal{I}$.*
3. *A segmentation of $\mathcal{I}$ is called* appropriate, *iff every segment J of $\mathcal{I}$ either consists of a single point or is right–open, i.e., for all $j \in J$ there is a $j' \neq j$ in J such that $j \leq j'$.*

Subsequently we will have to consider segmentations of the linearly ordered set $\mathcal{C} := (C_s, \preceq)$, where $C_s := \{[s]\} \cup \{[t] \mid t \in C \text{ and } [s] \prec [t]\}$ and

$$[t] \preceq [u] :\iff [t] = [u] \text{ or } [t] \prec [u], \text{ for all } t, u \in C^s,$$

such that the truth value of a given formula remains unaltered on every segment. (Note that both C^s and $\prec$ have been defined in Section 3; moreover, $\preceq$ is in fact a linear ordering because of Proposition 5.) The next definition says what this means precisely.

Definition 4. *Let $\widetilde{\mathcal{M}} = (C^s, \{\xrightarrow{L} |_{C^* \times C^*}, \xrightarrow{\diamond} |_{C^* \times C^*}\}, \sigma)$ be the submodel of the canonical model of* **CG** *generated by $s \in C$, I an indexing set and $\mathcal{P} := \{\mathcal{P}_\iota \mid \iota \in I\}$ a segmentation of $\mathcal{C}$.*

1. *For every $\iota \in I$ we define an equivalence relation $\sim_\iota$ on $\mathcal{P}_\iota$ by*

$$x \sim_\iota y :\iff \begin{cases} \text{there is some } z \in \bigcup \mathcal{P}_\iota \text{ such that } x \xrightarrow{\diamond} z \text{ and } y \xrightarrow{\diamond} z, \\ \text{or } x = y, \end{cases}$$

for all $x, y \in \bigcup \mathcal{P}_\iota$. Every $\sim_\iota$–class is called a cone *of $\mathcal{P}_\iota$, and the set of all cones of $\mathcal{P}_\iota$ is designated S_ι.*
2. *Let $\alpha \in \mathcal{F}$ be a formula. Then α is called* stable on $\mathcal{P}$, *iff for all $\iota \in I$ and cones $S \in S_\iota$*

$$\widetilde{\mathcal{M}} \models \alpha[y] \text{ for all } y \in S, \text{ or } \widetilde{\mathcal{M}} \models \neg\alpha[y] \text{ for all } y \in S.$$

The relation $\sim_\iota$ is in fact transitive because $\xrightarrow{\diamond}$ is transitive and weakly connected, by Proposition 2(b). — We can always achieve a *finite appropriate* segmentation of $\mathcal{C}$ on which a given formula is stable.

Proposition 8. *Let $\alpha \in \mathcal{F}$ be a formula and $\mathcal{C}$ as above. Then there exists a finite appropriate segmentation $\mathcal{P}_\alpha := \{\mathcal{P}_1, \ldots, \mathcal{P}_n\}$ of $\mathcal{C}$ such that α is stable on $\mathcal{P}_\alpha$. Moreover, $\mathcal{P}_\alpha$ can be chosen such that it refines $\mathcal{P}_\beta$ for every subformula β of α.*

It should be remarked that proving the proposition one proceeds inductively and starts with the trivial segmentation $\{\mathcal{C}\}$ in case α a propositional variable. Only the cases $\alpha = (\beta \wedge \gamma)$ and $\alpha = K\beta$ contribute to a refinement of the actually obtained segmentation.

Now we define an intermediate model of 'finite depth' depending on a given formula α. We let $\mathcal{P}_\alpha := \{\mathcal{P}_1, \ldots, \mathcal{P}_n\}$ be the finite segmentation of $\mathcal{C}$ according to Proposition 8 on which α is stable and $X := \{S \mid S \in S_i \text{ for some } 1 \leq i \leq n\}$. Furthermore, we define binary relations $\xmapsto{L}, \xmapsto{\diamond}$ on X by

$$S \xmapsto{L} T :\iff \text{ for some } i \in \{1, \ldots, n\} \text{ both } S \text{ and } T \text{ are cones of } \mathcal{P}_i$$
$$S \xmapsto{\diamond} T :\iff \text{ there exist } x \in S \text{ and } y \in T \text{ such that } x \xrightarrow{\diamond} y,$$

for all $S, T \in X$. Finally, we let a valuation τ on X be defined by $\tau(A, S) := \sigma(A, x)$ for some $x \in S$, for all $A \in PV$ and $S \in X$. Then we have the following

Lemma 4. *The structure* $\mathcal{M} := (X, \{\overset{L}{\longmapsto}, \overset{\diamond}{\longmapsto}\}, \tau)$ *is a model such that*

(a) *all properties stated in Proposition 2 and Proposition 4 are also valid for* $\overset{L}{\longmapsto}$ *and* $\overset{\diamond}{\longmapsto}$ *respectively, and*

(b) *for all subformulas* β *of* α, *points* $t \in C^s$ *and cones* $S \in X$ *containing* t *it holds that* $\widetilde{\mathcal{M}} \models \beta[t]$ *iff* $\mathcal{M} \models \beta[S]$.

The model $\mathcal{M}$ can be 'vertically decomposed' into *slices*, i.e., sequences

$$\Xi: \quad S_{n-m+1} \overset{\diamond}{\longmapsto} S_{n-m+2} \overset{\diamond}{\longmapsto} \ldots \overset{\diamond}{\longmapsto} S_n$$

of maximal length such that $S_i \in \mathcal{S}_i$ for all $n - m + 1 \leq i \leq n$ ($m \in \mathbb{N}$, $m \leq n$). Now let Ξ be of length n and $S_{i_1}, \ldots, S_{i_k}$ the reflexive cones in Ξ ($i_1, \ldots, i_k \in \{1, \ldots, n\}$). Consider any right–open interval $J = [r, p)$ of either $\mathbb{Q}$ or $\mathbb{R}$ (p may be ∞). Decompose J into k right–open subintervals $J_1 = [r_1, p_1), \ldots, J_k = [r_k, p_k)$ (in ascending order). Because of the next lemma one can p–morphically map J onto Ξ, thereby assigning subintervals to reflexive cones and right end-points to a following non–reflexive cone, if need be (see [9], p. 57).

Lemma 5. *For any slice of* $\mathcal{M}$, *two non–reflexive cones are non–adjacent.*

Since the segmentation of $\mathcal{C}$ was chosen to be appropriate, *all* cones of fixed index $i \in \{1, \ldots, n\}$ are either reflexive or irreflexive. So, we may define a *bimodal* model $\mathcal{M}'$ in which the equivalence classes belonging to the operator K are indexed by J; moreover, the above p–morphism extends suitably to that model. To be more precise, let for every $j \in \{1, \ldots, n\}$

$$J'_j := \begin{cases} J_{i_l} & \text{if } j = i_l \text{ for some } 1 \leq l \leq k \text{ and } (j = 1 \text{ or } j - 1 = i_{l-1}) \\ (r_{i_l}, q_{i_l}) & \text{if } j = i_l \text{ for some } 2 \leq l \leq k \text{ and } i_{l-1} < j - 1 \\ p_{i_l} & \text{if } j - 1 = i_l \text{ for some } 1 \leq l \leq k \text{ and } j \neq i_{l+1} \\ r_1 & \text{if } j = 1 \text{ and } j \neq i_1. \end{cases}$$

Note that exactly one of the cases on the right–hand side occurs, according to our discussion above. Using this we define

$$X' := \{(q, S) \mid q \in J'_j \text{ and } S \in \mathcal{S}_j \text{ for some } 1 \leq j \leq n\}.$$

Furthermore, let accessibility relations on X' be given by

$$(q, S) \overset{L}{\rightharpoonup} (q', S') :\Longleftrightarrow q = q' \text{ and } S \overset{L}{\longmapsto} S'$$
$$(q, S) \overset{\diamond}{\rightharpoonup} (q', S') :\Longleftrightarrow q < q' \text{ and } S \overset{\diamond}{\longmapsto} S',$$

for all $q, q' \in J$ and $S, S' \in X$. Letting finally the valuation of $\mathcal{M}'$ be the one induced on X' by that of $\mathcal{M}$ we obtain:

Lemma 6. *The relations $\xrightarrow{L}$ and $\xrightarrow{\diamond}$ fulfill all the respective properties stated in Proposition 6, and for all $\beta \in \mathcal{F}$ and $(q, S) \in X'$ we have $\mathcal{M}' \models \beta[(q, S)]$ iff $\mathcal{M} \models \beta[S]$.*

The interval J (together with the restriction of $<$ to J) is nothing but a generated substructure of $(\mathbb{Q}, <)$ (and of $(\mathbb{R}, <)$, respectively). Thus, by the standard submodel lemma of modal logic ([9], 1.7, e.g.), we may assume that q varies over $\mathbb{Q}$ in the above lemma (and over $\mathbb{R}$, respectively).

The situation now is obviously the same as that after Lemma 2, except for the fact that the set indexing the $\xrightarrow{L}$ –equivalence classes equals $\mathbb{Q}$ here ($\mathbb{R}$, respectively). Consequently, carrying out the construction of the final model as a suitable space of functions in the same way as above we get:

Theorem 2. *The system **CG** is complete w.r.t. the classes of rational and continuous models, respectively.*

5   Decidability

Our examination of the previous section also leads to *decidability* of the set of formulas holding in every densely ordered (rational, continuous) model and, equivalently, of the set of **CG**–derivable formulas. Our starting point now is Lemma 4, from which the following definition is derived.

Definition 5. *Let $\mathcal{M} := (W, \{R, S\}, \sigma)$ be a bimodal Kripke model; i.e., W is a non–empty set, R and S are binary relations on W, and σ is a valuation. Then $\mathcal{M}$ is called a CG–model, iff*

- *R is an equivalence relation on W,*
- *S serial, antisymmetric, transitive, weakly dense and weakly connected,*
- *for all $s, t, u \in W$: if $s R t$ and $t S u$, then there exists an element $v \in W$ such that $s S v$ and $v R u$,*
- *for all $s, t \in W$ such that $s S t$ and every $A \in PV$ it holds that $\mathcal{M} \models A[s]$ iff $\mathcal{M} \models A[t]$.*

In the above definition the relation R corresponds with the modality K; accordingly, S and $\Box$ are related. — It turns out that the system **CG** is sound and complete w.r.t. the class of CG–models as well.

Theorem 3. *A formula $\alpha \in \mathcal{F}$ is **CG**–derivable, iff it holds in every CG–model.*

Proving soundness of the system w.r.t. CG–models is straightforward, but concerning completeness one has to utilize Theorem 1.

Because of this theorem it suffices to consider CG–models in order to falsify a given formula α that is not derivable in the system **CG**. Revisiting Lemma 4 shows that even CG–models of 'finite depth' are sufficient for this purpose. But we can go one step further: we may in fact confine ourselves to *finite CG–* models. We do not carry out this in detail presently, but mention that one can use standard methods of 'topological' modal logic to this end; see [8], e.g. — All in all, this gives decidability of our logic.

Theorem 4. *The set of formulas derivable in the system **CG** is decidable.*

6   Prospect

The given system describes formally the continuous growth of sets in a modal setting. As it stands, it is a very general framework, last but not least due to the fact that it deals with the propositional case only. However, because of the results obtained so far we feel that it could be a good basis for qualitative (spatio–)temporal reasoning in contexts where such continuous phenomena have to be modelled.

It was already indicated in the introduction that a *discrete version* of the logic has been worked out and will appear elsewhere. Incorporating among other things the common *nexttime* operator, it generalizes propositional temporal logic of linear time correspondingly. Apart from this a formalism expressing both increasing and shrinking of sets, and their acting in combination, is desirable. If one confines oneself to *nexttime* one can get such a system which is clearly rather weak. So, further research has to be concerned with more expressive logical languages being tailor–made for the present context.

References

1. Artemov, S., Davoren, J., Nerode, A.: Topological Semantics for Hybrid Systems. Lecture Notes in Computer Science, Vol. 1234 (1997) 1–8
2. Chagrov, A., Zakharyaschev, M.: Modal Logic. Oxford (1997)
3. Chellas, B. F.: Modal Logic: An Introduction. Cambridge (1980)
4. Davoren, J.: Modal Logics for Continuous Dynamics. PhD dissertation, Cornell University (1998)
5. Dabrowski, A., Moss, L. S., Parikh, R.: Topological Reasoning and The Logic of Knowledge. Annals of Pure and Applied Logic **78** (1996) 73–110
6. Fagin, R., Halpern, J. Y., Moses, Y., Vardi, M. Y.: Reasoning about Knowledge. Cambridge(Mass.) (1995)
7. Georgatos, K.: Knowledge Theoretic Properties of Topological Spaces. Lecture Notes in Computer Science, Vol. 808 (1994) 147–159
8. Georgatos, K.: Knowledge on Treelike Spaces. Studia Logica **59** (1997) 271–301
9. Goldblatt, R.: Logics of Time and Computation. Stanford (1987)
10. Halpern, J.Y., Moses,Y.: Knowledge and Common Knowledge in a Distributed Environment. Journal of the ACM **37** (1990) 549–587
11. Halpern, J.Y., Vardi, M.Y.: The Complexity of Reasoning about Knowledge and Time. I. Lower Bounds. Journal of Computer and System Sciences **38** (1989) 195–237
12. Hansen, M. R., Chaochen, Z.: Duration Calculus: Logical Foundations. Formal Aspects of Computing **9** (1997) 283–330
13. Heinemann, B.: A Topological Generalization of Propositional Linear Time Temporal Logic. Lecture Notes in Computer Science, Vol. 1295 (1997) 289–297
14. Heinemann, B.: Separating Sets by Modal Formulas. Lecture Notes in Computer Science, Vol. 1548 (1999) 140–153
15. Prade, H. (ed.): ECAI 98. 13th European Conference on Artificial Intelligence. Chichester (1998)

Model Checking Knowledge and Time in Systems with Perfect Recall*

(Extended Abstract)

Ron van der Meyden[1] and Nikolay V. Shilov[2]

[1] School of Computer Science and Engineering,
University of New South Wales, Sydney 2052, Australia.
meyden@cse.unsw.edu.au
[2] Institute of Informatics Systems, Novosibirsk
6, Lavrent'ev av., Novosibirsk, 630090, Russia
shilov@iis.nsk.su

Abstract. This paper studies model checking for the modal logic of knowledge and linear time in distributed systems with perfect recall. It is shown that this problem (1) is undecidable for a language with operators for *until* and *common knowledge*, (2) is PSPACE-complete for a language with *common knowledge* but without *until*, (3) has non-elementary upper and lower bounds for a language with *until* but without *common knowledge*. Model checking *bounded knowledge depth* formulae of the last of these languages is considered in greater detail, and an automata-theoretic decision procedure is developed for this problem, that yields a more precise complexity characterization.

1 Introduction

Modal logics have been found to be convenient formalisms for reasoning about distributed systems [MP91], in large part because such logics enable automated verification by model checking of specifications [CGP99]. This involves constructing a model of the system to be verified, and then testing that this model satisfies a formula specifying the system. Frequently, the model of the system is finite state, but model checking of infinite state systems is an emerging area of research.

Epistemic logic, or the *logic of knowledge* [HM90,FHMV95], is a recent addition to the family of modal logics that have been applied to reasoning about distributed systems. This logic allows one to express that an agent in the system *knows* (has the information) that some fact holds. This expressiveness is particularly useful for reasoning about distributed systems with unreliable components or communication media. In such settings, information arises in subtle ways, and it can be difficult to express the precise conditions under which an agent

---

* Work supported by an Australian Research Council Large Grant, done while the authors were employed in the School of Computing Sciences, University of Technology, Sydney. The first author acknowledges the hospitality of the Department of Computer Science, Utrecht University while revising this paper.

C. Pandu Rangan, V. Raman, R. Ramanujam (Eds.): FSTTCS'99, LNCS 1738, pp. 432–445, 1999.

has certain knowledge. On the other hand, the behavior of agents is often a simple function of their state of knowledge. Examples of knowledge-level analysis of systems illustrating this claim are given in [FHMV95].

A topic of interest for logics of knowledge is the extent to which they (like other modal logics) allow for automated analysis of designs and specifications. Combinations of temporal and epistemic logics are especially significant, since a frequent concern in applications is how knowledge changes over time. A number of papers have studied the problem of model checking logics of knowledge and time in *finite* state systems [HV91,FHMV95,Var96]. However, much of the literature on applications of logics of knowledge assumes that agents have *perfect recall*, i.e., remember all their past states, and this results in infinite state systems. Model checking of the logic of knowledge with respect to the perfect recall semantics has been considered by van der Meyden [Mey98], but this work deals with a language that does not include temporal operators.

In the present paper, we study model checking a combined logic of knowledge and linear time in *synchronous* systems with perfect recall. Like van der Meyden [Mey98], we assume that agents operate in a finite state environment, but we extend this framework to allow Büchi fairness constraints. Since the perfect recall assumption generates an infinite Kripke structure from the environment, the problem we study is an example of infinite state model checking.

Formal definitions of synchronous systems with perfect recall and the model checking problem are presented in Section 2. While model checking a logic with operators for knowledge and *common knowledge* is decidable [Mey98], the addition of the linear time temporal operators *next* and *until* makes the problem undecidable (Section 3). However, decidability is retained for two fragments of this extended language: the fragments in which we (1) omit the *until* operator (this case is PSPACE complete) or (2) omit the *common knowledge* operator (this case is non-elementary). The latter result may be obtained by means of reductions to and from various powerful logics already known to be decidable (weak S1S and Chain Logic with an equal level predicate [Tho92]). However, we also present (in Section 4) an alternative proof of this latter result, using novel automata-theoretic constructions, that provides a more informative complexity characterization. Section 5 discusses related work and topics for further research.

2   Basic Definitions

This section further develops the definition of environments of [Mey98] by adding fairness constraints, and defines the model checking problem we study.

Let *Prop* be a set of atomic propositional constants, $n \geq 0$ be a natural number and $\mathcal{O}$ be a set. Define a *finite interpreted environment for n agents* to be a tuple E of the form $\langle S, I, T, O, \pi, \alpha \rangle$ where the components are as follows:

1. S is a finite set of *states* of the environment,
2. I is a subset of S, representing the possible *initial states*,
3. $T \subseteq S^2$ is a *transition relation*,

4. O is a tuple $(O_1, \ldots, O_n)$ of functions, where for each $i \in \{1 \ldots n\}$
   the component $O_i : S \to \mathcal{O}$ is called the *observation function of agent i,*
5. $\pi : S \to \{0, 1\}^{Prop}$ is an *interpretation,*
6. $\alpha \subseteq S$ is an *acceptance condition.*

Intuitively, an environment is a finite-state transition system where states encode values of local variables, messages in transit, failure of components, etc. For states s, s' the relation $s\,T\,s'$ means that if the system is in state s, then at the next tick of the clock it could be in state s'. If s is a state and i an agent then $O_i(s)$ represents the observation agent i makes when the system is in state s, i.e., the information about the state that is accessible to the agent. The interpretation maps each state to an assignment of truth values to the atomic propositional constants in *Prop*. The acceptance conditions are standard Büchi conditions which are used to model fairness requirements on evolutions of the environment.

A *trace* of an environment E is a finite sequence of states $s_0 s_1 \ldots s_m$ such that $s_0 \in I$ and $s_j\,T\,s_{j+1}$ for all $j < m$. A *run* of an environment E is an infinite sequence $r : \mathbb{N} \to S$ of states of E such that every finite prefix of r is a trace of E and there exists a state $s \in \alpha$ that occurs infinitely often in r. We say that the acceptance condition of E is *trivial* if $\alpha = S$. A *point* of E is a tuple (r, m), where r is a run of E and m a natural number. Intuitively, a point identifies a particular instant of time along the history described by the run.

Individual runs of an environment provide sufficient structure for the interpretation of formulae of linear temporal logic. To interpret formulae involving knowledge, we use the agents' observations to determine the points they consider possible. There are many ways one could do this. The particular approach used in this paper models a *synchronous perfect-recall* semantics of knowledge. Given a run r of an environment for n agents with observation functions $O_1, \ldots, O_n$, we define the *local state of agent i at time* $m \geq 0$ to be the sequence $r_i(m) = O_i(r(0)) \ldots O_i(r(m))$. That is, the local state of an agent at a point in a run consists of a complete record of the observations the agent has made up to that point. These local states may be used to define for each agent i a relation $\sim_i$ of *indistinguishability* on points $(r, m), (r', m')$ of E, by $(r, m) \sim_i (r', m')$ if $r_i(m) = r'_i(m')$. Intuitively, when $(r, m) \sim_i (r', m')$, agent i has failed to receive enough information at these points to determine whether it is in one situation or the other. Clearly, each $\sim_i$ is an equivalence relation. The use of the term "synchronous" above reflects the fact that if $(r, m) \sim_i (r', m')$, then we must have $m = m'$. The relations $\sim_i$ will be used to define the semantics of knowledge for individual agents. We will also consider an operator for common knowledge, a kind of group knowledge, for which we use another relation. If $G \subseteq \{1 \ldots n\}$ is a *group* of agents (i.e., two or more) then we define the relation $\sim_G$ on points to be the reflexive transitive closure of the union of all indistinguishability relations $\sim_i$ for $i \in G$, i.e., $\sim_G = (\bigcup_{i \in G} \sim_i)^*$.

We will be concerned with model checking a propositional multi-modal language for knowledge and linear time based on a set *Prop* of atomic propositional constants, with formulae generated by the modalities $\bigcirc$ (next), $\mathcal{U}$ (until), a knowledge operator K_i for each agent $i \in \{1 \ldots n\}$, and a common knowledge operator C_G for each group of agents $G \subseteq \{1 .. n\}$. Formulae of the language are defined as follows: each atomic propositional constant $p \in Prop$ is a formula, and if φ and ψ are formulae, then so are $\neg\varphi$, $\varphi \wedge \psi$, $\bigcirc\varphi$, $\varphi \mathcal{U} \psi$, $K_i\varphi$ and $C_G\varphi$ for each $i \in \{1 .. n\}$ and group $G \subseteq \{1 .. n\}$. We write $\mathcal{L}_{\{\bigcirc, \mathcal{U}, K_1, ..., K_n, C\}}$ for the set of formulae. We will refer to sublanguages of this language by a similar expression that lists the operators generating the language. For example, $\mathcal{L}_{\{K_1, ..., K_n, C\}}$ refers to the language of the logic of knowledge (without time). As usual, we use the abbreviations $\Diamond\varphi$ for $\mathbf{true}\,\mathcal{U}\,\varphi$, and $\Box\varphi$ for $\neg\Diamond\neg\varphi$.

The semantics of this language is defined as follows. Suppose we are given an environment E with interpretation π. We define satisfaction of a formula φ at a point (r, m) of a run of E, denoted $E, (r, m) \models \varphi$, inductively on the structure of φ. The cases for the temporal fragment of the language are standard:

$$E, (r, m) \models p \qquad \text{if } \pi(r(m))(p) = 1, \text{ where } p \in Prop,$$
$$E, (r, m) \models \varphi_1 \wedge \varphi_2 \quad \text{if } E, (r, m) \models \varphi_1 \text{ and } E, (r, m) \models \varphi_2,$$
$$E, (r, m) \models \neg\varphi \qquad \text{if not } E, (r, m) \models \varphi,$$
$$E, (r, m) \models \bigcirc\varphi \qquad \text{if } E, (r, m + 1) \models \varphi,$$
$$E, (r, m) \models \varphi_1 \mathcal{U} \varphi_2 \quad \text{if there exists } m'' \geq m \text{ such that } E, (r, m'') \models \varphi_2$$
$$\text{and } E, (r, m') \models \varphi_1 \text{ for all } m' \text{ with } m \leq m' < m''.$$

The semantics of the knowledge and common knowledge operators is defined by

$$E, (r, m) \models K_i\varphi \text{ if } E, (r', m') \models \varphi \text{ for all points } (r', m') \text{ of } E$$
$$\text{satisfying } (r', m') \sim_i (r, m)$$

$$E, (r, m) \models C_G\varphi \text{ if } E, (r', m') \models \varphi \text{ for all points } (r', m') \text{ of } E$$
$$\text{satisfying } (r', m') \sim_G (r, m)$$

This definition can be viewed as an instance of the general framework for the semantics of knowledge proposed in [HM90]. Intuitively, an agent knows a formula to be true if this formula holds at all points that the agent is unable to distinguish from the actual point. Common knowledge may be understood as follows. For G a group of agents, define the operator E_G, read "everyone in G knows" by $E_G\varphi \equiv \bigwedge_{i \in G} K_i\varphi$. Then $C_G\varphi$ is equivalent to the infinite conjunction of the formulae $E_G^k\varphi$ for $k \geq 1$. That is, φ is common knowledge if everyone knows φ, everyone knows that everyone knows φ, etc. We refer the reader to [HM90,FHMV95] for further motivation and background.

We may now define the model checking problem we consider in this paper. Say that a formula φ is *realized* in the environment E if for all runs r of E, we have $E, (r, 0) \models \varphi$. We are interested in the following problem, which we call the *realization problem*: given an an environment E and a formula φ of a language $\mathcal{L}$, determine if φ is realized in E. We will consider this problem with respect to several sublanguages of $\mathcal{L}_{\{\bigcirc, \mathcal{U}, K_1, ..., K_n, C\}}$.

3   Complexity Bounds

We now present a number of results on the complexity of the realization problem for various fragments of the language, and briefly sketch their proofs. First, we consider the most expressive language $\mathcal{L}_{\{\bigcirc, \mathcal{U}, K_1, \ldots, K_n, C\}}$, containing all the modal operators we have defined. Here the outcome of our investigation is negative:

Theorem 1. *There exist a class of finite environments for two agents with trivial acceptance conditions and a formula of $\mathcal{L}_{\{\bigcirc, \mathcal{U}, K_1, \ldots, K_n, C\}}$ such that it is undecidable whether the formula is realized in a given environment of this class.*

That is, even a restricted case of the realization problem for $\mathcal{L}_{\{\bigcirc, \mathcal{U}, K_1, \ldots, K_n, C\}}$ is undecidable. The proof of Theorem 1 employs ideas from concerning *model checking at a trace* for the language $\mathcal{L}_{\{K_1, \ldots, K_n, C\}}$. Stated in terms of our current terms and notations, this problem is to determine, *given an environment E with a trivial acceptance condition, a trace t of E and a formula φ of $\mathcal{L}_{\{K_1, \ldots, K_n, C\}}$, whether $E, (r, |t|) \models \varphi$ for all runs r of E extending t.*[1] This model checking problem was studied in [Mey98] for both the synchronous and an *asynchronous* perfect recall semantics of knowledge. The following two results are proved in [Mey98].[2]

Theorem 2. *With respect to the synchronous perfect recall semantics, the problem of model checking at a trace is in PSPACE for the language $\mathcal{L}_{\{K_1, \ldots, K_n, C\}}$. It is PSPACE hard for the fixed formula $C_{\{1,2\}}p$.*

Theorem 3. *There exists an environment for two agents such that with respect to the asynchronous perfect recall semantics, the problem of model checking the fixed formula $C_{\{1,2\}}p$ at a given trace of the environment is undecidable.*

The proof of the lower bound in Theorem 2 involved showing that the synchronous semantics can simulate PSPACE computations, with Turing machine configurations represented as traces and the step relation on configurations represented by the composition $\sim_1 \circ \sim_2$. The common knowledge operator then allows us to represent the transitive closure of the step relation, enabling a formula to refer to the result of a PSPACE computation. The proof of Theorem 3 used a similar representation of Turing machine computations, but first uses asynchrony to "guess" the amount of space required by the computation. To prove Theorem 1 we reuse this approach to representation of Turing machine computations. However, instead of asynchrony, we now use the temporal operators to refer to a sufficiently long configuration. As before, we then describe the outcome of the computation starting at that configuration using the common knowledge operator.

In the language $\mathcal{L}_{\{\bigcirc, \mathcal{U}, K_1, \ldots, K_n, C\}}$ we have two operators, the until operator and the common knowledge operator, whose semantics allows an arbitrary reach through two orthogonal dimensions in our semantic structures. In other contexts, these operators are individually tractable, e.g., the validity problem for both

---

[1] It can be shown that if r and r' are two runs extending t and $\varphi \in \mathcal{L}_{\{K_1, \ldots, K_n, C\}}$ then $E, (r, |t|) \models \varphi$ iff $E, (r', |t|) \models \varphi$.

[2] In both these results $\{1, 2\}$ is a set of agents while p is a propositional constant.

the logic of knowledge and common knowledge $\mathcal{L}_{\{K_1,\ldots,K_n,C\}}$ and temporal logic $\mathcal{L}_{\{\bigcirc,\mathcal{U}\}}$ are known to be decidable. It therefore makes sense to study the result of eliminating one of these operators from our language. For the language obtained by excluding the until operator we have:

Theorem 4. *The realization problem for $\mathcal{L}_{\{\bigcirc,K_1,\ldots,K_n,C\}}$ is PSPACE complete.*

The proof of Theorem 4 is similar to the proof in [Mey98] of Theorem 2, and exploits the fact, for checking realization of a formula φ, instead of the infinite set of runs we can confine our attention to the finite set of traces of length at most $|\varphi|$, which is, intuitively, the furthest that the temporal operators can reach. This set of traces has exponentially many elements, but since each trace is of linear size we may do model checking within polynomial space using techniques of [Mey98].

For the language without common knowledge, some new techniques are required. Here we obtain the following.

Theorem 5. *The realization problem for $\mathcal{L}_{\{\bigcirc,\mathcal{U},K_1,\ldots,K_n\}}$ is decidable, with non-elementary upper and lower bounds.*

The proof for both the upper and the lower bound can be obtained by reductions from variants of SnS, the Monadic Second Order Logic of n Successors [Tho92], which is interpreted over the infinite tree $\{1 \ldots n\}^*$. The proof of the lower bound in Theorem 5 is by a reduction from WS1S, a version of S1S in which the second order quantifiers are restricted to range over finite sets. It is known [Mey74] that WS1S is decidable with lower bound non-elementary in the size of the formula. The upper bound result can be established by a translation to the problem of checking the validity of a formula of *Chain Logic with the Equal Level predicate* [Tho92] (or CLE) on tree structures. The logic CLE is an extension of a restriction of SnS. Chain Logic (CL) is obtained from SnS by restricting the interpretation of the second order quantifiers to sets that are chains, i.e., are totally ordered by the prefix relation. The logic CLE is obtained by adding to this restriction of SnS the equal level predicate, defined on $u, v \in \{1 \ldots n\}^*$ by $E(u,v)$ if $|u| = |v|$. This approach to the upper bound for the realization problem for $\mathcal{L}_{\{\bigcirc,\mathcal{U},K_1,\ldots,K_n\}}$ from this proof is rather indirect, however, as decidability of CLE is proved by Thomas [Tho92] using a translation to S1S. The logic S1S in turn is known to be decidable using automata theoretic arguments [Büc60]. Thus, we have gone from automata (in the definition of environments) to logic, and back to automata. In the following section we present a proof of Theorem 5 that is based directly on automata theoretic constructions, and which yields a more informative complexity characterization.

4 An Algorithm for Bounded Knowledge Depth Formulae

Our more informative characterization of the complexity of realization for the language $\mathcal{L}_{\{\bigcirc,\mathcal{U},K_1,\ldots,K_n\}}$ is cast in terms of the *knowledge depth* of formulae, i.e., the maximal depth of nesting of knowledge operators in a formula. For example, $depth(K_1(\bigcirc K_2(q \wedge K_2 r))) = 3$. Our approach to the decidability result will exploit k-*trees*, a data structure that has previously been used in the literature [Mey98] to represent depth k formulae of $\mathcal{L}_{\{K_1,\ldots,K_n\}}$ holding at a point of an environment. We show in this section that k-trees also encode enough information to interpret formulae of $\mathcal{L}_{\{\bigcirc,\mathcal{U},K_1,\ldots,K_n\}}$ with knowledge depth at most k. Throughout this section we assume a fixed finite environment E. We assume without loss of generality that every trace of E can be extended to a run of E.[3]

4.1   Trees

Intuitively, a k-tree, for $k \geq 0$, is a type of finite tree of height k in which vertices are labelled by states of the environment and edges are labelled by agents. It is convenient to represent these trees as follows.[4] For numbers $k \geq 0$ we define by mutual recursion the set $\mathcal{T}_k$ of k-*trees over* E, and the set $\mathcal{F}_k$ of *forests of k-trees over* E. Define $\mathcal{T}_0$ to be the set of tuples of the form $\langle s, \emptyset, \ldots, \emptyset \rangle$ where s is a state of E and the number of copies of the empty set $\emptyset$ is equal to the number of agents n. Once $\mathcal{T}_k$ has been defined, let $\mathcal{F}_k$ be the set of all subsets of $\mathcal{T}_k$. Now, define $\mathcal{T}_{k+1}$ to be the set of all tuples of the form $\langle s, U_1, \ldots, U_n \rangle$, where s is a state and U_i is in $\mathcal{F}_k$ for each $i \in \{1 .. n\}$. We denote $\bigcup_{k \geq 0} \mathcal{T}_k$ by $\mathcal{T}$.

Intuitively, in a tuple $\langle s, U_1, \ldots, U_n \rangle$, the state s represents the actual state of the environment, and for each $i \in \{1 .. n\}$ the set U_i represents the knowledge of agent i. Identifying a 0-tree $\langle s, \emptyset, \ldots, \emptyset \rangle$ with the state s, note that each component U_i in a 1-tree is simply a set of states: intuitively, those states agent i considers possible. For higher k, the set U_i represents agent i's knowledge both about the universe and other agents' knowledge, up to depth k.

The elements of $\mathcal{T}_k$ correspond in an obvious way to trees of height k, with edges labelled by agents and nodes labelled by states. If $w = \langle s, U_1, \ldots, U_n \rangle$ we define $root(w)$ to be the state s. If w' is an element of U_i, then we say that w' is an i-*child* of w. When $w \in \mathcal{T}_0$ the labelled tree corresponding to w consists of just the root, labelled $root(w)$. The tree corresponding to $w \in \mathcal{T}_{k+1}$ has root labelled with $root(w)$, and for each i-child $w' \in \mathcal{T}_k$ of w there is an i-labelled edge from the root to a vertex at which the labelled subtree is that corresponding to w'. The following result characterises C_k, the number of k-trees over E.

Lemma 1. *Let $k \geq 0$ be a natural number and E be a finite environment for n agents with l states. Then C_k is not greater than $\exp(n \times l, k)/n$, where $\exp(a, b)$ is the function defined by $\exp(a, 0) = a$ and $\exp(a, b+1) = a2^{\exp(a,b)}$.*

---

[3] An environment not satisfying this condition can easily by modified (without changing its realization properties) by eliminating states that do not belong to any run.

[4] The definitions we give here are (for reasons of clarity and space) a slight simplification of those in [Mey98], which add some complications to enable k-trees to be used to interpret formulae of *alternation depth* at most k, a slightly larger class than the class of formulae of knowledge depth at most k.

For each $k \geq 0$ we may associate with each point (r, m) of E a k-tree $F_k(r, m)$, which captures some of the structure of the indistinguishability relations of the environment around that point. We proceed inductively. For $k = 0$ we define $F_0(r, m) = \langle r(m), \emptyset, \ldots, \emptyset \rangle$. For $k > 0$ we define $F_k(r, m) = \langle r(m), U_1, \ldots, U_m \rangle$, where for each agent i we have U_i equal to the set of $k - 1$-trees $F_{k-1}(r', m')$ where (r', m') is a point of E with $(r', m') \sim_i (r, m)$.

For each point (r, m) of E let $\tau(r, m)$ be the trace $r(0) \ldots r(m)$. It is not difficult to see that for all $k \geq 0$ and for all points (r, m) and (r', m') with $\tau(r, m) = \tau(r', m')$ we have $F_k(r, m) = F_k(r', m')$. Thus, we may also view F_k as a function mapping traces of E to k-trees, and write $F_k(\tau)$ where τ is a trace of E. Note that we exploit the fact that every trace may be extended to a run here.

We now recall from [Mey98] some functions that may be used to update k-trees. These functions were used in [Mey98] to provide an algorithm for the problem of model checking at a trace (see above). We will use these functions below to define a sequence of Büchi automata for the realization problem. Let S, T and $\mathcal{O}$ be the set of states, the transition relation, and the set of observations of the environment E, respectively. We define for each number $k \geq 0$ the function $G_k : \mathcal{T}_k \times S \rightarrow \mathcal{T}_k$. The definition of G_k will be by mutual recursion with the functions $H_{k,i} : \mathcal{F}_k \times \mathcal{O} \rightarrow \mathcal{F}_k$, where $i \in \{1 .. n\}$ and $k \geq 0$. Intuitively, if agent i's state of knowledge (to depth k) is represented by the the set of k-trees U, then $H_{k,i}(U, o)$ represents the agent's revised state of knowledge after it makes the observation $o \in \mathcal{O}$. We define $G_0(w, s) = \langle s, \emptyset, \ldots, \emptyset \rangle$. Once G_k has been defined, we define for each $i \in \{1 .. n\}$ the function $H_{k,i}$ by taking $H_{k,i}(U, o)$ to be the set of k-trees $G_k(w, s)$ where $w \in U$ and $O_i(s) = o$ and $root(w)Ts$, i.e., there exists a transition of E from $root(w)$ to s. Using the functions $H_{k,i}$ we may now define G_{k+1} by setting $G_{k+1}(\langle s, U_1, \ldots, U_n \rangle, s')$ to be $\langle s', H_{k,1}[U_1, O_1(s')], \ldots, H_{k,n}[U_n, O_n(s')] \rangle$.

For our definitions (which are a slight variant of those in [Mey98]), we may establish the following theorem, essentially the same as a result proved in [Mey98].

Theorem 6. *For each $k \geq 0$, and for every finite trace $\tau \cdot s$ of E with final state s and prefix τ, we have the incremental update property $F_k(\tau \cdot s) = G_k(F_k(\tau), s)$.*

The definition of the function F_k above is not effective, ranging over the possibly infinite set of runs of E. In the case of traces of length 0 is it easily seen how to make it effective, obtaining functions that represent agents' knowledge in the *initial* states of the environment as a k-tree. We inductively define mappings $f_k : I \rightarrow \mathcal{T}_k$ for $k \geq 0$. In the base case, we put $f_0(s) = \langle s, \emptyset, \ldots, \emptyset \rangle$. For $k \geq 0$ we define $f_{k+1}(s)$ to be the $k + 1$-tree with root s and an i-child $f_k(s')$ for each initial state s' with $O_i(s') = O_i(s)$.

Lemma 2. *For all runs r of E we have $F_k(r, 0) = f_k(r(0))$.*

Note that, here again, we rely on the fact that every trace (of length 0) can be extended to a run.

4.2   An Automaton Theoretic Characterization

We now give an automata-theoretic characterization of realization that forms the basis for the algorithm discussed below.

We begin by defining a type of Büchi automata [Büc60]. The specific variety of automata we need are tuples of the form $A = \langle S, I, T, \alpha \rangle$, where S is a finite set of (control) states, $I \subseteq S$ is the set of initial states of the automaton, $T \subseteq S^2$ is a a transition relation, and $\alpha \subseteq S$ is its acceptance condition.[5] An *execution* of A is an infinite sequence $e : \mathbb{N} \rightarrow S$ of states of S such that for all $m \geq 0$ we have $e(m) \, T \, e(m+1)$. An execution e is said to be *properly initialised* if $e(0) \in I$. An execution is said to be *fair* if some state in α occurs infinitely often in the execution. A fair, properly initialised execution is called *accepting*. We also call accepting executions *runs*. The language accepted by the automaton A is the set $\mathcal{L}(A)$ of all runs of A.

Given the environment E fixed above, we now define an infinite sequence of Büchi automata $A_0(E), \ldots, A_k(E), \ldots$. Each automaton $A_k(E)$ defines a language consisting of infinite sequences of k-trees over that environment. These automata will be crucial to the algorithm we develop.

Let E be $\langle S, I, T, O, \pi, \alpha \rangle$ and $k \geq 0$. Define $A_k(E) = \langle S_k, I_k, T_k, \alpha_k \rangle$ to be the Büchi automaton with

1. S_k equal to the set T_k of k-trees over E,
2. initial states I_k equal to the set of k-trees $f_k(s)$ where $s \in I$,
3. transition relation T_k defined by $w T_k w'$ when there exists state $s \in S$ such that $root(w) T s$ and $w' = G_k(w, s)$,
4. acceptance condition α_k defined by $\alpha_k = \{ w \in S_k : root(w) \in \alpha \}$.

Since $A_k(E)$ is a Büchi automaton on infinite words, the notions of an execution, a fair execution, a properly intialised execution and a run of $A_k(E)$ are meaningful. We define a *projection* operation $Proj_k$ mapping runs r of the automata $A_k(E)$ to infinite sequences of states of E, by $Proj_k(r)(m) = root(r(m))$. Conversely, there exists a *lift* operation $Lift_k$ mapping runs r of the environment E to sequences of k-trees, defined by $Lift_k(r)(m) = F_k(r(0) \ldots r(m))$. The proof of the following is straightforward from the definitions, Theorem 6 and Lemma 2:

Lemma 3. *For each $k \geq 0$ the mappings $Proj_k$ and $Lift_k$ are inverse functions; $Proj_k$ maps the set of runs of $A_k(E)$ onto the set of runs of E while $Lift_k$ maps the set of runs of E onto the set of runs of $A_k(E)$.*

We now show that the automata $A_k(E)$ adequately capture the depth k formulae of $\mathcal{L}_{\{\bigcirc, \mathcal{U}, K_1, \ldots, K_n\}}$ holding at points of E. To do so, we define for each k a relation $\models_k$ between points in executions of $A_k(E)$ and formulae of knowledge depth $\leq k$. The definition of $\models_k$ is by induction on k, as follows. For the basic propositions and the temporal operators, the definition is much like the standard semantics. Thus, for an execution e of $A_k(E)$ and $m \geq 0$,

---

[5] These are slightly less general than usual, in that the input alphabet and control states coincide, and the transition function has a specific form that is derived from the transition relation.

$$E, (e, m) \models_k p \qquad \text{if } \pi(root(e(m)))(p) = 1, \text{ where } p \in Prop,$$
$$E, (e, m) \models_k \varphi_1 \wedge \varphi_2 \qquad \text{if } E, (e, m) \models_k \varphi_1 \text{ and } E, (e, m) \models_k \varphi_2,$$
$$E, (e, m) \models_k \neg\varphi \qquad \text{if not } E, (e, m) \models_k \varphi,$$
$$E, (e, m) \models_k \bigcirc\varphi \qquad \text{if } E, (e, m + 1) \models_k \varphi,$$
$$E, (e, m) \models_k \varphi_1 \, \mathcal{U} \, \varphi_2 \qquad \text{if there exists } m'' \geq m \text{ such that}$$
$$E, (e, m'') \models_k \varphi_2 \text{ and } E, (e, m') \models_k \varphi_1 \text{ for all}$$
$$m' \text{ with } m \leq m' < m''.$$

The interesting case concerns the knowledge operators, where we make use of the fact that we are dealing with a sequence of k-trees. It is convenient to first define $\models_k$ not on points, but on k-trees w, for formulae $K_i\varphi$ of $\mathcal{L}_{\{\bigcirc, \mathcal{U}, K_1, ..., K_n\}}$ of knowledge depth at most k, by

$$E, w \models_k K_i\varphi \text{ if for all } k - 1\text{-trees } w' \text{ that are } i\text{-children of } w, \text{ and for all fair}$$
executions e of $A_{k-1}(E)$ such that $e(0) = w'$, we have $E, (e, 0) \models_{k-1} \varphi$.

Note that we consider fair executions e rather than runs in this definition because w' is not necessarily an initial state of $A_{k-1}(E)$. We then define $\models_k$ on points by

$$E, (e, m) \models_k K_i\varphi \text{ if } E, e(m) \models_k K_i\varphi.$$

The following result establishes a connection between the relations $\models_k$ and the semantics of $\mathcal{L}_{\{\bigcirc, \mathcal{U}, K_1, ..., K_n\}}$ in an environment E.

Lemma 4. *For every natural number $k \geq 0$, every formula φ in $\mathcal{L}_{\{\bigcirc, \mathcal{U}, K_1, ..., K_n\}}$ of knowledge depth at most k, for every environment E, every run r of E and $m \geq 0$ we have $E, (r, m) \models \varphi$ iff $E, (Lift_k(r), m) \models_k \varphi$.*

This equivalence forms the basis for our decision procedure for realization.

4.3   The Algorithm

We are now in a position to present the algorithm for the realization problem for $\mathcal{L}_{\{\bigcirc, \mathcal{U}, K_1, ..., K_n\}}$. Let us first note that in the special case where $depth(\varphi) = 0$, i.e., formulae not containing the knowledge operators, testing realization amounts to a problem of temporal logic to which well-known techniques may straightforwardly be applied (see below). To generalize to formulae of greater depth, we show how to decide the relation $E, w \models_k K_i\varphi$. We achieve this by factoring formulae into their temporal and knowledge components. To represent the temporal components, define a *context* to be just like a formula of $\mathcal{L}_{\{\bigcirc, \mathcal{U}, K_1, ..., K_n\}}$, but with additional propositional variables from a special separate countable set Var. If β is a context then we denote by $Var(\beta)$ the set of all variables which occur in β. A *pure temporal context* is a context not containing any occurrences of the knowledge operators.

We separate the temporal and knowledge aspects of formulae by means of the following way of "exploding" a formula. Define a *puff* to be a finite sequence of pairs $(set_0, map_0) \ldots (set_m, map_m)$ where the set_j are finite sets of pure temporal contexts, such that $Var(set_0) = \emptyset$, and for $j \neq j'$ the sets $Var(set_j)$

and $Var(set_{j'})$ are disjoint, and the map_j are mappings $map_j : Var(set_j) \rightarrow \{K_1, \ldots, K_n\} \times set_{j-1}$. (Thus, $map_0 = \emptyset$.) The result $\varphi \cdot map_j$ of applying a mapping map_j to a context φ is defined to be the context obtained by simultaneously substituting for each occurrence in φ of a variable $x \in Var(set_j)$ the formula $K_i \psi$ such that $map_j(x) = (K_i, \psi)$. A puff is said to be a *complete separation* of a formula φ if set_m contains a single context β, such that $\varphi = \beta \cdot map_m \cdot \ldots \cdot map_1$. Let cep be a function from formulae to puffs such that $cep(vp)$ is a complete separation of φ. We call the unique pure temporal context β in the top level set_m of a complete separation of φ the *temporal skeleton* of φ.

Example 1. The puff

$$set_0 = \{p\} \qquad\qquad map_0 = \emptyset$$
$$set_1 = \{\bigcirc x, \Box x\} \qquad map_1 : x \mapsto (K_2, p)$$
$$set_2 = \{y \, \mathcal{U} \, z\} \qquad\quad map_2 : y \mapsto (K_1, \bigcirc x)$$
$$map_2 : z \mapsto (K_2, \Box x)$$

is a complete separation of the formula $(K_1 \bigcirc K_2 p) \, \mathcal{U} \, (K_2 \Box K_2 p)$. The temporal skeleton of this formula is $y \, \mathcal{U} \, z$.

A *valuation* is a partial mapping $\upsilon : Var \rightarrow \mathcal{P}(\mathcal{T})$. This associates with each propositional variable the set of trees at which it is true. For each valuation υ one can extend the relation $\models_k$ on formulae to a relation $\models_k^\upsilon$ on temporal contexts by simply adding the superscript υ to $\models_k$ throughout the clauses above, and by adding the following clause:

- $E, (e, m) \models_k^\upsilon x$, where $x \in Var$ is a variable, iff $e(m) \in \upsilon(x)$.

A valuation υ is said to be *consistent* with a puff $(set_0, map_0) \ldots (set_m, map_m)$ if for each $j \in \{1 .. m\}$, every variable $x \in Var(set_j)$ and every $w \in \mathcal{T}_j$, if $map_j(x) = (K_i, \beta)$ then $w \in \upsilon(x)$ iff $E, w \models_j^\upsilon K_i \beta$.

Lemma 5. *Suppose φ is a formula of $\mathcal{L}_{\{\bigcirc, \mathcal{U}, K_1, \ldots, K_n\}}$ of knowledge depth k. Let β be the temporal skeleton of φ. Suppose that υ is a valuation consistent with $cep(\varphi)$. Then for all fair executions e of $A_k(E)$ and for all $m \geq 0$, we have $E, (e, m) \models_k \varphi$ iff $E, (e, m) \models_k^\upsilon \beta$.*

By Lemma 4, to determine $E, (r, m) \models \varphi$ for formulae of depth φ it suffices to decide $E, (Lift_k(r), m) \models_k \varphi$. The effect of Lemma 5 is to reduce the complicated recursion through k-trees required to evaluate the latter to the problem $E, (Lift(r), m) \models_k^\upsilon \beta$, whose determination involves only temporal steps.

Thus, we obtain the following approach to deciding realization: (1) represent a formula as a puff, (2) construct a consistent valuation for the puff, (3) evaluate the temporal skeleton of the formula with respect to this valuation and (4) check that the skeleton is valid for all initial states. This approach is formalized in the algorithm in Figure 1. By construction and in accordance with the definition of consistency, the assertion "υ is consistent with $(set_0, map_0) \ldots (set_j, map_j)$" is an invariant of the loop of the algorithm. Combining this with the results above

INPUT: a finite environment E and a formula φ
OUTPUT: Y if φ is realized in E, N otherwise.

PROCEDURE:

1. Let $k := depth(\varphi)$ and suppose $cep(\varphi) = (set_0, map_0)\dots(set_k, map_k)$.
   Let $\upsilon := \emptyset$.
2. For $j := 0$ to $k - 1$ do:
   (a) For all $\beta \in set_j$, let $[\beta] :=$
       $\{ w \in T_j : (e, 0) \models_j^{\upsilon} \beta$ for every fair execution e of $A_j(E)$ with $e(0) = w \}$.
   (b) For all $x \in Var(set_{j+1})$, if $map_{j+1}(x) = (K_i, \beta)$,
       let $[x] := \{w \in T_{j+1} : w' \in [\beta]$ for all i-children w' of $w\}$.
   (c) Let $\upsilon := \upsilon \cup (\bigcup_{x \in Var(set_{j+1})}\{(x, [x])\})$.
3. For temporal skeleton β of the formula φ let $[\beta] :=$
   $\{ w \in T_k : (e, 0) \models_k^{\upsilon} \beta$ for every fair execution e of $A_k(E)$ with $e(0) = w \}$.
4. If $w \in [\beta]$ for all $w \in \mathcal{I}_k$ then output(Y)   else output(N).

Fig. 1. An algorithm for realization

we conclude that the algorithm is correct. As presented, the algorithm is not yet fully operational: we still need to show how it is possible to compute $[\beta]$ at steps 2(a) and 3. This can be done in space polynomial in the size of $A_k(E)$ and β using known techniques [SC85].

Theorem 7. *The problem of determining if a formula φ of $\mathcal{L}_{\{\bigcirc, \mathcal{U}, K_1, \dots, K_n\}}$ is realized in an environment E is decidable in space polynomial in $|\varphi| \cdot \exp(depth(\varphi), O(|E|))$.*

5   Conclusion

It is interesting to note that for each of the languages we have considered, the complexity we have obtained for the realization problem is the same as the complexity obtained by Halpern and Vardi [HV88,HV89] for the validity problem. While there are some commonalities in the proof techniques used, we are not aware of any straightforward reductions between the two problems.

   The problem we have studied here, of checking whether a formula is *realized* in a given environment, is closely related to a problem studied by van der Meyden and Vardi [MV98]. This work also concerns the synchronous perfect recall semantics. They deal with a notion of environment in which agents are able to choose their actions based on their local state, and the choice of action determines the state transitions. They consider the *realizability* question, of whether it is possible to decide the existence of (and if so, construct) a protocol (a function from local state to actions) for an agent such that running this protocol in a given environment generates a system realizing a given specification in the logic of knowledge and time. By contrast with our results in this paper, however, realizability is decidable only in the single agent case, even for environments with trivial acceptance condition.

Model checking a logic of knowledge and time has also been studied by Clarke et al. [CJM98] in the context of verification of cryptographic protocols. Their work assumes a semantics of knowledge very different from ours, based on explicit computation rather than the information theoretic notion we have studied. An interesting topic for further research is the applicability of our results, or adaptations thereof, to verification of cryptographic protocols.

Several dimensions of generalisation of our results are worth considering. In particular, it would be desirable to know if our results generalise to the case of languages with past time or branching time operators — this generalisation is able to express the problem of checking that a given finite state protocol implements a given *knowledge based program* [FHMV95,FHMV97] in a given environment, in such a way that agents operate as if they had perfect recall. A knowledge based program is a type of specification that describes how an agent's actions relate to its state of knowledge. Verification of knowledge based programs for finite state definitions of knowledge has been shown decidable by Vardi [Var96]. The perfect recall case remains to be addressed, although the linear time case we have presented is already able to yield this result in the special case of *deterministic* knowledge-based programs [MV98]. Also of interest is the complexity of realization with respect to other natural definitions of knowledge, such as the *asynchronous* perfect recall semantics.

References

Büc60.   J.R. Büchi. On a decision method in restricted second order arithmetic. In *Proc. Internat. Congr. on Logic, Methodology and Philosophy of Science*, pages 1–11, Stanford, CA, 1960. Stanford Univ. Press.

CGP99.   E.M. Clarke, O. Grumberg, and D. Peled. *Model Checking*. MIT Press, Cambridge, MA, 1999.

CJM98.   E. Clarke, S. Jha, and W. Marrero. A machine checkable logic of knowledge for specifying security properties of electronic commerce protocols. In *LICS Workshop on Formal Methods and Security Protocols*, 1998.

FHMV95. R. Fagin, J. Y. Halpern, Y. Moses, and M. Y. Vardi. *Reasoning about Knowledge*. MIT Press, 1995.

FHMV97. R. Fagin, J. Y. Halpern, Y. Moses, and M. Y. Vardi. Knowledge-based programs. *Distributed Computing*, 10(4):199–225, 1997.

HM90.   J. Y. Halpern and Y. Moses. Knowledge and common knowledge in a distributed environment. *Journal of the ACM*, 37(3):549–587, 1990.

HV88.   J. Y. Halpern and M. Y. Vardi. The complexity of reasoning about knowledge and time: synchronous systems. Research Report RJ 6097, IBM, 1988.

HV89.   J. Y. Halpern and M. Y. Vardi. The complexity of reasoning about knowledge and time, I: lower bounds. *Journal of Computer and Systems Science*, 38(1):195–237, 1989.

HV91.   J. Y. Halpern and M. Y. Vardi. Model checking vs. theorem proving: a manifesto. In V. Lifschitz, editor, *Artificial Intelligence and Mathematical Theory of Computation (Papers in Honor of John McCarthy)*, pages 151–176. Academic Press, San Diego, Calif., 1991.

Mey74.   A. R. Meyer. The inherent complexity of theories of ordered sets. In *Proc. of the Int. Congr. of Mathematics*, volume 2, pages 477–482, Vancouver, 1974. Canadian Mathematical Congress.

Mey98.   R. van der Meyden. Common knowledge and update in finite environments. *Information and Computation*, 140(2):115–157, 1998.

MP91.   Z. Manna and A. Pnueli. *The Temporal logic of Reactive and Concurrent Systems*. Springer-Verlag, Berlin, 1991.

MV98.   R. van der Meyden and M. Y. Vardi. Synthesis from knowledge-based specifications. In *Proc. CONCUR'98, 9th International Conf. on Concurrency Theory*, pages 34–49. Springer LNCS No. 1466, 1998.

SC85.   A. P. Sistla and E. M. Clark. The complexity of propositional linear temporal logic. *Journal of the ACM*, 32(3):733–749, 1985.

Tho92.   W. Thomas. Infinite trees and automaton-definable relations over ω-words. *Theoretical Computer Science*, 103:143–159, 1992.

Var96.   M. Y. Vardi. Implementing knowledge-based programs. In *Proc. of the Conf. on Theoretical Aspects of Rationality and Knowledge*, pages 15–30, San Mateo, CA, 1996. Morgan Kaufmann.

The Engineering of Some Bipartite Matching Programs

Kurt Mehlhorn

Max-Planck-Institut für Informatik,
Im Stadtwald, 66123 Saarbrücken, Germany,
www.mpi-sb.mpg.de/~mehlhorn

Over the past years my research group was involved in the development of three algorithm libraries:

- LEDA, a library of efficient data types and algorithms [LED]
- CGAL, a computational geometry algorithms library [CGA], and
- AGD, a library for automatic graph drawing [AGD].

In this talk I will discuss some of the lessons learned from this work. I will do so on the basis of the LEDA-implementations of bipartite cardinality matching algorithms. The talk is based on Section 7.6, pages 360–392, of [MN99]. In this book Stefan Näher and I give a comprehensive treatment of the LEDA system and its use. We treat the architecture of the system, we discuss the functionality of the data types and algorithms available in the system, we discuss the implementation of many modules of the system, and we give many examples for the use of LEDA.

My personal level of involvement was very different in the three projects: The LEDA project was started in 89 by Stefan Näher and myself and I have been involved as a designer and system architect, implementer of algorithms and tools, writer of documentation and tutorials, and user of the system. For CGAL I acted as an advisor and for AGD my involvement was marginal.

The bipartite cardinality matching problem asks for the computation of a maximum cardinality matching M in a bipartite graph $(A \cup B, E)$. A *matching* in a graph G is a subset M of the edges of G such that no two share an endpoint.

I will discuss the following points:

- Specification: We discuss several specifications of the problem and discuss their relative merits, in particular, with respect to verification and flexibility.
- Checking and verification: We discuss how a matching algorithm can justify its answers and how answers can be checked.
- Representations of matchings: We discuss how matchings can be represented and what the relative merits of the representations are.
- Reinitialization in iterative algorithms: Most matching algorithms work in phases. We discuss how to reinitialize data structures in a cost-effective way.
- Search for augmenting paths by depth-first search or by breadth-first search: We discuss the relative merits of the two methods.

C. Pandu Rangan, V. Raman, R. Ramanujam (Eds.): FSTTCS'99, LNCS 1738, pp. 446–449, 1999.

n	m	k	FFB	dfs−	dfs+	bfs−	bfs+	HK−	HK+	AB−	AB+	Check
2	4	1	311.1	1.63	1.14	1.08	0.93	1.5	1.42	0.94	0.96	0.09
2	4	50	319.2	1.24	0.65	0.71	0.58	1.09	0.99	0.7599	0.77	0.07001
2	4	2500	316.1	0.35	0.32	0.32	0.3	1.01	0.92	0.69	0.68	0.07001
2	6	1	404	24.06	6.72	18.97	7.1	2.35	2.24	1.29	1.26	0.1001
2	6	50	397	71	15.22	12.29	7.57	1.76	1.67	0.97	0.95	0.08997
2	6	2500	313.9	3.15	1.12	0.6902	0.6299	2.37	2.29	0.78	0.76	0.08008
2	8	1	364.8	7.42	2.61	7.56	3.71	2.71	2.55	2.84	2.95	0.1199
2	8	50	360	34.15	12.47	9.35	7.59	2.19	2.1	1.68	1.64	0.09985
2	8	2500	360.2	42.8	10.47	1.9	1.76	2.9	2.74	1.23	1.22	0.08984
4	8	1	−	4.43	3.23	3.03	2.64	3.3	3.05	2.47	2.47	0.1699
4	8	50	−	2.95	1.84	1.76	1.52	2.84	2.62	1.92	1.91	0.1501
4	8	2500	−	0.9202	0.6599	0.71	0.6599	2.38	2.18	1.69	1.66	0.1501
4	12	1	−	108	27.77	87.78	29.59	5.49	5.22	3.44	3.43	0.21
4	12	50	−	317.5	67.39	57.71	34.76	3.86	3.65	2.62	2.53	0.1699
4	12	2500	−	291.2	77.17	29.31	22.05	4.19	3.92	2.09	2.05	0.1699
4	16	1	−	23.81	9.01	26.91	10.93	5.3	4.93	2.66	2.66	0.25
4	16	50	−	205.2	59.7	46.32	38.94	4.92	4.62	2.27	2.31	0.2002
4	16	2500	−	470.3	105.3	16.62	14.71	4.46	4.23	2.09	2.1	0.1699
8	16	1	−	−	−	6.27	7.57	8.28	7.75	6.76	6.61	0.37
8	16	50	−	−	−	4.5	4.22	5.8	5.55	5.13	5.13	0.31
8	16	2500	−	−	−	1.66	1.4	4.69	4.42	4.33	4.26	0.28
8	24	1	−	−	−	378.3	116.5	12.54	12.01	9.77	9.52	0.45
8	24	50	−	−	−	248.2	152.6	9.94	9.51	7.39	7.32	0.36
8	24	2500	−	−	−	118.2	82.81	6.8	6.41	5.48	5.42	0.3301
8	32	1	−	−	−	109.6	39.3	12.2	11.63	9.87	9.84	0.39
8	32	50	−	−	−	181.9	157.6	10.47	10.08	7.4	7.36	0.39
8	32	2500	−	−	−	63.56	50.97	9.9	9.54	5.53	5.48	0.37

Table 1. Running times of our matching algorithms. The first columns show the values of $n/10^4$ and $m/10^4$, respectively. The meaning of the other columns is explained in the text. A dash indicates that the program was not run on the instance.

- The use of heuristics to find an initial solution: Matching algorithms can either start from an empty matching or can use a heuristic to construct an initial matching.
- Simultaneous search for augmenting paths: The fastest matching algorithms search for augmenting paths in order of increasing length.
- Documentation: We discuss the merits of literate programming for documentation and why we use it document our implementations.

Figure 1 shows the running times of our bipartite matching algorithms; the source code of all our implementations can be found in [MN99]. A plus sign indicates the use of the greedy heuristic for finding an initial matching and a minus sign indicates that the algorithm starts with the empty matching. The algorithms HK [HK73] and AB [ABMP91] have a worst case running time of $O(\sqrt{n}m)$ and the other algorithms have a worst case running time of $O(nm)$. FFB stands for the basic version of the Ford and Fulkerson algorithm [FF63]. It runs in n phases, uses depth-first-search for finding augmenting paths and uses $\Theta(n)$ time at the beginning of each phase for initialization. Its best case running time is $\Theta(n^2)$. The algorithms dfs and bfs are variants of the Ford and Fulkerson algorithm. They avoid the costly initialization at the beginning of each phase and use depth-first and breadth-first search, respectively. The algorithms HK and AB use breadth-first search and search for augmenting paths in order of increasing length. The last column shows the time to check the result of the computation.

We used bipartite group graphs $G_{n,m,k}$, as suggested by [CGM$^+$97] in their experimental study of bipartite matching algorithms, for our experiments. A graph $G_{n,m,k}$ has n nodes on each side. On each side the nodes are divided into k groups of size n/k each (this assumes that k divides n). Each node in A has degree $d = m/n$ and the edges out of a node in group i of A go to random nodes in groups $i+1$ and $i-1$ of B.

The running times our algorithms differ widely. We observe (the book attempts to explain the observations, but we will not do so here) that the program with the quadratic best case running time is much slower than the other implementations, dfs is almost always slower than bfs and frequently much slower, that the use of the heuristic helps and the advantage is more prominent for the slower algorithms, and that the asymptotically better algorithms are never much slower than the asymptotically slower algorithms and sometimes much better. We also see that the time for checking the result of the computation is negligable.

Table 1 is a strong case for algorithm engineering and its interplay with the theoretical investigation of algorithms. We have algorithms with the same asymptotic bounds and widely differing observed behavior. The differences can be explained, sometimes analytically and sometimes heuristically, coined into implementation principles, and applied to other algorithms. See Sections 7.7 on maximum cardinality matching in general graphs, 7.8 on weighted matchings in bipartite graphs, and 7.9 on weighted matchings in general graphs of [MN99] to see how we applied the lessons learned from bipartite cardinality matchings to other matching problems.

References

ABMP91. H. Alt, N. Blum, K. Mehlhorn, and M. Paul. Computing a maximum cardinality matching in a bipartite graph in time $O(n^{1.5}\sqrt{m/\log n})$. *Information Processing Letters*, 37:237–240, 1991.

AGD.   The AGD graph drawing library. http://www.mpi-sb.mpg.de/AGD/.

CGA.   CGAL   (Computational   Geometry   Algorithms   Library). www.cs.ruu.nl/CGAL.

CGM$^+$97. B. Cherkassky, A. Goldberg, P. Martin, J. Setubal, and J. Stolfi. Augment or relabel? A computational study of bipartite matching and unit capacity maximum flow algorithms. Technical Report TR 97-127, NEC Research Institute, 1997.

FF63.   L.R. Ford and D.R. Fulkerson. *Flows in Networks*. Princeton University Press, Princeton, NJ, 1963.

HK73.   J.E. Hopcroft and R.M. Karp. An $n^{5/2}$ algorithm for maximum matchings in bipartite graphs. *SIAM Journal of Computing*, 2(4):225–231, 1973.

LED.   LEDA (Library of Efficient Data Types and Algorithms). www.mpi-sb.mpg.de/LEDA/leda.html.

MN99.   K. Mehlhorn and S. Näher. *The LEDA Platform for Combinatorial and Geometric Computing*. Cambridge University Press, 1999.

Author Index

Abadi, Martín 122
Aluru, Srinivas 21
Amadio, Roberto M. 304
Amo, Sandra de 245

Baaz, Matthias 258
Becchetti, Luca 201
Bidoit, Nicole 245
Boudol, Gérard 304

Castellani, Ilaria 219
Chandran, L. Sunil 283
Chaudhuri, Jeet 34
Ciabattoni, Agata 258
Corradini, Flavio 381

D'Souza, Deepak 60
Daley, Mark 269
De Nicola, Rocco 381
Demaine, Erik D. 84
Di Ianni, Miriam 201
Dix, Jürgen 142

Ésik, Zoltán 316

Fagerberg, Rolf 72
Fermüller, Christian 258
Fournet, Cédric 122

Garg, Naveen 213
Gloor, Greg 269
Gonthier, Georges 122

Heinemann, Bernhard 420
Herrmann, Philippe 47
Huth, Michael 369

Jain, Sachin 213
Jakoby, Andreas 394
Janardan, Ravi 291
Johnson, Eric 291

Kari, Lila 269

Labella, Anna 381
Landweber, Laura F. 269
Lévy, Jean-Jacques 181
Lhoussaine, Cédric 304
Löding, Christof 97
Lu, James J. 155

Majhi, Jayanth 291
Maranget, Luc 181
Marchetti-Spaccamela, Alberto 201
Mehlhorn, Kurt 446
Meyden, Ron van der 432
Mostéfaoui, Achour 329
Mukhopadhyay, Supratik 232
Mukund, Madhavan 219
Munro, Ian J. 84
Murray, Neil V. 155

Nandy, Subhas C. 34

Oheimb, David von 168
Okawa, Satoshi 316

Podelski, Andreas 232

Raynal, Michel 329
Rosenthal, Erik 155

Sauerhoff, Martin 342
Schindelhauer, Christian 394

Schlechta, Karl 142
Schwerdt, Jörg 291
Sevilgen, Fatih E. 21
Sharir, Micha 1
Shilov, Nikolay V. 432
Sieling, Detlef 356
Siromoney, Rani 269
Smid, Michiel 291
Swamy, Chaitanya 213

Thiagarajan, P. S. 60, 219
Toda, Seinosuke 341

Veith, Helmut 258

Wilke, Thomas 110

Yamakami, Tomoyuki 407

Lecture Notes in Computer Science

For information about Vols. 1–1663
please contact your bookseller or Springer-Verlag

Vol. 1664: J.C.M. Baeten, S. Mauw (Eds.), CONCUR'99.Concurrency Theory. Proceedings, 1999. XI, 573 pages. 1999.

Vol. 1665: P. Widmayer, G. Neyer, S. Eidenbenz (Eds.), Graph-Theoretic Concepts in Computer Science. Proceedings, 1999. XI, 414 pages. 1999.

Vol. 1666: M. Wiener (Ed.), Advances in Cryptology – CRYPTO '99. Proceedings, 1999. XII, 639 pages. 1999.

Vol. 1667: J. Hlavička, E. Maehle, A. Pataricza (Eds.), Dependable Computing – EDCC-3. Proceedings, 1999. XVIII, 455 pages. 1999.

Vol. 1668: J.S. Vitter, C.D. Zaroliagis (Eds.), Algorithm Engineering. Proceedings, 1999. VIII, 361 pages. 1999.

Vol. 1669: X.-S. Gao, D. Wang, L. Yang (Eds.), Automated Deduction in Geometry. Proceedings, 1998. VII, 287 pages. 1999. (Subseries LNAI).

Vol. 1670: N.A. Streitz, J. Siegel, V. Hartkopf, S. Konomi (Eds.), Cooperaltive Buildings. Proceedings, 1999. X, 229 pages. 1999.

Vol. 1671: D. Hochbaum, K. Jansen, J.D.P. Rolim, A. Sinclair (Eds.), Randomization, Approximation, and Combinatorial Optimization. Proceedings, 1999. IX, 289 pages. 1999.

Vol. 1672: M. Kutylowski, L. Pacholski, T. Wierzbicki (Eds.), Mathematical Foundations of Computer Science 1999. Proceedings, 1999. XII, 455 pages. 1999.

Vol. 1673: P. Lysaght, J. Irvine, R. Hartenstein (Eds.), Field Programmable Logic and Applications. Proceedings, 1999. XI, 541 pages. 1999.

Vol. 1674: D. Floreano, J.-D. Nicoud, F. Mondada (Eds.), Advances in Artificial Life. Proceedings, 1999. XVI, 737 pages. 1999. (Subseries LNAI).

Vol. 1675: J. Estublier (Ed.), System Configuration Management. Proceedings, 1999. VIII, 255 pages. 1999.

Vol. 1676: M. Mohania, A M. Tjoa (Eds.), Data Warehousing and Knowledge Discovery. Proceedings, 1999. XII, 400 pages. 1999.

Vol. 1677: T. Bench-Capon, G. Soda, A M. Tjoa (Eds.), Database and Expert Systems Applications. Proceedings, 1999. XVIII, 1105 pages. 1999.

Vol. 1678: M.H. Böhlen, C.S. Jensen, M.O. Scholl (Eds.), Spatio-Temporal Database Management. Proceedings, 1999. X, 243 pages. 1999.

Vol. 1679: C. Taylor, A. Colchester (Eds.), Medical Image Computing and Computer-Assisted Intervention – MICCAI'99. Proceedings, 1999. XXI, 1240 pages. 1999.

Vol. 1680: D. Dams, R. Gerth, S. Leue, M. Massink (Eds.), Theoretical and Practical Aspects of SPIN Model Checking. Proceedings, 1999. X, 277 pages. 1999.

Vol. 1681: D. A. Forsyth, J. L. Mundy, V. di Gesú, R. Cipolla (Eds.), Shape, Contour and Grouping in Computer Vision. VIII, 347 pages. 1999.

Vol. 1682: M. Nielsen, P. Johansen, O.F. Olsen, J. Weickert (Eds.), Scale-Space Theories in Computer Vision. Proceedings, 1999. XII, 532 pages. 1999.

Vol. 1683: J. Flum, M. Rodríguez-Artalejo (Eds.), Computer Science Logic. Proceedings, 1999. XI, 580 pages. 1999.

Vol. 1684: G. Ciobanu, G. Păun (Eds.), Fundamentals of Computation Theory. Proceedings, 1999. XI, 570 pages. 1999.

Vol. 1685: P. Amestoy, P. Berger, M. Daydé, I. Duff, V. Frayssé, L. Giraud, D. Ruiz (Eds.), Euro-Par'99. Parallel Processing. Proceedings, 1999. XXXII, 1503 pages. 1999.

Vol. 1686: H.E. Bal, B. Belkhouche, L. Cardelli (Eds.), Internet Programming Languages. Proceedings, 1998. IX, 143 pages. 1999.

Vol. 1687: O. Nierstrasz, M. Lemoine (Eds.), Software Engineering – ESEC/FSE '99. Proceedings, 1999. XII, 529 pages. 1999.

Vol. 1688: P. Bouquet, L. Serafini, P. Brézillon, M. Benerecetti, F. Castellani (Eds.), Modeling and Using Context. Proceedings, 1999. XII, 528 pages. 1999. (Subseries LNAI).

Vol. 1689: F. Solina, A. Leonardis (Eds.), Computer Analysis of Images and Patterns. Proceedings, 1999. XIV, 650 pages. 1999.

Vol. 1690: Y. Bertot, G. Dowek, A. Hirschowitz, C. Paulin, L. Théry (Eds.), Theorem Proving in Higher Order Logics. Proceedings, 1999. VIII, 359 pages. 1999.

Vol. 1691: J. Eder, I. Rozman, T. Welzer (Eds.), Advances in Databases and Information Systems. Proceedings, 1999. XIII, 383 pages. 1999.

Vol. 1692: V. Matoušek, P. Mautner, J. Ocelíková, P. Sojka (Eds.), Text, Speech and Dialogue. Proceedings, 1999. XI, 396 pages. 1999. (Subseries LNAI).

Vol. 1693: P. Jayanti (Ed.), Distributed Computing. Proceedings, 1999. X, 357 pages. 1999.

Vol. 1694: A. Cortesi, G. Filé (Eds.), Static Analysis. Proceedings, 1999. VIII, 357 pages. 1999.

Vol. 1695: P. Barahona, J.J. Alferes (Eds.), Progress in Artificial Intelligence. Proceedings, 1999. XI, 385 pages. 1999. (Subseries LNAI).

Vol. 1696: S. Abiteboul, A.-M. Vercoustre (Eds.), Research and Advanced Technology for Digital Libraries. Proceedings, 1999. XII, 497 pages. 1999.

Vol. 1697: J. Dongarra, E. Luque, T. Margalef (Eds.), Recent Advances in Parallel Virtual Machine and Message Passing Interface. Proceedings, 1999. XVII, 551 pages. 1999.

Vol. 1698: M. Felici, K. Kanoun, A. Pasquini (Eds.), Computer Safety, Reliability and Security. Proceedings, 1999. XVIII, 482 pages. 1999.

Vol. 1699: S. Albayrak (Ed.), Intelligent Agents for Telecommunication Applications. Proceedings, 1999. IX, 191 pages. 1999. (Subseries LNAI).

Vol. 1700: R. Stadler, B. Stiller (Eds.), Active Technologies for Network and Service Management. Proceedings, 1999. XII, 299 pages. 1999.

Vol. 1701: W. Burgard, T. Christaller, A.B. Cremers (Eds.), KI-99: Advances in Artificial Intelligence. Proceedings, 1999. XI, 311 pages. 1999. (Subseries LNAI).

Vol. 1702: G. Nadathur (Ed.), Principles and Practice of Declarative Programming. Proceedings, 1999. X, 434 pages. 1999.

Vol. 1703: L. Pierre, T. Kropf (Eds.), Correct Hardware Design and Verification Methods. Proceedings, 1999. XI, 366 pages. 1999.

Vol. 1704: Jan M. Żytkow, J. Rauch (Eds.), Principles of Data Mining and Knowledge Discovery. Proceedings, 1999. XIV, 593 pages. 1999. (Subseries LNAI).

Vol. 1705: H. Ganzinger, D. McAllester, A. Voronkov (Eds.), Logic for Programming and Automated Reasoning. Proceedings, 1999. XII, 397 pages. 1999. (Subseries LNAI).

Vol. 1706: J. Hatcliff, T. Æ. Mogensen, P. Thiemann (Eds.), Partial Evaluation – Practice and Theory. 1998. IX, 433 pages. 1999.

Vol. 1707: H.-W. Gellersen (Ed.), Handheld and Ubiquitous Computing. Proceedings, 1999. XII, 390 pages. 1999.

Vol. 1708: J.M. Wing, J. Woodcock, J. Davies (Eds.), FM'99 – Formal Methods. Proceedings Vol. I, 1999. XVIII, 937 pages. 1999.

Vol. 1709: J.M. Wing, J. Woodcock, J. Davies (Eds.), FM'99 – Formal Methods. Proceedings Vol. II, 1999. XVIII, 937 pages. 1999.

Vol. 1710: E.-R. Olderog, B. Steffen (Eds.), Correct System Design. XIV, 417 pages. 1999.

Vol. 1711: N. Zhong, A. Skowron, S. Ohsuga (Eds.), New Directions in Rough Sets, Data Mining, and Granular-Soft Computing. Proceedings, 1999. XIV, 558 pages. 1999. (Subseries LNAI).

Vol. 1712: H. Boley, A Tight, Practical Integration of Relations and Functions. XI, 169 pages. 1999. (Subseries LNAI).

Vol. 1713: J. Jaffar (Ed.), Principles and Practice of Constraint Programming – CP'99. Proceedings, 1999. XII, 493 pages. 1999.

Vol. 1714: M.T. Pazienza (Eds.), Information Extraction. IX, 165 pages. 1999. (Subseries LNAI).

Vol. 1715: P. Perner, M. Petrou (Eds.), Machine Learning and Data Mining in Pattern Recognition. Proceedings, 1999. VIII, 217 pages. 1999. (Subseries LNAI).

Vol. 1716: K.Y. Lam, E. Okamoto, C. Xing (Eds.), Advances in Cryptology – ASIACRYPT'99. Proceedings, 1999. XI, 414 pages. 1999.

Vol. 1717: Ç. K. Koç, C. Paar (Eds.), Cryptographic Hardware and Embedded Systems. Proceedings, 1999. XI, 353 pages. 1999.

Vol. 1718: M. Diaz, P. Owezarski, P. Sénac (Eds.), Interactive Distributed Multimedia Systems and Telecommunication Services. Proceedings, 1999. XI, 386 pages. 1999.

Vol. 1719: M. Fossorier, H. Imai, S. Lin, A. Poli (Eds.), Applied Algebra, Algebraic Algorithms and Error-Correcting Codes. Proceedings, 1999. XIII, 510 pages. 1999.

Vol. 1720: O. Watanabe, T. Yokomori (Eds.), Algorithmic Learning Theory. Proceedings, 1999. XI, 365 pages. 1999. (Subseries LNAI).

Vol. 1721: S. Arikawa, K. Furukawa (Eds.), Discovery Science. Proceedings, 1999. XI, 374 pages. 1999. (Subseries LNAI).

Vol. 1722: A. Middeldorp, T. Sato (Eds.), Functional and Logic Programming. Proceedings, 1999. X, 369 pages. 1999.

Vol. 1723: R. France, B. Rumpe (Eds.), UML'99 – The Unified Modeling Language. XVII, 724 pages. 1999.

Vol. 1725: J. Pavelka, G. Tel, M. Bartošek (Eds.), SOFSEM'99: Theory and Practice of Informatics. Proceedings, 1999. XIII, 498 pages. 1999.

Vol. 1726: V. Varadharajan, Y. Mu (Eds.), Information and Communication Security. Proceedings, 1999. XI, 325 pages. 1999.

Vol. 1727: P.P. Chen, D.W. Embley, J. Kouloumdjian, S.W. Liddle, J.F. Roddick (Eds.), Advances in Conceptual Modeling. Proceedings, 1999. XI, 389 pages. 1999.

Vol. 1728: J. Akoka, M. Bouzeghoub, I. Comyn-Wattiau, E. Métais (Eds.), Conceptual Modeling – ER '99. Proceedings, 1999. XIV, 540 pages. 1999.

Vol. 1729: M. Mambo, Y. Zheng (Eds.), Information Security. Proceedings, 1999. IX, 277 pages. 1999.

Vol. 1730: M. Gelfond, N. Leone, G. Pfeifer (Eds.), Logic Programming and Nonmonotonic Reasoning. Proceedings, 1999. XI, 391 pages. 1999. (Subseries LNAI).

Vol. 1732: S. Matsuoka, R.R. Oldehoeft, M. Tholburn (Eds.), Computing in Object-Oriented Parallel Environments. Proceedings, 1999. VIII, 205 pages. 1999.

Vol. 1734: H. Hellwagner, A. Reinefeld (Eds.), SCI: Scalable Coherent Interface. XXI, 490 pages. 1999.

Vol. 1564: M. Vazirgiannis, Interactive Multimedia Documents. XIII, 161 pages. 1999.

Vol. 1591: D.J. Duke, I. Herman, M.S. Marshall, PREMO: A Framework for Multimedia Middleware. XII, 254 pages. 1999.

Vol. 1735: J.W. Amtrup, Incremental Speech Translation. XV, 200 pages. 1999. (Subseries LNAI).

Vol. 1736: L. Rizzo, S. Fdida (Eds.): Networked Group Communication. Proceedings, 1999. XIII, 339 pages. 1999.

Vol. 1738: C. Pandu Rangan, V. Raman, R. Ramanujam (Eds.), Foundations of Software Technology and Theoretical Computer Science. Proceedings, 1999. XII, 452 pages. 1999.

Vol. 1740: R. Baumgart (Ed.): Secure Networking – CQRE [Secure] '99. Proceedings, 1999. IX, 261 pages. 1999.

Vol. 1742: P.S. Thiagarajan, R. Yap (Eds.), Advances in Computing Science – ASIAN'99. Proceedings, 1999. XI, 397 pages. 1999.